METHODS IN MOLECULAR BIOLOGY

Series Editor
John M. Walker
School of Life and Medical Sciences
University of Hertfordshire
Hatfield, Hertfordshire, UK

For further volumes:
http://www.springer.com/series/7651

For over 35 years, biological scientists have come to rely on the research protocols and methodologies in the critically acclaimed *Methods in Molecular Biology* series. The series was the first to introduce the step-by-step protocols approach that has become the standard in all biomedical protocol publishing. Each protocol is provided in readily-reproducible step-by-step fashion, opening with an introductory overview, a list of the materials and reagents needed to complete the experiment, and followed by a detailed procedure that is supported with a helpful notes section offering tips and tricks of the trade as well as troubleshooting advice. These hallmark features were introduced by series editor Dr. John Walker and constitute the key ingredient in each and every volume of the *Methods in Molecular Biology* series. Tested and trusted, comprehensive and reliable, all protocols from the series are indexed in PubMed.

SARS-CoV-2

Methods and Protocols

Edited by

Justin Jang Hann Chu

Department of Microbiology and Immunology; Infectious Disease TRD; BSL3 Core Facility, Yong Loo Lin School of Medicine, National University of Singapore, Singapore, Singapore

Bintou Ahmadou Ahidjo

The University of British Columbia, Vancouver, BC, Canada

Chee Keng Mok

BSL3 Core Facility, Yong Loo Lin School of Medicine, National University of Singapore, Singapore, Singapore

Humana Press

Editors
Justin Jang Hann Chu
Department of Microbiology and Immunology; Infectious Disease TRD; BSL3 Core Facility, Yong Loo Lin School of Medicine
National University of Singapore
Singapore, Singapore

Bintou Ahmadou Ahidjo
The University of British Columbia
Vancouver, BC, Canada

Chee Keng Mok
BSL3 Core Facility, Yong Loo Lin School of Medicine
National University of Singapore
Singapore, Singapore

ISSN 1064-3745 ISSN 1940-6029 (electronic)
Methods in Molecular Biology
ISBN 978-1-0716-2110-3 ISBN 978-1-0716-2111-0 (eBook)
https://doi.org/10.1007/978-1-0716-2111-0

Cover Illustration Caption: This image was taken by Mr. Wong Yi Hao, MSc, Science and Service Support Team, Biosafety Level 3 Core Facility, Yong Loo Lin School of Medicine, National University of Singapore, Singapore.

This Humana imprint is published by the registered company Springer Science+Business Media, LLC, part of Springer Nature.
The registered company address is: 1 New York Plaza, New York, NY 10004, U.S.A.

Preface

Severe Acute Respiratory Syndrome Coronavirus 2 (SARS-CoV-2) is a person-to-person transmission, single-stranded, positive-sense RNA virus belonging to the *Betacoronavirus* genus of the family *Coronaviridae*. SARS-CoV-2 infections result in the novel coronavirus disease, COVID-19, which in 2020 has resulted in a global pandemic with significant economic burden. *SARS-CoV-2: Methods and Protocols* provides the increasing number of SARS-CoV-2 researchers with a useful handbook covering multidisciplinary approaches on various aspects of SARS-CoV-2 research brought together by leading laboratories across the globe. Topics covered include techniques in clinical and diagnostic virology, basic protocols in cell and virus culture, as well as bioinformatics and proteomics approaches in cellular response studies. This comprehensive handbook also covers methods in immunology, animal models, antivirals and vaccine development strategies, as well as biorisk and mitigation measurements for SARS-CoV-2 research.

Singapore, Singapore — *Justin Jang Hann Chu*
Vancouver, BC, Canada — *Bintou Ahmadou Ahidjo*
Singapore, Singapore — *Chee Keng Mok*

Contents

Contributors

ANWARI AKHTER • *Disease Intervention & Prevention and Population Health Programs, Texas Biomedical Research Institute, San Antonio, TX, USA*
KENDRA J. ALFSON • *Texas Biomedical Research Institute, San Antonio, TX, USA*
XAVIER ALVAREZ • *Southwest National Primate Research Center, San Antonio, TX, USA; Disease Intervention & Prevention and Population Health Programs, Texas Biomedical Research Institute, San Antonio, TX, USA*
ENYIA R. ANDERSON • *Department of Vector Biology, Liverpool School of Tropical Medicine, Liverpool, UK; Department of Tropical Disease Biology, Liverpool School of Tropical Medicine, Liverpool, UK; Centre for Neglected Tropical Disease, Liverpool School of Tropical Medicine, Liverpool, UK*
JEAN ARMENGAUD • *Département Médicaments et Technologies pour la Santé (DMTS), Université Paris-Saclay, CEA, INRAE, SPI, Bagnols-sur-Cèze, France*
MUHAMMAD AZHAR • *Department of Biosciences, COMSATS University Islamabad, Sahiwal, Pakistan*
COREY BALINSKY • *Henry M. Jackson Foundation for the Advancement of Military Medicine, Bethesda, MD, USA*
CHRISTOPHER F. BASLER • *Department of Microbiology, Icahn School of Medicine at Mount Sinai, New York, NY, USA*
SHABANA BIBI • *Yunnan Herbal Laboratory, School of Ecology and Environmental Sciences, Yunnan University, Kunming, Yunnan, China*
MATTIA BUGATTI • *Section of Pathology, Department of Molecular and Translational Medicine, University of Brescia, Spedali Civili di Brescia, Brescia, Italy*
C. KORIN BULLEN • *Department of Medicine, Johns Hopkins University School of Medicine, Baltimore, MD, USA*
RICARDO CARRION JR • *Texas Biomedical Research Institute, San Antonio, TX, USA*
SHENG CHEN • *Department of Infectious Diseases and Public Health, City University of Hong Kong, Kowloon, Hong Kong SAR; Jockey Club College of Veterinary Medicine and Life Sciences, City University of Hong Kong, Kowloon, Hong Kong SAR*
JUSTIN JANG HANN CHU • *Laboratory of Molecular RNA Virology and Antiviral Strategies, Department of Microbiology and Immunology, Yong Loo Lin School of Medicine, National University of Singapore, Singapore, Singapore; Infectious Disease Translational Research Programme, Yong Loo Lin School of Medicine, National University of Singapore, Singapore, Singapore; Biosafety Level 3 Core Facility, Yong Loo Lin School of Medicine, National University of Singapore, Singapore, Singapore; Collaborative and Translation Unit for HFMD, Institute of Molecular and Cell Biology, Agency for Science, Technology and Research (A*STAR), Singapore, Singapore*
ASHLEY CLIFF • *The Bredesen Center for Interdisciplinary Research and Graduate Education, University of Tennessee Knoxville, Knoxville, TN, USA*
JOURNEY COLE • *Southwest National Primate Research Center, San Antonio, TX, USA; Texas Biomedical Research Institute, San Antonio, TX, USA*
ANURA DAVID • *Department of Molecular Medicine and Haematology, School of Pathology, Faculty of Health Sciences, University of the Witwatersrand, Johannesburg, South Africa*

Stephanie L. Davis • *Department of Medicine, Johns Hopkins University School of Medicine, Baltimore, MD, USA*

Gabriel Defang • *Viral and Rickettsial Diseases Department, Naval Medical Research Center, Silver Spring, MD, USA*

Ning Dong • *Department of Infectious Diseases and Public Health, City University of Hong Kong, Kowloon, Hong Kong SAR*

Ruby A. Escobedo • *Southwest National Primate Research Center, San Antonio, TX, USA; Texas Biomedical Research Institute, San Antonio, TX, USA*

Fabio Facchetti • *Section of Pathology, Department of Molecular and Translational Medicine, University of Brescia, Spedali Civili di Brescia, Brescia, Italy*

Akshamal M. Gamage • *Programme in Emerging Infectious Diseases, Duke-NUS Medical School, Singapore, Singapore*

Shashank R. Ganatra • *Southwest National Primate Research Center, San Antonio, TX, USA; Texas Biomedical Research Institute, San Antonio, TX, USA*

Michael R. Garvin • *Oak Ridge National Laboratory, Computational Systems Biology, Oak Ridge, TN, USA; National Virtual Biotechnology Laboratory, US Department of Energy, Washington, DC, USA*

Joao Gabriel Felipe Machado Gazolla • *Oak Ridge National Laboratory, Computational Systems Biology, Oak Ridge, TN, USA; National Virtual Biotechnology Laboratory, US Department of Energy, Washington, DC, USA*

Yu-Nong Gong • *Research Center for Emerging Viral Infections, College of Medicine, Chang Gung University, Taoyuan, Taiwan; Department of Laboratory Medicine, Linkou Chang Gung Memorial Hospital, Taoyuan, Taiwan*

Margarida Goulart • *Joint Research Centre, European Commission, Brussels, Belgium*

Duarte Gouveia • *Département Médicaments et Technologies pour la Santé (DMTS), Université Paris-Saclay, CEA, INRAE, SPI, Bagnols-sur-Cèze, France*

Lucia Grenga • *Département Médicaments et Technologies pour la Santé (DMTS), Université Paris-Saclay, CEA, INRAE, SPI, Bagnols-sur-Cèze, France*

Qinxue Hu • *State Key Laboratory of Virology, Wuhan Institute of Virology, Center for Biosafety Mega-Science, Chinese Academy of Sciences, Wuhan, China; Institute for Infection and Immunity, St. George's, University of London, London, UK*

Linda Hueston • *Arbovirus Emerging Diseases Unit, NSW Health Pathology, Institute of Clinical Pathology and Medical Research (ICPMR), Westmead, NSW, Australia; Menzies Health Institute Queensland (MHIQ), Griffith University, Qld, Australia*

Grant L. Hughes • *Department of Vector Biology, Liverpool School of Tropical Medicine, Liverpool, UK; Department of Tropical Disease Biology, Liverpool School of Tropical Medicine, Liverpool, UK; Centre for Neglected Tropical Disease, Liverpool School of Tropical Medicine, Liverpool, UK*

Awais Ihsan • *Department of Biosciences, COMSATS University Islamabad, Sahiwal, Pakistan*

Daniel Jacobson • *Oak Ridge National Laboratory, Computational Systems Biology, Oak Ridge, TN, USA; National Virtual Biotechnology Laboratory, US Department of Energy, Washington, DC, USA; The Bredesen Center for Interdisciplinary Research and Graduate Education, University of Tennessee Knoxville, Knoxville, TN, USA; Genome Science and Technology, University of Tennessee Knoxville, Knoxville, TN, USA; Department of Psychology, NeuroNet Research Center, University of Tennessee Knoxville, Knoxville, TN, USA*

Vihasi Jani • *Henry M. Jackson Foundation for the Advancement of Military Medicine, Bethesda, MD, USA*

Piet Jones • *The Bredesen Center for Interdisciplinary Research and Graduate Education, University of Tennessee Knoxville, Knoxville, TN, USA*

Tessy Joseph • *BSL-3 Core Facility, Yong Loo Lin School of Medicine, National University of Singapore, Singapore, Singapore*

Alexander S. Jureka • *Centers for Disease Control and Prevention, Atlanta, GA, USA*

Amy Justice • *VA Connecticut Healthcare/General Internal Medicine, West Haven, CT, USA; Yale University School of Medicine, New Haven, CT, USA*

Takashi Kajiya • *Clinical Research Center, Tenyoukai Central Hospital, Kagoshima, Japan*

Deepak Kaushal • *Southwest National Primate Research Center, San Antonio, TX, USA; Texas Biomedical Research Institute, San Antonio, TX, USA*

Vidya Keshav • *Department of Molecular Medicine and Haematology, School of Pathology, Faculty of Health Sciences, University of the Witwatersrand, Johannesburg, South Africa*

Muhammad Saad Khan • *Department of Biosciences, COMSATS University Islamabad, Sahiwal, Pakistan*

Jörg Kriegsmann • *Molecular Pathology Trier, Trier, Germany; Danube Private University Krems, Krems, Austria; Proteopath Trier, Trier, Germany*

Katharina Kriegsmann • *Department of Hematology, Oncology and Rheumatology, University Hospital Heidelberg, Heidelberg, Germany*

Mark Kriegsmann • *Institute of Pathology, University Hospital Heidelberg, Heidelberg, Germany; Member of the German Centre for Lung Research (DZL), Translational Lung Research Centre Heidelberg, Heidelberg, Germany*

Matthew Lane • *The Bredesen Center for Interdisciplinary Research and Graduate Education, University of Tennessee Knoxville, Knoxville, TN, USA*

Kuo-Ming Lee • *Research Center for Emerging Viral Infections, College of Medicine, Chang Gung University, Taoyuan, Taiwan; Division of Infectious Diseases, Department of Pediatrics, Linkou Chang Gung Memorial Hospital, Taoyuan, Taiwan*

Tae-Hyung Lee • *Southwest National Primate Research Center, San Antonio, TX, USA; Texas Biomedical Research Institute, San Antonio, TX, USA*

Changchang Li • *Key Laboratory of Special Pathogens and Biosafety, Center for Biosafety Mega-Science, Wuhan Institute of Virology, Chinese Academy of Sciences, Wuhan, Hubei, China; International College, University of Chinese Academy of Sciences, Beijing, China*

Xiohong Li • *Key Laboratory of Special Pathogens and Biosafety, Center for Biosafety Mega-Science, Wuhan Institute of Virology, Chinese Academy of Sciences, Wuhan, Hubei, China*

Jing Liu • *Department of Otolaryngology, Yong Loo Lin School of Medicine, National University of Singapore, Singapore, Singapore*

Silvia Lonardi • *Section of Pathology, Department of Molecular and Translational Medicine, University of Brescia, Spedali Civili di Brescia, Brescia, Italy*

Monika M. Looney • *Department of Medicine, Johns Hopkins University School of Medicine, Baltimore, MD, USA*

Rolf Marschalek • *Institute of Pharmaceutical Biology, Goethe University, Frankfurt am Main, Germany*

Luis Martinez-Sobrido • *Disease Intervention & Prevention and Population Health Programs, Texas Biomedical Research Institute, San Antonio, TX, USA*

Elizabeth Mayne • *Department of Immunology, School of Pathology, Faculty of Health Sciences, University of the Witwatersrand, Johannesburg, South Africa; National Health Laboratory Service, Johannesburg, Gauteng, South Africa*

J. Izaak Miller • *Oak Ridge National Laboratory, Computational Systems Biology, Oak Ridge, TN, USA; National Virtual Biotechnology Laboratory, US Department of Energy, Washington, DC, USA*

Chee Keng Mok • *Laboratory of Molecular RNA Virology and Antiviral Strategies, Department of Microbiology and Immunology, Yong Loo Lin School of Medicine, National University of Singapore, Singapore, Singapore; Biosafety Level 3 Core Facility, Yong Loo Lin School of Medicine, National University of Singapore, Singapore, Singapore*

Caroline Mwaliko • *International College, University of Chinese Academy of Sciences, Beijing, China; Sino-Africa Joint Research Center, Nairobi, Kenya; CAS Key Laboratory of Molecular Virology & Immunology, Institut Pasteur of Shanghai, Chinese Academy of Sciences, Wuhan, Hubei, China*

Sabari Nath Neerukonda • *Office of Vaccines Research and Review, Center for Biologics Evaluation and Research and Review, US Food and Drug Administration, Silver Spring, MD, USA*

Yan Ling Ng • *Laboratory of Molecular RNA Virology and Antiviral Strategies, Department of Microbiology and Immunology, Yong Loo Lin School of Medicine, National University of Singapore, Singapore, Singapore*

Jun-ichiro Nishi • *Department of Microbiology, Graduate School of Medical and Dental Sciences, Kagoshima University, Kagoshima, Japan*

Lara Noble • *Department of Molecular Medicine and Haematology, School of Pathology, Faculty of Health Sciences, University of the Witwatersrand, Johannesburg, South Africa*

Raphael Nyaruaba • *Key Laboratory of Special Pathogens and Biosafety, Center for Biosafety Mega-Science, Wuhan Institute of Virology, Chinese Academy of Sciences, Wuhan, Hubei, China; International College, University of Chinese Academy of Sciences, Beijing, China; Sino-Africa Joint Research Center, Nairobi, Kenya*

Faith Ogolla • *Key Laboratory of Special Pathogens and Biosafety, Center for Biosafety Mega-Science, Wuhan Institute of Virology, Chinese Academy of Sciences, Wuhan, Hubei, China; International College, University of Chinese Academy of Sciences, Beijing, China; Sino-Africa Joint Research Center, Nairobi, Kenya*

Christiane Pallas • *Institute for Medical Virology, University Hospital Frankfurt am Main, Goethe University, Frankfurt am Main, Germany*

Jun-Gyu Park • *Disease Intervention & Prevention and Population Health Programs, Texas Biomedical Research Institute, San Antonio, TX, USA*

Edward I. Patterson • *Department of Biological Sciences, Brock University, St. Catharines, Canada*

Mirko Pavicic • *Oak Ridge National Laboratory, Computational Systems Biology, Oak Ridge, TN, USA; National Virtual Biotechnology Laboratory, US Department of Energy, Washington, DC, USA*

Sabai Phyu • *Laboratory Biorisk Consultancy & Training Pte. Ltd., Singapore, Singapore; European Union Chemical, Biological, Radiological and Nuclear Risk Mitigation Centres of Excellence Regional Secretariat—South East Asia/B&S Europe, Manila, Philippines*

Paula A. Pino • *Disease Intervention & Prevention and Population Health Programs, Texas Biomedical Research Institute, San Antonio, TX, USA*

Kevin R. Porter • *Viral and Rickettsial Diseases Department, Naval Medical Research Center, Silver Spring, MD, USA*

Erica T. Prates • *Oak Ridge National Laboratory, Computational Systems Biology, Oak Ridge, TN, USA; National Virtual Biotechnology Laboratory, US Department of Energy, Washington, DC, USA*

TESSA PRINCE • *NIHR Health Protection Unit in Emerging and Zoonotic Infections, Department of Clinical Infection, Microbiology and Immunology, University of Liverpool, Liverpool, UK*

HOLGER F. RABENAU • *Institue of Medical Viroogy, University of Frankfurt, Goethe-University, Frankfurt, Germany*

CHRISTOPHER T. RENTSCH • *Faculty of Epidemiology and Population Health, London School of Hygiene and Tropical Medicine, London, UK; VA Connecticut Healthcare/General Internal Medicine, West Haven, CT, USA*

JONATHON ROMERO • *The Bredesen Center for Interdisciplinary Research and Graduate Education, University of Tennessee Knoxville, Knoxville, TN, USA*

LESLEY SCOTT • *Department of Molecular Medicine and Haematology, School of Pathology, Faculty of Health Sciences, University of the Witwatersrand, Johannesburg, South Africa*

SU YUN SE-THOE • *Biosafety Branch, Public Health Group, Ministry of Health, Singapore, Singapore*

MANESH B. SHAH • *Genome Science and Technology, University of Tennessee Knoxville, Knoxville, TN, USA*

SHIN-RU SHIH • *Research Center for Emerging Viral Infections, College of Medicine, Chang Gung University, Taoyuan, Taiwan; Department of Laboratory Medicine, Linkou Chang Gung Memorial Hospital, Taoyuan, Taiwan; Department of Medical Biotechnology and Laboratory Science, College of Medicine, Chang Gung University, Taoyuan, Taiwan; Research Center for Chinese Herbal Medicine, Research Center for Food and Cosmetic Safety, and Graduate Institute of Health Industry Technology, College of Human Ecology, Chang Gung University of Science and Technology, Taoyuan, Taiwan*

BINDU SINGH • *Southwest National Primate Research Center, San Antonio, TX, USA; Texas Biomedical Research Institute, San Antonio, TX, USA*

DHIRAJ K. SINGH • *Southwest National Primate Research Center, San Antonio, TX, USA; Texas Biomedical Research Institute, San Antonio, TX, USA*

WENDY STEVENS • *Department of Molecular Medicine and Haematology, School of Pathology, Faculty of Health Sciences, University of the Witwatersrand, Johannesburg, South Africa; National Priority Programme, National Health Laboratory Service, Johannesburg, Gauteng, South Africa*

YASUO SUDA • *Department of Chemistry and Biotechnology, Graduate School of Science and Engineering, Kagoshima University, Kagoshima, Japan*

KYLE A. SULLIVAN • *Oak Ridge National Laboratory, Computational Systems Biology, Oak Ridge, TN, USA; National Virtual Biotechnology Laboratory, US Department of Energy, Washington, DC, USA*

PEIFANG SUN • *Viral and Rickettsial Diseases Department, Naval Medical Research Center, Silver Spring, MD, USA*

YASUHISA TAJIMA • *Division of Infectious Disease, Hamamatsu Medical Center, Shizuoka, Japan*

KAI SEN TAN • *Biosafety level 3 Core Facility, Yong Loo Lin School of Medicine, National University Health System, National University of Singapore, Singapore, Singapore*

JORDI B. TORRELLES • *Disease Intervention & Prevention and Population Health Programs, Texas Biomedical Research Institute, San Antonio, TX, USA*

LANCE TURTLE • *NIHR Health Protection Unit in Emerging and Zoonotic Infections, Department of Clinical Infection, Microbiology and Immunology, University of Liverpool, Liverpool, UK; Tropical and Infectious Disease Unit, Liverpool University Hospitals Foundation NHS Trust (member of Liverpool Health Partners), Liverpool, UK*

ARIANNA VALZELLI • *Section of Pathology, Department of Molecular and Translational Medicine, University of Brescia, Spedali Civili di Brescia, Brescia, Italy*

RUSSELL VASSELL • *Office of Vaccines Research and Review, Center for Biologics Evaluation and Research and Review, US Food and Drug Administration, Silver Spring, MD, USA*

ANGELICA M. WALKER • *The Bredesen Center for Interdisciplinary Research and Graduate Education, University of Tennessee Knoxville, Knoxville, TN, USA*

PETRA WANDERNOTH • *Molecular Pathology Trier, Trier, Germany*

WEI WANG • *Office of Vaccines Research and Review, Center for Biologics Evaluation and Research and Review, US Food and Drug Administration, Silver Spring, MD, USA*

HONPING WEI • *Key Laboratory of Special Pathogens and Biosafety, Center for Biosafety Mega-Science, Wuhan Institute of Virology, Chinese Academy of Sciences, Wuhan, Hubei, China*

CAROL D. WEISS • *Office of Vaccines Research and Review, Center for Biologics Evaluation and Research and Review, US Food and Drug Administration, Silver Spring, MD, USA*

SANDRA WESTHAUS • *Institue of Medical Viroogy, University of Frankfurt, Goethe-University, Frankfurt, Germany*

MAREK WIDERA • *Institute for Medical Virology, University Hospital Frankfurt am Main, Goethe University, Frankfurt am Main, Germany*

ALEXANDER WILHELM • *Institute for Medical Virology, University Hospital Frankfurt am Main, Goethe University, Frankfurt am Main, Germany; Institute of Pharmaceutical Biology, Goethe University, Frankfurt am Main, Germany*

MAYA WILLIAMS • *Chemistry Division, US Naval Research Laboratory, Washington, DC, USA*

YAN YAN • *Infection and Immunity Laboratory, The Fifth People's Hospital of Wuxi, Affiliated Hospital of Jiangnan University, Wuxi, China*

HANG YANG • *Key Laboratory of Special Pathogens and Biosafety, Center for Biosafety Mega-Science, Wuhan Institute of Virology, Chinese Academy of Sciences, Wuhan, Hubei, China*

XUEMEI YANG • *Department of Infectious Diseases and Public Health, City University of Hong Kong, Kowloon, Hong Kong SAR*

QUDSIA YOUSAFI • *Department of Biosciences, COMSATS University Islamabad, Sahiwal, Pakistan*

JUNPING YU • *Key Laboratory of Special Pathogens and Biosafety, Center for Biosafety Mega-Science, Wuhan Institute of Virology, Chinese Academy of Sciences, Wuhan, Hubei, China*

LU ZHAO • *Key Laboratory of Special Pathogens and Biosafety, Center for Biosafety Mega-Science, Wuhan Institute of Virology, Chinese Academy of Sciences, Wuhan, Hubei, China; International College, University of Chinese Academy of Sciences, Beijing, China*

Part I

Clinical and Diagnostic Virology

Chapter 1

Evolution and Epidemiology of SARS-CoV-2 Virus

Yu-Nong Gong, Kuo-Ming Lee, and Shin-Ru Shih

Abstract

A novel coronavirus (CoV) that emerged in Wuhan, Hubei province in China, in December 2019, has rapidly spread worldwide. Named as the severe acute respiratory syndrome coronavirus 2 (SARS-CoV-2), this virus has been responsible for infecting about 153 million people and causing 3 million deaths by May 2021. There is obvious interest in gaining novel insights into the epidemiologic evolution of this virus; however, inappropriate application and interpretation of genomic and phylogenetic analyses has led to dangerous outcomes and misunderstandings. This chapter focuses on not only introducing this virus, its genomic characteristics and molecular mechanisms but also describing the application and interpretation of phylogenetic tree analyses, in order to provide useful information to better understand the evolution and epidemiology of this virus. In addition, recombinant region and genetic ancestry of SARS-CoV-2 remain unknown. It is urgently required to collect samples and obtain related viral genetic data from animal sources for identifying the intermediate host of SARS-CoV-2 that is responsible for its cross-species transmission.

Key words Coronavirus, SARS-CoV-2, Evolution, Epidemiology, Phylogenetic tree, Recombination

1 Introduction

In December 2019, a novel coronavirus (CoV) emerged in Wuhan, Hubei province in China, and rapidly spread worldwide [1]. The virus was designated as the severe acute respiratory syndrome coronavirus 2 (SARS-CoV-2), and its associated disease was named coronavirus disease 2019 (COVID-19) [2]. The World Health Organization declared this disease a pandemic on March 11, 2020. As of May 2021, COVID-19 resulted in about 153 million confirmed cases and 3 million deaths worldwide [3]. However, the real number of infected cases is probably underestimated since many patients have very mild or asymptomatic infections and are usually excluded from these surveys. To date, more than one million genomic data on SARS-CoV-2 infection cases are available on the Global Initiative on Sharing All Influenza Data (GISAID), a global initiative and primary source of open-access genomic data [4]. Nonetheless, the challenges of biased data sampling and low

Justin Jang Hann Chu et al. (eds.), *SARS-CoV-2: Methods and Protocols*,
Methods in Molecular Biology, vol. 2452, https://doi.org/10.1007/978-1-0716-2111-0_1,

phylogenetic resolution still remain due to high similarities of genomic sequences at an early stage of this outbreak. Herein, the current knowledge on SARS-CoV-2 evolution and epidemiology are summarized, highlighting the gap within the origin and intermediate host of this virus. Moreover, the issues related to the collection and analysis of SARS-CoV-2 sequences through computational technologies are also discussed.

2 Genomic Characteristics and Molecular Mechanisms of SARS-CoV-2

CoVs are classified into four genera: *Alphacoronavirus*, *Betacoronavirus*, *Gammacoronavirus*, and *Deltacoronavirus* [5]. Prior to 2019, six CoVs were known to infect human, including the human CoV 229E and NL63 belonging to the genera *Alphacoronavirus*, as well as the human CoV OC43, HKU1, Severe Acute Respiratory Syndrome-related CoV (SARS-CoV), and Middle East Respiratory Syndrome CoV (MERS-CoV) belonging to the genera *Betacoronavirus* [6–9]. SARS-CoV-2 is the seventh CoV identified to infect human and is a member of the genus *Betacoronavirus* within the subgenus *Sarbecovirus*, along with SARS-CoV, whereas MERS-CoV belongs to the separate subgenus *Merbecovirus*.

CoV is the longest RNA virus and shares the characteristic of high mutation rate with other RNA viruses. Nonetheless, the mutation rate of SARS-CoV-2 is of approximately two mutations per month, which is considered lower than most RNA viruses due to the presence of proofreading activity in the replication machinery [10]. The first complete genome of SARS-CoV-2 (strain name Wuhan-Hu-1), published in January, 2020, comprised a sequence of 29,903 bp in length [11], with six evolutionarily conserved open reading frames (ORFs) including the nonstructural ORF1a and 1b and four structural proteins, as well as several accessory proteins that assemble in a virus-specific manner (Fig. 1a). The ORF1a/b occupies two thirds of the viral genome, and its overlapping ORFs can produce two polypeptides (pp1a and pp1ab) due to a programmed −1 ribosomal frameshift at the boundary of ORF1a and ORF1b [12–14]. Further proteolytic processes mediated by two cysteine proteases that are also encoded in ORF1a, a papain-like protease (PLpro or nsp3) and a 3C-like protease (3CLpro or nsp5), generate a total of 16 nonstructural proteins (nsps) (Fig. 1b). Of note, PLpro is responsible for the cleavage of nsp1, nsp2, and nsp3 while the remaining nsps are subjected to 3CLpro processing [15, 16].

Most nsps act as viral replicases and are mainly involve in the viral genome replication [17]. Nsp9 forms homodimers and binds to the single stranded RNA [18], whereas Nsp12 functions as a RNA-dependent RNA polymerase that plays a central role in viral transcription and replication in coordination with the cofactor

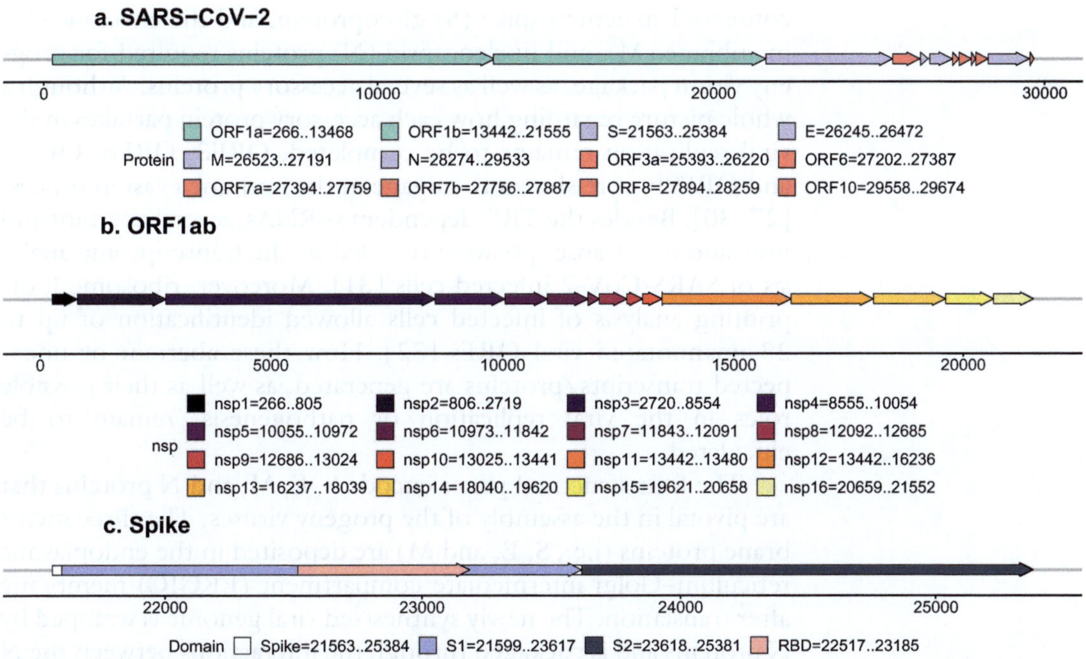

Fig. 1 Genomic coordinates of SARS-CoV-2 proteins. (**a**) Two polyproteins, and four structural and six accessory proteins, (**b**) 16 nonstructural proteins, and (**c**) two major domains including S1 and S2, and a receptor binding domain (RBD) within S1 in the spike protein are illustrated

proteins, nsp7 and nsp8 [18–20]. In addition to the primase activity, nsp8 from 229E is associated with RNA 3′terminal adenylyltransferase activity, thus might involve in the 3′ end processing of viral transcripts [21]. Moreover, the holo-RdRp complex binds to two nsp13 helicases, which not only unwind the template but also regulate replication-transcription6 complex backtracking, a key process that affects template switching and proofreading [22]. The CoV-specific proofreading activity is mediated by nsp14 that can excise the misincorporated nucleotides through its exonuclease (ExoN) domain [23]. Replication of CoVs is also characterized by its own capping machinery operated by the RNA 5′-triphosphatase activity of nsp13, and the guanine-*N7* methyltransferase and 2′-*O*-methyltransferase activity of nsp14 and nsp16, respectively. The enzymatic activities of both nsp14 and nsp16 are boosted by the cofactor nsp10 [24, 25]. During CoV replication, the replicase produces the full-length genomic RNAs, but also generates a series of discontinuous subgenomic RNAs (sgRNAs). These sgRNAs, which are nidovirus-specific, are characterized by the coterminal sequences at both 5′- and 3′-ends, and are synthesized through long- and/or short-range RNA-RNA interactions mediated by transcription-regulatory sequences (TRS) located at the 5′-untranslated region (TRS-L) and regions upstream of most ORFs (TRS-body), except ORF1ab [26]. SgRNAs encode the conserved structural spike (S) glycoprotein, and the envelope (E), membrane (M), and nucleocapsid (N) proteins required for progeny virion package, as well as several accessory proteins. Although a whole picture regarding how each accessory protein partakes in the viral replication remains to be completed, ORF3, ORF6, ORF8, and ORF9b were shown to engage in the immune evasion process [27–30]. Besides the TRS-dependent sgRNAs, several aberrant and noncanonical transcripts were revealed by the transcriptome analysis of SARS-CoV-2 infected cells [31]. Moreover, ribosome footprinting analysis of infected cells allowed identification of up to 23 unannotated viral ORFs [32]. How these aberrant or unexpected transcripts/proteins are generated, as well as their possible roles in the viral replication or pathogenesis, remain to be elucidated.

The four structural genes encode S, E, M, and N proteins that are pivotal in the assembly of the progeny viruses. The three membrane proteins (i.e., S, E, and M) are deposited in the endoplasmic reticulum–Golgi intermediate compartment (ERGIC) membrane after translation. The newly synthesized viral genome is wrapped by N proteins and is packaged through the interactions between the N and M proteins. The structural proteins are also involved in CoV pathogenesis through other functions. The small E protein forms a homopentameric cation channel that may activate the host inflammasome through Ca^{2+} release from the ERGIC lumen [33]. The SARS-CoV-2 N and SARS-CoV M proteins are able to suppress the host innate immunity by targeting STAT1/2 and TRAF3, respectively [34, 35]. In turn, the S protein has been widely studied given its important role in binding to the host receptor ACE2 and thereby promoting cell entry [35–37]. The S proteins can be divided into the receptor binding domain (RBD)-harboring S1 and fusion peptide-containing S2 domain responsible for the receptor binding and membrane fusion, respectively (Fig. 1c). The trimeric S proteins on the viral surface suffer two proteolytic cleavages accompanied by conformational changes that will activate the membrane fusion with the host cell [38, 39]. Thus, genetic traits that might affect functions of the S protein attract extensive attention worldwide as a strategy to prevent CoV infections.

3 Genomic Mutation and Deletion of SARS-CoV-2

Mutations commonly happen after RNA virus infects the host and replicates. Indeed, mutation (genetic drift) and recombination (genetic shift) are critical strategies used by viruses to achieve host adaptation, or even cross-species transmission. Phylogeny lineages are also inferred from specific mutation transmission. A study analyzing 30,983 complete SARS-CoV-2 genomes collected globally from December 24, 2019, to May 13, 2020, identified several geographically shared mutations, including nsp2 T265I, nsp6 L3606F, nsp12 T4847I, S D614G, N R203K-G204R, ORF3a Q57H–G251V, and ORF8 L84S [40]. Many of these mutations occur on proteins important for viral replication; however, the impact of each mutation on the virulence and pathogenesis of SARS-CoV-2 remains to be explored and carefully examined. Nonetheless, genomic changes can help us better understand this emerging virus and its transmission; although sporadic mutations usually do not dramatically impact on the outcome of the ongoing outbreak [41].

The S D614G mutation serves as a striking example. The D614 residue forms an inter-protomer hydrogen bond with T859 of the other promoter; thus, the D614G substitution disrupts the hydrogen bond leading to a more open conformation. This particular change exposes the RBD, which in turn will increase the ACE2-binding capacity of the S protein [42]. Given the advantage that the D614G variant provides to the virus, this mutation has replaced the ancestral sequence since May 2000, with the infection mediated by the new variant resulting in higher upper respiratory tract viral loads than the previous strain. In addition, the G-type pseudotype and nano-luciferase reporter viruses infected cell lines more efficiently than D-type viruses [43, 44]. More importantly, competition analyses showed that the G-type virus outcompeted the D-type virus in human airway epithelial cell culture models [44, 45]. Infection of G-type viruses in *Syrian* hamsters also showed increased transmissibility and higher viral titers in nasal washes and trachea. However, no obvious changes in disease severity regarding the body weight loss and pathological analyses were described. In fact, both the D- and G-type viruses can be efficiently neutralized by convalescent serums and neutralizing antibodies [44, 45]. Overall, these results suggest that the D614G substitution might be important for the human adaptation and for viral competition fitness. Similar analyses on other mutations should be considered to conclude on their impact on SARS-CoV2 replication.

Higher frequency of viral replication is associated to higher chance of genetic mutation, as well as for genetic deletions. In addition to tracking genomic mutations, deletion events in SARS-CoV-2 genome have also been reported, which are summarized in Fig. 2. Firstly, the longest deletion detected was 382-bp long and was reported in different countries, including Singapore, Bangladesh, Australia, and Spain [46], and Taiwan [47]. This deletion, which corresponds to the genomic positions from 27,848 to 28,229, according to Wuhan-Hu-1, covers from ORF7b and the TRS of ORF8 (Fig. 2a). Importantly, this long deletion might be a candidate to improve the development of vaccines or drug treatment targeting SARS-CoV-2, due to its clinical effects reported in patients from Singapore [30]. For example, the ORF8 deletion was clinically associated with a mild infection and reduced systemic

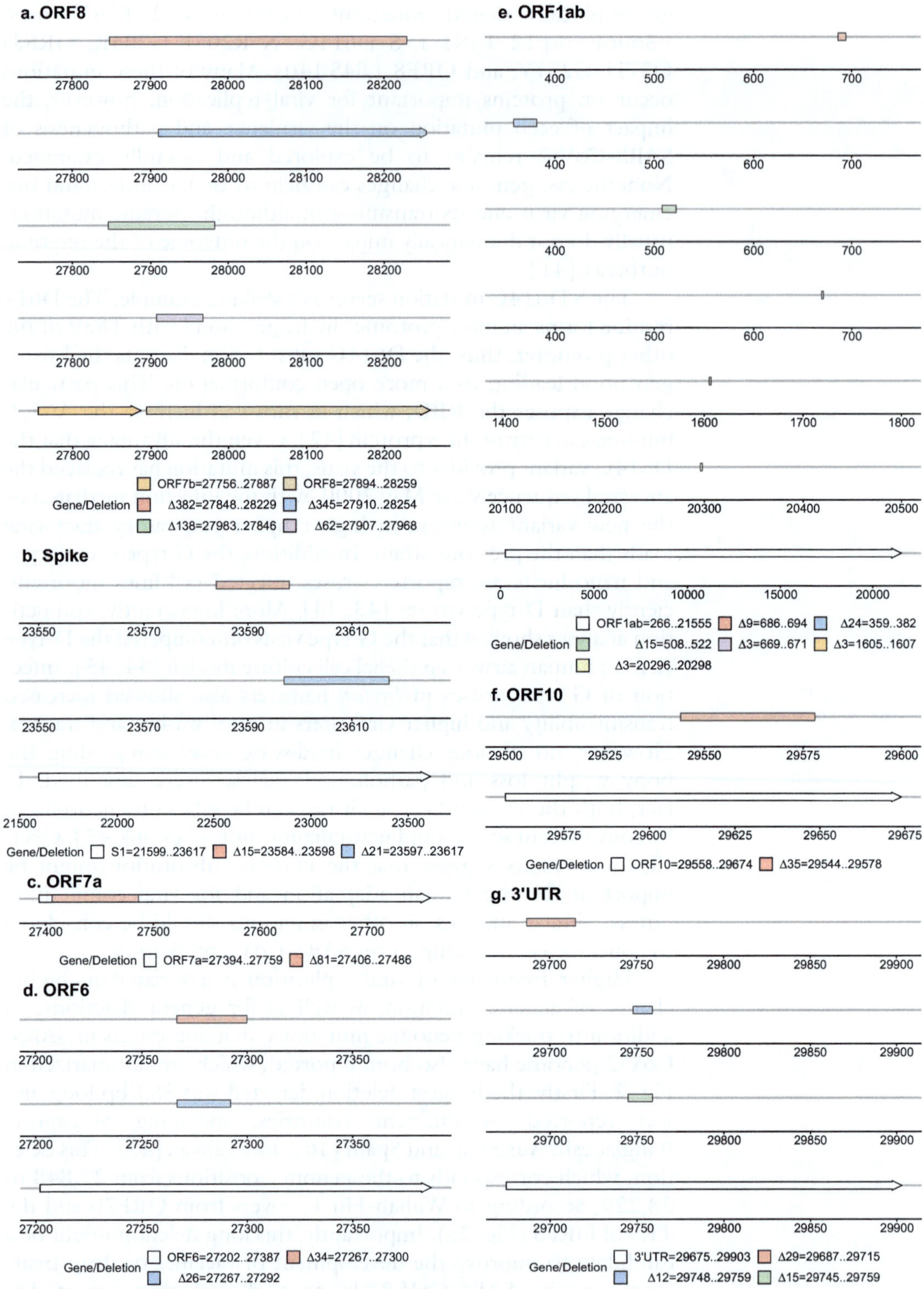

Fig. 2 Genomic deletions in SARS-CoV-2. Summary of four, two, one, two, six, one, and three deletions located in (**a**) ORF8, (**b**) spike, (**c**) ORF7a, (**d**) ORF6, (**e**) ORF1ab, (**f**) ORF10, and (**g**) 3′-untranslated region (UTR), respectively, and their genomic coordinates and lengths

release of proinflammatory cytokines. Although three different variations of the deletion were identified within Bangladesh, Australia, and Spain (345-, 138-, and 62-bp long, respectively) no specific clinical data are provided [46]. Secondly, Liu et al. identified two commonly seen 15- and 21-bp long deletions at the S1/S2 junction of the S protein (Fig. 2b) [48]. Lau et al. further indicated that a unique cleavage motif promoting viral infection, which will be discussed in detail in the "Evolution of SARS-CoV-2" section below, could be under strong selective pressure, thus the replication of this virus in Vero-E6 cells led to loss of adaptive function [49]. This finding suggested that deletion variants in *S* gene could hold potential for designing a vaccine. Thirdly, one 81-bp deletion in ORF7a was detected in the AZ-ASU2923 genome from a surveillance program in Arizona, USA, on March 16–19, 2020, which covered the putative signal peptide as well as two beta strands according to the protein structure [50] (Fig. 2c). Fourthly, two frameshifting deletions in 34- and 26-bp in ORF6 were found in Lyon, France, from March to early April, 2020. These deletions generated premature stop codons leading to ORF6 truncations [51] (Fig. 2d). Fifthly, a very small deletion of 9-bp, translated to three amino acids as KSF in nsp1 (position 686–694) of the ORF1ab polyprotein was identified, which could be associated with the C-terminal tail structure as determined using structural prediction modelling [52] (Fig. 2e). Sixthly, another six deletions in the ORF1ab polyprotein (Fig. 2e), one deletion of 35-bp in ORF10 (Fig. 2f), and three deletions of 29-, 12-, and 15-bp in the 3′-untranslated region (UTR) (Fig. 2g) were summarized [53, 54] (Fig. 2e). Lastly, this 15-bp deletion in the 3′-UTR disrupts a conserved secondary structure important for the replication of several animal CoV, suggesting the loading of replication complex might be altered [54] (Fig. 2g). Overall, most deletions identified to date tend to attenuate the virus. However, whether these isolates were collected from asymptomatic and/or mild COVID-19 patients, or collected at the later stages of COVID-19 disease progression remain an open question.

4 Evolution of SARS-CoV-2

There are considerable discussions and arguments on the origin of SARS-CoV-2 isolated from Wuhan, Hubei province, China, after the first reports of the virus [1]. At an early stage of this pandemics, SARS-CoV-2 sequences analyzed showed approximately 96% identity with a bat CoV (RaTG13 strain isolated in 2013), highlighting the capacity of viral spill-over from animals to humans [55]. Although the difference of their genome was of only 4%, there were about 1200 nucleotide changes, according to almost 30,000-bp of full lengths sequence analysis. Thus, these results suggested that many questions about viral transmissions across hosts and species still remain to be answered. Interestingly and surprisingly, Lam et al. reported a SARS-CoV-2-related coronavirus identified in Malayan pangolins (*Manis javanica*) during anti-smuggling operations in southern China, and possible recombination among SARS-CoV-2, and bat and pangolin CoVs [56]. Although pangolin-associated SARS-CoV-2-related coronavirus showed lower similarity with SARS-CoV-2 than the bat RaTG13 strain, these pangolin-associated strains exhibited strong similarity with the S protein RBD of SARS-CoV-2. This discovery suggested that pangolins could be considered potential hosts responsible for the emergence of new CoVs. Despite the lower genomic identity of the pangolin strain with SARS-CoV-2 compared with the bat RaTG13 virus, amino acids of pangolin RBD region shared 97.4% with SARS-CoV-2, including five key residues involved in the ACE2 binding process [56, 57]. Similarity plot analysis also suggested that the emerging SARS-CoV-2 could have resulted from the recombination of the bat and pangolin CoVs [58, 59].

Two notable features of SARS-CoV-2 genome are believed to demonstrate that this virus was not manmade [57]. Specifically, it is the optimized binding ability to the human receptor ACE2 [60–62], and its functional polybasic furin cleavage site (residues PRRA) resulted from 12-nucleotide insertion at the S1/S2 boundary of the S protein which in turn might affect the glycosylation status of S proteins [62]. This unique furin cleavage site of SARS-CoV-2 is absent in related *Betacoronaviruses*, even in the closest pangolin S protein [57, 62], leading to the precleavage of S during the biogenesis and helping the subsequent fusion process upon receptor binding [62, 63]. As a consequence, the highly efficient infection of SARS-CoV-2 in humans as compared with SARS and MERS is partly attributed to this unique feature [57, 64]. Moreover, an insertion event (residues "PAA") at the S1/S2 cleavage site was identified in a bat CoV (RmYN02) strain isolated in 2019 in Yunnan province, over 1500 km far from Wuhan, which reflects a natural evolution of this insertion in CoVs [65]. Although bats are important reservoir species for CoVs, the recombinant region and genetic ancestry of this bat strain or other SARS-CoV-2-related strains remain unknown. It would be urgent to collect samples and obtain related viral data from animal sources for identifying intermediate host of SARS-CoV-2 involved in the zoonotic transmission.

5 Epidemiology and Transmission of SARS-CoV-2

With rapidly increasing number of infections, the GISAID [4] provides a platform for sharing SARS-CoV-2 genetic information, such as footprint data, thereby providing valuable information for global spatiotemporal tracking of the virus, as well as clinical and epidemiologic data related to COVID-19 [66]. Viral phylodynamic analyses using genetic data have successfully inferred the transmission and evolutionary patterns of SARS-CoV-2 in Washington [10], California [67], and the East Coast [68] of USA, as well as in the Netherlands [69] in Europe. With increasing viral genomic data becoming available, it was noticed that the evolutionary rate of this virus is slower than its transmission rate. In other words, many genomes are identical and rapidly spreading worldwide, which makes difficult to infer viral transmissions. Furthermore, low availability of genomic data from some countries, for example, the Middle Eastern countries, introduces unrepresentative sampling and promotes misinterpretation of the phylogenetic tree [47]. To address these issues and find strategies to tackle the ongoing outbreak, it is important to accurately investigate and explain the global spreading of these emerging viruses. Worobey et al. revealed possible networks of SARS-CoV-2 transmissions in Europe and North America at the early stages of the global outbreak, showing introduction of the virus from China into both Italy and the Washington State, USA [70]. They also highlighted the value of establishing intensive and community-level virus surveillance during the ongoing outbreak, as well as for combining it with the efforts of genomic surveillance for closing the gap between and community-level transmission and global-level mitigation.

Due to the increasing number of SARS-CoV-2 sequences, a nomenclature system is urgently needed. Rambaut et al. presented a rational, dynamic, and tractable nomenclature strategy with hierarchical labels, named Pango lineages, which uses a phylogenetic framework and machine learning technology to describe the initial lineages and their members based on collection dates of sampling and geographical locations [71, 72]. A total of 81 viral lineages based on 27,767 complete SARS-CoV-2 genomes as of May 18, 2020 were identified, majorly belonging to A (including six lineages), B (16 lineages), and B.1 (>70 sublineages). Although Pango nomenclature was designed for facilitating scientific discussion, in particular by describing local and short-term genetic diversity, it generated large and long labels in a hierarchical structure. To address this, Nextstrain [73] provides a year-letter nomenclature to present dominantly and genetically well-defined clades, for example, 19A, 19B, 20A, 20B, and 20C. Taken together, uniformed nomenclatures can assist to track and better understand the patterns and determinants of globally spreading viruses, as well as providing assistance toward responding measures against this pandemic.

Phylogenetic approaches are capable of inferring viral transmission within travelers, of which travel paths represent an important source. However, this information was usually not fully incorporated into the phylogeographic inference process. In particular, discrete traits of phylogeographic inference were confounded due to spatiotemporal bias and unbalanced data in genome sampling. For example, genomic surveillances in Taiwan [47] and Australia [74] highlighted the viral diversity driving the epidemic of Iran and Turkey in the Middle Eastern countries, from which none or very few sequences have been reported. To address this issue, Lemey et al. incorporated individual travel history and unsampled diversity in Bayesian phylogeographic inference of SARS-CoV-2 [75]. Importantly, the imbalance of genome information incorporated into GIASID platform may have prevent a more comprehensively understanding of the viral transmission and epidemiology during this pandemic, since none or very few sequences have been reported from some countries. This phylogeographic analysis comprising lineages from undersampled locations is capable of improving the impact of sampling bias. Moreover, the Bayesian phylogeographic inference may help construct and/or reinforce specific transmission hypotheses using the metadata of travel history, as well as to better explain plausible routes of viral transmission within the epidemiological context, which are not represented in public sampling efforts.

To conclude, with the rapid and advanced sequencing technologies, genomic surveillance can help find the link and explain the clinical and epidemiological features of this pandemic for better responses to emerging infectious diseases. Following the first release of the SARS-CoV-2 genome [11], research laboratories globally have rapidly shared new sequences (more than one million sequences) on GISAID, as well as on GenBank from the National Center for Biotechnology Information, which currently provided more than 300,000 nucleotide sequences and more than 500,000 runs in the Sequence Read Archive (SRA) database by May 2021. Public analytic and visualization tools, such as Nextstrain [73], also provide near to real-time snapshots of the phylogenetic tree of global strains. Mutations are important evidences to trace the transmission of this virus; however, those public database strains might be generated from different sequencing platforms or technologies, as well as from different sample handling and preprocessing procedures, which might introduce technical artifacts rather than displaying real biological mutations. To overcome this technical challenge, a comprehensive pipeline is needed for rapid assembling and assessment of SARS-CoV-2 genomes, as well as for verifying mutations using different computational approaches.

6 Concluding Remarks

Genomic sequence data not only help trace the origin and evolution of SARS-CoV-2 but also explain its transmission.

Comprehensive phylodynamic inference may pave the way for a better understanding and verification of the transmission hypotheses, in particular to explain plausible routes of viral transmission within the ongoing epidemiological context. Considering the ability of CoV to spread through zoonotic transmission, it would be urgent to setup a global surveillance network for collecting samples and obtaining related viral genetic data from animal sources, in order to identify the intermediate host of SARS-CoV-2 that is responsible for its cross-species transmission.

Acknowledgments

This work was financially supported by the Research Center for Emerging Viral Infections from The Featured Areas Research Center Program within the framework of the Higher Education Sprout Project by the Ministry of Education (MOE) in Taiwan and the Ministry of Science and Technology (MOST), Taiwan (MOST 110-2222-E-182-004, 110-2634-F-182-001, 109-2320-B-182-045-MY2, and MOST 109-2327-B-182-002).

References

1. Zhu N, Zhang D, Wang W, Li X, Yang B, Song J, Zhao X, Huang B, Shi W, Lu R, Niu P, Zhan F, Ma X, Wang D, Xu W, Wu G, Gao GF, Tan W, China Novel Coronavirus Investigating and Research Team (2020) A novel coronavirus from patients with pneumonia in China, 2019. N Engl J Med 382(8): 727–733. https://doi.org/10.1056/NEJMoa2001017
2. Coronaviridae Study Group of the International Committee on Taxonomy of Viruses (2020) The species severe acute respiratory syndrome-related coronavirus: classifying 2019-nCoV and naming it SARS-CoV-2. Nat Microbiol 5(4):536–544. https://doi.org/10.1038/s41564-020-0695-z
3. Dong E, Du H, Gardner L (2020) An interactive web-based dashboard to track COVID-19 in real time. Lancet Infect Dis 20(5):533–534. https://doi.org/10.1016/S1473-3099(20)30120-1
4. Shu Y, McCauley J (2017) GISAID: global initiative on sharing all influenza data - from vision to reality. Euro Surveill 22(13):30494. https://doi.org/10.2807/1560-7917.ES.2017.22.13.30494
5. Lefkowitz EJ, Dempsey DM, Hendrickson RC, Orton RJ, Siddell SG, Smith DB (2018) Virus taxonomy: the database of the international committee on taxonomy of viruses (ICTV). Nucleic Acids Res 46(D1):D708–D717. https://doi.org/10.1093/nar/gkx932
6. Drosten C, Gunther S, Preiser W, van der Werf S, Brodt HR, Becker S, Rabenau H, Panning M, Kolesnikova L, Fouchier RA, Berger A, Burguiere AM, Cinatl J, Eickmann M, Escriou N, Grywna K, Kramme S, Manuguerra JC, Muller S, Rickerts V, Sturmer M, Vieth S, Klenk HD, Osterhaus AD, Schmitz H, Doerr HW (2003) Identification of a novel coronavirus in patients with severe acute respiratory syndrome. N Engl J Med 348(20):1967–1976. https://doi.org/10.1056/NEJMoa030747
7. van der Hoek L, Pyrc K, Jebbink MF, Vermeulen-Oost W, Berkhout RJ, Wolthers KC, Wertheim-van Dillen PM, Kaandorp J, Spaargaren J, Berkhout B (2004) Identification of a new human coronavirus. Nat Med 10(4): 368–373. https://doi.org/10.1038/nm1024
8. Woo PC, Lau SK, Chu CM, Chan KH, Tsoi HW, Huang Y, Wong BH, Poon RW, Cai JJ, Luk WK, Poon LL, Wong SS, Guan Y, Peiris JS, Yuen KY (2005) Characterization and complete genome sequence of a novel coronavirus, coronavirus HKU1, from patients with pneumonia. J Virol 79(2):884–895. https://doi.org/10.1128/JVI.79.2.884-895.2005
9. de Groot RJ, Baker SC, Baric RS, Brown CS, Drosten C, Enjuanes L, Fouchier RA,

Galiano M, Gorbalenya AE, Memish ZA, Perlman S, Poon LL, Snijder EJ, Stephens GM, Woo PC, Zaki AM, Zambon M, Ziebuhr J (2013) Middle East respiratory syndrome coronavirus (MERS-CoV): announcement of the coronavirus study group. J Virol 87(14): 7790–7792. https://doi.org/10.1128/JVI.01244-13

10. Bedford T, Greninger AL, Roychoudhury P, Starita LM, Famulare M, Huang ML, Nalla A, Pepper G, Reinhardt A, Xie H, Shrestha L, Nguyen TN, Adler A, Brandstetter E, Cho S, Giroux D, Han PD, Fay K, Frazar CD, Ilcisin M, Lacombe K, Lee J, Kiavand A, Richardson M, Sibley TR, Truong M, Wolf CR, Nickerson DA, Rieder MJ, Englund JA, Seattle Flu Study I, Hadfield J, Hodcroft EB, Huddleston J, Moncla LH, Muller NF, Neher RA, Deng X, Gu W, Federman S, Chiu C, Duchin JS, Gautom R, Melly G, Hiatt B, Dykema P, Lindquist S, Queen K, Tao Y, Uehara A, Tong S, MacCannell D, Armstrong GL, Baird GS, Chu HY, Shendure J, Jerome KR (2020) Cryptic transmission of SARS-CoV-2 in Washington state. Science 370(6516):571–575. https://doi.org/10.1126/science.abc0523
11. Wu F, Zhao S, Yu B, Chen YM, Wang W, Song ZG, Hu Y, Tao ZW, Tian JH, Pei YY, Yuan ML, Zhang YL, Dai FH, Liu Y, Wang QM, Zheng JJ, Xu L, Holmes EC, Zhang YZ (2020) A new coronavirus associated with human respiratory disease in China. Nature 579(7798):265–269. https://doi.org/10.1038/s41586-020-2008-3
12. Boursnell ME, Brown TD, Foulds IJ, Green PF, Tomley FM, Binns MM (1987) Completion of the sequence of the genome of the coronavirus avian infectious bronchitis virus. J Gen Virol 68(Pt 1):57–77. https://doi.org/10.1099/0022-1317-68-1-57
13. Brierley I, Boursnell ME, Binns MM, Bilimoria B, Blok VC, Brown TD, Inglis SC (1987) An efficient ribosomal frame-shifting signal in the polymerase-encoding region of the coronavirus IBV. EMBO J 6(12): 3779–3785
14. Brierley I, Digard P, Inglis SC (1989) Characterization of an efficient coronavirus ribosomal frameshifting signal: requirement for an RNA pseudoknot. Cell 57(4):537–547. https://doi.org/10.1016/0092-8674(89)90124-4
15. Chan JF, Kok KH, Zhu Z, Chu H, To KK, Yuan S, Yuen KY (2020) Genomic characterization of the 2019 novel human-pathogenic coronavirus isolated from a patient with atypical pneumonia after visiting Wuhan. Emerg Microbes Infect 9(1):221–236. https://doi.org/10.1080/22221751.2020.1719902
16. Harcourt BH, Jukneliene D, Kanjanahaluethai A, Bechill J, Severson KM, Smith CM, Rota PA, Baker SC (2004) Identification of severe acute respiratory syndrome coronavirus replicase products and characterization of papain-like protease activity. J Virol 78(24):13600–13612. https://doi.org/10.1128/JVI.78.24.13600-13612.2004
17. V'Kovski P, Kratzel A, Steiner S, Stalder H, Thiel V (2021) Coronavirus biology and replication: implications for SARS-CoV-2. Nat Rev Microbiol 19(3):155–170. https://doi.org/10.1038/s41579-020-00468-6
18. Littler DR, Gully BS, Colson RN, Rossjohn J (2020) Crystal structure of the SARS-CoV-2 - non-structural protein 9, Nsp9. iScience 23(7): 101258. https://doi.org/10.1016/j.isci.2020.101258
19. Yin W, Mao C, Luan X, Shen DD, Shen Q, Su H, Wang X, Zhou F, Zhao W, Gao M, Chang S, Xie YC, Tian G, Jiang HW, Tao SC, Shen J, Jiang Y, Jiang H, Xu Y, Zhang S, Zhang Y, Xu HE (2020) Structural basis for inhibition of the RNA-dependent RNA polymerase from SARS-CoV-2 by remdesivir. Science 368(6498):1499–1504. https://doi.org/10.1126/science.abc1560
20. Hillen HS, Kokic G, Farnung L, Dienemann C, Tegunov D, Cramer P (2020) Structure of replicating SARS-CoV-2 polymerase. Nature 584(7819):154–156. https://doi.org/10.1038/s41586-020-2368-8
21. Tvarogova J, Madhugiri R, Bylapudi G, Ferguson LJ, Karl N, Ziebuhr J (2019) Identification and characterization of a human coronavirus 229E nonstructural protein 8-associated RNA 3′-terminal adenylyltransferase activity. J Virol 93(12):e00291-19. https://doi.org/10.1128/JVI.00291-19
22. Chen J, Malone B, Llewellyn E, Grasso M, Shelton PMM, Olinares PDB, Maruthi K, Eng ET, Vatandaslar H, Chait BT, Kapoor TM, Darst SA, Campbell EA (2020) Structural basis for helicase-polymerase coupling in the SARS-CoV-2 replication-transcription complex. Cell 182(6):1560–1573.e13. https://doi.org/10.1016/j.cell.2020.07.033
23. Robson F, Khan KS, Le TK, Paris C, Demirbag S, Barfuss P, Rocchi P, Ng WL (2020) Coronavirus RNA proofreading: molecular basis and therapeutic targeting. Mol Cell 79(5):710–727. https://doi.org/10.1016/j.molcel.2020.07.027
24. Krafcikova P, Silhan J, Nencka R, Boura E (2020) Structural analysis of the SARS-CoV-2 methyltransferase complex involved in RNA

cap creation bound to sinefungin. Nat Commun 11(1):3717. https://doi.org/10.1038/s41467-020-17495-9

25. Ma Y, Wu L, Shaw N, Gao Y, Wang J, Sun Y, Lou Z, Yan L, Zhang R, Rao Z (2015) Structural basis and functional analysis of the SARS coronavirus nsp14-nsp10 complex. Proc Natl Acad Sci U S A 112(30):9436–9441. https://doi.org/10.1073/pnas.1508686112
26. Sola I, Almazan F, Zuniga S, Enjuanes L (2015) Continuous and discontinuous RNA synthesis in coronaviruses. Annu Rev Virol 2(1):265–288. https://doi.org/10.1146/annurev-virology-100114-055218
27. Lei X, Dong X, Ma R, Wang W, Xiao X, Tian Z, Wang C, Wang Y, Li L, Ren L, Guo F, Zhao Z, Zhou Z, Xiang Z, Wang J (2020) Activation and evasion of type I interferon responses by SARS-CoV-2. Nat Commun 11(1):3810. https://doi.org/10.1038/s41467-020-17665-9
28. Miorin L, Kehrer T, Sanchez-Aparicio MT, Zhang K, Cohen P, Patel RS, Cupic A, Makio T, Mei M, Moreno E, Danziger O, White KM, Rathnasinghe R, Uccellini M, Gao S, Aydillo T, Mena I, Yin X, Martin-Sancho L, Krogan NJ, Chanda SK, Schotsaert M, Wozniak RW, Ren Y, Rosenberg BR, Fontoura BMA, Garcia-Sastre A (2020) SARS-CoV-2 Orf6 hijacks Nup98 to block STAT nuclear import and antagonize interferon signaling. Proc Natl Acad Sci U S A 117(45):28344–28354. https://doi.org/10.1073/pnas.2016650117
29. Jiang HW, Zhang HN, Meng QF, Xie J, Li Y, Chen H, Zheng YX, Wang XN, Qi H, Zhang J, Wang PH, Han ZG, Tao SC (2020) SARS-CoV-2 Orf9b suppresses type I interferon responses by targeting TOM70. Cell Mol Immunol 17(9):998–1000. https://doi.org/10.1038/s41423-020-0514-8
30. Young BE, Fong SW, Chan YH, Mak TM, Ang LW, Anderson DE, Lee CY, Amrun SN, Lee B, Goh YS, Su YCF, Wei WE, Kalimuddin S, Chai LYA, Pada S, Tan SY, Sun L, Parthasarathy P, Chen YYC, Barkham T, Lin RTP, Maurer-Stroh S, Leo YS, Wang LF, Renia L, Lee VJ, Smith GJD, Lye DC, Ng LFP (2020) Effects of a major deletion in the SARS-CoV-2 genome on the severity of infection and the inflammatory response: an observational cohort study. Lancet 396(10251):603–611. https://doi.org/10.1016/S0140-6736(20)31757-8
31. Kim D, Lee JY, Yang JS, Kim JW, Kim VN, Chang H (2020) The architecture of SARS-CoV-2 transcriptome. Cell 181(4):914–921.e10. https://doi.org/10.1016/j.cell.2020.04.011
32. Finkel Y, Mizrahi O, Nachshon A, Weingarten-Gabbay S, Morgenstern D, Yahalom-Ronen Y, Tamir H, Achdout H, Stein D, Israeli O, Beth-Din A, Melamed S, Weiss S, Israely T, Paran N, Schwartz M, Stern-Ginossar N (2020) The coding capacity of SARS-CoV-2. Nature 589(7840):125–130. https://doi.org/10.1038/s41586-020-2739-1
33. Mandala VS, McKay MJ, Shcherbakov AA, Dregni AJ, Kolocouris A, Hong M (2020) Structure and drug binding of the SARS-CoV-2 envelope protein transmembrane domain in lipid bilayers. Nat Struct Mol Biol 27(12):1202–1208. https://doi.org/10.1038/s41594-020-00536-8
34. Siu KL, Chan CP, Kok KH, Chiu-Yat Woo P, Jin DY (2014) Suppression of innate antiviral response by severe acute respiratory syndrome coronavirus M protein is mediated through the first transmembrane domain. Cell Mol Immunol 11(2):141–149. https://doi.org/10.1038/cmi.2013.61
35. Mu J, Fang Y, Yang Q, Shu T, Wang A, Huang M, Jin L, Deng F, Qiu Y, Zhou X (2020) SARS-CoV-2 N protein antagonizes type I interferon signaling by suppressing phosphorylation and nuclear translocation of STAT1 and STAT2. Cell Discov 6:65. https://doi.org/10.1038/s41421-020-00208-3
36. Wang Q, Zhang Y, Wu L, Niu S, Song C, Zhang Z, Lu G, Qiao C, Hu Y, Yuen KY, Wang Q, Zhou H, Yan J, Qi J (2020) Structural and functional basis of SARS-CoV-2 entry by using human ACE2. Cell 181(4):894–904.e9. https://doi.org/10.1016/j.cell.2020.03.045
37. Lan J, Ge J, Yu J, Shan S, Zhou H, Fan S, Zhang Q, Shi X, Wang Q, Zhang L, Wang X (2020) Structure of the SARS-CoV-2 spike receptor-binding domain bound to the ACE2 receptor. Nature 581(7807):215–220. https://doi.org/10.1038/s41586-020-2180-5
38. Benton DJ, Wrobel AG, Xu P, Roustan C, Martin SR, Rosenthal PB, Skehel JJ, Gamblin SJ (2020) Receptor binding and priming of the spike protein of SARS-CoV-2 for membrane fusion. Nature 588(7837):327–330. https://doi.org/10.1038/s41586-020-2772-0
39. Shang J, Wan Y, Luo C, Ye G, Geng Q, Auerbach A, Li F (2020) Cell entry mechanisms of SARS-CoV-2. Proc Natl Acad Sci U S A 117(21):11727–11734. https://doi.org/10.1073/pnas.2003138117
40. Alouane T, Laamarti M, Essabbar A, Hakmi M, Bouricha EM, Chemao-Elfihri MW, Kartti S, Boumajdi N, Bendani H, Laamarti R, Ghrifi F,

Allam L, Aanniz T, Ouadghiri M, El Hafidi N, El Jaoudi R, Benrahma H, Attar JE, Mentag R, Sbabou L, Nejjari C, Amzazi S, Belyamani L, Ibrahimi A (2020) Genomic diversity and hotspot mutations in 30,983 SARS-CoV-2 genomes: moving toward a universal vaccine for the "confined virus"? Pathogens 9(10):829. https://doi.org/10.3390/pathogens9100829

41. Grubaugh ND, Petrone ME, Holmes EC (2020) We shouldn't worry when a virus mutates during disease outbreaks. Nat Microbiol 5(4):529–530. https://doi.org/10.1038/s41564-020-0690-4

42. Yurkovetskiy L, Wang X, Pascal KE, Tomkins-Tinch C, Nyalile TP, Wang Y, Baum A, Diehl WE, Dauphin A, Carbone C, Veinotte K, Egri SB, Schaffner SF, Lemieux JE, Munro JB, Rafique A, Barve A, Sabeti PC, Kyratsous CA, Dudkina NV, Shen K, Luban J (2020) Structural and functional analysis of the D614G SARS-CoV-2 spike protein variant. Cell 183(3):739–751.e8. https://doi.org/10.1016/j.cell.2020.09.032

43. Korber B, Fischer WM, Gnanakaran S, Yoon H, Theiler J, Abfalterer W, Hengartner N, Giorgi EE, Bhattacharya T, Foley B, Hastie KM, Parker MD, Partridge DG, Evans CM, Freeman TM, de Silva TI, Sheffield C-GG, McDanal C, Perez LG, Tang H, Moon-Walker A, Whelan SP, LaBranche CC, Saphire EO, Montefiori DC (2020) Tracking changes in SARS-CoV-2 spike: evidence that D614G increases infectivity of the COVID-19 virus. Cell 182(4): 812–827.e19. https://doi.org/10.1016/j.cell.2020.06.043

44. Hou YJ, Chiba S, Halfmann P, Ehre C, Kuroda M, Dinnon KH 3rd, Leist SR, Schafer A, Nakajima N, Takahashi K, Lee RE, Mascenik TM, Graham R, Edwards CE, Tse LV, Okuda K, Markmann AJ, Bartelt L, de Silva A, Margolis DM, Boucher RC, Randell SH, Suzuki T, Gralinski LE, Kawaoka Y, Baric RS (2020) SARS-CoV-2 D614G variant exhibits efficient replication ex vivo and transmission in vivo. Science 370(6523):1464–1468. https://doi.org/10.1126/science.abe8499

45. Plante JA, Liu Y, Liu J, Xia H, Johnson BA, Lokugamage KG, Zhang X, Muruato AE, Zou J, Fontes-Garfias CR, Mirchandani D, Scharton D, Bilello JP, Ku Z, An Z, Kalveram B, Freiberg AN, Menachery VD, Xie X, Plante KS, Weaver SC, Shi PY (2020) Spike mutation D614G alters SARS-CoV-2 fitness. Nature 592(7852):116–121. https://doi.org/10.1038/s41586-020-2895-3

46. Su YCF, Anderson DE, Young BE, Linster M, Zhu F, Jayakumar J, Zhuang Y, Kalimuddin S, Low JGH, Tan CW, Chia WN, Mak TM, Octavia S, Chavatte JM, Lee RTC, Pada S, Tan SY, Sun L, Yan GZ, Maurer-Stroh S, Mendenhall IH, Leo YS, Lye DC, Wang LF, Smith GJD (2020) Discovery and genomic characterization of a 382-nucleotide deletion in ORF7b and ORF8 during the early evolution of SARS-CoV-2. mBio 11(4):e01610-20. https://doi.org/10.1128/mBio.01610-20

47. Gong YN, Tsao KC, Hsiao MJ, Huang CG, Huang PN, Huang PW, Lee KM, Liu YC, Yang SL, Kuo RL, Chen KF, Liu YC, Huang SY, Huang HI, Liu MT, Yang JR, Chiu CH, Yang CT, Chen GW, Shih SR (2020) SARS-CoV-2 genomic surveillance in Taiwan revealed novel ORF8-deletion mutant and clade possibly associated with infections in Middle East. Emerg Microbes Infect 9(1): 1457–1466. https://doi.org/10.1080/22221751.2020.1782271

48. Liu Z, Zheng H, Lin H, Li M, Yuan R, Peng J, Xiong Q, Sun J, Li B, Wu J, Yi L, Peng X, Zhang H, Zhang W, Hulswit RJG, Loman N, Rambaut A, Ke C, Bowden TA, Pybus OG, Lu J (2020) Identification of common deletions in the spike protein of severe acute respiratory syndrome coronavirus 2. J Virol 94(17): e00790-20. https://doi.org/10.1128/JVI.00790-20

49. Lau SY, Wang P, Mok BW, Zhang AJ, Chu H, Lee AC, Deng S, Chen P, Chan KH, Song W, Chen Z, To KK, Chan JF, Yuen KY, Chen H (2020) Attenuated SARS-CoV-2 variants with deletions at the S1/S2 junction. Emerg Microbes Infect 9(1):837–842. https://doi.org/10.1080/22221751.2020.1756700

50. Holland LA, Kaelin EA, Maqsood R, Estifanos B, Wu LI, Varsani A, Halden RU, Hogue BG, Scotch M, Lim ES (2020) An 81-nucleotide deletion in SARS-CoV-2 ORF7a identified from sentinel surveillance in Arizona (January to March 2020). J Virol 94 (14):e00711-20. https://doi.org/10.1128/JVI.00711-20

51. Queromes G, Destras G, Bal A, Regue H, Burfin G, Brun S, Fanget R, Morfin F, Valette M, Trouillet-Assant S, Lina B, Frobert E, Josset L (2021) Characterization of SARS-CoV-2 ORF6 deletion variants detected in a nosocomial cluster during routine genomic surveillance, Lyon, France. Emerg Microbes Infect 10(1):167–177. https://doi.org/10.1080/22221751.2021.1872351

52. Benedetti F, Snyder GA, Giovanetti M, Angeletti S, Gallo RC, Ciccozzi M, Zella D (2020) Emerging of a SARS-CoV-2 viral strain

with a deletion in nsp1. J Transl Med 18(1): 329. https://doi.org/10.1186/s12967-020-02507-5

53. Islam MR, Hoque MN, Rahman MS, Alam A, Akther M, Puspo JA, Akter S, Sultana M, Crandall KA, Hossain MA (2020) Genome-wide analysis of SARS-CoV-2 virus strains circulating worldwide implicates heterogeneity. Sci Rep 10(1):14004. https://doi.org/10.1038/s41598-020-70812-6
54. McNamara RP, Caro-Vegas C, Landis JT, Moorad R, Pluta LJ, Eason AB, Thompson C, Bailey A, Villamor FCS, Lange PT, Wong JP, Seltzer T, Seltzer J, Zhou Y, Vahrson W, Juarez A, Meyo JO, Calabre T, Broussard G, Rivera-Soto R, Chappell DL, Baric RS, Damania B, Miller MB, Dittmer DP (2020) High-density amplicon sequencing identifies community spread and ongoing evolution of SARS-CoV-2 in the Southern United States. Cell Rep 33(5):108352. https://doi.org/10.1016/j.celrep.2020.108352
55. Zhou P, Yang XL, Wang XG, Hu B, Zhang L, Zhang W, Si HR, Zhu Y, Li B, Huang CL, Chen HD, Chen J, Luo Y, Guo H, Jiang RD, Liu MQ, Chen Y, Shen XR, Wang X, Zheng XS, Zhao K, Chen QJ, Deng F, Liu LL, Yan B, Zhan FX, Wang YY, Xiao GF, Shi ZL (2020) A pneumonia outbreak associated with a new coronavirus of probable bat origin. Nature 579(7798):270–273. https://doi.org/10.1038/s41586-020-2012-7
56. Lam TT, Jia N, Zhang YW, Shum MH, Jiang JF, Zhu HC, Tong YG, Shi YX, Ni XB, Liao YS, Li WJ, Jiang BG, Wei W, Yuan TT, Zheng K, Cui XM, Li J, Pei GQ, Qiang X, Cheung WY, Li LF, Sun FF, Qin S, Huang JC, Leung GM, Holmes EC, Hu YL, Guan Y, Cao WC (2020) Identifying SARS-CoV-2-related coronaviruses in Malayan pangolins. Nature 583(7815):282–285. https://doi.org/10.1038/s41586-020-2169-0
57. Andersen KG, Rambaut A, Lipkin WI, Holmes EC, Garry RF (2020) The proximal origin of SARS-CoV-2. Nat Med 26(4):450–452. https://doi.org/10.1038/s41591-020-0820-9
58. Xiao K, Zhai J, Feng Y, Zhou N, Zhang X, Zou JJ, Li N, Guo Y, Li X, Shen X, Zhang Z, Shu F, Huang W, Li Y, Zhang Z, Chen RA, Wu YJ, Peng SM, Huang M, Xie WJ, Cai QH, Hou FH, Chen W, Xiao L, Shen Y (2020) Isolation of SARS-CoV-2-related coronavirus from Malayan pangolins. Nature 583(7815): 286–289. https://doi.org/10.1038/s41586-020-2313-x
59. Li X, Giorgi EE, Marichannegowda MH, Foley B, Xiao C, Kong XP, Chen Y, Gnanakaran S, Korber B, Gao F (2020) Emergence of SARS-CoV-2 through recombination and strong purifying selection. Sci Adv 6(27): eabb9153. https://doi.org/10.1126/sciadv.abb9153
60. Wan Y, Shang J, Graham R, Baric RS, Li F (2020) Receptor recognition by the novel coronavirus from Wuhan: an analysis based on decade-long structural studies of SARS coronavirus. J Virol 94(7):e00127-20. https://doi.org/10.1128/JVI.00127-20
61. Wrapp D, Wang N, Corbett KS, Goldsmith JA, Hsieh CL, Abiona O, Graham BS, McLellan JS (2020) Cryo-EM structure of the 2019-nCoV spike in the prefusion conformation. Science 367(6483):1260–1263. https://doi.org/10.1126/science.abb2507
62. Walls AC, Park YJ, Tortorici MA, Wall A, McGuire AT, Veesler D (2020) Structure, function, and antigenicity of the SARS-CoV-2 spike glycoprotein. Cell 181:281–292.e6. https://doi.org/10.1016/j.cell.2020.02.058
63. Jaimes JA, Millet JK, Whittaker GR (2020) Proteolytic cleavage of the SARS-CoV-2 spike protein and the role of the novel S1/S2 site. iScience 23(6):101212. https://doi.org/10.1016/j.isci.2020.101212
64. Zhang YZ, Holmes EC (2020) A genomic perspective on the origin and emergence of SARS-CoV-2. Cell 181(2):223–227. https://doi.org/10.1016/j.cell.2020.03.035
65. Zhou H, Chen X, Hu T, Li J, Song H, Liu Y, Wang P, Liu D, Yang J, Holmes EC, Hughes AC, Bi Y, Shi W (2020) A novel bat coronavi rus closely related to SARS-CoV-2 contains natural insertions at the S1/S2 cleavage site of the spike protein. Curr Biol 30(11): 2196–2203.e3. https://doi.org/10.1016/j.cub.2020.05.023
66. Lam TT (2020) Tracking the genomic footprints of SARS-CoV-2 transmission. Trends Genet 36(8):544–546. https://doi.org/10.1016/j.tig.2020.05.009
67. Deng X, Gu W, Federman S, du Plessis L, Pybus OG, Faria NR, Wang C, Yu G, Bushnell B, Pan CY, Guevara H, Sotomayor-Gonzalez A, Zorn K, Gopez A, Servellita V, Hsu E, Miller S, Bedford T, Greninger AL, Roychoudhury P, Starita LM, Famulare M, Chu HY, Shendure J, Jerome KR, Anderson C, Gangavarapu K, Zeller M, Spencer E, Andersen KG, MacCannell D, Paden CR, Li Y, Zhang J, Tong S, Armstrong G, Morrow S, Willis M, Matyas BT, Mase S, Kasirye O, Park M, Masinde G, Chan C, Yu AT, Chai SJ, Villarino E, Bonin B, Wadford DA, Chiu CY (2020) Genomic surveillance reveals multiple introductions of

SARS-CoV-2 into Northern California. Science 369(6503):582–587. https://doi.org/10.1126/science.abb9263
68. Fauver JR, Petrone ME, Hodcroft EB, Shioda K, Ehrlich HY, Watts AG, Vogels CBF, Brito AF, Alpert T, Muyombwe A, Razeq J, Downing R, Cheemarla NR, Wyllie AL, Kalinich CC, Ott IM, Quick J, Loman NJ, Neugebauer KM, Greninger AL, Jerome KR, Roychoudhury P, Xie H, Shrestha L, Huang ML, Pitzer VE, Iwasaki A, Omer SB, Khan K, Bogoch II, Martinello RA, Foxman EF, Landry ML, Neher RA, Ko AI, Grubaugh ND (2020) Coast-to-coast spread of SARS-CoV-2 during the early epidemic in the United States. Cell 181(5):990–996.e5. https://doi.org/10.1016/j.cell.2020.04.021
69. Oude Munnink BB, Nieuwenhuijse DF, Stein M, O'Toole A, Haverkate M, Mollers M, Kamga SK, Schapendonk C, Pronk M, Lexmond P, van der Linden A, Bestebroer T, Chestakova I, Overmars RJ, van Nieuwkoop S, Molenkamp R, van der Eijk AA, GeurtsvanKessel C, Vennema H, Meijer A, Rambaut A, van Dissel J, Sikkema RS, Timen A, Koopmans M, Dutch-Covid-19 response team (2020) Rapid SARS-CoV-2 whole-genome sequencing and analysis for informed public health decision-making in the Netherlands. Nat Med 26(9):1405–1410. https://doi.org/10.1038/s41591-020-0997-y
70. Worobey M, Pekar J, Larsen BB, Nelson MI, Hill V, Joy JB, Rambaut A, Suchard MA, Wertheim JO, Lemey P (2020) The emergence of SARS-CoV-2 in Europe and North America. Science 370(6516):564–570. https://doi.org/10.1126/science.abc8169
71. Rambaut A, Holmes EC, O'Toole A, Hill V, McCrone JT, Ruis C, du Plessis L, Pybus OG (2020) A dynamic nomenclature proposal for SARS-CoV-2 lineages to assist genomic epidemiology. Nat Microbiol 5(11):1403–1407. https://doi.org/10.1038/s41564-020-0770-5
72. Rambaut A, Holmes EC, O'Toole A, Hill V, McCrone JT, Ruis C, du Plessis L, Pybus OG (2021) Addendum: a dynamic nomenclature proposal for SARS-CoV-2 lineages to assist genomic epidemiology. Nat Microbiol 6(3):415. https://doi.org/10.1038/s41564-021-00872-5
73. Hadfield J, Megill C, Bell SM, Huddleston J, Potter B, Callender C, Sagulenko P, Bedford T, Neher RA (2018) Nextstrain: real-time tracking of pathogen evolution. Bioinformatics 34(23):4121–4123. https://doi.org/10.1093/bioinformatics/bty407
74. Eden JS, Rockett R, Carter I, Rahman H, de Ligt J, Hadfield J, Storey M, Ren X, Tulloch R, Basile K, Wells J, Byun R, Gilroy N, O'Sullivan MV, Sintchenko V, Chen SC, Maddocks S, Sorrell TC, Holmes EC, Dwyer DE, Kok J, 2019-nCoV Study Group (2020) An emergent clade of SARS-CoV-2 linked to returned travellers from Iran. Virus Evol 6(1):veaa027. https://doi.org/10.1093/ve/veaa027
75. Lemey P, Hong SL, Hill V, Baele G, Poletto C, Colizza V, O'Toole A, McCrone JT, Andersen KG, Worobey M, Nelson MI, Rambaut A, Suchard MA (2020) Accommodating individual travel history and unsampled diversity in Bayesian phylogeographic inference of SARS-CoV-2. Nat Commun 11(1):5110. https://doi.org/10.1038/s41467-020-18877-9

Chapter 2

Molecular Epidemiology of SARS-CoV-2 by Sequencing

Yan Yan and Qinxue Hu

Abstract

Sequences of severe acute respiratory syndrome coronavirus 2 (SARS-CoV-2) contains preliminary information on the phylodynamics and phylogeography of this new virus. A maximum clade credibility tree (MCCT) was constructed using available whole genome sequences of SARS-CoV-2 and highly similar whole genome sequences from bat SARS-like coronavirus, which are available in GenBank. In this chapter, we describe the molecular epidemiology of SARS-CoV-2 by sequencing the viral genomes from confirmed COVID-19 patients, utilizing methods such as target fragment amplification, sequencing, alignment, and maximum similarity mapping.

Key words SARS-CoV-2, COVID-19, Molecular epidemiology, Sequencing, Genotype

1 Introduction

Severe acute respiratory syndrome coronavirus-2 (SARS-CoV-2), a novel human enveloped positive-strand RNA virus in the β genus of the coronavirus family, was first reported in late December 2019 in Wuhan, China [1–3]. SARS-CoV-2 may be derived from an animal host, and it may have infected humans after undergoing genetic evolution [4]. The molecular epidemiology of SARS-CoV-2 concerns the molecular similarity and divergence between SARS-CoV-2 and other related coronaviruses or coronaviruses originating in different countries [2, 5]. Therefore, sequence analysis is helpful in determining the virology, transmission rules, and virulence of SARS-CoV-2 according to the sequence information of related viruses [6].

SARS-CoV-2 sequence information is also useful in molecular evolutionary analyses investigating transmission among species and evolutionary status among affected countries; these findings are in agreement with epidemiological observations and may be employed to determine the most likely geographic origin of the virus according to molecular clock analysis [7]. SARS-CoV-2 sequences contain information on the genetic evolutionary

Justin Jang Hann Chu et al. (eds.), *SARS-CoV-2: Methods and Protocols*,
Methods in Molecular Biology, vol. 2452, https://doi.org/10.1007/978-1-0716-2111-0_2,

dynamics underlying the epidemic, which demonstrates the urgent importance of developing effective molecular surveillance strategies for SARS-CoV-2 among patients; these strategies are important for prevention efforts, as they may enable researchers to trace the source of the infection, evaluate control effectiveness and develop vaccines. SARS-CoV-2 genomes continue to be generated at a rate far greater than those of any other pathogens [8], with more than 200,000 complete genomes being available on GISAID as of December 2020. In this chapter, we introduce protocols for describing the molecular epidemiology of SARS-CoV-2 by sequencing. Phylogenetic network analyses of SARS-CoV-2 genomes from around the world may similarly be employed to help trace undocumented COVID-19 infection sources [7, 9].

2 Materials

2.1 Epidemiological Data

The epidemiological data of target strains were collected from local medical or epidemiological network databases; these data included demographic information, travel history, date of symptom onset and laboratory confirmation for local SARS-CoV-2-infected patients.

2.2 Open Access Websites of SARS-CoV-2 Sequences

SARS-CoV-2 epidemic and available full genome sequences were obtained from the following open access websites (*see* **Note 1**):

1. Global Initiative on Sharing All Influenza Database (GISAID, https://www.gisaid.org/).
2. BLAST (NCBI, https://www.ncbi.nlm.nih.gov/).
3. Coronavirus Global Shared Database (CGSD, http://nmdc.cn/coronavirus).

2.3 Sample Handling and SARS-CoV-2 RNA Detection

1. Viral transport medium (VTM) (BD Diagnostic systems, USA). Samples were stored at room temperature for a short period of time and transported at low temperature (~4 °C).
2. QIAamp viral RNA Mini Kit. Reagents were stored at room temperature.
3. The primer and probe sequences for the open reading frame (ORF) 1b gene assay are as follow [3]: 5′-TGG GGY TTT ACR GGT AAC CT-3′ (Forward; Y is C or T, R is A or G), 5′-AAC RCG CTT AAC AAA GCA CTC-3′ (Reverse; R is A or G) and 5′-TAG TTG TGA TGC WAT CAT GAC TAG-3′ (probe is in 5′-FAM/ZEN/3′-IBFQ format; W is A or T), while primer and probe sequences for the N gene assay are 5′-TAA TCA GAC AAG GAA CTG ATT A-3′ (Forward), 5′-CGA AGG TGT GAC TTC CAT G-3′ (Reverse) and 5′-GCA AAT TGT GCA ATT TGC GG-3′ (probe is in 5′-FAM/ZEN/3′-IBFQ

format). Primer and probe products were supplied by Sangon Biotech (Shanghai, China). The primers were prepared in a stock of 100 μM, and probes were prepared in a stock of 50 μM; primer and probe working solutions of 20 μM and 10 μM were prepared for use in real-time RT-PCR, respectively. Reagents were stored at −20 °C until use.

4. SARS-CoV-2 primers [10] (*see* **Note 2**): Partial ORF 1a: 5′-AAT AAT TGG TTG AAG CAG C-3′ (Forward) and 5′-TCT ATA AGT TTT GAT GGT-3′ (Reverse); Full ORF 8: 5′-CTT ATT ATC TTT TGG TTC TCC-3′ (Forward) and 5′-GGG GTC CAT TAT CAG ACA TTT T-3′ (Reverse) (Sangon Biotech, China). The primers were prepared in a stock of 100 μM. A working solution of 20 μM was prepared for use in one-step RT-PCR. Reagents were stored at −20 °C until use.
5. One Step PrimeScript™ RT-PCR Kit (Perfect Real Time) (TaKaRa, China) containing 2× One Step RT-PCR Buffer III, TaKaRa Ex Taq HS, PrimeScript RT Enzyme Mix II, RNase-Free dH_2O. Reagents were stored at −20 °C.
6. PrimeScript™ One Step RT-PCR Kit Ver. 2 (TaKaRa, China) containing PrimeScript One Step Enzyme Mix, 2× Step buffer, and RNase Free dH_2O. Reagents were stored at −20 °C.
7. Applied Biosystems ABI 7500 PCR system (Thermo Fisher Scientific, USA).
8. 1× Tris–acetate–EDTA (TAE) buffer: dilute 10 × TAE buffer 1:10 with Milli-Q water. Reagents were stored at room temperature.
9. 1% agarose gels: Prepare the gels by diluting 1 g agarose gel in 100 ml 1× TAE buffer; next, add 0.6 μl GelGreen (*see* **Note 3**). Reagents were stored at room temperature.
10. DL2,000 DNA Marker. Reagents were stored at 4 °C.

3 Methods

1. Posterior oropharyngeal mucosal or nasopharyngeal swab specimens of suspected COVID-19 patients were collected by swabbing the posterior oropharynx or nasopharyngeal areas, and the samples were preserved in VTM.
2. Inactivation of the virus: specimens were incubated in a 56 °C water bath for 30 min before specimen repacking and nucleic acid detection.
3. After vortexing and squeezing the swab tip, the original specimens were divided, and 200 μl of the solution was taken for RNA extraction according to the manufacturer's instructions using the Spin Protocol (Qiagen QIAamp viral RNA mini kit

handbook). The aliquot samples were stored at −80 °C, and total RNA was extracted (60 μl).

4. Qualitative testing of SARS-CoV-2 nucleic acid in patient specimens was performed using ORF 1b/N gene primers, probes and commercial reagents (One Step PrimeScript™ RT-PCR Kit (Perfect Real Time) by real-time RT-PCR following the manufacturer's instructions (TaKaRa, China). The products of the ORF 1b and N genes are 132 bp and 110 bp, respectively [3]. Briefly, 20 μl reaction volume reagent contained 10 μl of 2× One Step RT-PCR Buffer III, 0.4 μl of PrimeScript RT Enzyme Mix II, 0.4 μl of PCR Forward Primer (10 μM), 0.4 μl of PCR Reverse Primer (10 μM), 0.8 μl of Probe (5 μM), 3 μl of total RNA (10 pg ~ 100 ng), and 4.6 μl of dH_2O. Real-time RT-PCR was performed with the following steps: 42 °C for 5 min; 94 °C for 10 s (1 cycle of amplification); 95 °C for 5 s, 60 °C for 20 s (40 cycles of amplification) detected in an Applied Biosystems ABI 7500 PCR system. A cycle threshold value (Ct value) less than 37 was defined as a positive record, and a Ct value exceeding 40 was defined as a negative record.

3.1 Complete Genome Sequence of Positive SARS-CoV-2 and Phylogenetic Analyses

1. The SARS-CoV-2-positive total RNA extracted from the original specimen was employed for full genome sequencing.
2. Complete genome sequencing was performed using the Illumina MiSeq system with the Burrows–Wheeler Aligner MEM algorithm (BWAMEM) 0.7.5a-r405 assembly method.
3. The existing SARS-CoV-2 complete genomes were collected using online GISAID, NCBI BLAST or CGSD.
4. To increase the reliability of the genotyping, the complete genomes have to be at least 29,000 bp in length and have fewer than 1% "N"s. The sequenced SARS-CoV-2 complete genomes and reference genomes were trimmed and aligned with BioEdit 7.02 software (*see* **Note 4**). Finally, consensus sequences with no gap and high average coverage were generated (*see* **Note 5**).
5. The trimmed query SARS-CoV-2 sequences were registered and deposited in GenBank, and the accession number was obtained; conversely, on the GISAID EpiCoV newly emerging coronavirus SARS-CoV-2 platform, the information of genotypes and mutation sites was obtained (*see* Fig. 1).
6. Recently emerged variants analysis: The eleven ORFs in SARS-CoV-2 genome were identified in the genomes using BioEdit 7.02 software (*see* Fig. 2).
7. For phylogenetic analyses, trimmed query SARS-CoV-2 sequences were aligned with Molecular Evolutionary Genetics Analysis (MEGA, version 6.06) by the maximum likelihood method (MLM) with 1000 replicates (*see* **Note 6**). Trimmed

A

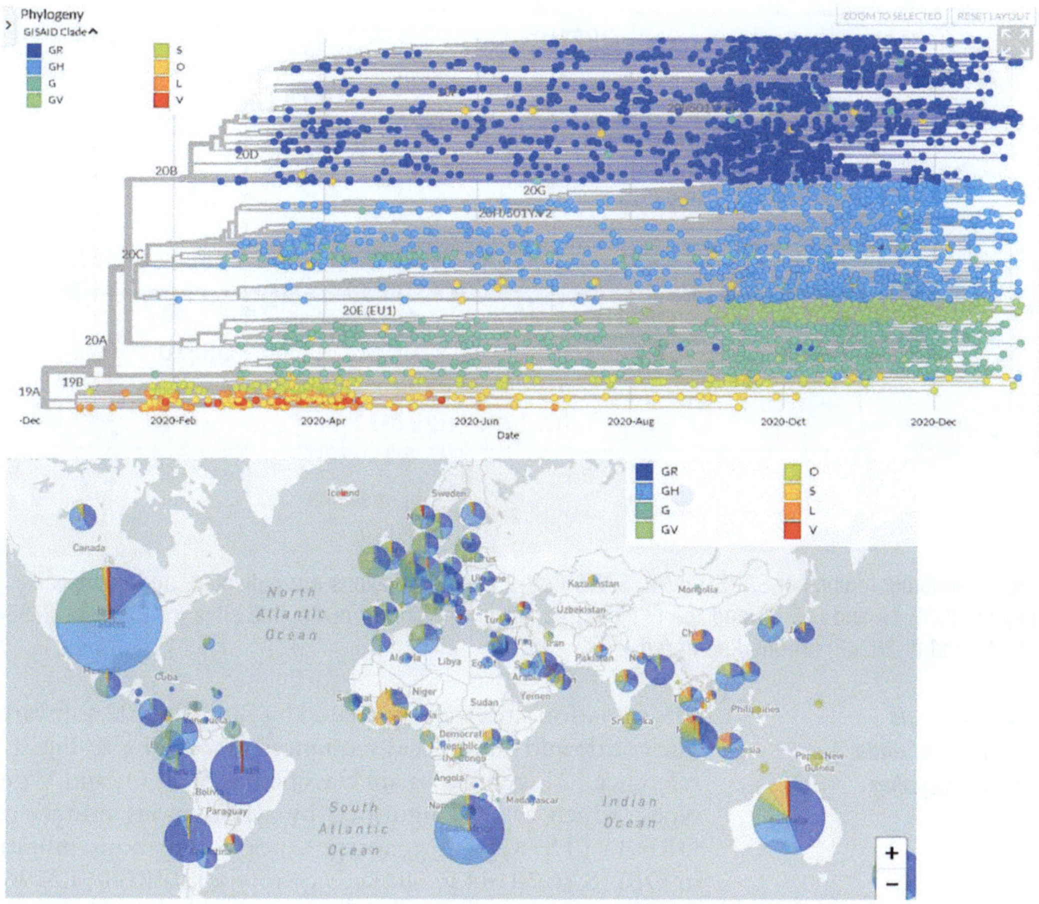

B

Fig. 1 Phylogenetic overview and genomic epidemiology of global SARS-CoV-2 variants. (**a**) Phylogeny of global SARS-CoV-2 variants. (**b**) Geography of global SARS-CoV-2 variants. Metadata were downloaded from the GISAID website (https:/www.gisaid.org/phylodynamics/global/nextstrain/) for 3953 samples grouped by their 557 authors

SARS-CoV-2 sequences were included in the phylogenetic analysis, and bootstrap values are shown. Gene accession numbers in GenBank for corresponding viral sequences are provided. The local or national molecular epidemiology of SARS-CoV-2 was determined by sequence analysis, including examination of subtypes, time, and geographical distributions according the cluster relationship (*see* Fig. 3). As shown in Fig. 3, the query SARS-CoV-2 sequences of the subtypes to be determined have temporal aggregation and are corresponding to GISAID phylogenetic classification in the context of the epidemic (*see* Table 1 and https://www.gisaid.org/phylodynamics/global/nextstrain/).

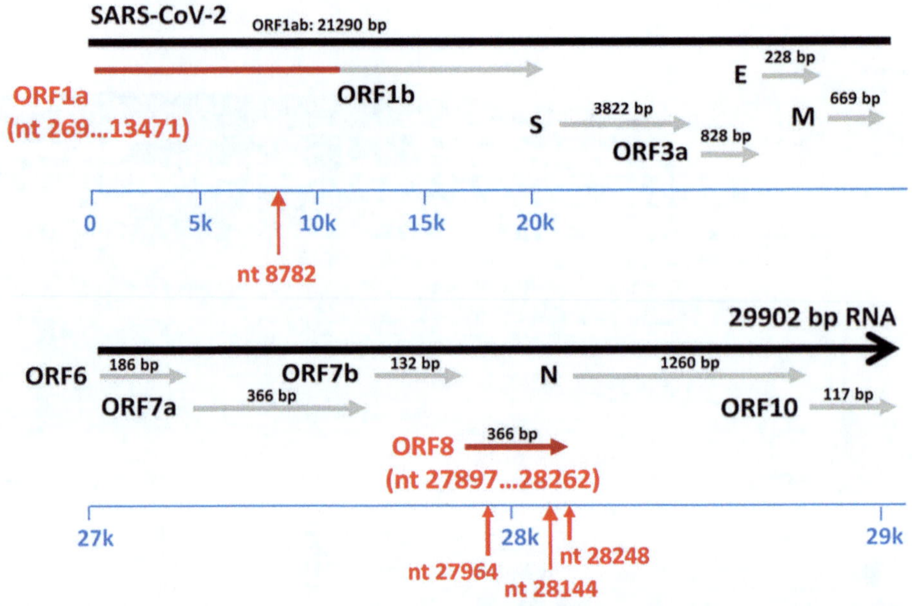

Fig. 2 Genomic regions of SARS-CoV-2. SARS-CoV-2 genomic regions are indicated, and the specific gene regions (ORF 1a and ORF 8) and the highest frequency mutation of nucleotide sites (nt 8782, nt 27964, nt 28144 and nt 28248) are marked in red

3.2 Analysis of Mutation Sites and Genotyping

1. The mutation sites were estimated by BioEdit 7.02 software for nucleotide and amino acid sequences. According to Fig. 1 and reference [2], genotypes are classified as S, L, O, and V genotypes, which are distinguished by synonymous mutations in ORF 1a (T8782C, L genotype), nonsynonymous mutations in ORF 8 (C28144T, S84L, L genotype), ORF8 (T27964C, S24L, O genotype) and ORF 8 (C28248A, L128M, V genotype). Here, S genotype genes are regarded as reference genes. These four areas are high-frequency mutation areas and are described at CGSD (http://nmdc.cn/coronavirus) (*see* Fig. 4). According to this classification method, the remaining RNAs of clinical specimens were amplified by the target genes of SARS-CoV-2 ORF 1a and ORF 8.
2. Simplification was employed in the detection of the high-frequency mutation sequences [2, 10], that is, partial ORF 1a and full ORF 8, which covered ORF 1a (nt 8782), ORF 8 (nt 27964), ORF 8 (nt 28144), and ORF 8 (nt 28248), respectively.
3. Specific genes were amplified by a PrimeScript™ One Step RT-PCR Kit Ver. 2 following the manufacturer's instructions. Briefly, a 25-μl reaction volume contained 12.5 μl of 2× Step buffer, 1 μl of PrimeScript One Step Enzyme Mix, 0.5 μl of PCR Forward Primer (10 μM), 0.5 μl of PCR Reverse Primer (10 μM), 5 μl of total RNA (10 pg ~ 100 ng), and 6.5 μl of

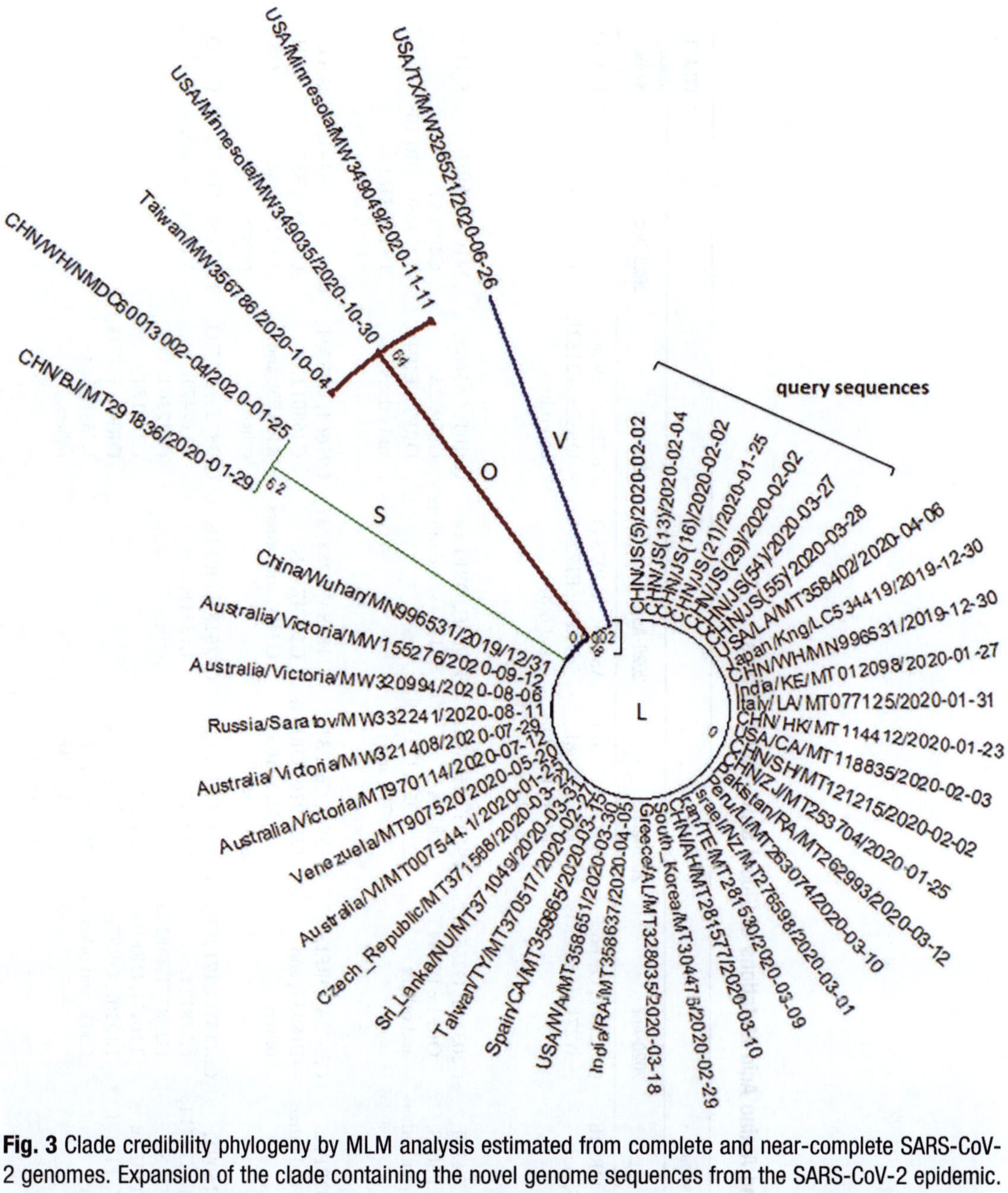

Fig. 3 Clade credibility phylogeny by MLM analysis estimated from complete and near-complete SARS-CoV-2 genomes. Expansion of the clade containing the novel genome sequences from the SARS-CoV-2 epidemic. Clade posterior probabilities are shown at well-supported nodes. Colors represent different genotypes. Sequenced SARS-CoV-2 complete query genomes: CHN/JS(5)/2020-02-02, CHN/JS(13)/2020-02-04, CHN/JS(16)/2020-02-02, CHN/JS(29)/2020-02-02, CHN/JS(54)/2020-03-27, and CHN/JS(55)/2020-03-28. This figure displays a small number of global SARS-CoV-2 strains throughout years 2019 to 2020. The green clade is S genotype, the black clade is L genotype, the red clade is O genotype, and the blue clade is V genotype

Table 1
Representative Nucleotide and Amino Acid Mutations of Global SARS-CoV-2 Strains[a]

Variant name	Representative mutation sites							GISAID Clade Name
	2019-Dec	2020-Feb	2020-Apr	2020-June	2020-Aug	2020-Oct	2020-Dec	
19A	T6996C, ORF1a: I2244T, and others	G17814T, ORF1b: L1449F, and others	G27281T, ORF6: W27L, and others	G2867T, A17637T, ORF1a: V868L, ORF1b: K1390N, and others	A5436C, G29734T, ORF1a: E1724A, and others	C6720T, T29685C, ORF1a: T2152I, and others	[b]	L, V, O
19B	C24034T, ORF1a: S1561P, and others	C1190T, C9438T, ORF1a: P309S, T3058I, and others	C20316T, G25500T, ORF1a: P2144T, and others	C1912T, G19398T, G24574A, C29095T, ORF1b: E1977D, and others	T3906C, ORF1a: F1214S, and others	G210T, C6568T, G25972A, ORF3a: E194K, and others	T241C, T14408C, C20964T, T27945C, ORF1b: L314P, ORF8: *18Q, and others	S, O
20A	[b]	G1685A, ORF1a: A474T, and others	C5672A, ORF1a: P1803T, and others	A1558G, C6573T, T22192C, ORF1a: I431M, S2103F, and others	T21808A, G28373T, G29747A, N: G34W, and others	C9166T, G2529T, C26801T, S: K1245N, and others	G1335A, C9344T, ORF1a: C357Y, L3027F, and others	G, O, GH, GR
20B	[b]	C593T, C1609T, C3315T, C7471T, C28677T, N: T135I, ORF1a: H11OY, T1017I, and others	C5115T, C17125T, C27812T, G27877T, ORF1a: T1617I, ORF1b: L1220F, ORF7b: C41F, and others	C106T, G4160T, C7011T, G19086T, C24023T, C26527T, M: A2V, ORF1a: V1299L, A2249V, ORF1b: K1837N, and others	C17518T, ORF1b: L1351F, and others	C1385T, A2475G, C11747T, C23230T, C26013T, ORF1a: H374Y, K737R, and others	T27716C, ORF7a: V108A, and others	GR, O, G

20C	b	C13620T, C21595T, and others	C335T, T19839C, T22965C, ORFa: R24C, S: I468T, and others	A21299C, A22666G, A28070C, G29705T, ORF1b: K2611T, ORF8: E59D, and others	T1623C, A5725T, A8446G, C15141T, G19539T, ORF1a: I453T, ORF1b: M2024I, and others	C28826T, N: R185C, and others	C10202T, C16716T, A27510G, ORF1a: L3313F, and others	GH, G, O, GR
20D	b	T17784C, and others	C13225T, C14919T, C21575T, S: L5F, and others	C2455T; T10009C, T19765C, and others	C16575T; C21595T, T21769A, S: H69Q, and others	C5079T, G29711T, ORF1a: T1605I, and others	G26718T, G27382T, M: V66L, ORF6: D61Y, and others	GR, O, G
20E	b	G20275A, ORF 1b: D2270N, and others	G410T, ORF1a: G49C, and others	C13862T, C15212T, ORF1b: T132I, T582I, and others	C3130T, G3403T, and others	G18651T, G25266T, ORF1b: E1728D, S: C1235F, and others	C2005T, T13258C, G20580T, and others	G, O, GR
20F	b	b	T12697C, and others	C1519T, and others	C2667T, A6601G, G21718T, ORF1a: A801V, S: Q52H, and others	G3340T, and others	G11335T, C12076T, and others	GR, G
20G	b	C19264T, ORF1b: L1933F, and others	G29402T, N: D377Y, and others	C3411T, T6394C, T7098G, G8179A, A15873G, C15924T, G18462A, ORF1a: A1049V, I2278S, and others	C8139T, G12028A, C23896T, ORF1a: S2625F, M3921I, and others	C106T, C5284T, T8871C, C12020T, ORF1a: L2869S, L3919F, and others	C446T, G15372T, C17004T, and others	GH, G, O
20H	b	C14805T, and others	G7473T, C26636T, C27213T, ORF1a: G2403V, and others	C12880T, G29553A, and others	C15024T, C27879T, ORF7b: H42Y, and others	T14222G, ORF1b: L252, and others	C29108T, N: P279S, and others	GH, G, O

(continued)

Table 1 (continued)

Variant name	Representative mutation sites							GISAID Clade Name
	2019-Dec	2020-Feb	2020-Apr	2020-June	2020-Aug	2020-Oct	2020-Dec	
20I	[b]	[b]	A6252G, ORF1a: N1996S, and others	A26162G, ORF3a: N257S, and others	C8299T, C15237T, C23929T, and others	G28362C, N: G30A, ORF9b: E27Q, and others	T6489G, G21600T, G23012A, C23604A, C28603T, C29253T, N: S327L, ORF1a: I2075S, S: S13I, E484K, P681H, and others	GR, O, G

[a]Nonsynonymous mutations with both nucleotide variation information and amino acid variation information; Synonymous mutations with only nucleotide variation information and no amino acid variation information. The information of the mutation sites of SARS-CoV-2 strains was quoted from the GISAID website (https://www.gisaid.org/phylodynamics/global/nextstrain/)

[b]No strain

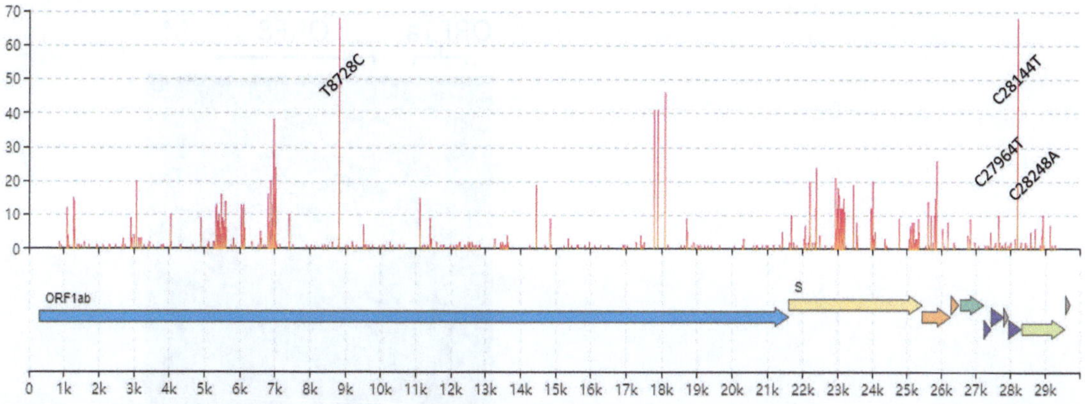

Fig. 4 Genomic structure of SARS-CoV-2 and the genomic location and frequency of single nucleotide polymorphisms (compared with global reference genomes). Download from CGSD (http:/nmdc.cn/coronavirus)

dH$_2$O. One-step RT-PCR was performed with the following steps: 50 °C for 30 min, 94 °C for 2 min (1 cycle of amplification); 94 °C for 30 s, 50 °C for 30 s, 72 °C for 30 s (35 cycles of amplification); and extension at 72 °C for 10 min in an Applied Biosystems ABI PCR system (*see* **Note 7**).

4. Next, 5 μl of RT-PCR product mixed with 1 μl 6× loading buffer was loaded onto a 1% agarose gel. Also, 3 μl of DNA Marker was loaded onto the side of the gel. The voltage used for gel electrophoresis was 220 V and 200 mA for 30 min. The bands were visualized by illuminating the gel with UV light on the Bio-Rad system.
5. SARS-CoV-2 ORF 1a and ORF 8 yielded 378-bp and 465-bp DNA products, respectively (*see* Fig. 5) (*see* **Note 8**). The target fragment sizes were determined by comparing the length of the bands to the DL2,000 DNA marker, and sequencing steps and trimming primer areas (5′/3′) were performed according to the above description.

3.3 Molecular Epidemiology of SARS-CoV-2 Genomes

1. For a specific segment comparison with the global epidemic reference genes, the available reference genomes were downloaded from the GISAID, GenBank (BLAST) and CGSD websites for alignment. Strains without sample date, location and host source were excluded.
2. The nucleotide and amino acid sequences of partial ORF 1a (nt 8573–8938, 366 bp) and full ORF 8 (nt 27881–28246, 366 bp) regions were trimmed and aligned using BioEdit 7.02 software.
3. Phylogenetic trees for targeted genomes were constructed using MEGA 6.06 software by the maximum likelihood method (MLM) with 1000 replicates. The classification results are consistent with Fig. 3.

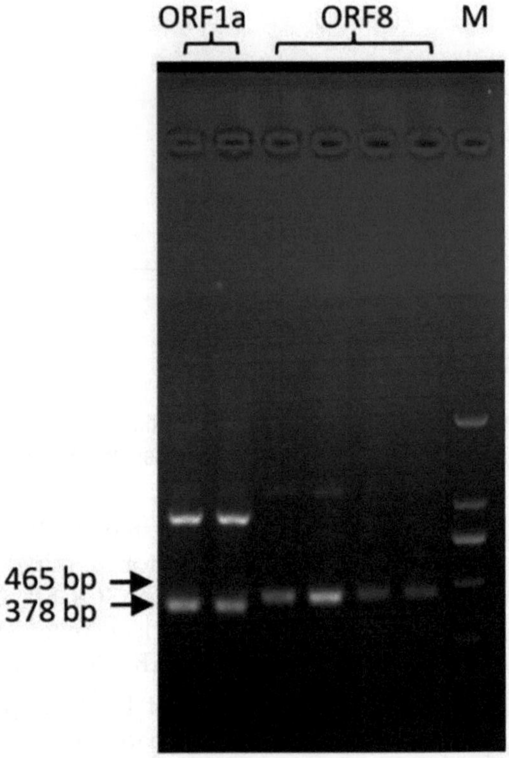

Fig. 5 Gel electrophoresis after RT-PCR. Partial ORF 1a (378 bp) and ORF 8 (465 bp) genomic DNA bands in clinical SARS-CoV-2 nucleic acid–positive specimens are indicated. DL2,000 DNA Marker (up to down, 2000 bp, 1000 bp, 750 bp, 500 bp, 250 bp, 100 bp). M: DNA Marker

4. The SARS-CoV-2 mutation sites of target genes (ORF1a and ORF 8) were analyzed with BioEdit 7.02 software. The local or national molecular epidemiology was analyzed depending on nucleotide and amino acid sequences.

4 Notes

1. Detailed information on SARS-CoV-2 strains was collected on NCBI or other websites and listed in a table, including simplified name, accession number, collection date, country, and isolation source.
2. PCR primers for SARS-CoV-2 ORF 1a and ORF 8 were designed referring to the reference strains downloaded from the Coronavirus Global Shared Database website (CGSD, http://nmdc.cn/coronavirus) using Primer Premier 5.0 software.

3. The agarose gel was dissolved in 1× TAE buffer in a microwave oven by boiling and shaking 3 times. The gel solution was cooled, and nucleic acid dye (GelGreen) was added until it reached approximately 70 °C.
4. BioEdit 7.02 software was used to assess the raw sequence quality before trimming and after alignment, which can prevent potential errors. This software can identify the different sites between test SARS-CoV-2 sequences and reference genomes. The primer binding sites at the 5′ and 3′ ends were removed, as they are usually high-frequency error areas. Normally, a shorter genome than a reference genome is obtained in GenBank.
5. The consensus sequences were saved as a "*.text" files by BioEdit 7.02 software, and the *.text files were used to register and deposit the sequences in GenBank. In the text file, the name of each sequence to be registered or analyzed by the software was listed after the symbol ">", and the sequence was listed on a new line.
6. The consensus sequences were saved as "*.fas" files by BioEdit 7.02 software, and the *.fas files were opened in MEGA 6.06 software; next, they were saved as "*. mas" files in MEGA 6.06 software, which were also used for phylogenetic analysis of the main interface in MEGA 6.06.
7. As shown in Table 1, RT-PCR can be used to identify SARS-CoV-2 mutations of partial ORFs in the viral genomes to determine the subtypes of the newly discovered strains. The complete sequences can be uploaded to the GISAID website after high-throughput sequencing to analyze their GISAID clades, subtypes and geography.
8. When the RT-PCR product was not a single band after electrophoresis, the gel of the target band was cut out and used for sequencing.

Acknowledgments

This work was supported by the Top Talent Support Program for Yong and Middle-aged People of Wuxi Health Committee (BJ2020094), the Chinese Public Health Alliance Program (GWLM202005), the Wuxi Key Medical Talents Program (ZDRC024), the National Natural Science Foundation of China (81701550), the National Mega-Projects against Infectious Diseases (2018ZX10301406-002), the Emergency Prevention and Control Capacity Program for New Severe Infectious Diseases of the National Institute for Viral Disease Control and Prevention, and the 135 Strategic Program of Chinese Academy of Sciences.

References

1. Zhu N, Zhang D, Wang W, Li X, Yang B, Song J, Zhao X, Huang B, Shi W, Lu R, Niu P, Zhan F, Ma X, Wang D, Xu W, Wu G, Gao GF, Tan W, China Novel Coronavirus Investigating and Research Team (2020) A novel coronavirus from patients with pneumonia in China, 2019. N Engl J Med 382(8): 727–733. https://doi.org/10.1056/NEJMoa2001017
2. Tang XL, Wu CC, Li X, Song YH, Yao XM, Wu XK, Duan YG, Hong Z, Yirong Wang ZQ, Cui J, Jian L (2020) On the origin and continuing evolution of SARS-CoV-2. Natl Sci Rev 7(6):1012–1023. https://doi.org/10.1093/nsr/nwaa036/5775463
3. Chu DKW, Pan Y, Cheng SMS, Hui KPY, Krishnan P, Liu Y, Ng DYM, Wan CKC, Yang P, Wang Q, Peiris M, Poon LLM (2020) Molecular diagnosis of a novel coronavirus (2019-nCoV) causing an outbreak of pneumonia. Clin Chem 66(4):549–555. https://doi.org/10.1093/clinchem/hvaa029
4. Andersen KG, Rambaut A, Lipkin WI, Holmes EC, Garry RF (2020) The proximal origin of SARS-CoV-2. Nat Med 26(4):450–452. https://doi.org/10.1038/s41591-020-0820-9
5. Benvenuto D, Giovanetti M, Salemi M, Prosperi M, De Flora C, Junior Alcantara LC, Angeletti S, Ciccozzi M (2020) The global spread of 2019-nCoV: a molecular evolutionary analysis. Pathog Glob Health 114(2): 64–67. https://doi.org/10.1080/20477724.2020.1725339
6. Lu J, du Plessis L, Liu Z, Hill V, Kang M, Lin H, Sun J, François S, Kraemer M, Faria NR MJ, Peng J, Xiong Q, Yuan R, Zeng L, , Zhou P, Liang C, Yi L, Liu J, Xiao J, Hu J, Liu T, Ma W, Li W, Su J, Zheng H, Peng B, Fang S, Su W, Li K, Sun R, Bai R, Tang X, Liang M, Quick J, Song T, Rambaut A, Loman N, Raghwani J, Pybus OG, Ke C (2020) Genomic epidemiology of SARS-CoV-2 in Guangdong province, China, Cell 181(5): 997-1003.e9. doi:https://doi.org/10.1016/j.cell.2020.04.023
7. Forster P, Forster L, Renfrew C, Forster M (2020) Phylogenetic network analysis of SARS-CoV-2 genomes. Proc Natl Acad Sci U S A 117(17):9241–9243. https://doi.org/10.1073/pnas.2004999117
8. Nadeau S, Vaughan TG, Crawford KHD, Althaus CL, Reichmuth ML, Bowen JE, Walls AC, Corti D, Bloom JD, Veesler D, Mateo D, Hernando A, Comas I, González-Candelas F; SeqCOVID-SPAIN consortium, Stadler T, Neher RA (2020) Spread of a SARS-CoV-2 variant through Europe in the summer of 2020. Nature 595(7869):707–712. https://doi.org/10.1038/s41586-021-03677-y
9. Giovanetti M, Benvenuto D, Angeletti S, Ciccozzi M (2020) The first two cases of 2019-nCoV in Italy: where they come from? J Med Virol 92(5):518–521. https://doi.org/10.1002/jmv.25699
10. Yan Y, Liu B, Ding H, Wang X, Dai Y, Ding D, Yu H, Sha M, Lui C, Chantsalmaa D, Qiu Y, Huang L, Hu Q (2020) Characterizing COVID-19 severity, epidemiology and SARS-CoV-2 genotypes in a regional business hub of China. J Infect 82(2):282–327. https://doi.org/10.1016/j.jinf.2020.08.031

Chapter 3

Advanced Genetic Methodologies in Tracking Evolution and Spread of SARS-CoV-2

Xuemei Yang, Ning Dong, and Sheng Chen

Abstract

A newly emerged coronavirus, SARS-CoV-2, caused severe pneumonia outbreaks in China in December 2019 and has since spread to various countries around the world. Here, we describe genetic methods to trace the evolution route and probe the transmission dynamics of this virus.

Key words SARS-CoV-2, COVID-19, Transmission, Phylodynamic analysis, Super-spreader

1 Introduction

A number of newly emerged coronaviruses such as the highly pathogenic severe acute respiratory syndrome coronavirus (SARS-CoV) and Middle East respiratory syndrome coronavirus (MERS-CoV) have caused serious respiratory and intestinal infections in human within the past two decades [1]. In December 2019, another new coronavirus, SARS-CoV-2, has emerged and caused outbreaks of lower respiratory tract infections, often with poor clinical outcome, in Wuhan, China. The virus, which has since spread to other cities in China and various countries worldwide [2], exhibited a high potential to undergo human-to-human transmission [3]. Global initiative on sharing all influenza data (GISAID) is a platform for sharing genetic data of influenza. Currently, a rapidly increasing number of SARS-CoV-2 genomic sequences are being deposited into this database from laboratories around the world [4]. With more sequences being released, we can obtain a more comprehensive view on the genomic features of this virus through in-depth sequence analysis. Here, we describe

---

Supplementary Information The online version of this chapter (https://doi.org/10.1007/978-1-0716-2111-0_3) contains supplementary material, which is available to authorized users.

Justin Jang Hann Chu et al. (eds.), *SARS-CoV-2: Methods and Protocols*,
Methods in Molecular Biology, vol. 2452, https://doi.org/10.1007/978-1-0716-2111-0_3,

genetic methods to retrieve and analyze the publicly shared genome sequences of SARS-CoV-2 to investigate the genetic diversity and phylodynamic of these viruses.

2 Materials

2.1 Equipment

1. Databases: the GenBank database (https://www.ncbi.nlm.nih.gov/genbank/), the GISAID database [5] (https://www.gisaid.org/). Example SARS-CoV-2 genome sequences for use in this protocol are available in supplementary file, see Subheading 2.2 for details).
2. MAFFT software [6] (https://mafft.cbrc.jp/alignment/software/).
3. IQ-TREE software [7] (http://www.iqtree.org/).
4. Nextstrain software [8] (https://nextstrain.org/).
5. TreeTime software [9] (https://github.com/neherlab/treetime).
6. Snippy software [10] (https://github.com/tseemann/snippy).
7. Online phylogenetic tree annotating tool: iTOL [11] (https://itol.embl.de/).
8. Graphic editing tool: Inscape [12] (https://inkscape.org/).

2.2 Equipment Setup

2.2.1 Critical

The commands used in the protocol should all be run from the Unix shell prompt within a terminal window. Commands meant to be executed from the Unix shell (e.g., bash or csh) are prefixed with a "$" character.

2.2.2 Data

This protocol is illustrated with a demo to track evolution and spread of SARS-CoV-2 by analyzing 247 genome sequences (*see* Electronic Supplementary Material). All the sequences were downloaded from the GISAID database, a platform for sharing genetic data of influenza viruses and the SARS-CoV-2 coronavirus [5]. This platform provides information regarding accession ID, virus name, host, location, collection date, submission date, sequence length, origin lab, and submission lab (Fig. 1). Some are with patient status and additional information, such as clinical and epidemiological data. SARS-CoV-2 isolate Wuhan-Hu-1 (also referred to as "2019-nCoV") was isolated from bronchoalveolar lavage fluid collected from a patient on 31 December 2019 [13]. This patient was a worker at the Huanan Seafood Wholesale Market and was admitted to the Central Hospital of Wuhan on 26 December 2019 while experiencing a severe respiratory syndrome that included fever, dizziness and a cough. Complete genomic sequence of SARS-CoV-2 isolate Wuhan-Hu-1 (29,903 nucleotides) was downloaded from GenBank (GenBank

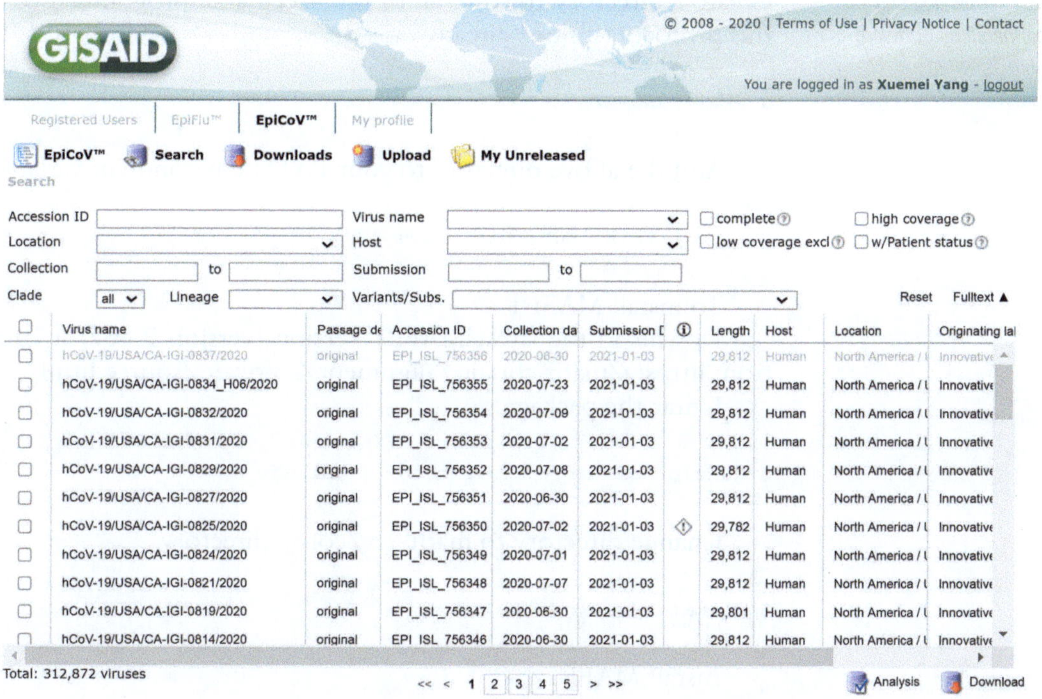

Fig. 1 Web page of GISAID database

accession No: NC_045512.2, GISAID ID: EPI_ISL_402125) and regarded as reference in this protocol.

For phylogenetic analysis, high quality sequences are required. Only nearly full-length ($\geq$ 29 kb) genome sequences of SARS-CoV-2 are included. Sequences with >5% Ns should be excluded. These two conditions can be accomplished by choose the two options, complete and low coverage excl, when downloading from the GISAID platform (Fig. 1). For phylodynamic analysis, information regarding the date and country of isolation are required which can be retrieved by downloading the Acknowledgement table. Sequences containing little temporal signal which are unsuitable for inference using phylogenetic molecular clock models should be excluded when performing phylodynamic analysis. All data used in this protocol is available in the file demo.tar.gz, described below.

2.2.3 Downloading and Organizing Required Data

- Download and unpack the data file demo.tar.gz (**Electronic Supplementary Material**) and inspect the contents.
- $ tar xvzf demo.tar.gz.
- The package expands to contain a folder demo/, which has the following directories: genome / phylogenetic / phylodynamic / SNP /.

2.2.4 Downloading and Installing Software

Create a directory to store all of the executable programs used in this protocol (if none already exists):

```
$ mkdir $HOME/bin
```

Add the above directory to your PATH environment variable:

```
$ export PATH=$HOME/bin:$PATH
```

To install MAFFT.

Download the latest MAFFT version (version 7.467 or later) from https://mafft.cbrc.jp/alignment/software/source.html.

Untar the package.

```
$ gunzip -cd mafft-x.x-src.tgz | tar xfv -
```

Change directory to mafft-x.x/core/ directory.

```
$ cd mafft-x.x/core/
```

Install MAFFT.

```
$ make clean
$ make
$ su
$ make install
```

To install IQ-TREE.

Download the latest IQ-TREE version for COVID-edition from http://www.iqtree.org/#download.

```
Unpack iqtree-2.x.x-Linux.tar.gz
$tar -zxvf iqtree-2.x.x-Linux.tar.gz
Copy iqtree2 to $HOME/bin
$ copy iqtree-2.1.2-Linux/bin/iqtree2 $HOME/bin
```

To install Snippy.

Install Snippy direct from Github (https://github.com/tseemann/snippy).

```
Change directory to $HOME/bin
$ cd $HOME/bin
```

Install Snippy

$ git clone https://github.com/tseemann/snippy.git

Add Snippy's bin directory to your $PATH.

```
$ export PATH = snippy/bin:$PATH
```

To install TreeTime

```
$pip install phylo-treetime
```

To install Nextstrain.

```
$ python3 -m pip install nextstrain-cli
```

3 Methods

3.1 Multiple Sequence Alignment

Perform multiple alignment using MAFFT (*see* **Note 1**):

```
$ mafft --auto --addfragments input.fasta Wuhan-Hu-1.fasta >
alignment.txt
```

This command line provides rapid calculation of full-length multiple sequence alignment of closely related viral genomes under MAFFT version 7.467 or later. Earlier versions (≤7.458) had the same options but were inefficient for this purpose. An online version for multiple sequence alignment is provided by MAFFT which supports more than 20,000 sequences of SARS-CoV-2. (https://mafft.cbrc.jp/alignment/software/closelyrelatedviralgenomes.html).

3.2 Phylogenetic Analysis

1. Build phylogenetic tree of SARS-CoV-2 sequences with IQ-TREE:

```
$ iqtree -s alignment.txt
```

   This command infers a maximum-likelihood tree from the sequence alignment with the best-fit model automatically selected by ModelFinder.

2. Tree Annotation.

   Upload the IQ-TREE generated treefile to iTOL (https://itol.embl.de/).

   Annotate the tree by drag-dropping the dataset files onto the tree display.

   Export the tree to a PNG file or a SVG file which can be further edited by Inkscape [12].

Here we annotated the tree with identified mutation sites (see below), location and isolation date (Fig. 2).

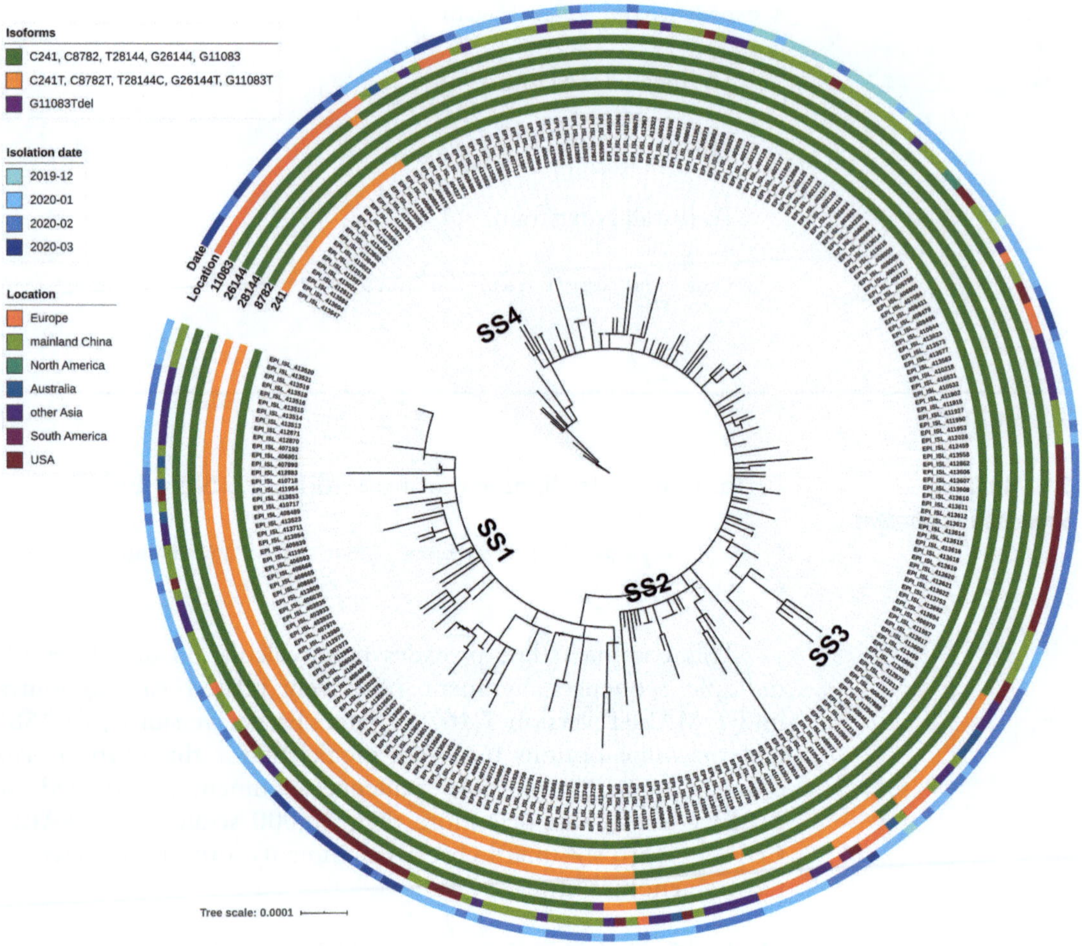

Fig. 2 Phylogenetic analysis of 247 SARS-CoV-2 genomes

3.3 Phylodynamic Analysis

Global genomic surveillance of SARS-CoV-2 was implemented by means of the automated phylogenetic analysis pipeline Nextstrain, which generates an interactive visualization integrating a phylogeny with sample metadata such as geographic location or isolation date [8]. The pipeline involved the sequence alignment module with MAFFT [6], phylogenetic analysis with IQ-TREE [7], maximum-likelihood phylodynamic analysis with Treetime [9], identification of nucleotide and amino acid mutations with Augur, and result visualization with Auspice [9].

1. Phylodynamic analysis with Treetime.

```
$ treetime --aln alignment.txt --tree iqtree.newick --dates
metadata.tsv --plot-tree tree.svg --plot-rtt rtt.svg
```

TreeTime provides routines for ancestral sequence reconstruction and inference of molecular-clock phylogenies. In this command, metadata.tsv is a tsv file with dates and/or other meta data. Tree.svg and rtt.svg are output files of treetime.

2. Augur: sequence alignment and phylogeny construction.

```
$ mkdir -p results/
```

This command line makes a folder "results" for all outputs of Augur.

```
$ augur filter -sequences data/sequences.fasta -metadata
data/metadata.tsv --exclude config/dropped_strains.txt --out-
put results/filtered.fasta --group-by country year month --
sequences-per-group 20 --min-date 2018
```

This command filters the parsed sequences and metadata to exclude strains from subsequent analysis and subsample the remaining strains to a fixed number of samples per group.

```
$ augur align --sequences results/filtered.fasta --reference-
sequence config/Wuhan-Hu-1.gb
 --output results/aligned.fasta --fill-gaps
```

This command creates a multisequence alignment using a custom reference. After this alignment, columns with gaps in the reference are removed. Additionally, the --fill-gaps flag fills gaps in nonreference sequences with "N" characters. These modifications force all sequences into the same coordinate space as the reference sequence.

```
$ augur tree --alignment results/aligned.fasta --output
results/tree_raw.nwk
```

This command infers a phylogenetic tree from the multisequence alignment. The resulting tree is stored in Newick format. Branch lengths in this tree measure nucleotide divergence.

```
$augur refine --tree results/tree_raw.nwk --alignment re-
sults/aligned.fasta --metadata data/metadata.tsv
 --output-tree results/tree.nwk --output-node-data results/
branch_lengths.json --timetree --coalescent opt --date-confi-
dence --date-inference marginal --clock-filter-iqd 4
```

Run the refine command to apply TreeTime to the original phylogenetic tree and produce a "time tree."

```
$ augur traits --tree results/tree.nwk --metadata data/
metadata.tsv --output-node-data results/traits.json
--columns region country --confidence
```

This command infers the region and country of all internal nodes from the time tree and original strain metadata. As with the refine command, the resulting JSON output is indexed by strain or internal node name.

```
$ augur ancestral --tree results/tree.nwk --alignment re-
sults/aligned.fasta --output-node-data results/nt_muts.json
--inference joint
```

This command infers the ancestral sequence of each internal node and identify any nucleotide mutations on the branches leading to any node in the tree.

```
$ augur translate --tree results/tree.nwk --ancestral-se-
quences results/nt_muts.json --reference-sequence config/ Wu-
han-Hu-1.gb --output-node-data results/aa_muts.json
```

This command identifies amino acid mutations from the nucleotide mutations and a reference sequence with gene coordinate annotations. The resulting JSON file contains amino acid mutations indexed by strain or internal node name and by gene name. To export a FASTA file with the complete amino acid translations for each gene from each node's sequence, specify the --alignment-output parameter in the form of results/aligned_aa_%GENE.fasta.

```
$ augur export v2 --tree results/tree.nwk --metadata data/
metadata.tsv --node-data results/branch_lengths.json results/
traits.json results/nt_muts.json results/aa_muts.json --col-
ors config/colors.tsv --lat-longs config/lat_longs.tsv --aus-
pice-config config/auspice_config.json --output auspice/SARS-
CoV-2.json
```

This command collects all node annotations and metadata and export it in Auspice's JSON format. This refers to three config files to define colors via config/colors.tsv, latitude and longitude coordinates via config/lat_longs.tsv, as well as page title, maintainer, filters present, etc., via config/auspice_config.json. The resulting tree and metadata JSON files are the inputs to the Auspice visualization tool.

3. Visualization of phylodynamic analysis results.

```
$ auspice view --datasetDir results
```

Finally, we will get the root-to-tip regression scatter plots and distribution of mutations across the 247 SAR-CoV-2 genomes (Fig. 3).

3.4 Genome-Wide Single-Nucleotide Variations

1. Call SNPs by Snippy.

```
$ snippy --cpus 16 --outdir sequence1 --ref Wuhan-Hu-1.gb --
ctgs sequence1.fasta
```

SNPs are called by using the annotated genome sequence of the SARS-CoV-2 isolate Wuhan-Hu-1 (GenBank accession No: NC_045512.2) as reference. Run the command one by one for all the sequences. Here, we write a sh file to put commands for all sequences and run it.

2. Extract "core SNPs" of all genome sequences.

```
$ snippy-core --ref Wuhan-Hu-1.fasta --prefix core sequence1
sequence2 sequence3...
```

You will get a tab file which is a tab-separated columnar list of core SNP sites with alleles.

3. Analyze nucleotide mutations of all genome.

Within the tab file, you will get the SNP sites of all analyzed genome. The amino acid mutations of separate proteins will be present according to the reference sequence.

4. Classify sequences with super-spreader clusters according relative variants.

   In this protocol, we identified 4 clusters. The first cluster contained two mutations, $C^{8782}T$ and $T^{28144}C$; the second cluster contained the mutation $G^{26144}T$; the third cluster contained the mutation $G^{11083}T$; the fourth cluster contained three mutations, $C^{241}T$, $C^{3037}T$ and $A^{23403}G$.

3.5 Quick Identification of the Types of SARS-CoV-2 Genome Sequences

1. Identify mutations of newly released sequences by aligning to the reference genome using snippy as described above.
2. Classify the new sequences into the above identified super-spreaders.
3. Trace the evolution route of new SARS-CoV-2 isolates through sequence alignment and probe the transmission dynamics of this virus.

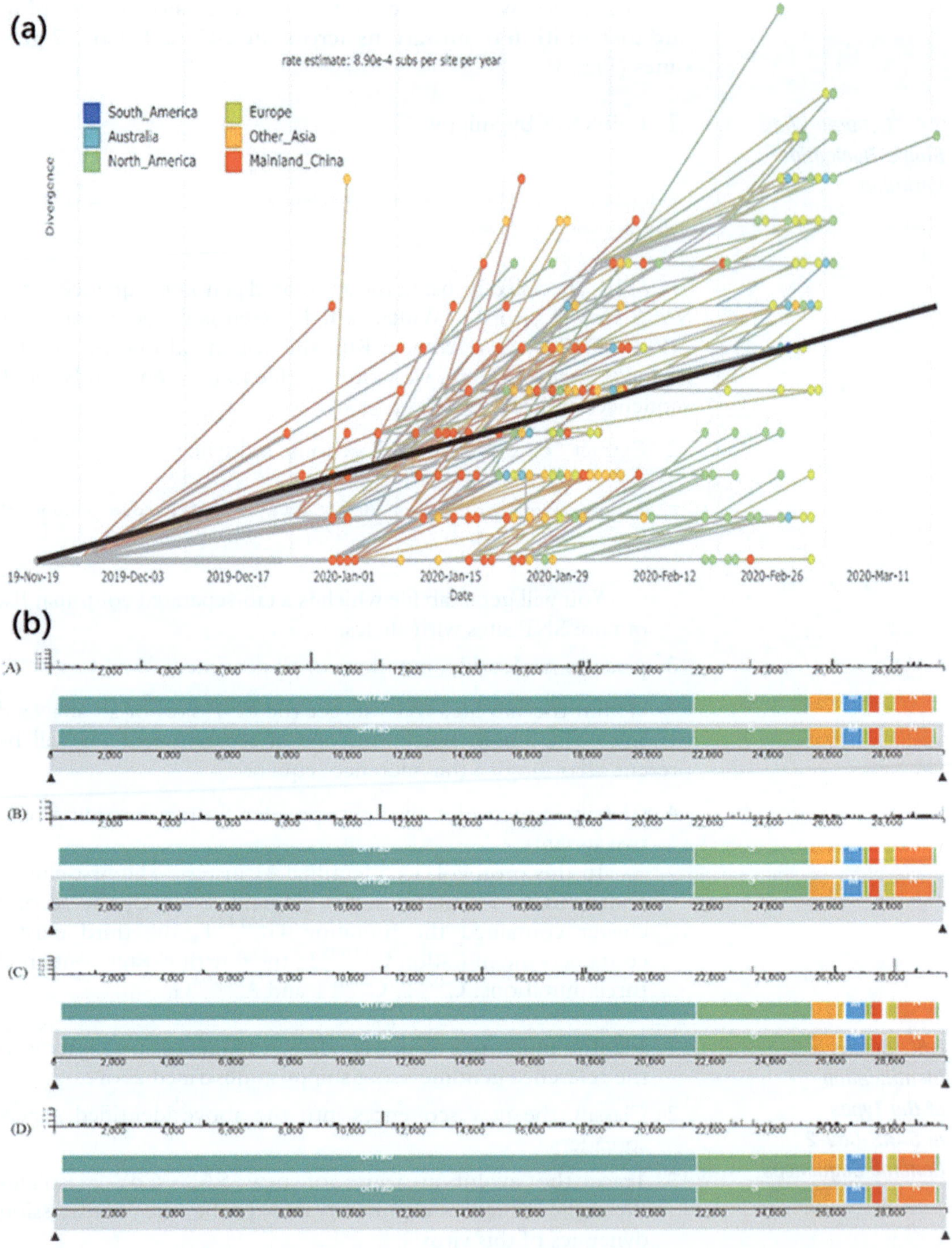

Fig. 3 Root-to-tip regression scatter plots and distribution of mutations across the 247 SAR-CoV-2 genomes

4 Notes

1. There are multiple alignment, phylogenetic analysis, phylodynamic analysis, and tree editing software. Users can choose proper ones depending the sample loads and preference.

Acknowledgments

This work was supported by Research Grant from City University of Hong Kong, SGP/CityU / 9380110.

References

1. Cui J, Li F, Shi Z-L (2019) Origin and evolution of pathogenic coronaviruses. Nat Rev Microbiol 17(3):181–192
2. Gorbalenya AE, Baker SC, Baric RS, de Groot RJ, Drosten C, Gulyaeva AA, Haagmans BL, Lauber C, Leontovich AM, Neuman BW, Penzar D, Perlman S, LLM P, Samborskiy DV, Sidorov IA, Sola I, Ziebuhr J, Coronaviridae Study Group of the International Committee on Taxonomy of V (2020) The species severe acute respiratory syndrome-related coronavirus: Classifying 2019-nCoV and naming it SARS-CoV-2. Nat Microbiol 5(4): 536–544. https://doi.org/10.1038/s41564-020-0695-z
3. Guan WJ, Ni ZY, Hu Y, Liang WH, Ou CQ, He JX, Liu L, Shan H, Lei CL, Hui DSC, Du B, Li LJ, Zeng G, Yuen KY, Chen RC, Tang CL, Wang T, Chen PY, Xiang J, Li SY, Wang JL, Liang ZJ, Peng YX, Wei L, Liu Y, Hu YH, Peng P, Wang JM, Liu JY, Chen Z, Li G, Zheng ZJ, Qiu SQ, Luo J, Ye CJ, Zhu SY, Zhong NS (2020) Clinical characteristics of coronavirus disease 2019 in China. N Engl J Med 382(18):1708–1720. https://doi.org/10.1056/NEJMoa2002032
4. Shu Y, McCauley J (2017) GISAID: Global initiative on sharing all influenza data–from vision to reality. Eur Secur 22(13):30494
5. Shu Y, McCauley J (2017) GISAID: Global initiative on sharing all influenza data - from vision to reality. Euro Surveill 22(13):30494. https://doi.org/10.2807/1560-7917.ES.2017.22.13.30494
6. Katoh K, Standley DM (2013) MAFFT multiple sequence alignment software version 7: Improvements in performance and usability. Mol Biol Evol 30(4):772–780. https://doi.org/10.1093/molbev/mst010
7. Nguyen L-T, Schmidt HA, von Haeseler A, Minh BQ (2015) IQ-TREE: A fast and effective stochastic algorithm for estimating maximum-likelihood phylogenies. Mol Biol Evol 32(1):268–274. https://doi.org/10.1093/molbev/msu300
8. Hadfield J, Megill C, Bell SM, Huddleston J, Potter B, Callender C, Sagulenko P, Bedford T, Neher RA (2018) Nextstrain: real-time tracking of pathogen evolution. Bioinformatics 34(23):4121–4123. https://doi.org/10.1093/bioinformatics/bty407
9. Sagulenko P, Puller V, Neher RA (2018) TreeTime: Maximum-likelihood phylodynamic analysis. Virus evolution 4 (1):vex042
10. Seemann T (2020) Snippy: rapid haploid variant calling and core genome alignment. https://github.com/tseemann/snippy
11. Letunic I, Bork P (2016) Interactive tree of life (iTOL) v3: An online tool for the display and annotation of phylogenetic and other trees. Nucleic Acids Res 44(W1):W242–W245. https://doi.org/10.1093/nar/gkw290
12. Bah T (2009) Inkscape: Guide to a vector drawing program. Prentice Hall Press, Hoboken, NJ
13. Wu F, Zhao S, Yu B, Chen YM, Wang W, Song ZG, Hu Y, Tao ZW, Tian JH, Pei YY, Yuan ML, Zhang YL, Dai FH, Liu Y, Wang QM, Zheng JJ, Xu L, Holmes EC, Zhang YZ (2020) A new coronavirus associated with human respiratory disease in China. Nature 579(7798):265–269. https://doi.org/10.1038/s41586-020-2008-3

Chapter 4

Antigen-Based Point of Care Testing (POCT) for Diagnosing SARS-CoV-2: Assessing Performance

Vidya Keshav, Lesley Scott, Anura David, Lara Noble, Elizabeth Mayne, and Wendy Stevens

Abstract

Currently, the most accurate way to diagnose an active SARS-CoV-2 (COVID-19) infection is through detection of viral RNA using reverse transcription polymerase chain reaction (RT-PCR) test. While RT-PCR tests are the most sensitive for identifying infection, there are significant limitations, such as global access to sufficient test kits, turnaround times (TAT) from specimen collection to test result is often greater than 24 h and the need for skilled operators in accredited laboratories requiring specialized equipment. A rapid test performed at the point of care (POC) could provide a result within an approximate time of 30 min post specimen collection, be performed by a health care worker and comprise a simple workflow, improving both turnaround time and potentially decreasing costs (e.g., transport, cold-chain, skilled laboratory staff, complex equipment). Determining the performance of SARS-CoV-2 RT-PCR tests is, however, easier to assess than antigen-based POCT, as residual clinical specimens (swabs in universal transport media [UTM]) are readily available in laboratory environments, and do not require patient informed consent. Evaluating the performance of POCT requires informed-consent driven studies, with patients required to provide a standard of care specimen as well as study evaluation specimens, which is often not acceptable as nasopharyngeal swabbing can be invasive, clinical field trials are costly and time consuming. Many institutions and regulatory bodies also require preliminary data prior to use in field settings. Therefore, we have developed a method to determine the performance of antigen based POCT that can be used by implementers in national healthcare programs, regulators and rapid test developers. The method investigates both quantitative and qualitative parameters, with the latter providing insights into the capability for implementation and national program uptake.

Key words SARS-CoV-2 diagnostics, SARS-CoV-2 rapid antigen test, SARS-CoV-2 point of care, Lateral flow assay, Nucleocapsid protein, Performance, Regulation, Implementation readiness

1 Introduction

SARS-CoV-2, the causative agent of COVID-19, was first identified in Wuhan in December 2019 [1]. Since the World Health Organization (WHO) deemed COVID-19 a global pandemic on 11 March 2020 > 100 million cases of infection and over two

Justin Jang Hann Chu et al. (eds.), *SARS-CoV-2: Methods and Protocols*,
Methods in Molecular Biology, vol. 2452, https://doi.org/10.1007/978-1-0716-2111-0_4,

million deaths have been reported globally 1 year later [2]. During this 1 year period the diagnostic pipeline has witnessed huge development of manual and automated diagnostic tests to efficiently detect SARS-CoV-2 infections. These diagnostic tests were mainly categorized in two groups: molecular test (RNA extractions, RT-PCR) and immunoassays (chromatographic and fluorescent lateral flow assays [LFA], and enzyme linked immunosorbent assays [ELISAs]). After extensive laboratory evaluations, RT-PCR was determined to be the gold standard for SARS-CoV-2 detection based on analytical performance (sensitivity and specificity) [3, 4]. However, with RT-PCR, there is a trade-off between improved assay performance and result turnaround time and availability of consumables [5]. As travel restrictions eased, and social activities resumed, there was an urgent need for the implementation of a reliable and rapid diagnostic test (RDT) for mass screening and management of SARS-CoV-2 infections. Consequently, several rapid antigen tests were made commercially available for SARS-CoV-2 diagnosis. Whereas the basic principles of the tests remained similar across various manufacturers [LFA containing immobilized monoclonal antibodies against SARS-CoV-2 surface proteins (nucleocapsid protein, spike glycoprotein) or glycans (sialic acid)] [6], analytical performance has varied significantly [7, 8]. In this regard, it became apparent that the use of a standardized protocol is critical to facilitate decision-making when considering the regulatory environment for implementation of a rapid diagnostic antigen test for SARS-CoV-2 detection for use in patient care. A key hurdle facing evaluators and regulators is the limitation of specimens available for both standard of care testing and evaluation purposes, as a single nasopharyngeal, mid-turbinate, oropharyngeal or nasal swab is required for SARS-CoV-2 antigen testing. This test is also designed to be performed at the POC, which is a difficult environment for performing test evaluation and equivalency testing. A further difficulty is knowing that antigen tests generally demonstrate a lower sensitivity than RT-PCR, and specimens should hence be collected from individuals when they are in the early stages of infection (<7 days from symptom onset) and when the viral load is generally highest (Ct <25) [9] would be best to determine an assays performance. This is not known prior to specimen collection and hence larger numbers of individuals would be required for screening to identify enough specimens for evaluation in the high viral load range. In addition, depending on epidemic status, enrolling a large number of participants may not be feasible further delaying the evaluation, and only a limited number of assays evaluated at any one time. Owing to these difficulties, our group were tasked with developing a suitable and standardized method (protocol) to evaluate SARS-Cov2 rapid antigen assays that could be performed under emergency use approval. Although bodies such as the US Food and Drug Administration (FDA) provide

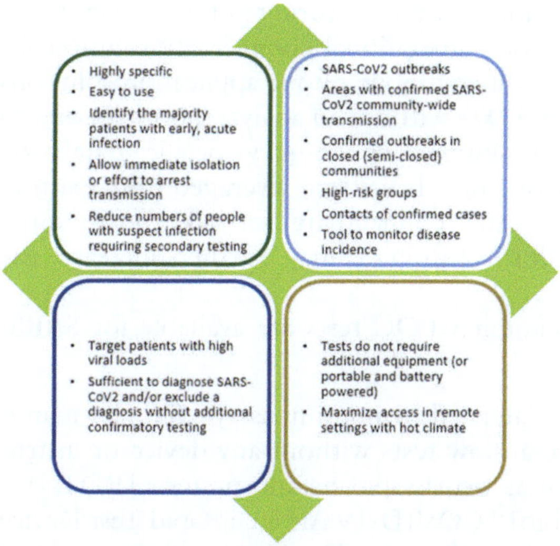

Criteria	Acceptable	Desirable
Target molecule:	RNA / protein / antigens detectable in 1 week of infection	RNA / protein / antigens detectable in 2 weeks of infection
Analytical sensitivity:	10^6 copies; (Ct=25); ≥70%	10^4 copies; (Ct=35); ≥80%
Analytical specificity:	Detects SARS-CoV2 with no cross-reactivity ≥97%	Detects SARS-CoV2 with no cross-reactivity ≥99%
Type of analysis:	Yes / No Visual or reader	
Specimen:	Swab – saliva (sputum)	Swab – saliva (sputum) (self-collection desirable)
Time to result:	<40mins	<20mins
Invalid rate:	<5%	<1%
Other:	Minimize the need for biosafety	
Cost:	$2 - $20	

Fig. 1 Summary of WHO TPP for POCT where reference testing is unavailable or TAT obviate its clinical utility

guidance [10] to developers on assay registration requirements which include the testing of clinical specimens from 30 positive and 30 negative patients, these guidelines were only available in October 2020, and do not provide guidance to implementers that require decisions regarding test uptake. The WHO also only provided a draft target product profile (TPP) for test developers in July 2020 [11]. The TPP summary relevant to POCT, where reference testing is unavailable or TAT limits its clinical utility, is outlined in Fig. 1, but limited guidance is provided to downstream end-users on how to best evaluate these POCT, especially where multiple tests are flooding the market. As of March 2021, we have evaluated 18 rapid POC SARS-CoV-2 antigen tests to determine their performance within the laboratory to assist in-country regulators with decision-making on product import.

The overall advantages of a SARS-CoV-2 antigen test is they would be cost-effective, easy to use, with a rapid turnaround time (< 15–20 min) [12] which minimizes the marginal differences on assay performance compared to RT-PCR. In addition, since no specialized equipment or accredited laboratory is required; antigen tests can be implemented for rapid screening at POC rather than requiring specimen transport to centralized laboratory facilities [13]. Furthermore, the advantages of testing at POC include nurse or on-site technician testing (minimal staff training) and the ability to report the result to a patient who remains onsite, limiting loss to follow-up, allowing for patient triage, and enabling rapid contact screening, which assists healthcare workers to make informed rapid treatment and/or isolation decisions [14, 15]. A disadvantage of rapid antigen tests at POC is the lack of connectivity to the centralized data bases that are critical to public health interventions and epidemiological monitoring [7]. However, many manufacturers now provide an optional connectivity solution for RDT analyzers and innovative mobile phone application solutions are being developed for RDTs without an analyzer, both linking to laboratory information systems or country-specific databases. These solutions, many of which have been leveraged from existing programs, may resolve the connectivity setback, but require specialized IT skills for setup and may further increase the implementation cost.

To date, multiple antigen POC tests are available for SARS-CoV-2 diagnostics:

(a) Chromatographic lateral flow immunoassay—Simple, manually run rapid lateral flow tests without any device or instrument (also known as visual rapid diagnostic tests [RDTs])—for example, Panbio™ COVID-19 Antigen Rapid Test Device (Abbott Laboratories, Chicago, IL, USA) and Standard Q COVID-19 Ag (SD Biosensor Inc., Suwon-si, South Korea).

(b) Chromatographic/fluorescent lateral flow immunoassay—Simple, manually run RDTs read with a reader—for example, BD Veritor™ Plus System (Becton Dickinson and Company, Franklin Lakes, NJ, USA), Sofia SARS Antigen Fluorescent Immunoassay (FIA) (Quidel Corporation, San Diego, CA, USA).

(c) Microfluidic immunofluorescence assay—Small, simple POC instrument rapid lateral flow tests that are both run and read by a dedicated device—for example, LumiraDx SARS-COV-2 Antigen Test (LumiraDx UK Limited, London, UK).

Preliminary data indicates that POC device-based assays have higher sensitivity compared to visual RDTs. However, as more sensitive antibodies against SARS-CoV-2 antigens become available, visual RDTs may achieve parity with device-based assays. Several rapid antigen assay instructions for use (IFU) focus only on testing for the presence of SARS-CoV-2 antigens directly off a nasal (or nasopharyngeal) swab, and the IFU seldom includes testing from viral transport media (VTM)/phosphate buffer saline (PBS) specimens. Of the 18 rapid Ag tests our group have received for evaluation in South Africa, only six tests included testing from a liquid specimen in their IFU. This is the greatest hurdle for evaluators as, for instance, in our experience 33 clinical trials is not feasible, too costly and above all, the waves of the current pandemic occur over fairly short time periods (in South Africa, the first infectious wave was in June/July 2020 and the second wave in December 2020/January 2021 [16].

This chapter therefore describes a detailed, impartial method based on extensive in-house troubleshooting/optimization and direct communication with several manufacturers that can be used

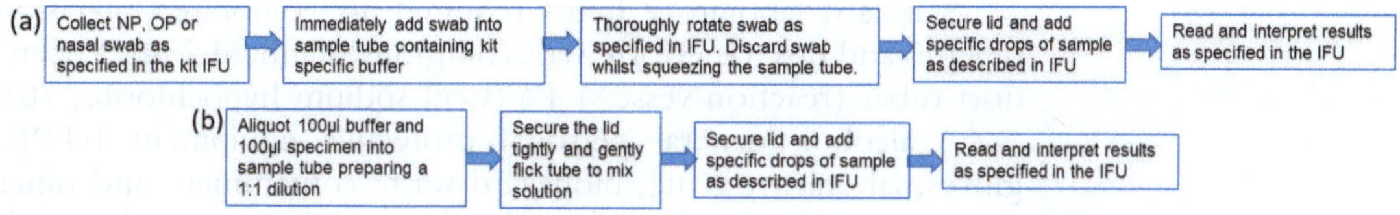

Fig. 2 Comparison of "clinical swab" and laboratory UVT/PBS antigen testing workflow. **(a)** Clinical specimen testing flow (as per the kit's Instructions for Use). **(b)** Specimen testing flow for evaluation of Rapid Test SARS-CoV-2 Ag using equal volumes of specimen to kit buffer

to evaluate any SARS-CoV-2 rapid antigen test in the absence of an IFU that caters for testing on VTM/PBS clinical specimens. As highlighted in Fig. 2, there are differences in the workflow for this laboratory-based protocol when compared to the "clinical swab" IFU but does enable continuous assay evaluation between waves of SARS-CoV-2 infection and without the expense or inconvenience to patients of a full clinical trial. An advantage of using residual VTM/PBS specimen for evaluation, is that several assays may be evaluated off a single specimen. This provides insight into multiple assays comparisons, which is not possible in a field trial as only two nasopharyngeal swabs are collected from each patient for a single use on rapid antigen test and standard of care (SOC) reference testing.

In this method's chapter, we describe a protocol for regulators and test implementers to rapidly assess the performance and implementation readiness of POC SARS-CoV-2 Ag tests under emergency use to screens assays prior to uptake. This approach can assist decision-makers, regulators and implementers faced with choosing suitable assays among multiple products flooding the market, address gaps in current evaluation guidelines and reduce time to market penetration, whilst also providing test developers the opportunity for rapid assessment by potential end-users.

2 Materials

2.1 Reagents Required to Evaluate POC SARS-CoV-2 Rapid Ag Assays in the Laboratory Environment

Suppliers are requested to provide a minimum of 250 SARS-CoV2 Ag kits for the evaluation. This may depend on individual country regulations and it may be necessary to purchase the required kits. This should include the IFU and all items required (especially the swab and test buffer) to perform the assay. The completeness of the kit components is assessed as part of the evaluation: Can the kit stand alone in field conditions at POC? If swabs are not provided by the supplier, we suggest purchasing Nasopharyngeal Specimen Collection nylon flocked swabs (WUXI Nest Biotechnology Co. Ltd., Jiangsu, China). A comprehensive evaluation of swab types are available [17]. The majority of SARS-CoV-2 antigen assays utilize a nasopharyngeal swab, although it is feasible that saliva testing may be implemented in the near-future [18].

Standard laboratory items required are: calibrated precision pipettes and tips to perform series dilutions, standard 2 mL Eppendorf tubes (reaction vessels) 4% (v/v) sodium hypochlorite, 70% (v/v) alcohol, general personal protective equipment [(PPE) gloves, laboratory coat], biohazard waste containment and timer to ensure test protocol is followed and record time to positive result (TPR). A relevant country-specific plug adaptor (where applicable) may be required for instrument-based technologies.

2.2 Challenge Panels

Viral cultures, commercially available reference materials and residual clinical patient specimens are suggested for inclusion in developing a comprehensive challenge panel. Examples are described below but can be adapted to reference materials available in-country. Two options for using the challenge panel are to spike (or submerse) the material onto the suppliers' swab or add the material directly to the test strip (or cartridge). Refer to notes for further information. A biosafety cabinet (BSC III) is required to ensure safe dilution of all infectious materials during the preparation of the challenge panels.

SARS-CoV-2 viral culture can be obtained from specialized virology laboratories with the capacity (BSL3 environment) to manufacture in bulk infectious virus. For our evaluations, SARS-CoV-2 viral culture supernatant was originally obtained through collaboration with Prof. Wolfgang Preiser (Stellenbosch University, Stellenbosch, Western Cape, South Africa) whose team grew culture isolates from clinical specimens by infecting Vero cells and harvesting supernatant. Viral culture growth was scaled for research in Johannesburg by Prof. Bavesh Kana and Dr. Bhavna Gordhan (Centre of Excellence for Biomedical Tuberculosis Research (CBTBR), University of the Witwatersrand, Johannesburg, Gauteng, South Africa). Viral supernatant was semiquantitated using reference RNA material (Dr. Christopher Ealand, CBTBR) and provided at ~log5 viral copies per microliter (cp/μL). For antigen studies, the viral culture supernatant was provided with live virus, as the effect of virus inactivation on protein structure is not known.

SARS-CoV-2 purified recombinant nucleocapsid (~60.68 kDa) and spike (~156.55 kDa) proteins were obtained from Prof. Jonathan Blackburn (University of Cape Town, Rondebosch, Cape Town, South Africa), whose team expressed SARS-CoV-2 nucleocapsid and spike proteins in Baculovirus expression system in insect cells using the patented KREX™ functional proteomics technology (Sengenics Co Pte Ltd., South Africa). The protein challenge panel was prepared in duplicates of protein concentrations (10 nM, 2.5 nM, 0.625 nM, 0.039 nM, 9.75 pM, and 2.4 pM) prepared in supplier test kit–specific buffer.

Residual clinical specimens of previously characterized SARS-CoV-2 positives and negative swab specimens immersed in either VTM, UTM, or PBS storage and transport media. Access to residual laboratory specimens may require institution review board

(IRB) approval based on local Human Research Ethics regulations within one's institution. All specimens should be deidentified for the evaluation. Specimens may be stored frozen (−80 °C) until evaluation and specimens should not be used post three freeze-thaw cycles.

Note: Comparator SARS-CoV-2 RT-PCR technology should be used to confirm the presence of SARS-CoV-2 and quantify the amount of SARS-CoV-2 through the proxy variable of the cycle threshold (Ct) [19, 20] *using standard of care (SOC) assays such as: RealTime SARS-CoV-2 (RT SARS-CoV-2) on the Abbott m2000 or Alinity M platforms Abbott Molecular, Abbott Park, IL, USA), cobas® SARS-CoV-2 on the cobas(r) 6800/8800 platforms (Roche Molecular, Pleasanton, CA, USA), TaqMan® TaqPath™ COVID-19 CE-IVD RT-PCR Kit (multiplex) on a suitable QuantStudio platform (both ThermoFisher Scientific, Waltham, MA, USA). RT-PCR was performed on the viral culture supernatant and residual clinical specimens.*

3 Methods

3.1 Investigate the SARS-CoV-2 Antigen Test IFU

1. Interrogate all the contents of the suppliers' kit and determine any additional items required to perform the test, such as power adaptor/plug, batteries, UV torch light, pipettes, buffers, and swabs.
2. Investigate the kit contents and test procedures. Rapid antigen tests are mainly intended for POC use and therefore should have all necessary materials required to perform the test on-site. In addition, the test should be rapid (turnaround time of <20–40 min), have no invalids/errors and should be relatively easy to use (does not require high level skill). Furthermore, the test procedure should be relatively safe to perform (no biohazards, toxic chemicals) for the operator, patient, and environment. During the evaluation process one should note the qualitative assay performance, which can be scored at the end of the evaluation using the Likert score as follows: 1 = very bad/difficult; 2 = bad/difficult; 3 = neutral; 4 = good/easy; and 5 = very good/easy to perform.

3.2 Determine the POC SARS-CoV-2 Rapid Antigen Performance

Determine the POC SARS-CoV-2 rapid antigen performance by measuring accuracy (sensitivity and specificity) and agreement (Cohen Kappa) compared to SOC RT-PCR.

3.2.1 Accuracy Evaluations

Accuracy evaluations should be performed using previously RT-PCR characterized SARS-CoV-2 specimens grouped into high viral load concentrations (Ct < 30), medium viral load concentrations (Ct ≤35) and negative specimens.

3.2.2 Clinical Specimens

A total of 60 high viral load and 20 medium viral load SARS-CoV-2 specimens and 30 negative specimens is recommended as a minimum sample size. The Ct threshold is selected based on current literature of measures of infectiousness [21] and the WHO antigen guidelines [9], and it is recommended to determine the in-country median Ct value, which will vary dependent on the positivity rate.

1. Perform all testing in a biosafety setting following standard laboratory protocols.
2. Test each specimen ($n = 110$) once, ensuring manufacturer's instructions are followed.
   (a) Defrost specimens at ambient temperature. Ensure that specimens are fully defrosted and warmed to ambient temperature before beginning processing.
   (b) Unless otherwise stated by a supplier, residual clinical specimen will be used in 1:1 ratio of specimen: kit buffer prior to applying to the rapid antigen test and perform the test following manufacturer's instructions. Record each specimen result including the time taken to generate a positive result as outlined in Fig. 3.

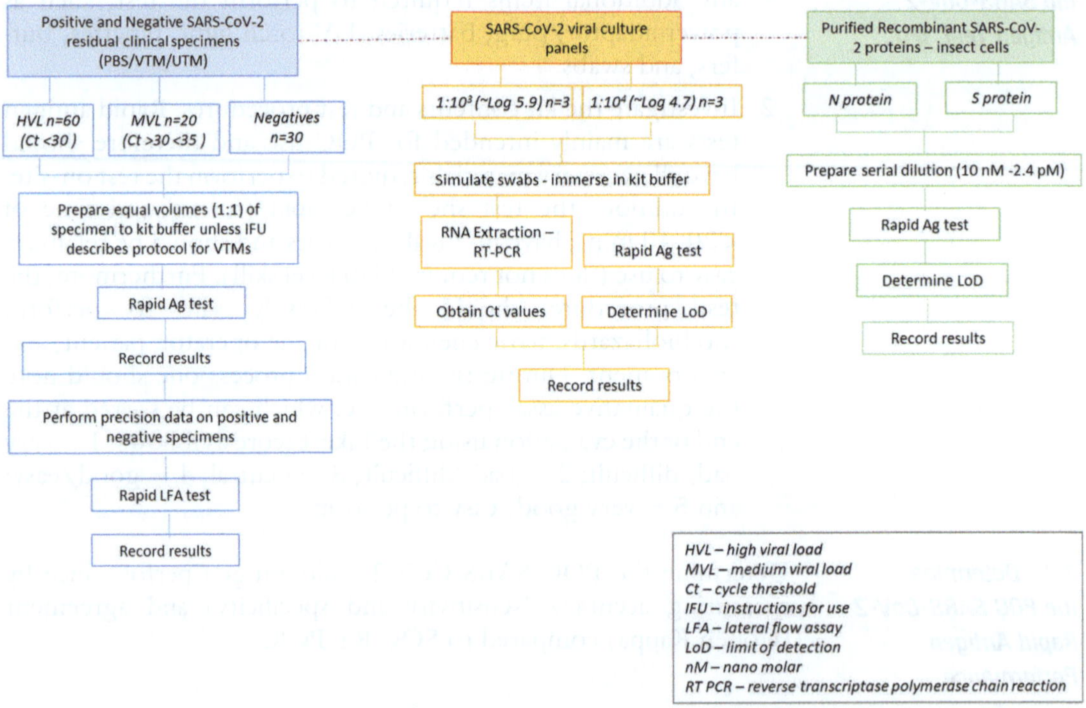

Fig. 3 Method comparison workflow to measure accuracy and precision (blue section), limit of detection using viral cultures (yellow section) and using recombinant protein reference material (green section)

3.2.3 Determine the Assay Precision (Reproducibility)

(a) To evaluate the assay precision (reproducibility), a minimum of two positive (Ct < 30) and two negative specimens should be tested in quintuplicates within the same day ($n = 20$).

(b) Select two SARS-CoV-2 specimens with Ct < 30 and two SARS-CoV-2 negative specimens and test as described in Subheading 3.2, by combining the residual specimen with the kit buffer in a 1:1 ratio.

(c) Ensure the manufacturer's instructions are followed and that all results are correctly recorded, including time to positivity if applicable.

3.2.4 Determine the Assay Limit of Detection (LoD)

SARS-CoV-2 Viral Culture

SARS-CoV-2 viral culture stock solution should be used to create standardized dilutions to verify the analytical sensitivity, in copies per reaction. The viral culture challenge panel should include a minimum of 4 replicates of 2 dilutions which approximate to log 5.0 and log 4.7 viral copies per milliliter (cp/mL) respectively. These generally produce Ct values between 25 and 30 on most assays, although this may vary based on SOC assays used. A process flow of viral culture testing is shown in Fig. 4.

Note: Substances present in certain viral transport media may obstruct the specimen flow; producing a false negative and/or invalid result. In this regard, viral culture diluent may be centrifuged (1200 × g for 2 min) and the supernatant can be applied to determine the assay LoD. However, due to a high number of variabilities, the viral culture LoD is not used to score the assay quality performance.

SARS-CoV-2 Recombinant Proteins

In addition to the viral culture supernatant, purified recombinant nucleocapsid (~60.68 kDa) and spike (~156.55 kDa) SARS-CoV-2 proteins that were expressed in Baculovirus expression system in insect cells were tested; protein size and availability may vary by country. Although most manufacturers state that the test line contains monoclonal antibodies against the N protein, the S protein was included to evaluate the specificity of the test. The recombinant protein challenge panel should include duplicates of each protein concentration (10 nM, 2.5 nM, 0.625 nM, 0.039 nM, 9.75 pM, and 2.4 pM) prepared in supplier test kit–specific buffer to a final volume of 150 μl. Purified (>90%) recombinant SARS-CoV-

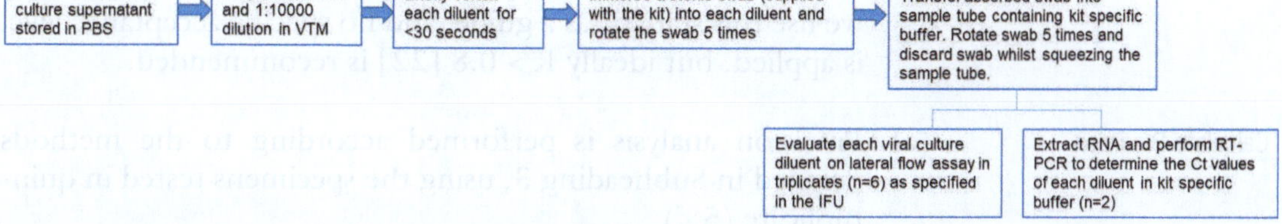

Fig. 4 Overview of viral culture supernatant processing

Table 1
Binary results to assess rapid antigen test performance

A: True positive (known positive PCR for COVID at least 14 days post-symptom onset)	B: False positive (negative COVID-19 test or symptoms or known contacts)
C: False negative (known positive PCR for COVID at least 14 days post-symptom onset)	D: True negative (negative COVID-19 test or symptoms or known contacts)

2 proteins are highly specific and therefore can be used to determine the assays LoD. However, at this stage, purified recombinant proteins are not easily accessible and for this reason was not included to score the overall test quality performance.

3.2.5 Perform the Statistical Analysis for Determining the Quantitative Assay Performance

Result Presentation

(a) Collate results in a MS Excel file or STATA (StataCorp. 2015. Stata Statistical Software: Release 14. College Station, TX: StataCorp LP.)
(b) Exclude inconclusive results from the assessment.
(c) Using the RT-PCR results as the reference, calculate the number of true positive, true negative, false positive and false negative results.
(d) Enter the binary results in Table 1 as shown below:

Calculate Accuracy

(a) Calculate sensitivity, specificity, and error rate as described above for the entire data set ($n = 150$).
(b) Calculate percentage sensitivity $= (A/(A + C)) \times 100$.
(c) Calculate percentage specificity $= (D/(D + B)) \times 100$.
(d) Calculate error rate: (inconclusive results/total specimens tested) × 100.
(e) Acceptance criteria for accuracy are sensitivity $\geq$80%, specificity $\geq$97% and error rate $<$ 5%.

Calculate Agreement

(a) Calculate method agreement using the Cohen kappa coefficient (K):
(b) Apply the Landis & Koch scale [22] to measure the degree of agreement according: $K < 0.2$ = Poor agreement; $0.2 < K < 0.4$ = Fair agreement; $0.4 < K < 0.6$ = Moderate agreement; $0.6 < K < 0.8$ = Substantial agreement; $0.8 < K < 1.0$ = Almost perfect agreement. In our experience we use this statistic as a guide, and no specific acceptance level is applied, but ideally $K > 0.8$ [22] is recommended.

Calculate Precision

(a) Precision analysis is performed according to the methods detailed in Subheading 3, using the specimens tested in quintuplicate (5×).
(b) All specimens tested must provide the correct result (positive or negative).

Table 2
Acceptable criteria when the assays are performed on laboratory obtained specimens

Assay name	Level achieved by the assay evaluated
≥97% specificity to minimize false-positive results is acceptable	*Provide yes/no*
≥80% sensitivity to identify virus in high concentration is acceptable	*Provide yes/no*
Identification of virus in low concentrations is desirable	*Provide yes/no*[a]
Assay generates a result within 40 min (acceptable)	*Provide yes/no*

[a]This may also be reported as "some detection"

3.2.6 Determine the Overall Acceptance Criteria

(a) Laboratory performance of antigen RDT assays in identifying SARS-CoV2 is determined through this protocol design (under emergency use and compared to existing in-country SOC technology) and is based on the criteria outlined in Table 2. *Note: In-country prevalence is not taken into account.*

(b) The most important criterion is specificity (>97%), followed by sensitivity (>80%) in specimens with high viral load (Ct < 30), a threshold also recommended by the WHO (interim guidelines) [9] as the minimum a SARS-CoV-2 antigen test should achieve. If these two criteria are met, the assay may proceed to further evaluation in field settings or used for patient care, depending on the country regulation.

Data Visualization

The raw data (positive or negative (binary) or invalid (excluded from assay performance analysis)) are entered into MS Excel spreadsheets (Microsoft Corporation, Redmond, Seattle, WA, USA) and can be visualized as a heat map as presented in the example in Fig. 5. Based on our experience, at least 3 POC rapid SARS-CoV-2 antigen tests can be evaluated off a single clinical specimen (VTM/PBS). The heat map provides a good visual to complement the accuracy analysis. For example: The RDT1 has better performance than RDT2 and 3 as evident from more red blocks (detected results) in both the high and medium viral load group than the other assay, which have more green (undetected results). It is also evident that all RDT assays 1, 2, and 3 show good specificity (i.e., no red blocks in the negative specimen group). Invalid specimens, while rare in our experience, are shown in yellow and contribute to the assay error rate (i.e., excessive yellow in a heat map may lead to an assay not meeting acceptance criteria based on a high error rate).

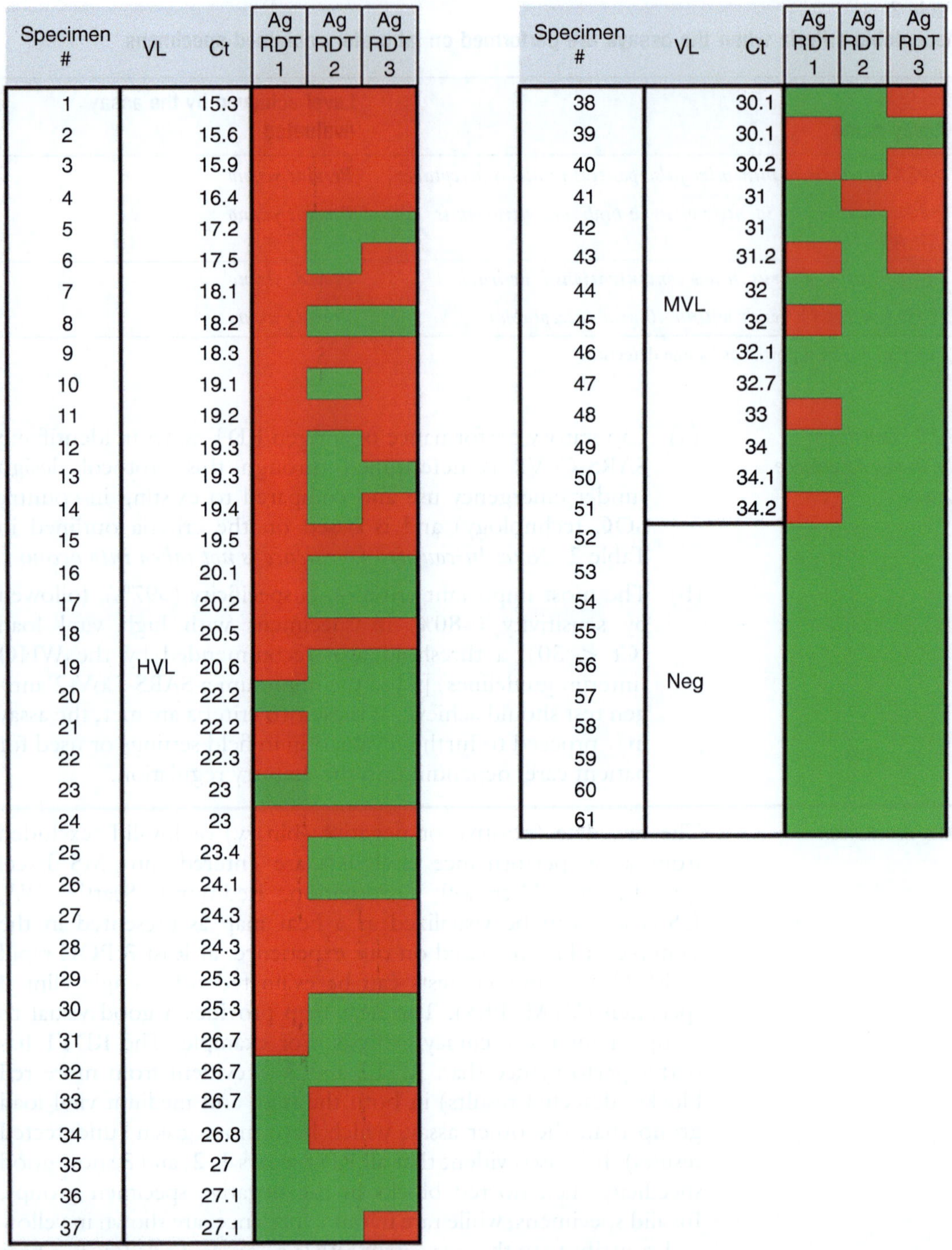

Fig. 5 An example (61 or 110 results shown) of a heat map illustrating the binary (positive [red], negative [green]) results from the evaluation of three POC rapid SARS-CoV-2 Antigen assays performed off the same residual clinical specimen. The Ct values from the SOC assay are included and sort the specimens into two groups: HVL (high viral load Ct < 30) and MVL (medium viral load Ct < 35) as well as negative specimens

Perform the Likert Score for Qualitative Measure of Assay Performance and Capability for Implementation

Answering the qualitative performance questions using the score system as outlined in Table 3, which shows two assays, one that performed well and one that performed less well. We recommend any score > 4 indicates an assay that may be suitable for use in the field.

4 Notes

4.1 Advise against Simulating "Clinical Swab" for the Evaluation to Prevent Dilution Effect

In our experience of evaluating over 18 SARS-CoV-2 POC rapid antigen assays to date using residual clinical specimens (PBS/VTM) collected between October to December 2020, simulating a "clinical swab" collected specimen by placing the supplier swab in the residual clinical patient VTM/PBS specimen prior to placing the swab in the supplier's buffer, to resemble the manufacturer's instructions for use (IFU) resulted in an increase in false negative results when compared to the Ct values obtained from a standard of care RT-PCR assay. This was due to specimen dilution compared to the original VTM/PBS specimen. To eliminate false negative results, the protocol was revised to use equal volumes of residual clinical specimen added directly to the manufacturer's supplied buffer (1:1) unless otherwise stipulated in their IFU. This strategy was based on extensive in-house troubleshooting/optimization, literature review and direct communication with several manufacturers to establish an impartial protocol for evaluation under laboratory-based conditions.

A laboratory-based evaluation (as described in this protocol) relies on the use of residual clinical specimens which are stored in either VTM, UTM, or PBS to preserve viral particles. However, several SARS-CoV-2 rapid antigen test IFU indicate possible interference/inhibition of test results when specimens are stored in transport media. In this regard, differentiating a false negative due to test performance vs. inhibition from substance(s) present in transport medium is vital. In addition, the brand and composition of storage media used in the field for transporting clinical patient swab specimens to the laboratory are is frequently unknown.

4.2 Post Market Surveillance

As the proposed protocol recommends performing the POC SARS-CoV-2 rapid Antigen assay evaluation in the laboratory setting, it is recommended that on-going quality monitoring of the assay within the field is monitored, that is, postmarket surveillance. In-country experiences from other RDT assays, notably the HIV diagnosis program, have highlighted variable technique (e.g., Imprecise specimen or diluent addition, reporting of result before recommended time) in the field, a relatively uncontrolled and frequently under-resourced environment [23].

Table 3
Qualitative performance analysis of POC SARS-CoV-2 rapid antigen tests

Assay characteristics	Availability of all required materials in kit	Ease of use	Time to result (acceptable $\leq$40 min, desirable $\leq$20 min)	Invalid (error rate)	Minimize the need for biosafety	Minimal training required
Score description	A score should be allocated by the operator to assess whether additional materials/ consumables (such as essential PPE, stopwatch, biohazard bins etc.) are required to complete the test process (i.e., vital components not included in the kit)	A score should be allocated based on the ease of use to perform a test (does the IFU provide clear instructions on the test process and result interpretation? are there any difficulties experienced by the operator during the test? Does the test require instruments/ analyzers to interpret results? Are the instruments/ analyzers user-friendly? Does it require frequent software updates? Are the instruments/ analyzers mobile?	A score should be allocated based on the overall process time (sample collection to test result) as well as an average detection time (observe a positive band/ result)	Majority of the rapid antigen test contains a control band which should be detected per test irrespective of the patient result. A score should be allocated based on the clear visibility of the control band per test card.	Although rapid antigen tests are intended to use at POC, SARS-CoV-2 is a highly contagious infection and therefore, all test procedures must be executed with highest safety measures and all protocols should be adhered to by the operator. A score is allocated based on the risk associated with the test process (does the kit buffer inactivates the viral particles? are the assay tubes slippery and may lead to a spill?)	A score is allocated based on the ability of the operator perform the test by following instructions provided on the IFU. If instruments/ analyzers are required, can they be operated without in-depth training? Are they additional connectivity and printer solutions? Can they be used with minimal training?

		Once fully charged, Is the battery life sufficient for at least 2 h?				
Ag RDT 1 score[a]	4	5	5	5	4	5
Comments	Does not include certain protective materials and consumables which will be required for field implementation	IFU provided with each kit describes a step by step process flow on specimen collection, specimen preparation, test assay and test interpretation	A mean TTP was observed within 4 min (range: 2–15 min).	Control line was clearly visible for each test device.	The lateral flow test is safe to use (buffer claims to inactivate SARS-CoV2) and can be executed without a biosafety laboratory; however, biosafety disposal and spill kit will be required and ALL safety precautions during specimen collection and addition of clinical specimen to the test buffer/sample tube need to be adhered	Basic standard procedure used for rapid antigen test. No instrument or specialized skills required.
Ag RDT 2 score[a]	4	3	4	5	4	4
Comments	The kit does not include certain protective materials, and consumables which will be	IFU provides a detailed description for specimen collection, specimen	Mean average time to result is not applicable with this assay as the prepared test device is incubated	No invalids were detected on the analyzer	The lateral flow test is relatively safe to use and can be executed without a biosafety laboratory;	Basic standard procedure used for rapid antigen test but requires the use of an analyzer and

(continued)

Table 3 (continued)

Assay characteristics	Availability of all required materials in kit	Ease of use	Time to result (acceptable $\leq$40 min, desirable $\leq$20 min)	Invalid (error rate)	Minimize the need for biosafety	Minimal training required
	required for field implementation. The kit contains a positive and negative control.	preparation, testing process and result interpretation. The assay result cannot be interpreted without the analyzer. Testing time is increased as only one test device can be inserted into the analyzer at one time. The connectivity solution (adaptor for barcode scanning, printing and Wi-Fi access) is not user friendly	for 15 min on benchtop and read immediately thereafter using the analyzer. However, the testing time is increased as only one test strip can be detected at a time (15 min per test)		however, biosafety disposal and spill kit will be required and ALL safety precautions during specimen collection and addition of clinical specimen to the test buffer/sample tube need to be adhered	specialized IT skills for connectivity solution

[a]An example of two assays (well performed and less well performed)

4.3 Instrument Based POC Assays

In our experience, a disadvantage to assays that rely on result interpretation using a fluorescent (or UV) torch is that batteries are required to operate torches may not always be supplied and requires replacement, usually an unbudgeted expense; the torch is easily misplaced; utilizing a torch with an incorrect wavelength can lead to errors; and if these assays are read under direct light, faint bands and/or discordant results may arise, and the use of the torch is not documented or captured along with the clinical result. POC tests that rely on instruments that require a power supply may be affected by unstable power. Systems with options for connectivity for result transmission may also be affected by unreliable networks, which is key if connectivity is required for the result to be provided.

4.4 Consideration for Assay Performance with the Introduction of SARS-CoV-2 Variant Strains Circulating in the Population

As early as October 2020, evidence was being provided of circulating SARS-CoV-2 variant strains globally [24]. Although the impact on diagnostics performance was first investigated in molecular assays, antigen-based assays also require monitoring. It is recommended that challenge panels sought for evaluation purposes do include current and relevant specimens, which optimally should be sequenced and characterized to determine possible false positive results reported by the POC SARS-CoV2 rapid Ag assays.

4.5 Quality Controls

Where possible, relevant SARS-CoV-2 positive and negative quality controls should be tested with each new kit lot number, to ensure ongoing test quality. Several suppliers provide these controls, and, where not provided, institutions may develop their own materials similarly to the residual clinical specimens applied in this protocol.

References

1. Zhu N, Zhang D, Wang W, Li X, Yang B, Song J, Zhao X, Huang B, Shi W, Lu R, Niu P, Zhan F, Ma X, Wang D, Xu W, Wu G, Gao GF, Tan W, China Novel Coronavirus I, Research T (2020) A novel coronavirus from patients with pneumonia in China, 2019. N Engl J Med 382(8):727–733. https://doi.org/10.1056/NEJMoa2001017
2. Wang C, Wang Z, Wang G, Lau JY, Zhang K, Li W (2021) COVID-19 in early 2021: current status and looking forward. Signal Transduct Target Ther 6(1):114. https://doi.org/10.1038/s41392-021-00527-1
3. Sethuraman N, Jeremiah SS, Ryo A (2020) Interpreting diagnostic tests for SARS-CoV-2. JAMA 323(22):2249–2251
4. Tahamtan A, Ardebili A (2020) Real-time RT-PCR in COVID-19 detection: issues affecting the results. Expert Rev Mol Diagn 20(5):453–454. https://doi.org/10.1080/14737159.2020.1757437
5. Corman VM, Landt O, Kaiser M, Molenkamp R, Meijer A, Chu DK, Bleicker T, Brunink S, Schneider J, Schmidt ML, Mulders DG, Haagmans BL, van der Veer B, van den Brink S, Wijsman L, Goderski G, Romette JL, Ellis J, Zambon M, Peiris M, Goossens H, Reusken C, Koopmans MP, Drosten C (2020) Detection of 2019 novel coronavirus (2019-nCoV) by real-time RT-PCR. Euro Surveill 25(3). https://doi.org/10.2807/1560-7917.ES.2020.25.3.2000045
6. Pohanka M, Denizli A (2021) Point-of-care diagnoses and assays based on lateral flow test. Int J Anal Chem 2021:1–9. https://doi.org/10.1155/2021/6685619
7. Dinnes J, Deeks JJ, Adriano A, Berhane S, Davenport C, Dittrich S, Emperador D, Takwoingi Y, Cunningham J, Beese S,

Dretzke J, Ferrante di Ruffano L, Harris IM, Price MJ, Taylor-Phillips S, Hooft L, Leeflang MM, Spijker R, Van den Bruel A, Cochrane C-DTAG (2020) Rapid, point-of-care antigen and molecular-based tests for diagnosis of SARS-CoV-2 infection. Cochrane DB Syst Rev 8:CD013705. https://doi.org/10.1002/14651858.CD013705

8. CDC (2020, December 16) Interim guidance for antigen testing for SARS-CoV-2. Centers for disease control and prevention. https://www.cdc.gov/coronavirus/2019-ncov/lab/resources/antigen-tests-guidelines.html. Accessed 30.03.2021 2021
9. WHO (2020, September 11) Antigen-detection in the diagnosis of SARS-CoV-2 infection using rapid immunoassays: interim guidance. Accessed 03.30.2021 2021
10. FDA (2021, March 26) In vitro diagnostics EUAs. https://www.fda.gov/medical-devices/coronavirus-disease-2019-covid-19-emergency-use-authorizations-medical-devices/in-vitro-diagnostics-euas. Accessed 30.03.2021 2021
11. WHO (2020, September 28) COVID-19 Target product profiles for priority diagnostics to support response to the COVID-19 pandemic v.0.1. Accessed 30.03.2021 2021
12. NICD (2020, December 11) The use of antigen testing for the diagnosis of SARS-CoV-2 in South Africa https://www.nicd.ac.za/diseases-a-z-index/covid-19/covid-19-guidelines/antigen-testing-guideline/. Accessed 30.03.2021 2021
13. Cerutti F, Burdino E, Milia MG, Allice T, Gregori G, Bruzzone B, Ghisetti V (2020) Urgent need of rapid tests for SARS CoV-2 antigen detection: evaluation of the SD-Biosensor antigen test for SARS-CoV-2. J Clin Virol 132:104654. https://doi.org/10.1016/j.jcv.2020.104654
14. Africa CDC (2021, February 10) Monitoring and evaluation of COVID-19 rapid antigen diagnostic test rollout in Africa. https://africacdc.org/download/monitoring-and-evaluation-of-covid-19-rapid-antigen-diagnostic-test-rollout-in-africa/. Accessed 30.03.2021 2021
15. Nagura-Ikeda M, Imai K, Tabata S, Miyoshi K, Murahara N, Mizuno T, Horiuchi, M, Kato K, Imoto Y, Iwata M, Mimura S, Ito T, Tamura K, Kato Y (2020) Clinical evaluation of self-collected saliva by quantitative reverse transcription-PCR (RT-qPCR), direct RT-qPCR, reverse transcription-loop-mediated isothermal amplification, and a rapid antigen test to diagnose COVID-19. J Clin Microbiol 58(9): e01438-20. https://doi.org/10.1128/JCM.01438-20. https://www.ncbi.nlm.nih.gov/pubmed/32636214
16. Jassat W, Mudara C, Ozougwu L, Tempia S, Blumberg L, Davies M, Pillay Y, Carter T, Morewane R, Wolmarans M, von Gottberg A, Bhiman JN, Walaza S, Cohen C (2021) Increased mortality among individuals hospitalised with COVID-19 during the second wave in South Africa. medRxiv. https://doi.org/10.1101/2021.03.09.21253184
17. Kahamba RT, Noble L, Stevens W, Scott L (2020) Comparison of three nasopharyngeal swab types and the impact of physiochemical properties for optimal SARS-CoV-2 detection. J Vaccines & Vaccination S6. https://doi.org/10.1101/2020.10.21.20206078
18. Agulló V, Fernández-González M, de la Tabla VO, Gonzalo-Jiménez N, García JA, Masiá M, Gutiérrez F (2020) Evaluation of the rapid antigen test Panbio COVID-19 in saliva and nasal swabs in a population-based point-of-care study. J Infect
19. PHE (2020, October) Understanding cycle threshold (Ct) in SARS-CoV-2 RT-PCR a guide for health protection teams. https://assets.publishing.service.gov.uk/government/uploads/system/uploads/attachment_data/file/926410/Understanding_Cycle_Threshold__Ct__in_SARS-CoV-2_RT-PCR_.pdf. Accessed 30.03.2021 2021
20. Rao SN, Manissero D, Steele VR, Pareja J (2020) A narrative systematic review of the clinical utility of cycle threshold values in the context of COVID-19. Infect Dis Ther 9:573–586. https://doi.org/10.1007/s40121-020-00324-3
21. La Scola B, Le Bideau M, Andreani J, Grimaldier C, Colson P, Gautret P, Raoult D (2020) Viral RNA load as determined by cell culture as a management tool for discharge of SARS-CoV-2 patients from infectious disease wards. Eur J Clin Microbiol Infect Dis 39: 1059–1061. https://doi.org/10.1007/s10096-020-03913-9
22. Landis JR, Koch GG (1977) The measurement of observer agreement for categorical data. Int Biometric Soc 33:159–174
23. Mwisongo A, Peltzer K, Mohlabane N, Tutshana B (2016) The quality of rapid HIV testing in South Africa: an assessment of testers' compliance. Afr Health Sci 16(3):646–654
24. PHE (2021, January, 15) Confirmed cases of COVID-19 variant from South Africa identified in UK. https://www.gov.uk/government/news/confirmed-cases-of-covid-19-variants-identified-in-uk

Chapter 5

Diagnostic Method for COVID-19 Using Sugar Chain–Immobilized Nanoparticles and Saliva Specimens

Yasuo Suda, Yasuhisa Tajima, Jun-ichiro Nishi, and Takashi Kajiya

Abstract

Identification of viruses that infects animals or plants, and determination of their quantity are essential for the diagnosis of infectious disease and for the determination of a strategy in the treatment of virus-derived diseases. However, the concentration of viruses existing in a living body (in bodily fluid), food, drinking water, river water, and so on. is not high enough to be detected using conventional diagnostic methods. For example, since the concentration of influenza virus released from an infected person is less than the detection limit of conventional simple examination kits (rapid kit) or even a PCR process at the initial stage of infection, it is difficult to detect the presence of influenza virus which will lead to influenza disease. Our technology allows for safe, efficient, and selective concentration of viruses without troublesome ultracentrifugation, using sugar chain–immobilized metal nanoparticles based on the binding interaction between viruses and sugar chains. For COVID-19, we have developed and commercialized two molecular diagnosis kits: SUDx SARS-CoV-2 detection kit, and SGNP nCoV/Flu PCR detection kit, for the Japanese market in 2020.

Key words Sugar chain, Ligand, Isolation, Spike domain, SGNP, SMGNP, Copy number

1 Introduction

Coronavirus disease 2019 (COVID-19) is a contagious disease caused by severe acute respiratory syndrome coronavirus 2 (SARS-CoV-2) [1]. While most people have mild symptoms, some people develop acute respiratory distress syndrome (ARDS). ARDS can be precipitated by cytokine storms, multiorgan failure, septic shock, and thrombosis. Influenza is an infectious disease caused by the influenza virus [2]. The most common symptoms include high fever, runny nose, sore throat, muscle and joint pain, headache, coughing, and fatigue. Complications of influenza may include viral pneumonia, secondary bacterial pneumonia, sinus infections, and worsening of previous health problems such as asthma or heart failure. Since it is very difficult to discriminate

Justin Jang Hann Chu et al. (eds.), *SARS-CoV-2: Methods and Protocols*,
Methods in Molecular Biology, vol. 2452, https://doi.org/10.1007/978-1-0716-2111-0_5,

between the symptoms of COVID-19 and those of influenza, a highly sensitive and accurate diagnosis method is important.

It is known that the concentration of viruses in saliva is lower than that in oropharyngeal swab, nasopharyngeal swab, or nasal swab [3–6], although it has been controversially reported that the concentration of SARS-CoV-2 may be higher in saliva compared with that found in an oropharyngeal or nasopharyngeal swab [7, 8]. Our method can concentrate the viruses in the specimen so that higher and more accurate molecular diagnosis is available using saliva, which is noninvasive and has a low risk of transmission during collection from patients. Till present, our SGNP nCoV/Flu PCR detection kit is the only kit which uses saliva to detect COVID-19 and/or Influenza at the same time in the world. The second benefit of our product is that it can indicate the concentration of virion (viral particle) which is crucial for the infection of virus to cells. Since our product first concentrates the virion, it can distinguish between infectious and free uninfectious RNA/DNA in the specimen. Therefore, our product is useful for the follow up of hospitalized patients or for preventing false-positive diagnostic results compared with the regular Qiagen or Boom methods [9], which cannot distinguish between infectious and free RNA/DNA.

Most of the surfaces of our cells are covered with sugar chains. It is known that viruses recognizes sugar chains existing on the surface of a cell, binding and migrating into cells using the cellular high-affinity receptor protein in some cases, and infect the cells. For example, the spike protein of SARS-CoV-2 possesses the binding domain for sulfated sugar chain, heparan sulfate, on one side and the binding domain for cell surface ACE-2 receptor protein on another side [10]. The spike protein of influenza viruses, called hemagglutinin, binds to cell surface sialic acids containing sugar chain, and the viruses migrate directly into the cells [11]. The infection of viruses into higher animals or plants causes diseases. Often, the infectious viruses tend to infect the same species of animals or plants. For example, avian influenza viruses regularly do not infect humans except in the case of mutation of viral RNA. In domestic animals, agricultural crops, and ornamental animals or plants, viruses that are capable of infecting such animals or plants are transmitted among the same species, causing disruption in production processes where these plants and animals are used for human consumption. Therefore, the identification of viruses capable of infecting such animals or plants is essential for early diagnosis, thus contributing to early treatment. This method builds on our previous work related to detecting virus in agricultural settings.

We have aimed to provide a method for concentrating target viruses in a specimen containing a number of contaminants and which can identify the virus in a short time. We have succeeded in developing a method using sugar chain–immobilized metal nanoparticles, prepared by immobilizing the ligand conjugate, in which a sugar chain is connected with a well-designed linker compound, and by using sugar chain linkages for viruses based on the binding interaction between sugar chains and viruses as shown in the figure below [12].

The size of viruses is approximately 30–300 nm, and recovery and concentration of viruses is generally carried out by high-speed centrifugation such as ultracentrifugation. However, with ultracentrifugation, an extremely large gravity has to be applied to a sample. Consequently, due to uneven balance in weight between samples, there is a danger that the samples may scatter. Since such scattering of samples must be avoided, there is a need for a method of concentrating viruses without using ultracentrifugation. Our technology is developed in view of the foregoing problems to provide a method for concentrating viruses, which is capable of concentrating viruses safely, efficiently, effectively, and in a short period of time. We have investigated methods that substitute for ultracentrifugation and tried first to use a bigger gold nanoparticle (around 15 nm size measured confirmed by Transmission electron microscopy, TEM) [13, 14]. We finally found that magnetizing the sugar chain–immobilized metal nanoparticles (less than 10 nm size by DLS) and collecting the sugar chain–immobilized metal nanoparticles bound to viruses with a magnetic force, exhibits collection efficiency almost equal to that of centrifugation, and we found that this method allows for safer and more rapid concentration of viruses than centrifugation [15].

2 Influenza

Figure 1 is a summary of our method for the pretreatment of molecular diagnosis of viruses. First, virus-binding sugar chains are determined using a Sugar Chip, in which structurally defined sugar chains are immobilized on a gold surface in a clustered form which mimics nature [12]. Sugar chains obtained from a natural source or via chemical synthesis are conjugated with our original linker molecule to prepare Ligand-conjugate and are then immobilized on a gold chip to prepare the Sugar Chip, or to gold nanoparticles to prepare sugar chain–immobilized gold nanoparticles (SGNP). The Sugar Chip can be used as a sensor chip for surface plasmon resonance apparatus for high-throughput analysis of virus-binding. The size of SGNP is smaller than that of the virus, so that lots of SGNP can easily bind to the spike proteins of the virus. The SGNP bound with virus becomes heavier and heavier, so it is easily precipitated by regular centrifugation ($10,000 \times g$) without using ultracentrifugation. In other words, viral particles are captured, concentrated and purified using SGNP. Then, the heat treatment or the addition of 0.1% SDS solution destroys the viral particle and viral DNA/RNA is released. This can then be analyzed

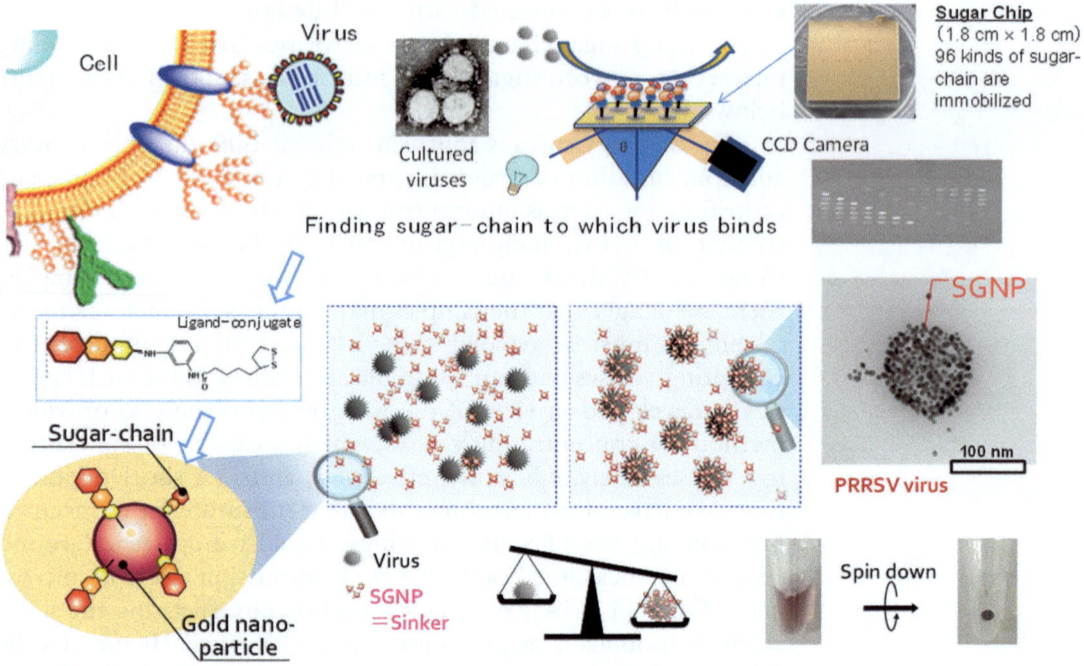

Fig. 1 Principle for the concentration and purification of viruses based on glycol-nanobiotechnology for the pretreatment of molecular diagnosis

Table 1
A result of the clinical research in 2011–2012 influenza season

			SGNP/PCR using saliva		
Patients	**Conventional kit**		**A**	**B**	**–**
Adults age ≥16 ($n = 74$)	A+	25	25	0	0
	B+	3	1	2	0
	–	46	24	0	22
Children age <16 ($n = 109$)	A+	56	54	0	2
	B+	7	1	5	1
	–	46	23	3	20

Comparison between conventional kit using nasal swab and immuno-chromatography and our SGNP/PCR method using saliva and RT-qPCR

using polymerase chain reaction (PCR) or reverse transcription polymerase chain reaction (RT-PCR). A result of clinical research on influenza during the 2011–2012 season using SGNP and RT-PCR (abbreviated SGNP/PCR method) is shown in Table 1, comparing diagnostic results between conventional nasal swab kits and SGNP/PCR using saliva specimens.

Conventional kits used are Quick Chaser Flu A, B(MIZUHO MEDY Co., Ltd., Tosu, Saga, Japan)and Quick Navi-Flu (DENKA

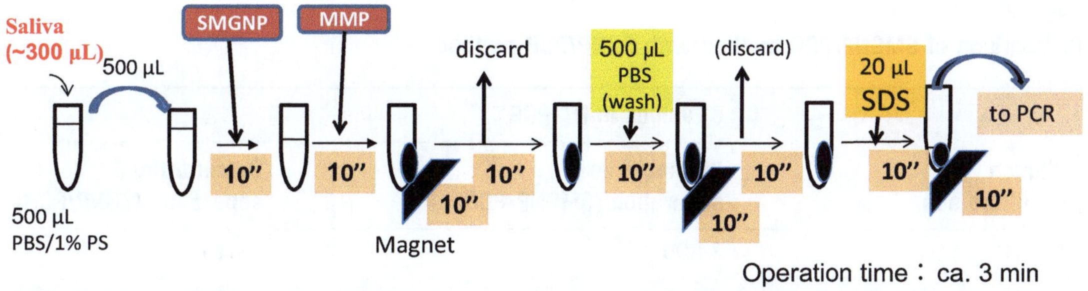

Fig. 2 Operation diagram of concentration and purification of viral particles, and extraction of DNA/RNA using the SMGNP/PCR method for RT-PCR

SEIKEN Co., Ltd., Tokyo, Japan). RT-qPCR is done with Thermal Cycler Dice Real Time II (Takara Bio Inc., Shiga, Japan) and One Step TB Green® PrimeScript™ RT-PCR Kit (Perfect Real Time) (Takara Bio). The primers for influenza type A (A) and type B (B) are as follows:

Type A, Forward primer 5′- CCMAGGTCGAAACG TAYGTTCTCTCTATC -3′, Reverse primer: 5′- TGACAGRA TYGGTCTTGTCTTTAGCCAYTCCA-3′;

Type B, Forward primer: 5′- AAATACGGTGGATTAAA TAAAAGCAA-3′,

Reverse primer: 5′-CCAGCAATAGCTCCGAAGAAA-3′

The positive results for the conventional kit are almost identical to those achieved using our SGNP/PCR method. However, the negative results for the conventional kit showed a big difference. In negative patients diagnosed using the conventional kit, it is found that 52% (24/46) of adults and 57% (26/46) of child patients possessed influenza viruses in their saliva. In other words, our method is more sensitive, and in addition, is better method due to the use of noninvasive saliva specimens.

Although we succeeded in developing a sensitive diagnostic method for influenza using noninvasive saliva specimens, it still needed low speed centrifugation (10,000 × g) and took a long time (over 30 min) to get viral RNA/DNA prior to RT-PCR. Therefore, we modified the SGNP/PCR method using sugar chain–immobilized magnetic gold nanoparticle (SMGNP) and magnetic microparticles (MMP) to collect viral particles, as shown in Fig. 2 [15].

When the improved magnetic method (abbreviated as SMGNP/PCR method) is compared with the original centrifugal method (SGNP/PCR method) using diluted porcine corona virus culture supernatants, the total operation time is about 1/10 of the original method, and the results of Ct value in RT-PCR is almost identical (Table 2).

Once the SMGNP/PCR method's viability is confirmed, several many clinical trials are performed. Figure 3 is an example of a

Table 2
Comparison of SMGNP/PCR method with SGNP/PCR method

Dilution of culture sup.	Ct values in RT-PCR	
	By magnetic separation (SMGNP/PCR)	By centrifugal separation (SGNP/PCR)
1/10	33.00	33.15
1/100	35.57	34.49
1/1000	38.32	37.76

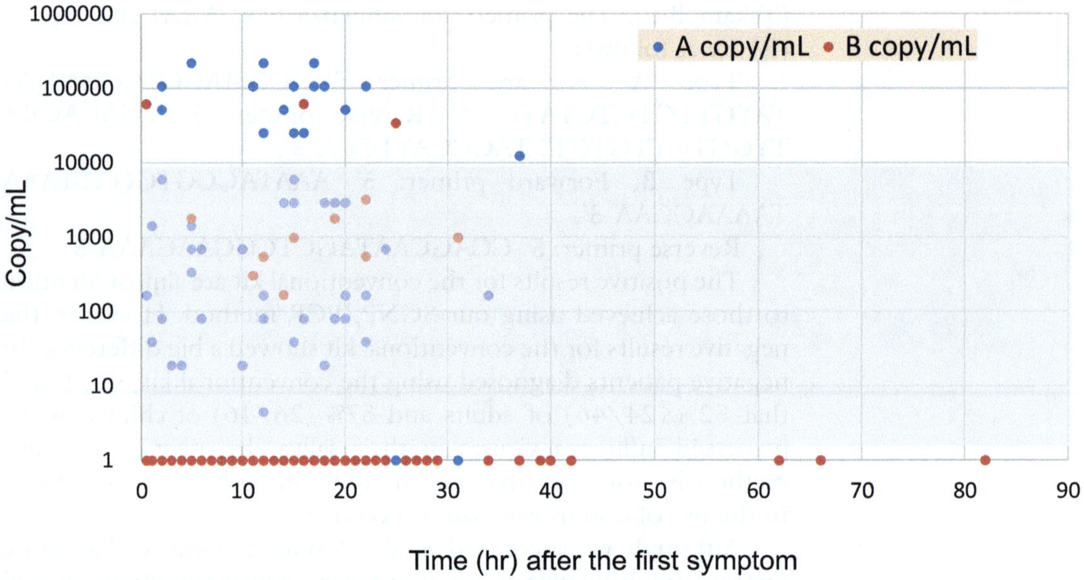

Fig. 3 Viral copy of influenza virus type A and B in the saliva of patients who had influenza-like symptoms. RT-qPCR was done with GeneSoc (J-Tas, Higashi-Osaka, Osaka, Japan). Reagent for reverse transcription and polymerase reaction used were Super Script (Thermo Fisher Scientific, Waltham, MA, USA) and SpeedSTAR™ HS DNA Polymerase (Takara Bio), respectively. The primers and probes for influenza type A (a) and type B (b) were as follows:

Type A, Forward primer 5′- CCMAGGTCGAAACGTAYGTTCTCTCTATC -3′, Revers primer: 5′- TGACAGRA TYGGTCTTGTCTTTAGCCAYTCCA-3′;

Prove: FAM-5′-ATYTCGGCTTTGAGGGGGCCTG-3′ MGB

Type B, Forward primer: 5′-AAATACGGTGGATTAAATAAAAGCAA-3′,

Revers primer: 5′-CCAGCAATAGCTCCGAAGAAA-3′

FAM-5′-CACCCATATTGGGCAATTTCCTATGGC-3′ MGB

trial done at a small clinic for children. Among saliva specimens from 194 patients who showed influenza-like symptoms, it is found that 64 patients possessed relatively high copies (over 10,000 copies/ml) of influenza viruses (type A: 55 patients, type B: 11, double infection: 2) in their saliva within 24 h after the first onset of influenza-like symptoms.

Then, we compared the SMGNP/PCR method results with virus culturing using MDCK cells. The clinical study using 30 saliva samples indicates 100% coincidence of sensitivity and specificity. A further clinical trial is performed following discussion with the Pharmaceuticals and Medical Devices Agency (PMDA). In this case, results of the SMGNP/PCR using saliva specimens of 463 patients who had influenza-like symptoms are compared with results by culturing using MDCK cells and nasal discharge specimens from the same patients. The sensitivity rate (comparison of positive results) is found to be 93.33% (SMGNP/PCR 98, culturing 105), the specificity rate (comparison of negative results) is 95.54% (SMGNP/PCR 342, culturing 358), and total match rate is 95.03% (440, 463), respectively. This data suggests that the SMGNP/PCR method can effectively detect infectious viruses.

3 COVID-19

The SMGNP/PCR method has been applied for the molecular diagnosis of SARS-CoV-2 since February 2020. Regarding the virus-binding sugar chain, we applied our previous data (Table 2) on porcine corona virus (porcine epidemic diarrhea virus, PEDV) where we used our sugar chip and partially purified PEDV culture supernatant combined with ultracentrifugation. Since PEDV bound to heparin or sulfated oligosaccharide (detail data will be reported elsewhere), we decided to use the same SMGNP that we used in the case of influenza [15]. Later in November 2020, Clausen et al., reported that the spike protein of SARS-CoV-2 possesses the binding site to heparan sulfate on the cell surface [10], confirming our strategy.

We first used the primer sets suggested in Centers for Disease Control and Prevention (CDC) [16] and National Institute of Infectious Diseases (NIID) [17] guidelines for SARS-CoV-2 molecular diagnosis. However, neither worked well because of high estimated self-complementary and 3′-self-complementary factors. Therefore, we tried to modify the primer sets and finally developed our own original primer sets as follows, using the information of N-protein gene of SARS-CoV-2/PC00101P/human/2020/USA:

mCDC set: Forward primer GACCCCAAAATCAGCGAAATG (21bp, 28280–28300); Reverse primer ATGTTGAGTGAGAGCGGTG (19bp, 28443–28425); Probe FAM-ACCCCGCATTACGTTTGGTGGACC -MGB (24bp, 28302–28325); PCR product 164 mer.

mNIID set: Forward primer ATTTTGGGGACCAGGAAC TAATC (23bp, 29120–29142); Reverse primer CGTTCCCGAA GGTGTGACTT (20bp, 29253–29234); Probe FAM-ATGTCGCGCATTGGCATGG -MGB (19bp, 29215–29233); PCR product 134 mer.

For the RT-PCR reagent, we used a commercially available reagent (One Step Prime Script RT-PCR Kit (Perfect Real Time), Takara Bio Inc., Shiga, Japan, code number RR064B). The reagent mixture for one RT-PCR reaction contained 2× One Step RT-PCR Buffer III 10 μl, 1×TaKaRa Ex Taq HS (5 U/μl) 0.4 μl, Prime-Script RT Enzyme Mix II 0.4 μl, PCR Primer Probe set (Primer: 0.2 μM, Probe: 0.4 μM) 1.6 μl, RNase Free distilled water 5.2 μl, and template solution 1 μl. Thermal Cycler Dice Real Time System II (Takara Bio Inc.) is used for RT-qPCR apparatus. The reaction protocol is as follows: 42 °C 300 s (reverse transcription reaction), 95 °C 10 s (hot start), 45 cycles of PCR: 95 °C 5 s, 60 °C 30 s.

Our SMGNP/PCR method is first used on a 71-year-old patient from a cruise ship docked at Yokohama, and this showed promising results using saliva to detect SARS-CoV-2 [18]. A further clinical study is performed using samples from 25 patients who had influenza-like symptoms. Nasopharyngeal swab and saliva specimens are collected from each patient. The nasopharyngeal swab specimen is tested according to the guidelines of the NIID at the local laboratory of Health and Welfare at Hamamatsu [17]. The saliva specimens are tested by our SMGNP/PCR method as described above. Results of both methods for the same patient are compared (Table 2). Based on the NIID method, the sensitivity and specificity are 90% (9/10) and 100% (15/15), respectively. From this data the SMGNP/PCR method is approved by the Japanese Ministry of Health, Labour and Welfare as one of the insurance-covered emergent diagnostic kits for SARS-CoV-2. The kit, named SUDx SARS-CoV-2 detection kit, has been on the Japanese market since June 10, 2020. See Table 3 for results of the clinical study of COVID-19 diagnosis.

The approved method is then used for the follow-up study of hospitalized patients after onset of COVID-19. Two typical examples are shown here. The clinical symptoms of one patient (TCH-59) were fever, cough, anosmia, and ageusia. These resolved on day 7 and the patient was discharged after 11 days hospitalization. The symptoms of another patient (TCH-61) were very mild; however, they grew worse after several days and the patient was transferred to the special hospital for severely affected patients. Saliva samples of patients were obtained in the early morning just after waking up. All samples were diluted with PBS (phosphate buffered saline, pH 7.4) containing 30 mM dithiothreitol and 1% antibacterial reagents (penicillin and streptomycin), and stored in the refrigerator before testing, which was done within 3 days. Five hundred micro liters (500 μl) of the diluted solution was used for

Table 3
Results of the clinical study of COVID-19 diagnosis

	Nasopharyngeal swab/NIID guideline					Saliva/SMGNP/PCR method		
		Primer/probe: NIID-1		Primer/probe: NIID-2			Primer/ probe: mCDC	Primer/ probe: mNIID
Patients	Judgment	Ct	Copy/μl (estimated)	Ct	Copy/μl (estimated)	Judgment	Copy/μl	Copy/μl
1	–					–		
2	–					–		
3	+	19.195	1.9.E+06	19.263	1.8.E+06	+	3.5.E+02	1.0.E+02
4	–					–		
5	–					–		
6	–					–		
7	–					–		
8	–					–		
9	+	–		40.202	2.7.E+00	+	8.8.E+01	
10	+	33.407	2.1.E+02	33.563	1.9.E+02	+	1.4.E+00	
11	+	38.641	7.4.E+00	–		+	3.9.E+02	
12	–					–		
13	+	40.088	2.9.E+00	–		+	4.5.E+01	
14	+	37.229	1.8.E+01	37.915	1.2.E+01	–		
15	–					–		
16	–					–		
17	+	25.995	2.5.E+04	25.988	2.5.E+04	+	2.0.E+01	
18	–					–		
19	–					–		
20	+	29.3586	2.8.E+03	29.6104	2.4.E+03	+	2.7.E+01	
21	–					–		
22	–					–		
23	–					–		
24	+	34.968	7.8.E+01	33.933	1.5.E+02	+	4.1.E+00	
25	+	35.921	4.2.E+01	36.21	3.5.E+01	+	4.4.E+00	

Copy/μl (estimated) was calculated according to the original calibration curve and Ct value obtained at the Laboratory of Health and Environment at Hamamatsu. Copy/μl was calculated according to the calibration curve obtained at the same time for the diagnosis by SMGNP/PCR method at Kagoshima University

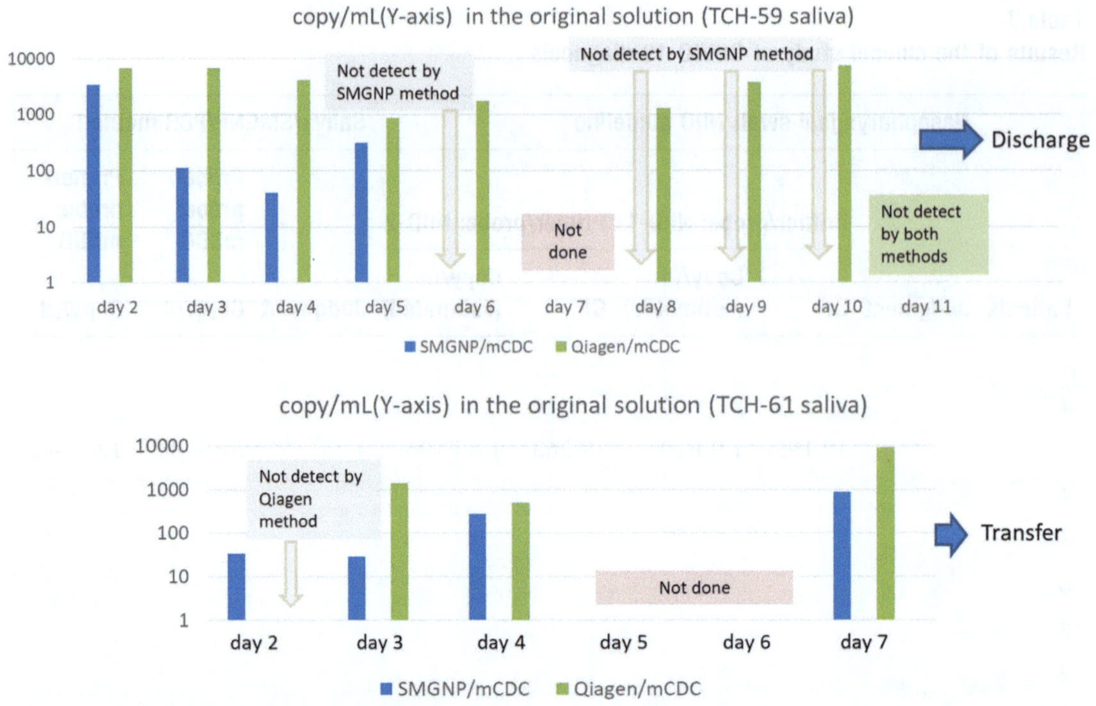

Fig. 4 Follow-up study of COVID-19 patients

SMGNP/PCR method, and 140 μl of the solution was applied to the Qiagen extraction kit to obtain RNA of SARS-CoV-2. The extracted RNA was then applied to the RT-PCR using the same RT-PCR reagent as for the SMGNP/PCR method and the same PCR apparatus was used. The results are shown in Fig. 4, where the copy number was calculated based on the calibration curve and volume of the mixture used for SMGNP or Qiagen extraction. In the case of THC-59, copy numbers of SARS-COV-2 RNA extracted by Qiagen kit did not change much from day 2 to day 10. However, the copy number determined by SMGNP/PCR method decreased day by day and was not detected after day 6. The symptoms of patient TCH-59 disappeared after day 7, and the patient was discharged on day 11. In comparison, in the case of TCH-61, a low number of RNA was found on day 2 using SMGNP/PCR and was not found with Qiagen extraction and RT-PCR. On day 3, the copy number detected by Qiagen extraction increased. Copy numbers obtained with both methods were increasing from day 4, and on day 7 more copy numbers were found with both methods. This corresponded with the worsening of symptoms in patient TCH-61, who was then moved to the special hospital. From these examples, it can be suggested that the follow up diagnosis using saliva is effective and the SMGNP/PCR method provides an accurate diagnosis from the standpoint of symptoms in patients, because SMGNP can selectively concentrate the virus particle which is crucial for the infection of virus. However, using Qiagen extraction, RNA from virus particles and from broken particles cannot be distinguished, therefore, over diagnosis or false positive results may occur.

Acknowledgements

We thank Eriko Arima and Bo Causer for support in preparing RT-PCR assay and English check, respectively.

References

1. CDC (2021) Symptoms of COVID-19. https://www.cdc.gov/coronavirus/2019-ncov/symptoms-testing/symptoms.html. Accessed 5/7/2021 2021
2. CDC (2021) Flu symptoms & complications. https://www.cdc.gov/flu/symptoms/symptoms.htm. 2021
3. Kim YG, Yun SG, Kim MY, Park K, Cho CH, Yoon SY, Nam MH, Lee CK, Cho YJ, Lim CS (2017) Comparison between saliva and nasopharyngeal swab specimens for detection of respiratory viruses by multiplex reverse transcription-PCR. J Clin Microbiol 55(1): 226–233. https://doi.org/10.1128/JCM.01704-16
4. Pere H, Podglajen I, Wack M, Flamarion E, Mirault T, Goudot G, Hauw-Berlemont C, Le L, Caudron E, Carrabin S, Rodary J, Ribeyre T, Belec L, Veyer D (2020) Nasal swab sampling for SARS-CoV-2: a convenient alternative in times of nasopharyngeal swab shortage. J Clin Microbiol 58(6). https://doi.org/10.1128/JCM.00721-20
5. Guclu E, Koroglu M, Yurumez Y, Toptan H, Kose E, Guneysu F, Karabay O (2020) Comparison of saliva and oro-nasopharyngeal swab sample in the molecular diagnosis of COVID-19. Rev Assoc Med Bras (1992) 66(8):1116–1121. https://doi.org/10.1590/1806-9282.66.8.1116
6. Williams E, Bond K, Zhang B, Putland M, Williamson DA (2020) Saliva as a noninvasive specimen for detection of SARS-CoV-2. J Clin Microbiol 58(8). https://doi.org/10.1128/JCM.00776-20
7. Yoon JG, Yoon J, Song JY, Yoon SY, Lim CS, Seong H, Noh JY, Cheong HJ, Kim WJ (2020) Clinical Significance of a High SARS-CoV-2 Viral Load in the Saliva. J Korean Med Sci 35(20):e195. https://doi.org/10.3346/jkms.2020.35.e195
8. Sakanashi D, Asai N, Nakamura A, Miyazaki N, Kawamoto Y, Ohno T, Yamada A, Koita I, Suematsu H, Hagihara M, Shiota A, Kurumiya A, Sakata M, Kato S, Muramatsu Y, Koizumi Y, Kishino T, Ohashi W, Yamagishi Y, Mikamo H (2021) Comparative evaluation of nasopharyngeal swab and saliva specimens for the molecular detection of SARS-CoV-2 RNA in Japanese patients with COVID-19. J Infect Chemother 27(1):126–129. https://doi.org/10.1016/j.jiac.2020.09.027
9. Boom R, Sol CJ, Salimans MM, Jansen CL, Wertheim-van Dillen PM, van der Noordaa J (1990) Rapid and simple method for purification of nucleic acids. J Clin Microbiol 28(3): 495–503. https://doi.org/10.1128/jcm.28.3.495-503.1990
10. Clausen TM, Sandoval DR, Spliid CB, Pihl J, Perrett HR, Painter CD, Narayanan A, Majowicz SA, Kwong EM, McVicar RN, Thacker BE, Glass CA, Yang Z, Torres JL, Golden GJ, Bartels PL, Porell RN, Garretson AF, Laubach L, Feldman J, Yin X, Pu Y, Hauser BM, Caradonna TM, Kellman BP, Martino C, Gordts P, Chanda SK, Schmidt AG, Godula K, Leibel SL, Jose J, Corbett KD, Ward AB, Carlin AF, Esko JD (2020) SARS-CoV-2 infection depends on cellular heparan sulfate and ACE2. Cell 183(4):1043–1057. e1015. https://doi.org/10.1016/j.cell.2020.09.033
11. Sriwilaijaroen N, Suzuki Y (2012) Molecular basis of the structure and function of H1 hemagglutinin of influenza virus. Proc Jpn Acad Ser B Phys Biol Sci 88(6):226–249. https://doi.org/10.2183/pjab.88.226
12. Suda Y, Zhang X, Takahashi Y, Yokoyama R, Nagatomo M, Aoyama K, Okuno T, Saito S, Morikawa S, Hiroi S, Kase T, Murakami N, Nishi J, Wakao M (2015) Discrimination of influenza virus strains and super high sensitive detection of viruses using sugar chip and sugar-chain immobilized gold nanoparticles. In:

Scholz C et al (ed) Tailored polymer architectures for pharmaceutical and biomedical applications, ACS Symposium Series Vol. 1135; American Chemical Society: Washington, DC. pp. 331–350. https://doi.org/10.1021/bk-2013-1135.ch020

13. Suda Y, Nagatomo M, Yokoyama R, Ohzono M, Aoyama K, Zhang X, Nakajima K, Murakami N, Shinoda T, Hirota T, Yanagihara S, Nishi JI (2015) Highly sensitive detection of influenza virus in saliva by real-time PCR method using sugar chain-immobilized gold nanoparticles; application to clinical studies. Biotechnol Rep (Amst) 7: 64–71. https://doi.org/10.1016/j.btre.2015.05.004
14. Saksono B, Dewi BE, Nainggolan L, Suda Y (2015) A highly sensitive diagnostic system for detecting dengue viruses using the interaction between a sulfated sugar chain and a virion. PLoS One 10(5):e0123981. https://doi.org/10.1371/journal.pone.0123981
15. Yasuo Suda MW, Takashi K (2010) Method for concentrating viruses, method for concentrating cells or bacteria, and magnetic composite. United States Patent 9464281
16. CDC (2020) Research use only 2019-novel coronavirus (2019-nCoV) real-time RT-PCR primers and probes. https://www.cdc.gov/coronavirus/2019-ncov/lab/rt-pcr-panel-primer-probes.html. Accessed 05/07/2021 2021
17. NIID (2020) Manual for the detection of pathogen 2019-nCoV. http://www.niid.go.jp/niid/images/epi/corona/2019-nCoVmanual20200217-en.pdf
18. Tajima Y, Suda Y, Yano K (2020) A case report of SARS-CoV-2 confirmed in saliva specimens up to 37 days after onset: proposal of saliva specimens for COVID-19 diagnosis and virus monitoring. J Infect Chemother 26(10): 1086–1089. https://doi.org/10.1016/j.jiac.2020.06.011

Chapter 6

Detection and Quantification of SARS-CoV-2 by Real-Time RT-PCR Assay

Alexander Wilhelm, Christiane Pallas, Rolf Marschalek, and Marek Widera

Abstract

The pandemic coronavirus SARS-CoV-2 is the causative agent of the acute respiratory disease COVID-19, which has become a global concern due to its rapid spread and high mortality rate. Detection and quantification of the (+) ssRNA virus, which has a genome size of 29,903 nucleotides, is commonly performed via reverse transcription quantitative polymerase chain reaction (RT-qPCR) targeting conserved sequences. Here, we describe a one-step RT-qPCR protocol for the quantitative detection of SARS-CoV-2 genomic RNA targeting M and RdRP genes, respectively, as well as active virus replication detecting subgenomic RNAs (sgRNA 4 and 8) that are formed by discontinuous transcription of the viral genome. Concomitantly, an input control targeting the human RNaseP gene (RPP30) was used in multiplex PCR to monitor the input of human nucleic acids. In vitro–transcribed RNA harboring the amplicon regions for M and RdRP regions served to set up a standard curve for absolute quantification.

In conclusion, the method described here allows for the detection and quantification of SARS-CoV-2 RNA isoforms for research by both using a probe-based or SYBR Green–based approach, but is also suitable for diagnostic purposes.

Key words Reverse transcription, Real time polymerase chain reaction, SARS-CoV-2, RdRP, COVID-19

1 Introduction

The pandemic coronavirus SARS-CoV-2 represents a global health threat. Beginning in December 2019 in Wuhan (China) the virus has since triggered an unprecedented burden on the global health and economic system [1]. SARS-CoV-2 infects susceptible cells expressing ACE2 and TMPRSS2, mainly in the respiratory tract [2]. After entry, the (+) ssRNA immediately serves as a template of polyprotein translation important for the initial infection [3]. During viral replication, the full-length viral genomic RNA (vRNA) serves as a template for the viral RNA dependent RNA-polymerase (RdRP) starting the complementary transcription of the (+)

Justin Jang Hann Chu et al. (eds.), *SARS-CoV-2: Methods and Protocols*,
Methods in Molecular Biology, vol. 2452, https://doi.org/10.1007/978-1-0716-2111-0_6,

ssRNA. In a process termed discontinuous transcription, during the synthesis of the (−) ssRNA, the polymerase pauses at transcription-regulating sequences (TRS) and translocates to downstream TRS. As a result, the viral subgenomic RNAs (sgRNAs) harbor a junction between the leader RNA and downstream genomic regions. Multiple sgRNAs are formed that are used as templates for the translation of all 13 SARS-CoV-2 ORFs, including the M gene (M) that is involved in the morphogenesis and assembly of the virus through its interactions with other viral proteins [3].

Thus, the vRNA deriving from cell culture supernatants or swab samples can be used for the quantification of genome copy equivalents. Of note, only the full-length mRNA is encapsidated into newly assembled virions. In addition, the detection of intracellular sgRNA junctions serves a as a suitable surrogate marker for active viral replication [4].

In this book section, we present methods that allow for the detection of SARS-CoV-2 RNA for research and diagnostic purposes. First, we describe methods for the detection of intracellular sgRNAs using SYBR Green (sgRNA 4) and a fluorescent probe (sgRNA 8). In addition, we demonstrate the quantitative detection of genome copy equivalents obtained from cell culture supernatants or from throat swab samples.

2 Materials

2.1 Preparation of RNA Standard Using In Vitro Transcription.

2.1.1 Linearization of pCR2.1 Plasmid Harboring RdRP- and M-Gene

1. Gloves.
2. Nuclease-free microtubes, pipettes, and pipette tips.
3. Heating block.
4. Ice bucket.
5. Microcentrifuge.
6. pCR2.1-based plasmid DNA harboring amplicon target regions of both SARS-CoV-2 M- and the RdRP gene sequence,
7. Restriction enzyme HindIII (20,000 U/ml).
8. CutSmart buffer (NEB).
9. Nuclease-free water.

2.1.2 Purification of Linearized DNA

1. Gloves.
2. Nuclease-free microtubes, pipettes, and pipette tips.
3. Heating block.
4. Ice bucket.
5. NucleoSpin Gel and PCR cleanup kit (Macherey-Nagel).

2.1.3 In Vitro Transcription

1. RNase-free bench area.
2. Gloves.
3. Nuclease-free microtubes, pipettes, and pipette tips.
4. Thermocycler.
5. HiScribe T7 High Yield kit (NEB).

2.1.4 DNase I Digestion

1. Gloves.
2. Nuclease-free microtubes, pipettes, and pipette tips.
3. RNase AWAY Reagent (Invitrogen).
4. Heating block.
5. Nuclease-free water.
6. 10× DNase I buffer (Promega)
7. RQ1 RNase-free DNase (Promega).

2.1.5 Purification and Quantification of RNA

1. Gloves.
2. Nuclease-free microtubes, pipettes, and pipette tips.
3. RNase AWAY Reagent (Invitrogen).
4. ≥ 95% EtOH (Industrial grade)
5. Microcentrifuge.
6. Monarch RNA cleanup kit (10 μg).
7. Nano Photometer N60 (Implen).

2.2 Sample Acquisition

2.2.1 Clinical Specimens

1. A BSL-2 facility for nonpropagative diagnostic laboratory work.
2. Class 2 laminar flow cabinet.
3. Gloves.
4. Individually wrapped sterile swabs of synthetic fiber origin (according to CDCs' Interim Guidelines for Collecting, Handling, and Testing Clinical Specimens from Persons for Coronavirus Disease 2019 (COVID-19) as of 08.10.2020) [5].
5. PBS.

2.2.2 SARS-CoV-2 RNA Derived from Viral Culture

1. A BSL-3 facility for propagative work with SARS-CoV-2 virus.
2. Certified personal protective equipment for BSL-3.
3. Class 2 laminar flow cabinet.
4. SARS-CoV-2 isolate.
5. Caco-2 cells (DSMZ, Braunschweig, Germany, no: ACC169).
6. Growth medium: Minimum Essential Medium supplemented with 10% fetal calf serum (FCS), 2% L-glutamine, 100 IU/ml

of penicillin and 100 g/ml streptomycin (all reagents purchased from Sigma).

7. Minimal Essential Medium (MEM) supplemented with 1% FCS, 2% L-glutamine, 100 IU/ml of penicillin, and 100 g/ml streptomycin.
8. Phosphate-buffered saline (PBS).
9. Gloves.
10. Nuclease-free pipettes and pipette tips.
11. Nuclease-free 96-well cell culture plate (flat bottom).
12. Standard light microscope.
13. Cell culture incubator.
14. Nunc cell scraper (Thermo Scientific).
15. RLT-buffer (Qiagen).
16. RNase-free DNase Set (Qiagen).

2.3 Viral RNA Isolation

1. Viral RNA Isolation derived from viral culture (see Subheadings 2.3.2 and 2.3.3 below) is performed under BSL-3 conditions until all infectious SARS-CoV-2 particles are inactivated. Nonpropagative diagnostic laboratory work handling clinical specimens may be performed under BSL-2 conditions with additional protective gear (see Subheading 2.3.1 below) if approved by institution.
2. Class 2 laminar flow cabinet.
3. Sterile filter tips, microtubes, and pipettes.
4. Gloves.
5. Sterillium Virugard (Paul Hartmann AG) or other approved disinfectant.
6. Plastic bags.
7. RNase AWAY Reagent (Invitrogen) or other reagent removing RNase contaminations.
8. Nuclease-free water.

2.3.1 Viral RNA Isolation from Swab Samples

1. 15 ml tube
2. PBS.
3. $\geq$ 95% EtOH (Industrial grade)
4. RNase-free 2.0 ml microtubes, pipette filter tips, and pipettes.
5. Microcentrifuge.
6. Vortexer.
7. QIAamp Viral RNA Mini Kit (Qiagen).

8. Optional: QIAcube HT (Qiagen) for high-throughput RNA Isolation in 96-well format.

2.3.2 Viral RNA Isolation from Supernatant

1. ≥ 95% EtOH (Industrial grade)
2. RNase-free 1.5 ml microtubes, pipette filter tips, and pipettes.
3. Microcentrifuge.
4. QIAamp Viral RNA Kit (Qiagen).
5. Optional: QIAcube for high-throughput RNA Isolation in 96-well format.

2.3.3 Viral RNA Isolation from Intracellular Lysates

1. Mercaptoethanol (β-ME).
2. RNase-free 1.5 ml microtubes, pipette filter tips, and pipettes.
3. Microcentrifuge.
4. ≥ 95% EtOH (Industrial grade)
5. QIAshredder homogenizer (Qiagen).
6. RNeasy Mini Kit (Qiagen).
7. Optional: QIAcube for high-throughput RNA Isolation in 96-well format.

2.4 Primer and Probes (Mixes for Multiplex)

1. RdRP and RNaseP (RPP30, *see* **Note 1**) primers are widely used in SARS-CoV-2 diagnostics and recommended by the WHO [6]. The M-gene [7], sgRNA4 [8], and sgRNA8 (original description in this protocol) primer and probes were designed in our lab and these are listed in Table 1.

2.5 Reagents for RT-qPCR (One-step)

1. PCR cabinet with incorporated UV light.
2. Nuclease-free microtubes, pipette tips, and pipettes.
3. RNase AWAY Reagent (Invitrogen).
4. Nuclease-free water.
5. Vortex mixer.
6. Microcentrifuge.
7. Ice bucket.
8. Luna Universal One-Step RT-qPCR Kit (NEB) for SYBR Green RT-qPCR or LightCycler Multiplex RNA Virus Master (Roche) for Probe RT-qPCR.
9. Target specific forward and reverse primer (see Table 1).
10. Target specific dual labeled probes with FAM,HEX, or Cy5, fluorophores conjugated at the 3′ end for probe RT-qPCR.
11. CFX96 Real-Time System, C1000 Touch Thermal Cycler (Bio-Rad) or other qPCR device.

Table 1
Primer and probes for detection and quantification of SARS-CoV-2 and RNaseP RNAs

Name	Sequence (5′-3′)	Length (nt)	Source
RdRP_SARSr-F2	GTGARATGGTCATGTGTGGCGG	22	[6, 9]
RdRP_SARSr-R1	CARATGTTAAASACACTATTAGCATA	26	[6, 9]
RdRP_SARSr-P2	6-Fam CAGGTGGAACCTCATCAGGAGATGC-BHQ1	25	[6, 9]
M-475-F	TGTGACATCAAGGACCTGCC	20	[7]
M-574-R	CTGAGTCACCTGCTACACGC	20	[7]
M-507-P	6-Fam TGTTGCTACATCACGAACGC-BHQ1	20	[7]
sgRNA4-F	AACGTACCTGTCTCTTCCGA	20	[8]
sgRNA4-R	CCAACCAACTTTCGATCTCTTGT	23	[8]
ACTB-F	CATCGAGCACGGCATCGTCA	20	[10, 11]
ACTB-R	TAGCACAGCCTGGATAGCAAC	21	[10, 11]
sgRNA8-F	ACCAACTTTCGATCTCTTGTAGA	23	[12]
sgRNA8-R	GCTCACAAGTAGCGAGTGTT	20	[12]
sgRNA8-P	HEX CTGTTCTCTAAACGAACATGAAAATTATTC-BHQ2	30	[12]
RPP30-F	AGATTTGGACCTGCGAGCG	19	[6]
RPP30-R	GAGCGGCTGTCTCCACAAGT	20	[6]
RPP30-P	Cy5 TTCTGACCTGAAGGCTCTGCGCG-BHQ3	23	[6]

IUPAC codes: S = G or C; R = A or G; o. d. = original description

2.6 Calculating Genome Copy Equivalents and X-fold Inhibition of Active Replication

1. Bio-Rad CFX Manager software, version 3.1 (Bio Rad Laboratories, Hercules, CA, USA).
2. Microsoft Excel.

3 Methods

3.1 Extraction of SARS-CoV-2 vRNA from Intracellular Samples

1. The SARS-CoV-2 isolate Frankfurt 1 (SARS-CoV-2 FFM1) obtained from the throat swab of a patient diagnosed with SARS-CoV-2 at the Frankfurt University Hospital (Germany) is used [7, 13] (*see* **Note 2**).
2. The analysis of active SARS-CoV-2 replication by the detection of sgRNAs requires the extraction of intracellular viral sgRNA. For a comprehensive analysis of a SARS-CoV-2 infection

experiment in cell culture it is suggested to also quantify extracellular genomic SARS-CoV-2 RNA of the same specimen. For this, harvest the supernatant and store it in a new plate at −80 °C prior to extract the extracellular viral RNA as described in Subheading 3.2: Viral RNA extraction from supernatant samples.

3. Carefully wash adherent cells with PBS (*see* **Note 3**).
4. Lyse cells directly grown in a monolayer by adding β-ME containing RLT buffer to the cell culture dish.
5. Use 160 μl RLT buffer in a 96-well plate format (6- to 24-well cell culture plates require 350 μl RLT buffer).
6. Collect the lysate by scratching the dish with a cell scraper provided with a soft and flexible blade followed by thoroughly pipetting up and down, taking care to minimize aerosol generation.
7. Optional: Optimize lysis by freezing cell culture plate at −80 °C and thaw plate at RT.
8. Optional: The QIAcube HT is well suited for quick mid- to high-throughput RNA isolation in 96-well format (*see* **Note 4**).
9. Resuspend thoroughly for optimal yield of RNA and virus inactivation.
10. Export lysed samples from the BSL-3 facility in compliance with corresponding safety regulations and perform all downstream processing in BSL-2 conditions. For more information on inactivating SARS-CoV-2 using a lysis buffer, please refer to Westhaus et al. 2020 [14] (*see* **Notes 5**–**7**).
11. Optional: To reduce viscosity homogenize with QIAshredder columns.
12. Optional: Transfer lysate to a QIAshredder column placed in a 2 ml microcentrifuge tube.
13. Optional: Centrifuge at 14,000 × *g* for 2 min in a microcentrifuge.
14. Optional: Discard the QIAshredder column.
15. Dilute the lysate 1:1 with 70% ethanol and mix well by pipetting. Transfer sample directly to an RNeasy spin column placed in a 2 ml microtube and close the lid. Centrifuge for 15 s at 8000 × *g* and discard the flow-through (*see* **Note 8**).
16. Wash the RNeasy spin column by adding 350 μl RW1 wash buffer and close the lid. Centrifuge for 15 s at 8000 × *g* and discard the flow-through.

17. Perform on-column DNase digestion using the RNase-free DNase Set by adding 80 μl of the DNase I incubation mix (well-mixed 10 μl DNase I stock solution with 70 μl RDD buffer). Incubate at ambient temperature for 15 min.
18. Wash the RNeasy spin column twice with 500 μl RPE buffer as indicated above. Centrifuge for 15 s at 8000 × *g* in the first and 2 min at 8000 × *g* in the second wash step to completely dry the spin column membrane before eluting the RNA.
19. Place the RNeasy spin column in a new 1.5 ml microtube and elute RNA by adding 30 μl RNase-free water directly onto the membrane of the RNeasy spin column. Close the lid and centrifuge for 1 min at 8000 × *g*.
20. Store the RNA at −80 °C.

3.2 Viral RNA Extraction from Supernatant Samples and Swab Samples

The reader is referred to specific literature and instructions to obtain an adequate sample. Compliance with biosafety regulations and hygiene standards based on the guidelines of the respective institution is essential to prevent infection with SARS-CoV-2. The nonpropagative diagnostic work is carried out under BSL-2 conditions wearing appropriate protective gear.

3.2.1 Sample Preparation for vRNA extraction from Patient Derived Swab Samples

1. For the extraction of vRNA dissolve swab derived patient material in 1000 μl PBS inside a sterile 15 ml tube. At least 500 μl should be used so that the swab is completely covered with PBS.
2. Stir the swab tip in PBS for 10 s and squeeze it against the side of the vial before removal.
3. Transfer 280 μl it into a 2 ml microtube.
4. Lyse the sample by adding 1120 μl of AVL buffer containing 1/100 diluted carrier RNA from the QIAamp Viral RNA Mini Kit and vortex for 15 s.
5. Incubate at room temperature for 10 min and proceed as indicated in Subheading 3.2.3 vRNA extraction using QIAamp Viral RNA Mini Kit.
6. Optional: The QIAcube HT is well suited for quick mid- to high-throughput RNA isolation in 96-well format. Perform the lysis by resuspending the sample in 160 μl of a mixture containing 146 μl ACL buffer, 9.2 μl PBS and 4.6 μl carrier RNA from the QIAamp 96 Virus QIAcube HT Kit. Incubate samples for 30 min in a 56 °C water bath and prepare the QIAcube HT for fully automated processing of 96 samples according to the manufacturer's instruction.

3.2.2 Sample Preparation for vRNA Extraction from Virus Culture

1. The SARS-CoV-2 isolate Frankfurt 1 (SARS-CoV-2 FFM1) obtained from the throat swab of a patient diagnosed with SARS-CoV-2 at the Frankfurt University Hospital (Germany) is used [7, 13] (*see* **Note 2**). This work must be carried out under BSL-3 conditions.
2. Harvest the supernatant from SARS-CoV-2 infected Caco-2 cells and store it in a microtube or new cell culture plate.
3. Optional: Supernatant can be stored at −80 °C.
4. A 96-well cell culture plate contains approximately 100–200 μl supernatant per well, which constitutes a suitable volume for the application of the following QIAamp Viral RNA Kit.
5. Optional: Use the QIAcube HT system for high-throughput processing of 96-samples using the QIAamp 96 Virus QIAcube HT Kit as described in Subheading 3.2.1 **step 6**.
6. Lyse the supernatant sample by adding 560 μl of AVL buffer containing 1/100 diluted carrier RNA and vortex for 15 s.
7. Incubate at room temperature for 10 min.
8. While incubating export lysed samples from the BSL-3 facility in compliance with corresponding safety regulations and perform all downstream processing in BSL-2 conditions. For more information on inactivating SARS-CoV-2 using a lysis buffer, please refer to Westhaus et al. 2020 [14] (*see* **Notes 6** and **9**).
9. Proceed the vRNA extraction as indicated in the next Subheading 3.2.3.

3.2.3 vRNA Extraction Using QIAamp Viral RNA Kit (Qiagen)

1. Shortly spin down the microtube to remove drops from the inside of the lid (*see* **Note** 7).
2. Add 560 μl ethanol (>96%) and vortex for 15 s.
3. Briefly centrifuge the microtube and transfer 630 μl of the solution to QIAamp Mini column placed in a 2 ml microcentrifuge tube.
4. Close the lid and centrifuge for 1 min at 6000 × *g*.
5. Discard the 2 ml microtube containing the flow-through and place the QIAamp Mini column in a clean 2 ml microtube.
6. Repeat the last two steps until all the lysate was loaded onto the QIAamp Mini column.
7. Add 500 μl of AW1 buffer and close the lid of the QIAamp Mini column. Centrifuge at 6000 × *g* for 1 min and place the QIAamp Mini Column in a new 2 ml microtube.
8. Add 500 μl of AW2 buffer and close the lid of the QIAamp Mini column. Centrifuge at 20,000 × *g* for 3 min and place the QIAamp Mini Column in a new 1.5 ml microtube.

9. Add 60 μl of AVE buffer and incubate for 1 min at room temperature. Centrifuge at 6000 × *g* for 1 min to collect the eluted RNA in the 1.5 ml microtube.
10. Store the RNA at −80 °C.

3.3 One Step Reverse Transcription Quantitative PCR for SARS-CoV-2 vRNA and sgRNA

3.3.1 Workflow for Multiplex Probe RT-qPCR for sgRNA8 Normalized to RNaseP

1. Use total SARS-CoV-2 vRNA extracted intracellularly from Caco-2 cells as described in Subheading 3.1.
2. The RT-qPCR is carried out in 20 μl reaction volume using the IVD-approved LightCycler Multiplex RNA Virus Master Kit. The One-step PCR system includes the cDNA synthesis step and allows for a significantly quicker procedure. By setting up the PCR reaction directly with the extracted RNA, the LightCycler Multiplex RNA Virus Master Kit significantly facilitates and accelerates the entire procedure (*see* **Note 10**).
3. Moreover, the multiplex approach allows for the quantification of both, the SARS-CoV-2 target, sgRNA8, as well as a reference gene, e.g. RNaseP, in a single multiplex PCR assay due to distinct fluorescently labelled probes (*see* **Note 11**).
4. Use 0.4 μM of each primer per reaction (detailed information of all primers used for this RT-qPCR are listed in Table 1).
5. Reaction Mix:

4.0 μl of RT-qPCR-Mix
0.8 μl of sgRNA8-F forward primer (0.4 μM)
0.8 μl of sgRNA8-R reverse primer (0.4 μM)
0.4 μl of sgRNA8-P probe (HEX-labelled) (0.2 μM)
0.8 μl of RPP30-F forward primer (0.4 μM)
0.8 μl of RPP30-R reverse primer (0.4 μM)
0.4 μl of RPP30-P probe (HEX-labeled) (0.2 μM)
0.1 μl of Enzyme Mix
6.9 μl of RNase-free H_2O
5.0 μl of Template RNA
20.0 μl Total Volume

6. Add a no-template control by replacing 5 μl RNA with RNase-free H_2O.
7. Program for the CFX96 Real-Time System, C1000 Touch Thermal Cycler:

50 °C	10 min	Reverse transcribe RNA	–
95 °C	30 s	Initial denaturation	–
95 °C	5 s	Denaturation	–

(continued)

60 °C	30 s	Annealing/Elongation	Repeat for 44 cycles
–	–	Read plate	–
40 °C	30 s	Final cooling	–

8. The obtained CT values can be evaluated with the Bio-Rad CFX Manager software, version 3.1 (Bio Rad Laboratories, Hercules, CA, USA) and exported to Excel for data analysis and relative quantification.

Relative Quantification Using the 2-ΔΔCT Method

1. The ΔΔ Ct method [15] is used for relative quantification of SARS-CoV-2 related transcripts compared to an endogenous reference or housekeeping gene. In contrast to absolute quantification based on quantification relative to a standard curve, it allows for the correlation of RNA copy numbers to a biological parameter and is based on the formula:

   $2^{-\Delta\Delta C_T}$.

2. The following calculation steps are required:
3. The calculation starts with the determination of ΔCT by subtracting the CT of SARS-CoV-2 target sample (e. g. M-gene) with CT of the endogenous control (e.g. RNaseP) (Table 2).
4. ΔΔCT is obtained by subtracting each calculated ΔCT with the average ΔCT of the untreated control samples, called calibrator samples.
5. In the last step, the ΔΔCT is preceded by a negative prefix: -ΔΔ CT; and subsequently potentiated to the base 2: 2-ΔΔCT.
6. The relative quantification approach successfully highlights the significant reduction of active viral replication in SARS-CoV-2 infected Caco-2 cells when pretreated for 48 h with 100 U interferon-β (Fig. 1).

3.3.2 Workflow for Multiplex Probe RT-qPCR for M-gene Normalized to RNaseP

1. Use SARS-CoV-2 vRNA extracted from the supernatant of Caco-2 cells as described in Subheading 3.2.
2. The RT-qPCR is carried out in 20 μl reaction volume using the IVD-approved LightCycler Multiplex RNA Virus Master Kit, which contains a special reaction mix for multiplex qPCRs. The One-step PCR system includes the cDNA synthesis step and allows for a significantly quicker procedure (see Subheading 3.3.1).
3. Use 0.4 μM of each primer per reaction (detailed information of all primers used for this RT-qPCR are listed in Table 1 in Subheading 2.5).
4. Reaction Mix:

4.0 μl of RT-qPCR-Mix
0.8 μl of M-475-F forward primer (0.4 μM)

(continued)

Table 2
Detailed calculations of $2^{-\Delta\Delta CT}$ method

Sample		C_T RNaseP	C_T sgRNA8	ΔC_T	$\Delta\Delta C_T$	$2^{-\Delta\Delta CT}$
Untreated	1	25.43	18.34	18.34–25.43 = −7.09	(−7.09) - (**−6.86**) = −0.23	$2^{-0.23} = 1.17 \times 10^{0}$
	2	26.06	19.47	19.47–26.06 = −6.58	(−6.58) - (**−6.86**) = 0.28	$2^{0.28} = 8.24 \times 10^{-1}$
	3	25.39	18.33	18.33–25.39 = −7.06	(−7.06) - (**−6.86**) = −0.20	$2^{-0.20} = 1.15 \times 10^{0}$
	4	25.22	18.51	18.51–25.22 = −6.71	(−6.71) - (**−6.86**) = 0.15	$2^{0.15} = 8.99 \times 10^{-1}$
IFN-β	1	25.56	32.62	32.62–25.56 = 7.06	7.06 - (**−6.86**) = 13.92	$2^{13.92} = 6.45 \times 10^{-5}$
	2	26.01	32.77	32.77–26.01 = 6.77	6.77 - (**−6.86**) = 13.63	$2^{13.63} = 7.90 \times 10^{-5}$
	3	25.20	33.19	33.19–25.20 = 7.99	7.99 - (**−6.86**) = 14.85	$2^{14.85} = 3.38 \times 10^{-5}$
	4	25.47	36.94	36.94–25.47 = 11.47	11.47 - (**−6.86**) = 18.33	$2^{18.33} = 3.03 \times 10^{-6}$

Average ΔC_T untreated samples: **−6.86**

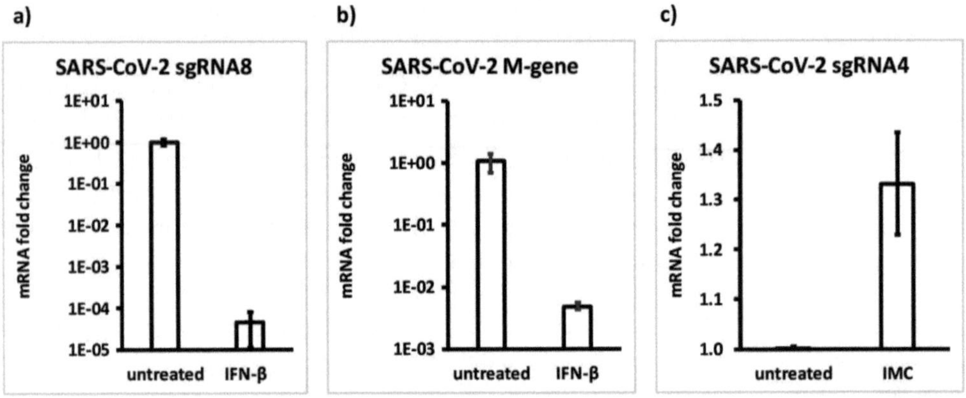

Fig. 1 Detection and relative quantification of SARS-CoV-2 specific mRNA isoforms. (**a**) Active SARS-CoV-2 replication intracellularly detected by sgRNA8 normalized to RNaseP. Caco-2 cells were pretreated for 48 h with PBS (untreated) or 100 U Interferon-β (IFN-β) prior to SARS-CoV-2 infection. (**b**) Quantitative Detection of SARS-CoV-2 genomic RNA targeting M-gene from supernatant normalized to RNaseP. (**c**) Active SARS-CoV-2 replication intracellularly detected by sgRNA4 normalized to ACTB in a low-cost SYBR Green RT-qPCR approach. Caco-2 cells pretreated with IMC or and infected with SARS-CoV-2 strain FFM1

0.8 μl of M-574-R reverse primer (0.4 μM)
0.4 μl of M-507-P probe (6-Fam labelled) (0.2 μM)
0.8 μl of RPP30-F forward primer (0.4 μM)
0.8 μl of RPP30-R reverse primer (0.4 μM)

(continued)

0.4 µl of RPP30-P probe (Cy5-labeled) (0.2 µM)
0.1 µl of Enzyme Mix
6.9 µl of RNase-free H_2O
5.0 µl of Template RNA
20.0 µl Total Volume

5. Add a no-template control by replacing 5 µl RNA with RNase-free H_2O.
6. Program for the CFX96 Real-Time System, C1000 Touch Thermal Cycler:

50 °C	10 min	Reverse transcribe RNA	–
95 °C	30 s	Initial denaturation	–
95 °C	5 s	Denaturation	–
60 °C	30 s	Annealing/Elongation	Repeat for 44 cycles
–	–	Read plate	–
40 °C	30 s	Final cooling	–

7. The obtained CT values can be evaluated with the Bio-Rad CFX Manager software and exported to Excel for data analysis and relative quantification.
8. Relative quantification using the 2-ΔΔCT Method is carried out accordingly to Subheading 3.3.1 (Table 3).
9. Calculate ΔΔCT values as described in Subheading 3.3, **step 9**.
10. Less SARS-CoV-2 viral genomic RNA targeting the M-gene is detected from the supernatant of Caco-2 cells with externally induced intracellular viral response prior to infection (Fig. 1b).

The supernatant is derived from the same specimen as used in the sgRNA8 PCR in Subheading 3.3.1 and highlights the sensitivity of both approaches shown in the well-studied activation of antiviral defenses upon interferon release (*see* **Note 12**).

3.3.3 Relative Quantification of SARS-CoV-2 sgRNA 4 Using SYBR Green

1. Use total SARS-CoV-2 vRNA extracted intracellularly from Caco-2 cells as described in Subheading 3.1.
2. The RT-qPCR is carried out in 20 µl reaction volume using the Luna One-Step Kit. The One-step PCR system includes the cDNA synthesis step and allows for a significantly quicker procedure (see Subheading 3.3.1).
3. Applying SYBR Green constitutes a suitable low-cost alternative to Probe PCRs for SARS-CoV-2 quantification (*see* **Note 13**).

Table 3
Calculations for relative expression of M-gene from supernatant samples using the 2-ΔΔCT method

Sample		C_T RNase P	C_T M-gene	ΔC_T	$\Delta\Delta C_T$	$2^{\Delta\Delta C_T}$
Untreated	1	28.65	19.48	19.48–28.65 = −9.17	(−9.17) - (−10.06) = 0.89	$2^{0.89} = 5.40 \times 10^{-1}$
	2	29.08	18.94	18.94–29.08 = −10.13	(−10.13) - (−10.06) = −0.07	$2^{-0.07} = 1.05 \times 10^{-1}$
	3	29.32	18.83	18.83–29.32 = −10.48	(−10.48) - (−10.06) = −0.43	$2^{-0.43} = 1.34 \times 10^{0}$
	4	29.61	19.16	19.16–29.61 = −10.45	(−10.45) - (−10.06) = −0.39	$2^{-0.39} = 1.31 \times 10^{0}$
IFN-β	1	26.61	24.44	24.44–26.61 = −2.17	(−2.17) - (−10.06) = 7.89	$2^{7.89} = 4.22 \times 10^{-3}$
	2	26.56	23.94	23.94–26.56 = −2.61	(−2.61) - (−10.06) = 7.44	$2^{7.44} = 5.74 \times 10^{-3}$
	3	26.15	23.70	23.70–26.15 = −2.45	(−2.61) - (−10.06) = 7.61	$2^{7.61} = 5.14 \times 10^{-3}$
	4	26.56	24.22	24.22–26.56 = −2.34	(−2.61) - (−10.06) = 7.72	$2^{7.72} = 4.75 \times 10^{-3}$

Average ΔC_T untreated samples: **−10.06**

4. Use 0.4 μM of sgRNA4 and β-Actin (ACTB) primer per reaction (detailed information of all primers used for this RT-qPCR are listed in Table 1 in Subheading 2.5).
5. Reaction Mix for sgRNA 4 PCR:

10.0 μl of Luna Universal One-Step Reaction Mix (2×)
1.0 μl of Luna WarmStart RT Enzyme Mix (20×)
0.8 μl of sgRNA4-F forward primer (0.4 μM)
0.8 μl of sgRNA4-R reverse primer (0.4 μM)
2.0 μl of Template RNA
5.4 μl of RNase-free H_2O
20.0 μl Total Volume

6. Add a no-template control by replacing 2 μl RNA with RNase-free H_2O.
7. Reaction Mix for ACTB PCR:

10.0 μl of Luna Universal One-Step Reaction Mix (2×)
1.0 μl of Luna WarmStart RT Enzyme Mix (20×)
0.8 μl of ACTB-F forward primer (0.4 μM)

(continued)

0.8 μl of ACTB-R reverse primer (0.4 μM)
2.0 μl of Template RNA
5.4 μl of RNase-free H_2O
20.0 μl Total Volume

8. Add a no-template control by replacing 2 μl RNA with RNase-free H_2O.
9. Program for the CFX96 Real-Time System, C1000 Touch Thermal Cycler:

55 °C	10 min	Reverse transcribe RNA	–
95 °C	30 s	Initial denaturation	–
95 °C	5 s	Denaturation	–
60 °C	30 s	Annealing/Elongation	Repeat for 44 cycles
–	–	Read plate	–

10. For a final melt-curve analysis subsequently to 44 cycles add the following steps to the program of the thermal cycler:

60 °C	**30 s**
+0.5 °C/cycle Ramp 0.5 °C/s	Repeat for 70 cycles

11. The obtained CT values can be evaluated with the Bio-Rad CFX Manager software and exported to Excel for data analysis and relative quantification.
12. Relative quantification using the 2-ΔΔCT method is carried out accordingly to Subheading 3.3.1, **step 3** (Table 4).
13. The mRNA levels of sgRNA4 are normalized to the housekeeping gene ACTB. The SYBR Green PCR shows a significant induction of active viral replication upon the treatment with an immune modulatory compound (IMC) currently under investigation in our lab (Fig. 1c).

3.4 Generation of RNA Standard Curve for Absolute Quantification of SARS-CoV-2 vRNA

3.4.1 Linearization of pCR2.1 Plasmid Harboring RdRP- and M-gene

1. Linearization Mix:

3.0 μg of pCR2.1 SARS-CoV-2 M-RdRP plasmid
1.0 μl of HindIII
2.0 μl of CutSmart buffer (NEB)
–
Fill up to 20 μl with Nuclease-free water
20.0 μl Total Volume

Table 4
Calculations for relative expression of sgRNA 4 from intracellular specimens using the 2-ΔΔCT method

Sample		C_T ACTB	C_T sgRNA4	ΔC_T	$\Delta\Delta C_T$	$2^{-\Delta\Delta C_T}$
Untreated	1	18.34	16.93	16.93–18.34 = −1.41	(−1.41) - (−1.42) = 0.01	$2^{0.01} = 1.00$
	2	18.13	16.71	16.71–18.13 = −1.42	(−1.42) - (−1.42) = −0.01	$2^{-0.01} = 1.00$
IMC	1	18.19	16.47	16.47–18.19 = −1.72	(−1.72) - (−1.42) = −0.30	$2^{-0.30} = 1.23$
	2	18.27	16.45	16.45–18.27 = −1.82	(−1.82) - (−1.42) = −0.40	$2^{-0.40} = 1.32$
	3	18.18	16.24	16.24–18.18 = −1.94	(−1.94) - (−1.42) = −0.52	$2^{-0.52} = 1.44$

Average ΔCq untreated control samples: **−1.42**

2. The pCR2.1 plasmid contains the PCR target sequence for the SARS-CoV-2 M gene (GenBank Accession number NC_045512 (475–574)) and the SARS-CoV-2 RdRP gene. A dual target RNA standard reduces the risk of a biased comparison by avoiding different RNA preparation steps.
3. Incubate at 37 °C for 1 h.
4. Store at −20 °C.

3.4.2 DNA Purification

1. Purify the linearized plasmid DNA with the NucleoSpin Gel and PCR Clean-Up Kit according to the manufacturer's instructions.
2. Carry out final elution step in 30 μl NE buffer (preheated in a heating block to 70 °C).

3.4.3 In Vitro Transcription

1. In-vitro transcribe with the HiScribeT7 High Yield kit.
2. Reaction Mix:

1.5 μl of ATP
1.5 μl of GTP
1.5 μ of UTP
1.5 μl of CTP
1.5 μl of 10× Reaction buffer (0.75× final)
1.5 μl T7 Polymerase Mix
11.0 μl linearized DNA (~1 μg)
20.0 μl Total Volume

3. Incubate 16 h at 37 °C.

3.4.4 DNase I Digestion and RNA Cleanup of Total RNA

1. Digest the template DNA by adding RQ1 RNase-free DNase.
2. Add to 20 μl IVT reaction mix:

68.0 μl of Nuclease-free water
10.0 μl of 10× DNase I buffer
–
2.0 μl of RQ1 RNase-free DNase
20.0 μl Total Volume

3. Incubate 15 min at 37 °C.
4. For RNA cleanup of total RNA use the Monarch RNA Cleanup Kit (10 μg) or alternatively the RNeasy Mini Kit according to the manufacturer's instructions.
5. Determine RNA yield using the Nano Photometer N60.

3.4.5 Generation of Serial Diluted RNA Standards

1. Dilute the linearized and purified RNA standards according to the following formula (*see* **Note 14**):

$$\frac{x\,\text{ng} \times 6.0221 \times 10^{23}\ \text{molecules/mole}}{(N \times 339.5\ \text{g/mole})1 \times 10^{9}\ \text{ng/g}}$$

2. Fill in the determined RNA stock concentration for "*x*" and the length (bp) of the template for "*N*" resulting in the required amount for the first dilution with a copy number of 1×10^8 molecules ("a" in Table 5).
3. Add nuclease-free water as Volume Diluent in "b" to obtain a total volume of 200 μl for Dilution 1.
4. The subsequent serial 1–10 dilutions are carried out by mixing 20 μl of the previous dilution step with 180 μl Nuclease-water.
5. Mix each dilution thoroughly before using it for the subsequent dilution step to obtain a consistent and accurate standard curve.
6. A total of 10 samples are used to calculate the standard curve including a no-template control "NTC" (Table 5).

3.4.6 Perform One-Step RT-qPCR for Standards and Negative Control Using the IVD-Approved LightCycler Multiplex RNA Virus Master Kit (Roche)

1. Add 10 standard samples to the RT-qPCR protocol from Subheading 3.3, **step 4** replacing 5 μl Template RNA (*see* **Note 15**) with the 9 dilutions of the RNA standards and a negative control (Table 5).
2. It is recommended to run the dilutions in addition to the RNA samples in a single PCR reaction.

Table 5
Serial dilutions for generating an RNA standard curve

Dilution	Copies/μl	Volume predilution	Volume diluent
1	1×10^8	a	b
2	1×10^7	20	180
3	1×10^6	20	180
4	1×10^5	20	180
5	1×10^4	20	180
6	1×10^3	20	180
7	1×10^2	20	180
8	1×10^1	20	180
9	1×10^0	20	180
NTC	0	0	200

3.4.7 Calculate Standard Curve

1. Plot in a base 10 semilogarithmic graph the obtained CT values from the RT-qPCR against the dilution factor (Fig. 2).
2. Generate a regression line estimated from the 9 plotted CT values. The equation of the regression line is as follows:

$$\text{M}: \quad y = 35.702 + 2.926; \quad R^2 = 0.995$$

$$\text{RdRP}: \quad y = 37.475 + 3.658; \quad R^2 = 0.999$$

3. The R^2 value is an indicator of accurate pipetting and should be close to 1. Hence, $R^2 = 0.995$ and 0.999, respectively, obtained from the 9 standards allow for an adequate downstream quantification (Fig. 2).
4. The absolute amount of genome copy equivalents per reaction from each patient sample can be calculated by rearrangement of the regression line's equation:
$M: \; x = \frac{y-35.702}{2.926}$
$\text{RdRP}: x = \frac{y-37.475}{3.658}$
5. The load of SARS-CoV-2 vRNA of each sample is defined in RNA molecules per reaction, but can be calculated to match copies/ml.

3.5 One Step Reverse Transcription Quantitative PCR for SARS-CoV-2 vRNA Derived from Patient Data—Absolute Quantification Using IVT RNA Standards

1. Detailed protocol for generating RNA standard curve for absolute quantification is described in Subheading 3.4: Generation of RNA standard curve for absolute quantification of SARS-CoV-2 vRNA.
2. Use SARS-CoV-2 vRNA extracted from cell culture supernatants of Caco-2 cells infected with swab derived SARS-CoV-2 following the protocol described in Subheading 3.2.

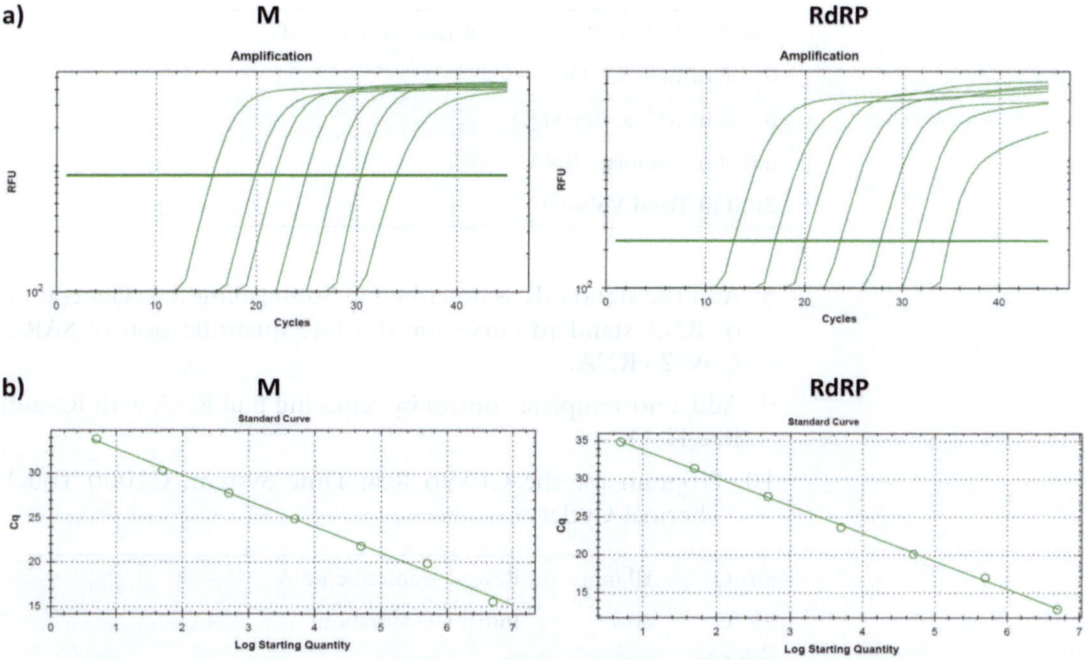

Fig. 2 Standard curves of SARS-CoV-2M and RdRP gene specific RT-qPCR using in vitro transcribed RNA templates. **(a)** Amplification curves and **(b)** log starting quantity (copies/reaction) is indicated and plotted against the Cq

3. Optional: the quantification can also be carried out directly from the swab sample derived material (e. g. PBS or lysis buffer).
4. The RT-qPCR is carried out in 20 μl reaction volume using the IVD-approved LightCycler Multiplex RNA Virus Master Kit optimized for multiplex qPCR approaches. The One-step PCR system includes the cDNA synthesis step and allows for a significantly quicker procedure (see Subheading 3.3.1).
5. Use 0.4 μM of each primer per reaction (detailed information of all primers used for this RT-qPCR are listed in Table 1 in Subheading 2.5).
6. Run two primer pairs and a probe, either for M and RNaseP or RdRP and RNaseP, respectively, in a single PCR reaction and take advantage of the cost-reducing multiplex approach.
7. Reaction Mixes for M- gene or RdRP-specific PCR are as follows:

4.0 μl of RT-qPCR-Mix
0.8 μl of M-475-F forward primer (0.4 μM)
0.8 μl of M-574-R reverse primer (0.4 μM)
0.4 μl of M-507-P probe (6-Fam labelled) (0.2 μM)
0.8 μl of RPP30-F forward primer (0.4 μM)
0.8 μl of RPP30-R reverse primer (0.4 μM)

(continued)

0.4 μl of RPP30-P probe (Cy5-labeled) (0.2 μM)
0.1 μl of Enzyme-Mix
6.9 μl of RNase-free H_2O
5.0 μl of Template RNA
20.0 μl Total Volume

8. Add the standards as described in Subheading 3.4. Generation of RNA standard curve for absolute quantification of SARS-CoV-2 vRNA.
9. Add a no-template control by replacing 5 μl RNA with RNase-free H_2O.
10. Program for the CFX96 Real-Time System, C1000 Touch Thermal Cycler:

50 °C	10 min	Reverse transcribe RNA	–
95 °C	30 s	Initial Denaturation	–
95 °C	5 s	Denaturation	
60 °C	30 s	Annealing/Elongation	Repeat for 44 cycles
–	–	Read Plate	–
40 °C	30 s	Final cooling	–

11. The obtained CT values can be evaluated with the Bio-Rad CFX Manager software (Tables 6 and 7). Quantification of SARS-CoV-2 with M-gene PCR is suitable for low and high

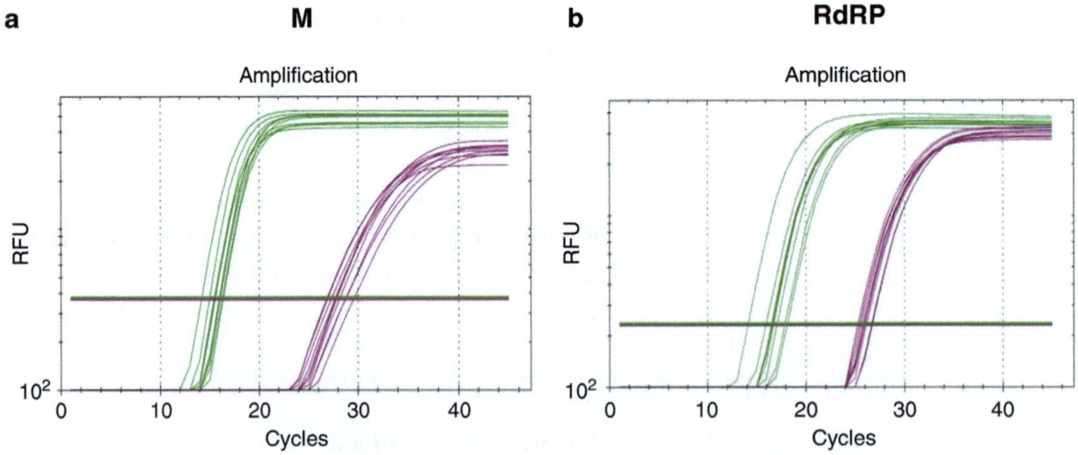

Fig. 3 Detection of SARS-CoV-2 RNAs. Caco-2 cells were infected with swab derived SARS-CoV-2 strains and 24 h post infection the RNA was subjected to (**a**) M and (**b**) RdRP-gene specific one step RT-qPCR-analysis. Amplification curves of SARS-CoV-2 RNAs (green) and RNaseP (purple) are shown as indicated

Table 6
Quantitative detection of SARS-CoV-2 genomic RNA from SARS-CoV-2 positive patient derived samples targeting M-gene and RNaseP

Patient#	Ct M	Copies/reaction	RNaseP
Pat#01	15.49	3.3E+06	27.66
Pat#02	15.38	3.6E+06	27.80
Pat#03	16.20	1.9E+06	26.84
Pat#04	15.41	3.5E+06	26.91
Pat#05	15.00	4.8E+06	28.15
Pat#06	14.16	9.3E+06	27.84
Pat#07	15.97	2.2E+06	27.29
Pat#08	16.00	2.2E+06	28.73
Pat#09	16.29	1.7E+06	28.77
Pat#10	16.05	2.1E+06	29.57

Table 7
Quantitative detection of SARS-CoV-2 genomic RNA from SARS-CoV-2 positive patient derived samples targeting RdRP gene and RNaseP

Patient#	Ct RdRP	Copies/reaction	RNaseP
Pat#01	16.57	5.2E+05	25.82
Pat#02	16.01	7.4E+05	26.01
Pat#03	16.72	4.7E+05	25.78
Pat#04	17.10	3.7E+05	25.55
Pat#05	16.59	5.1E+05	25.79
Pat#06	14.14	2.4E+06	25.22
Pat#07	16.49	5.4E+05	25.37
Pat#08	16.54	5.3E+05	26.59
Pat#09	18.13	1.9E+05	26.11
Pat#10	17.89	2.3E+05	26.68

viral loads and allows for a more sensitive viral detection in comparison to RdRP [7].

12. For absolute quantification and data analysis export the CT values to Excel.
13. Note that differences in the number of detected genome copy equivalents are due to the fact that the different viral RNA isoforms are present in different abundancies in cell culture supernatant.

14. The presence of human nucleic acids in the form of RNaseP can be used as a loading control for normalization purposes. It can also be used for quality control to assess the swab quality (*see* **Note 16**).

4 Notes

1. The RNaseP primers recommended from the CDC detect both, RNA and DNA, and therefore lack in discriminating background genomic DNA. Applying an alternative exon-exon junction reverse primer might prevent false-negative SARS-CoV-2 diagnoses [16].
2. Any SARS-CoV-2 isolate can be used for downstream analysis.
3. Consider pronounced effects of viral infection on adherence.
4. Using 160 μl is important for an optional QIAcube HT application to accurately transfer 150 μl from 96-well plate onto S-Block.
5. In case of multiwell plates, add lysis buffer into all wells and interspaces for proper inactivation of putative infectious material.
6. It is possible to transfer lysates already in the BSL-3 facility into the S-Block when using the QIAcube HT system. The S-Block is sealed with adhesive tape sheets and can be easily inactivated allowing for a straightforward export from the BSL-3 facility.
7. All subsequent steps are performed accordingly to the manufacturer's protocol.
8. The 2 ml microtube can be reused.
9. In case of multiwell plates, add lysis buffer into all wells and interspaces for proper inactivation of putative infectious material.
10. Importantly, one-Step PCR assays also avoid discrimination between the reverse transcription and the PCR reaction
11. In combination with the QIAcube HT technology, it is possible to obtain RT-qPCR data from up to 96-sample on the same day as sample acquisition.
12. Using the cell lysate with the corresponding supernatant for a double read-out is optional. Both PCR approaches individually allow for highly sensitive detection and quantification of SARS-CoV-2 and can be carried out independently from each other.
13. SYBR Green PCRs with the sgRNA 4 primers used in the protocol were found to be as sensitive as the Probe PCR with sgRNA 8 as described in Subheading 3.3.1.

14. http://www.scienceprimer.com/copy-number-calculator-for-realtime-pcr
15. Using 5 μl RNA standard per reaction will increase the copy number by the factor of 5 (e.g., 5×10^8 copies/reaction).
16. Reproducibility for RNaseP values for this cell culture supernatant derived samples is high; however, high variability is expected in swab samples.

Acknowledgments

The authors thank Denisa Bojkova, Jindrich Cinatl Jr., and Tuna Toptan-Grabmair for providing reagents and fruitful discussions. We are thankful for the numerous donations to the Goethe-Corona-Fond and for the support of our SARS-CoV-2 research. M.W. and R.M. were supported by the Deutsche Forschungsgemeinschaft (DFG, WI 5086/1–1, MA 1876/13-1, and MA 1876/12-1). All authors have read and agreed to the published version of the chapter.

References

1. Zhu N, Zhang D, Wang W, Li X, Yang B, Song J, Zhao X, Huang B, Shi W, Lu R, Niu P, Zhan F, Ma X, Wang D, Xu W, Wu G, Gao GF, Tan W, China Novel Coronavirus I, Research T (2020) A novel coronavirus from patients with pneumonia in China, 2019. N Engl J Med 382(8):727–733. https://doi.org/10.1056/NEJMoa2001017
2. Hoffmann M, Kleine-Weber H, Schroeder S, Kruger N, Herrler T, Erichsen S, Schiergens TS, Herrler G, Wu NH, Nitsche A, Muller MA, Drosten C, Pohlmann S (2020) SARS-CoV-2 Cell entry depends on ACE2 and TMPRSS2 and is blocked by a clinically proven protease inhibitor. Cell 181(2):271–280. e278. https://doi.org/10.1016/j.cell.2020.02.052
3. Kim D, Lee JY, Yang JS, Kim JW, Kim VN, Chang H (2020) The architecture of SARS-CoV-2 transcriptome. Cell 181(4):914–921 e910. https://doi.org/10.1016/j.cell.2020.04.011
4. Wolfel R, Corman VM, Guggemos W, Seilmaier M, Zange S, Muller MA, Niemeyer D, Jones TC, Vollmar P, Rothe C, Hoelscher M, Bleicker T, Brunink S, Schneider J, Ehmann R, Zwirglmaier K, Drosten C, Wendtner C (2020) Virological assessment of hospitalized patients with COVID-2019. Nature 581(7809):465–469. https://doi.org/10.1038/s41586-020-2196-x
5. CDC (2020) Interim guidelines for collecting, handling, and testing clinical specimens for COVID-19
6. WHO (2020) PCR protocol - World Health Organization
7. Toptan T, Hoehl S, Westhaus S, Bojkova D, Berger A, Rotter B, Hoffmeier K, Cinatl J Jr, Ciesek S, Widera M (2020) Optimized qRT-PCR approach for the detection of intra- and extra-cellular SARS-CoV-2 RNAs. Int J Mol Sci 21(12). https://doi.org/10.3390/ijms21124396
8. Shin D, Mukherjee R, Grewe D, Bojkova D, Baek K, Bhattacharya A, Schulz L, Widera M, Mehdipour AR, Tascher G, Geurink PP, Wilhelm A, van der Heden van Noort GJ, Ovaa H, Muller S, Knobeloch KP, Rajalingam K, Schulman BA, Cinatl J, Hummer G, Ciesek S, Dikic I (2020) Papain-like protease regulates SARS-CoV-2 viral spread and innate immunity. Nature 587: 657–662. https://doi.org/10.1038/s41586-020-2601-5
9. Corman VM, Landt O, Kaiser M, Molenkamp R, Meijer A, Chu DKW, Bleicker T, Brunink S, Schneider J, Schmidt ML, Mulders D, Haagmans BL, van der Veer B, van den Brink S, Wijsman L,

Goderski G, Romette JL, Ellis J, Zambon M, Peiris M, Goossens H, Reusken C, Koopmans MPG, Drosten C (2020) Detection of 2019 novel coronavirus (2019-nCoV) by real-time RT-PCR. Euro Surveill 25(3). https://doi.org/10.2807/1560-7917.ES.2020.25.3.2000045

10. Rho HW, Lee BC, Choi ES, Choi IJ, Lee YS, Goh SH (2010) Identification of valid reference genes for gene expression studies of human stomach cancer by reverse transcription-qPCR. BMC Cancer 10:240. https://doi.org/10.1186/1471-2407-10-240
11. Zhang X, Ding L, Sandford AJ (2005) Selection of reference genes for gene expression studies in human neutrophils by real-time PCR. BMC Mol Biol 6:4. https://doi.org/10.1186/1471-2199-6-4
12. Kohmer N, Toptan T, Pallas C, Karaca O, Pfeiffer A, Westhaus S, Widera M, Berger A, Hoehl S, Kammel M, Ciesek S, Rabenau HF (2021) The comparative clinical performance of Four SARS-CoV-2 rapid antigen tests and their correlation to infectivity in vitro. J Clin Med 10(2). https://doi.org/10.3390/jcm10020328
13. Hoehl S, Rabenau H, Berger A, Kortenbusch M, Cinatl J, Bojkova D, Behrens P, Boddinghaus B, Gotsch U, Naujoks F, Neumann P, Schork J, Tiarks-Jungk P, Walczok A, Eickmann M, Vehreschild M, Kann G, Wolf T, Gottschalk R, Ciesek S (2020) Evidence of SARS-CoV-2 infection in returning travelers from Wuhan, China. The New England J Med. https://doi.org/10.1056/NEJMc2001899
14. Widera M, Westhaus S, Rabenau HF, Hoehl S, Bojkova D, Cinatl J, Ciesek S (2021) Evaluation of stability and inactivation methods of SARS-CoV-2 in context of laboratory settings. Med Microbiol Immunol, 210(4):235–244. https://doi.org/10.1007/s00430-021-00716-3. Epub 2021 Jul 1.
15. Livak KJ, Schmittgen TD (2001) Analysis of relative gene expression data using real-time quantitative PCR and the 2(-Delta Delta C (T)) method. Methods 25(4):402–408. https://doi.org/10.1006/meth.2001.1262
16. Dekker RJ, Ensink WA, van Leeuwen S, Rauwerda H, Breit TM (2020) Overhauling a faulty control in the CDC-recommended SARS-CoV-2 RT-PCR test panel. bioRxiv.:2020.2006.2012.147819. https://doi.org/10.1101/2020.06.12.147819

Chapter 7

Immunofluorescent Antibody Techniques in the Diagnosis of SARS-CoV-2 Infection in Humans

Linda Hueston

Abstract

Immunofluorescence (IF) is an important technique used in the diagnosis of many infectious diseases. In virology, it has proven to be particularly suited to detecting antibody directed against newly emerging viruses able to be cultivated in cell culture. It permits visualization of antibody and allows for antibody class to be determined which is critical to understanding the timing of infection. The procedure used to determine IgG, IgA, and IgM antibody directed against SARS-CoV-2 in humans is described in this chapter. These methods were developed for routine diagnosis of SARS-CoV-2 infection in Australia at the start of the global pandemic in 2020.

Key words Immunofluorescence, Immunofluorescent antibody detection, SARS-CoV-2 immunofluorescent antibody detection, Fluorescence, FITC, SARS-CoV-2

1 Introduction

Coons et al. [1] initially described the immunofluorescent technique in 1941. It was later improved upon by Riggs et al. [2] who developed a stable linkage method to attach fluorescein to antibodies. Immunofluorescence has been an important technique in the diagnosis of numerous human and animal diseases. It is well suited to investigating newly emerging diseases and is often the technique against which ELISA and chemiluminescent techniques are evaluated. It allows rapid results (usually in 60–90 min), allowing for antibody classes and subclasses to be determined. It can also be performed qualitatively or quantitatively. Sensitive and specific FITC conjugates are widely available commercially, making the test more accessible.

There are however some concerns about the technique due to its subjective nature. Good training practices can significantly improve variation in reading the slides. Indeed, there are now robotic systems that can be "taught" to read fluorescent slides.

Justin Jang Hann Chu et al. (eds.), *SARS-CoV-2: Methods and Protocols*,
Methods in Molecular Biology, vol. 2452, https://doi.org/10.1007/978-1-0716-2111-0_7,

Preparation of the infected cells, inactivation and fixation methods can all affect the performance of the assay and each of these criteria need to be validated for each organism. Despite these issues, immunofluorescence remains an important diagnostic tool in infectious disease serology.

1.1 Safety

Interim recommendations published by the World Health Organization (WHO) classify SARS-CoV-2 as a Risk Group 3 biological agent [3]. This requires propagative work (such as viral culture, neutralization assays and preparation of immunofluorescent antibody slides) to be undertaken in PC3 or higher laboratories. The availability of such laboratories is limited and working in them is not suitable for large-scale antibody testing that a pandemic requires. However, once inactivated, materials can be handled safely in PC2 laboratories. Gamma radiation is an ionizing radiation often used by containment facilities to inactivate high-risk group organisms [4] as it is known to preserve viral morphology and protein integrity [5]. A 25kGY dose of gamma radiation was used to inactivate SARS-CoV-2 infected cells to allow IFA slides to be prepared in a PC2 laboratory.

2 Materials

2.1 Chemicals

- Bronidox (store at RT).
- Evans Blue (store at RT).
- Tween 20 (store at RT).
- Acetone (store at −20 °C).

2.2 Buffers and Solutions

Prepare all solutions using sterile endotoxin free water and analytical grade reagents. If a preservative is required Bronidox is the preferred preservative.

1. Evans Blue 0.5%. Dissolve 0.5 g Evans Blue dye in 100 ml of PBS. Store at RT.
2. Phosphate buffered saline (PBS) (10×). Store at RT.
3. Sample diluent. Store at 4 °C (*see* **Note 1**).
4. Conjugate diluent. Store at 4 °C (*see* **Note 2**).
5. Wash buffer: 2 l PBS (pH 7.2), 1 ml Tween 20. Mix. Store at RT.
6. Eagle's Minimum Essential Media (MEM) with Earle's salts, L-glutamine, and sodium bicarbonate.
7. MEM Growth Media (MEM + 4% FCS).
8. MEM Maintenance Media (MEM + 1% FCS).
9. Foetal calf serum (aliquot and store at −20 °C).

10. Trypsin–EDTA solution 1× (aliquot and store at −20 °C).
11. FITC F'ab 2Conjugates (IgG, IgA, IgM). Store at 4 °C (*see* **Note 3**).
12. Human IgG inactivation reagent. Store at 4 °C (*see* **Note 4**).
13. Mounting medium (1 part PBS + 9 parts glycerin pH 7.8).
14. Known positive and negative control sera.
15. Cryoprotectant media (*see* **Note 5**).

3 Methods

3.1 Cell Culture

Vero E6 cells (C1008; African Green Monkey Kidney cells) obtained from the ATCC (ATCC CRL-1586) were grown in 150 cm^3 flasks to 80% confluency in MEM Growth media at 37 °C with 5% CO_2. The flasks were then removed to the PC3 containment laboratory to enable infections with SARS-CoV-2 virus.

SARS-CoV-2 cultured from a nasopharyngeal swab from a symptomatic patient who returned from China in January 2020 was passaged a further 5 times in Vero E6 cells. The fifth passage of virus was cultured in Vero E6 cells with growth media at 37 °C with 5% CO_2 and was harvested 48 h postinoculation. Virus stocks were stored at −80 °C and this stock was used to infect cells to use in IFA slide production.

3.2 Infection and Preparation of Cells for Use in IFA

All infectious work is performed in a PC3 laboratory. Vero E6 cells were grown in 150cm^3 flasks to 80% confluency in MEM Growth media at 37 °C with 5% CO_2. Four flasks were used per batch. Growth media was removed from all four and replaced with 50 ml of maintenance media per flask. Three flasks were infected at 1/1000 with SARS-CoV-2 virus (titer $10^{-6.5}$), the fourth was kept as a negative control. Flasks were then incubated at 37 °C with 5% CO_2 for 36–40 h at which time the cells were harvested as follows.

Flasks were removed to a Class 2 Biosafety cabinet within the PC3 laboratory. Media was decanted into waste containers with 10% chlorize. Cell sheets were washed with 50 ml of sterile PBS 3 times to remove all trace of media (waste was decanted into containers with 10% chlorize). Five milliliters of trypsin–EDTA was added to each flask. The flask was recapped and gently rocked to allow the trypsin–EDTA to bathe the cell surface. Flasks were then placed at 37 °C for 5 min to allow the trypsin to dislodge the cells from the surface of the flask. Once the cells dislodged from the flask 2 ml of maintenance media was added to each flask to neutralize the trypsin–EDTA. Cells were gently mixed with a sterile pipette until no visible clumps remained. Cells from the infected and noninfected flasks were added to a sterile 50 ml centrifuge tube, volume made up to 50 ml with sterile PBS sealed and placed inside centrifuged buckets inside the biosafety cabinet sealed, surface decontaminated with chlorize before removing from the cabinet to the centrifuge. Tubes were spun at 800 × g for 10 min. The sealed buckets were returned to the biosafety cabinet to remove the tubes. The supernatant was carefully removed into chlorize waste containers and was replaced with fresh sterile PBS. The cell pellet was gently resuspended with a pipette until no visible clumps remained. This process was repeated 3 times.

Following the last PBS wash, the cell pellet was gently resuspended in 30 ml of Cryoprotectant media and dispensed in 1.5 ml aliquots into 2 ml sterile plastic tubes with an internal thread. The tubes were first stored for 2 h at 4 °C, then 4 h at −20 °C before placing at −80 °C until required for inactivation by gamma irradiation [6].

3.3 Inactivation by Gamma Irradiation

Frozen infected cells were placed inside a storage box, then sealed in a plastic bag whose surface was decontaminated with 10% chlorize. This was then placed in an Esky containing 5 kg dry ice. The bag was then covered with another 5 kg dry ice. The Esky was sealed with duct tape and the surface decontaminated with 10% chlorize before it was removed from the PC3 facility for irradiation affirming "chain of custody" procedures.

All irradiations were performed on dry ice inside a shielded irradiator with a cobalt-60 source. Appropriate dosimeters were attached to the Esky. Each run of irradiation was documented with actual absorbed dose of radiation. A dose of 25kGY was used to inactivate the SARS-CoV-2 infected cells.

Following irradiation, the material was returned to biocontainment to assess the inactivation by testing in tissue culture. A tube of irradiated cells was thawed, and the cells washed in 10 ml sterile PBS to remove the media, after this 1.5 ml of sterile PBS was added to the cell pellet, the cells were resuspended with gentle mixing. Vero E6 cells were infected with a 1:1 dilution of the cells in maintenance media and the culture was examined daily for cytopathic effect over a 7-day period. Negative cultures were passaged once more in freshly seeded Vero E6 cells and monitored as before for another 7-day period.

Once it was established that the material had been inactivated, it was removed to the PC2 laboratory to make the immunofluorescent slides.

3.4 Preparing Glass Slides

Glass slides suitable for IFA preparation are available from numerous commercial sources in a variety of formats. The number of wells per slide can be chosen to suit the laboratory workload, but the diameter of the well will influence the volume of cell suspension used. For this work, slides with three rows of 8 wells (4 mm

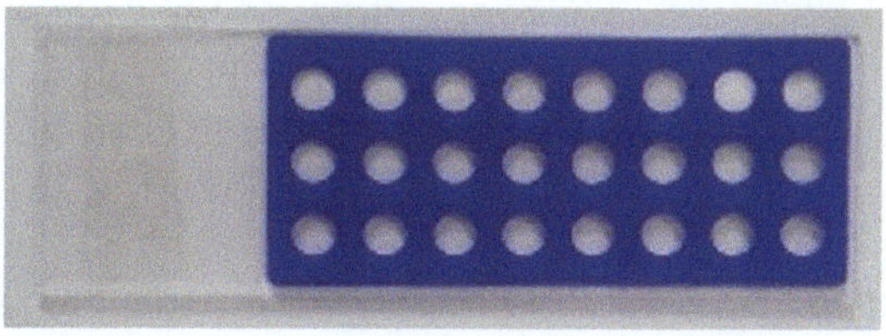

Fig. 1 IFA glass slide

diameter) were chosen for use (see Fig. 1). These were loaded with 5 μl of cell suspension per well.

Slides should always be cleaned thoroughly before use to ensure all residue from manufacture is removed. First, clean slides with hot soapy water, rinse well and dry individually, store until required. The day before slides are used, make up a solution of 1% Tween 20 in endotoxin-free water and dispense 5 μl of this solution to each well, allow it to sit for 30–60 s then remove and dry the slide. This will ensure that the cell suspension is evenly distributed across the well.

3.5 Preparing SARS-CoV-2 IF Slides

1. Place a beaker containing 200 ml of 37 °C water in a Class2 biosafety cabinet.
2. Place one vial of frozen, gamma irradiated SARS-CoV-2 infected cells into the water and allow to thaw.
3. Once thawed, working quickly, remove lid and using a small pipette gently resuspend the cell pellet.
4. Add the cell suspension to a 10 ml sterile centrifuge tube and add sterile PBS to the 10 ml line. Mix by gentle inversion.
5. Centrifuge cell suspension at 800 × *g* for 10 min. Aseptically decant the supernatant without disturbing the cell pellet.
6. Resuspend the cell pellet in 1 ml of sterile PBS, then make up the volume to 10 ml with PBS. Mix by gentle inversion.
7. Centrifuge cell suspension at 800 × *g* for 10 min. Aseptically decant the supernatant without disturbing the cell pellet.
8. Gently resuspend the pellet in 4 ml of sterile PBS to break up clumps. Add 5 μl of cell suspension to the wells of a glass slide. Allow to stand for 5 min, then observe under the microscope to determine if there are sufficient cells in the suspension so that cell coverage of the slide well is sufficient to cover the surface. Adjust the volume with PBS if required.
9. Dispense 5 μl of the cell suspension onto each well on the glass slide. Place the slide on a heating block or warming tray set to 40 °C for 30 min or until dry. If a heating block is not available, leave slides overnight in the biosafety cabinet to air-dry.

10. Once slides are dry, place into slide racks and immerse in ice cold acetone. Place in a −20 °C freezer and allow to fix for 15 min. Decant acetone and allow slides to air-dry before packaging and storing at −80 °C until required.

3.6 IFA Testing Procedure

3.6.1 Screening Protocol for IgG, IgA, and IgM

1. Dilute IgG controls and serum samples 1/10, that is, 10 μl control/sample + 90 μl sample diluent buffer. Mix.
2. Dilute IgA and IgM controls and serum samples 1/10, that is, 10 μl control/sample + 90 μl human IgG inactivation reagent. Mix well, allow samples to stand for 10 min then centrifuge at 13,000 × *g* in a microfuge for 10 min. The supernatant is used for IgA and IgM testing.
3. Dispense 15 μl of each control and patient sample into the appropriate wells of the IFA slide.
4. Incubate the slide for 30 min at 37 °C in a humidified chamber.
5. Wash slides with three changes of wash buffer, then fill the Coplin jar with wash buffer and allow to sit for 5 min. Decant wash buffer.
6. Rinse slide once with endotoxin free water to remove salts. Carefully wipe around wells with a cotton tip to remove excess water, do not touch the cells. Work quickly to ensure cells do not dry out.
7. Dispense 15 μl of predetermined dilution of either antihuman IgG, IgA, or IgM fluorescein labeled conjugate onto the appropriate well.
8. Incubate the slide for 30 min at 37 °C in a humidified chamber.
9. Wash slides with three changes of wash buffer then the fill the Coplin jar with PBS wash buffer and allow to sit for 5 min. Decant wash buffer.
10. Rinse slide once with endotoxin free water to remove salts. Air-dry the slide.
11. Add a small drop (5 μl) of mounting medium to each well and add coverslips.
12. Cells are examined for the presence of specific fluorescence. Bright apple green fluorescence in the cytoplasm of the infected cells denotes positivity. The intensity of the fluorescence is scored as 4+ (the most intense staining), 3+, 2+, and 1+ (the weakest intensity). The last well showing specific 1+ fluorescence is the endpoint.
13. All positives are titrated to determine an endpoint titer. Samples negative at the screening dilution are reported as <10.

3.7 IFA Titration Series Procedure

1. Dilute IgG controls and serum samples 1/10, that is, 10 μl control/sample + 90 μl sample diluent buffer. Mix. Use the supernatant in step 3.
2. Dilute IgA and IgM controls and serum samples 1/10, that is, 10 μl control/sample + 90 μl human IgG inactivation reagent. Mix well and allow samples to stand for 10 min then centrifuge at 13,000 × *g* in a microfuge for 5 min. Use the supernatant in step 3.
3. Serially dilute 1/10 samples (IgG, IgA, and IgM) in a low binding plastic 96 well plate (see Subheading 4) from 1/10 to 1/1280 as shown in Fig. 2.
4. Dispense 15 μl of each dilution of patient samples into the appropriate wells of the glass IFA slide. N.B. Controls are not diluted beyond 1/10 and are run on a separate slide.
5. Incubate the slides for 30 min at 37 °C in a humidified chamber.
6. Wash slides with three changes of wash buffer then fill the Coplin jar with wash buffer and allow to sit for 5 min. Decant wash buffer.
7. Rinse slide once with endotoxin free water to remove salts. Carefully wipe around wells with a cotton tip to remove excess

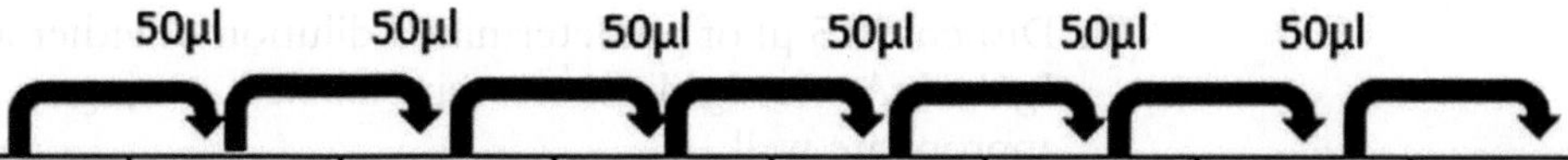

	A	B	C	D	E	F	G	H
1	100µl 1/10 sample	50µl sample diluent	50µl sample diluent	50µl sample diluent	50µl sample diluent	50µl sample diluent	50µl sample diluent	50µl sample diluent
2								
3								
4								
5								
6								
7								
8								
9								
10								
11								
12								

Fig. 2 Performing titration series on positive screening samples

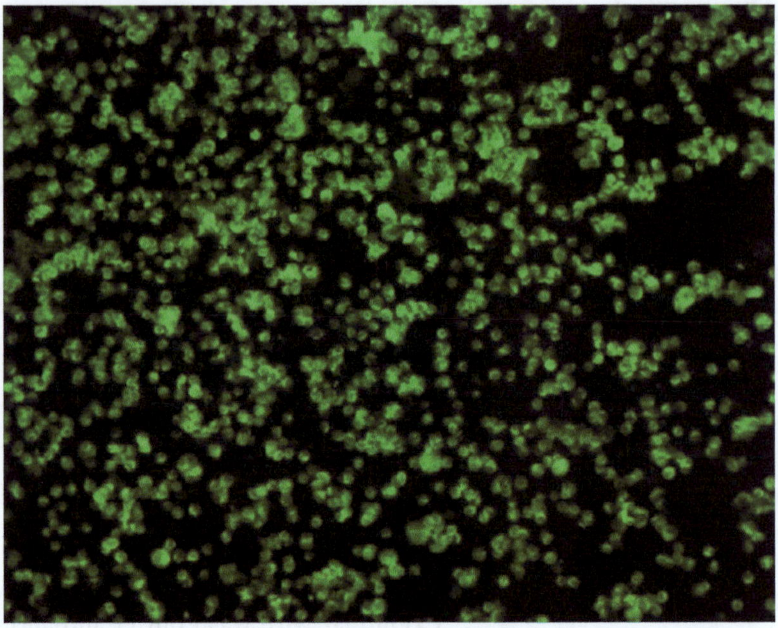

Fig. 3 Positive IFA for SARS-CoV-2 displaying cytoplasmic fluorescence

water, do not touch the cells. Work quickly to ensure cells do not dry out.

8. Dispense 15 μl of predetermined dilution of either antihuman IgG, IgA, or IgM fluorescein labeled conjugate onto the appropriate well.
9. Incubate the slide for 30 min at 37 °C in a humidified chamber.
10. Wash slides with three changes of wash buffer then fill the Coplin jar with wash buffer and allow to sit for 5 min. Decant wash buffer.
11. Rinse slide once with endotoxin free water to remove salts. Air-dry the slide.
12. Add a small drop (5 μl) of mounting medium to the wells and add coverslips.
13. Cells are examined for the presence of specific fluorescence. Bright apple green fluorescence in the cytoplasm of the infected cells denotes positivity (Fig. 3). The intensity of the fluorescence is scored as 4+ (the most intense staining), 3+, 2+, and 1+ (the weakest intensity). The last well showing specific 1+ fluorescence is the endpoint titer.

4 Notes

1. Sample diluent: To 95 ml wash buffer (pH 7.2) add 5 ml fetal calf serum and 0.1 ml Bronidox. Mix, filter through a 0.2 μm filter, dispense into 20 ml aliquots. Store at 4 °C. The presence of fetal calf serum and Tween 20 acts to reduce background.
2. Conjugate Diluent: To 95 ml sample diluent add 5 ml of 0.5% Evans Blue solution. Mix, filter through a 0.2 μm filter, dispense into 20 ml aliquots. Store at 4 °C. The presence of Evans Blue acts as a counterstain producing a red background to improve reading of the fluorescence.
3. Fab_2 IgG, IgA, and IgM FITC conjugates are commercially available from numerous sources. Fab_2 conjugates lack the Fc domain which reduces nonspecific staining. Conjugate should be made fresh daily and should be protected from light.
4. Human inactivation reagent is an anti-human IgG solution and is used as the sample diluent for IgA and IgM estimations. It binds to IgG in the patient sample precipitating it. The precipitate is removed by centrifugation. IgG is the most prevalent antibody and if it is not removed it will compete with IgA and IgM for binding sites on the infected cells potentially leading to false negative results. There are multiple commercial sources available.
5. Cryoprotectant media: MEM media with 10% FCS and 10% DMSO. This can be made in the lab or purchased from several commercial sources. It is usually sold as a recovery cell culture freezing medium. It protects the cell integrity during the freezing process.

References

1. Coons AH, Creech HJ, Jones RN (1941) Immunological properties of an antibody containing a fluorescent group. Proc Soc Exp Biol Med 47:200–202
2. Riggs JL, Seiwald RJ, Burckhalter JH, Downs CM, Metcalf TG (1958) Isothiocyanate compounds as fluorescent labelling agents for immune serum. Am J Pathol 34:1081–1097
3. WHO (2020) Laboratory biosafety guidance related to coronavirus Disease (COVID-19). WHO, Interim Guidance
4. Feldmann F, Lesley Shupert W, Haddock E, Twardoski B, Feldmann H (2019) Gamma irradiation as an effective method for inactivation of emerging viral pathogens. Am J Trop Med Hyg 100(5):1275–1277
5. Elliot LH, McCormick JB, Johnson KM (1982) Inactivation of Lassa, Marburg, and Ebola viruses by gamma irradiation. J Clin Microbiol 16(4):704–870
6. Leung A, Tran K, Audet J, Lavineway S, Bastien N, Krishnan J (2020) In vitro inactivation of SARS-CoV-2 using gamma radiation. Appl Biosafety:1–4

Part II

Cell Culture, Virus Replication, and Cellular Responses

Chapter 8

Propagation and Quantification of SARS-CoV-2

Alexander S. Jureka and Christopher F. Basler

Abstract

In late 2019, the novel coronavirus severe acute respiratory syndrome coronavirus 2 (SARS-CoV-2) emerged in Wuhan, China. Since its emergence, SARS-CoV-2 has been responsible for a world-wide pandemic resulting in over 80 million infections and over 1.8 million deaths. The severity of the pandemic has prompted widespread research efforts to more fully understand SARS-CoV-2 and the disease it causes, COVID-19. Research into this novel virus will be facilitated by the availability of clearly described and effective protocols that enable the propagation and quantification of infectious virus. Here, we describe protocols for the propagation of SARS-CoV-2 in Vero E6 cells as well as two human cells lines, the intestinal epithelial Caco-2 cell line and the respiratory epithelial Calu-3 cell line. Additionally, we provide protocols for the quantification of SARS-CoV-2 by plaque assays and immunofocus forming assays in Vero E6 cells utilizing liquid overlays. These protocols provide a foundation for laboratories acquiring the ability to study SARS-CoV-2 to address this ongoing pandemic.

Key words SARS-CoV-2, Coronavirus, COVID-19, Focus forming assay, Plaque assay, Virus

1 Introduction

The novel severe acute respiratory syndrome coronavirus 2 (SARS-CoV-2) belongs to the betacoronavirus genus which also includes the highly pathogenic SARS-CoV and Middle East respiratory syndrome coronavirus (MERS CoV). SARS-CoV-2 emerged in late 2019 in Wuhan, China and has been the cause of the most significant pandemic in the last century [1, 2]. The pandemic and the public response have severely impacted public health and resulted in unprecedented social and economic disruption. As of January 5, 2021, SARS-CoV-2 had caused more than 80 million infections and 1.8 million deaths worldwide [3]. The rapid emergence of SARS-CoV-2 has prompted urgent worldwide efforts to better characterize the virus for the development of novel countermeasures.

The expanded interest in studying SARS-CoV-2 to address the ongoing pandemic requires that laboratories acquire the capacity to

Justin Jang Hann Chu et al. (eds.), *SARS-CoV-2: Methods and Protocols*,
Methods in Molecular Biology, vol. 2452, https://doi.org/10.1007/978-1-0716-2111-0_8,

work with the virus. The ability to grow and accurately quantify infectious virus is critical for virological studies. SARS-CoV-2 has been demonstrated to grow well in only a limited number of cell lines due to its reliance on the angiotensin converting-enzyme 2 (ACE2) receptor for entry [4–7]. Vero cells, specifically the Vero E6 clone, are one of the most common cell lines used in virology research and are a preferred cell line for the propagation and quantification of SARS-CoV-2. However, the utility of Vero E6 cells is limited due to their nonhuman origin. For more detailed inquiries into SARS-CoV-2 biology, the intestinal epithelial Caco-2 and the bronchiolar epithelial line Calu3, are two human cell lines commonly in use. These support productive SARS-CoV-2 replication, and their intestinal and respiratory epithelial origins provide added relevance to human SARS-CoV-2 infections [8, 9].

Equally important to having the ability to grow SARS-CoV-2 is the ability to quantify its growth. Numerous studies have now determined that SARS-CoV-2 can be easily quantified using traditional plaque assay techniques with Vero E6 cells. SARS-CoV-2 infection results in readily identifiable and countable plaques after 72 h post-infection [9, 10]. While plaque assays have been perhaps the most commonly used technique for the quantification of SARS-CoV-2, immunohistochemical focus forming assays are another effective method for the quantification of SARS-CoV-2 [9, 11–14]. For SARS-CoV-2, we have found that plaque assays and immunofocus forming assays are complementary techniques with individual strengths. Traditional plaque assays provide high levels of accuracy at relatively low throughput, whereas focus forming assays provide sufficient sensitivity with higher-throughput capability.

Here, we describe detailed protocols for the propagation of SARS-CoV-2 in Vero E6 cells as well as protocols for accomplishing cell culture–based infections for the study of SARS-CoV-2. In addition, we provide protocols for the quantification of SARS-CoV-2-containing samples using plaque assays with a semiliquid overlay system as well as a SARS-CoV-2-specific immunofocus forming assay performed in 96-well plate format.

2 Materials

2.1 Cell Culture and Maintenance

Recovery, propagation, and quantification of SARS-CoV-2 is most efficiently carried out using Vero E6 cells. However, for the study of SARS-CoV-2 biology, several other human cell lines such as the Caco2 intestinal epithelial cell line and the Calu-3 bronchiolar epithelial cell line may also be required. The following materials are needed to culture and maintain these cell lines (*see* **Note 1**).

1. Frozen vials of cells or cells grown to near confluency in 75 cm^2 tissue culture flask (Vero E6 ATCC CRL1586, Caco-2 ATCC HTB37, Calu-3 ATCC HTB55).
2. Dulbecco's modified eagle medium (DMEM: 4.5 g/L glucose, 2 mM L-glutamine, 1 mM sodium pyruvate; Corning) supplemented with 10% fetal bovine serum (FBS; Gibco) and 1× antibiotic–antimycotic (Gibco; optional).
3. Cell culture grade phosphate-buffered saline (PBS) without calcium or magnesium.
4. Trypsin-EDTA (0.25%; Gibco).
5. Conical tubes (15 and 50 mL).
6. 75 cm^2 tissue culture treated flasks with straight necks.
7. Serological pipettes.
8. Aspirating pipettes.
9. 70% ethanol.
10. 37 °C water bath.
11. Cell culture incubator at 37°C, 5% CO_2.
12. Biosafety cabinet.

2.2 Preparation of SARS-CoV-2 Stock Virus and Cell Culture Infections

Several lineages of Vero cells are available (Vero, Vero 76, and Vero E6) and have been shown to be permissive to SARS-CoV-2 infection. However, Vero E6 cells are the preferred cell line for propagation of SARS-CoV-2 for generation of virus stocks [8–11]. All cell culture manipulations of SARS-CoV-2 should be performed at biosafety level 3 (BSL3) following institutional guidelines.

2.2.1 Preparation of SARS-CoV-2 Stock Virus

1. Vero E6 cells cultured to confluency in a 150 cm^2 or larger tissue culture flask with vented cap (*see* **Note 2**).
2. SARS-CoV-2 virus stock (ideally with known titer).
3. PBS without calcium or magnesium (sterile).
4. Opti-MEM (Gibco) supplemented with 2% FBS or 0.3% BSA (*see* **Note 3**) and 1× antibiotic–antimycotic (Gibco).
5. Conical tubes (15 or 50 mL).
6. Serological pipettes (sterile).
7. Aspirating pipettes (sterile).
8. Micropipettes (10, 20, 200, and 1000 μL).
9. Micropipette filter tips (sterile).
10. 1.5–2 mL externally threaded sterile cryogenic storage tubes.
11. Centrifuge capable of spinning conical tubes at ≥3000 × *g* with gasketed safety cups.
12. Cell culture incubator at 37 °C, 5% CO_2.

13. Class II biosafety cabinet.
14. −80 °C freezer
15. 70% ethanol or other chemical disinfectant (e.g., Microchem Plus).

2.2.2 Cell Culture SARS-CoV-2 Infections

1. Cells in tissue culture flask(s) or multiwell plate(s) cultured to desired confluency.
2. SARS-CoV-2 virus stock with known titer.
3. PBS without calcium or magnesium (sterile).
4. Dulbecco's modified eagle medium (DMEM: 4.5 g/L glucose, 2 mM L-glutamine, 1 mM sodium pyruvate; Corning) supplemented with 10% fetal bovine serum (FBS; Gibco) and 1× antibiotic–antimycotic (Gibco; optional).
5. Conical tubes (15 or 50 mL).
6. Serological pipettes (sterile).
7. Aspirating pipettes (sterile).
8. Micropipettes (10, 20, 200, and 1000 μL).
9. Micropipette filter tips (sterile).
10. Cell culture incubator at 37 °C, 5% CO_2.
11. Class II biosafety cabinet.
12. −80 °C freezer
13. 70% ethanol or other chemical disinfectant (e.g., Microchem Plus).

2.3 Quantification of SARS-CoV-2 by Plaque Assay

2.3.1 Preparation of 2.4% Microcrystalline Cellulose (MCC) Overlay Stock

Traditionally, plaque assays utilize either agar or carboxymethyl cellulose (CMC) as an overlay media. While these overlay medias generally work quite well, they both suffer from the significant drawbacks of requiring heating in their preparation for use and the manual removal of the agar or CMC "plugs" from the plaque assay plates. In contrast, microcrystalline cellulose (MCC) does not require heating at the time of use and is removed by simple aspiration prior to staining of the plaques. The following materials are required for the preparation of the MCC overlay media for the purposes of SARS-CoV-2 plaque assays.

1. Microcrystalline colloidal cellulose (MCC) powder (Sigma Aldrich; 435,244).
2. Cell culture grade water (Gibco; sterile).
3. DMEM powdered media (Gibco 12,800–017).
4. Cell culture grade sodium bicarbonate (Gibco; 7.5%).
5. 100× antibiotic–antimycotic (Gibco).
6. 2 L Erlenmeyer flask.

7. Large magnetic stirring bar.
8. Stir plate.
9. Autoclave.

2.3.2 SARS-CoV-2 Plaque Assay

1. Vero E6 cells plated to confluency in 6-, 12-, or 24-well tissue culture plates.
2. SARS-CoV-2 virus stock.
3. Dulbecco's modified eagle medium (DMEM: 4.5 g/L glucose, 2 mM L-glutamine, 1 mM sodium pyruvate; Corning) supplemented with 10% fetal bovine serum (FBS; Gibco) and 1× antibiotic–antimycotic (Gibco; optional).
4. Sterile dilution rack, deep-well 96-well plate, or standard 96-well plate for dilutions.
5. Serological pipettes (sterile).
6. Micropipettes (10, 20, 200, 1000 μL) and sterile filter tips.
7. Prepared 2.4% MCC overlay stock (preparation described in Subheading 3.3.1).
8. Absorbent pads.
9. Cell culture incubator at 37 °C, 5% CO_2.
10. 0.5% crystal violet staining solution. To make 1 L, add 5 g of crystal violet powder (Fisher Scientific) to 200 mL of methanol while stirring and allow to dissolve. Continue stirring and slowly add 800 mL of deionized water.
11. 10% neutral buffered formalin.

2.4 Quantification of SARS-CoV-2 by Immunofocus Forming Assay

Plaque assays are generally regarded as the standard for quantification of infectious virus within a sample. However, plaque assays require larger format cell culture dishes in order to properly visualize plaques. To accomplish higher throughput processing of samples, being able to quantify infectious virus in 96-well plate format is desirable. Immunofocus forming assays allow for the direct quantification of virus titers within a sample in 96-well plate format more rapidly and with similar effort to plaque forming assays. The following materials will be required for quantifying SARS-CoV-2 containing samples using immunofocus forming assays.

1. Vero E6 cells plated to confluency in 96-well tissue culture plates.
2. SARS-CoV-2 virus stock.
3. Dulbecco's modified eagle medium (DMEM: 4.5 g/L glucose, 2 mM L-glutamine, 1 mM sodium pyruvate; Corning) supplemented with 10% fetal bovine serum (FBS; Gibco) and 1× antibiotic–antimycotic (anti-anti; Gibco; optional).

4. Sterile dilution rack, deep-well 96 well plate, or standard 96 well plate for dilutions.
5. Serological pipettes (sterile).
6. Micropipettes (10, 20, 200, 1000 μL) and sterile filter tips.
7. Prepared 2.4% MCC overlay stock (preparation described in Subheading 3.3.1).
8. Absorbent pads.
9. Cell culture incubator at 37 °C, 5% CO_2.
10. 10% neutral buffered formalin.
11. Ice-cold methanol with 0.3% hydrogen peroxide. To prepare add 1 mL of cold 30% hydrogen peroxide to 100 mL of ice-cold 100% methanol.
12. Nonsterile PBS.
13. Nonsterile PBS–0.5% Tween® 20.
14. Nonfat dry milk.
15. Novus biologicals Rabbit anti-SARS-CoV N primary antibody (NB100–56576; *see* **Note 4**).
16. HRP conjugated anti-rabbit secondary antibody (Cell Signaling; 7074).
17. KPL TrueBlue peroxidase substrate (Sera Care; 5510-0030; *see* **Note 5**).

3 Methods

3.1 Cell Culture and Maintenance

1. Depending on the number of cells and size of the tissue culture flask, Vero cells should be passaged twice a week at a sub-cultivation ratio of 1:5. Caco-2 cells should be passaged once a week at a sub-cultivation ratio 1:5, and Calu3 cells should be passaged once a week at a sub-cultivation ratio of 1:2 (*see* **Note 6**). It is not recommended to grow Caco-2 or Calu3 cells in flasks larger than 75 cm^2.
2. When cells are nearly confluent, aspirate the growth media and wash the cells once with 5 mL of sterile PBS.
3. Add 3 mL of 0.25% trypsin to a 75 cm^2 flask or 5 mL to a 150 cm^2 flask and incubate at 37 °C/5% CO_2 for 5 min, or until cells begin to detach. Shaking or gently tapping the flask, especially for Calu3 cells, can help this process.
4. Add 5 mL of DMEM 10% FBS to the flask to inactivate the trypsin and gently pipet up and down several times to resolve any clumps.

5. Transfer the cells to a conical tube (either 15 or 50 mL depending on the total volume of cells) and spin at 300 × *g* for 5 min to pellet the cells.
6. Aspirate off the supernatant and resuspend the cells in 10 mL of DMEM 10% FBS.
7. Add cells back to a culture flask to the desired sub-cultivation ratio with a final volume of 12 mL for a 75 cm^2 culture flask or 25 mL for a 150 cm^2 culture flask. Gently rock the flask for even coverage.
8. Return the cells to the incubator at 37 °C/5% CO_2 and culture until they are 90% confluent. For Vero E6 this usually takes 2 days (for Caco-2: 3–4 days, and for Calu3 cells: 5–7 days). Replace media on cells every 3 days regardless of their confluency to maintain cell viability.

3.2 Preparation of SARS-CoV-2 Stock Virus and Cell Culture Infections

Producing virus stocks and performing cell culture infections with SARS-CoV-2 are similar processes. However, we have determined particular conditions that are optimal for producing high titer virus stocks. This is described in detail in Subheading 3.2.1. As a companion, we have also provided a more generic protocol for planning and executing cell culture infections with SARS-CoV-2 in the format required for a particular experimental design. Depending on the questions being addressed by an experiment and the supplies that are on hand, it may be necessary to use culture dishes of different sizes and varying multiplicities of infection (MOI). These details are described in Subheading 3.2.2 with recommendations for conditions we have found to work well.

3.2.1 Preparation of SARS-CoV-2 Stock Virus

1. Culture Vero E6 cells in a ≥ 150 cm^2 tissue culture flask to 80–90% confluency the day of infection.
2. All subsequent steps in this protocol should be performed within a biosafety level 3 (BSL3) laboratory unless otherwise indicated.
3. Thaw early passage SARS-CoV-2 stock virus (preferably of known titer).
4. Prepare the propagation inoculum (4–5 mL) to a multiplicity of infection (MOI) of 0.001 in Opti-MEM supplemented with 2% FBS or 0.3% BSA and 1× antibiotic-antimycotic (*see* **Note 7**).
5. Using an aspirating pipette, gently remove the growth media from the Vero E6 cells, taking care not to disturb the monolayer.
6. Wash the cell monolayer once with sterile PBS to remove residual growth media.

7. Pipet the prepared inoculum onto the cells and gently rock the flask to cover all of the cells.
8. Place the flask in a cell culture incubator at 37 °C/5% CO_2 and gently rock the flask every 15 min for 1 h to allow efficient virus adsorption.
9. After adsorption, return the flask to the biosafety cabinet and, without removing the initial inoculum, add a minimal amount of Opti-MEM 2% FBS that comfortably covers the cells. For a 150 cm^2 flask, addition of 21 mL of Opti-MEM 2% FBS to 4 mL of inoculum works well.
10. Incubate the flask at 37 °C/5% CO_2 for 72 h. Cytopathic effects, such as cell monolayer sloughing and a large number of floating cells, should be readily noticeable at 72 h post-infection. A mock-infected flask provides a useful comparison to help identify virus-induced cytopathic effects.
11. At harvest, transfer the virus containing supernatant to a 50 mL conical tube and centrifuge at $\geq$3000 × *g* for 15 min to remove cell debris.
12. Transfer the clarified supernatant to a new 50 mL conical tube.
13. Aliquot the clarified supernatant into sterile 1.5–2 mL externally threaded cryogenic storage tubes.
14. Store all virus stocks at −80°C and allow to remain at −80°C for at least 24 h before determining the titer (*see* **Note 8**).

3.2.2 Cell Culture SARS-CoV-2 Infections

1. Culture cells to be infected in a multiwell tissue culture plate or tissue culture flask should be at 80–90% confluent the day of infection.
2. All subsequent steps in this protocol should be performed within a biosafety level 3 (BSL3) laboratory unless otherwise indicated.
3. Prepare the virus inoculum to the desired MOI in a volume of DMEM 2% FBS that best suits the size of the cell culture plate/ flask being infected (Table 1).
4. Gently aspirate the growth media from the cells, taking care not to disturb the monolayer.
5. Wash the monolayer with PBS to remove residual growth media.
6. Add the inoculum to the cells using a micropipette with sterile filter tips or sterile serological pipette.
7. Gently rock the plate or flask to ensure even coverage of the monolayer.
8. Place cells in a cell culture incubator at 37 °C/5% CO_2 and gently rock every 10–15 min for 1 h to facilitate virus adsorption.

Table 1
Recommended SARS-CoV-2 inoculum volumes and final media volumes for different cell culture vessels

Culture vessel (cm^2)	Inoculum volume (mL)	Cells at confluency (for MOI calculation)	Recommended final volume (mL)
225	5	30×10^6	30
150	4	17.5×10^6	25
75	3	8.4×10^6	10
25	1	2.8×10^6	5
6 well plate	0.5	1.2×10^6	3
12 well plate	0.25	0.4×10^6	2
24 well plate	0.1	0.2×10^6	1
48 well plate	0.05	0.1×10^6	0.5
96 well plate	0.05	0.05×10^6	0.1

9. After adsorption, return the flask to the biosafety cabinet and remove the inoculum.
10. Add the appropriate amount of DMEM 2% FBS to the cell culture plate or flask (Table 1).
11. Incubate at 37 °C/5% CO_2 until the determined end point.
12. At the experimental endpoint, transfer supernatants to 1.5–2 mL externally threaded cryogenic storage tubes and centrifuge at $\geq 3000 \times g$ for 10 min to remove cell debris.
13. Transfer supernatants to new tubes and store at −80 °C for further analysis.

3.3 Quantification of SARS-CoV-2 by Plaque Assay

3.3.1 Preparation of 2.4% Microcrystalline Cellulose (MCC) Overlay Stock

1. Weigh out 24 g of dry MCC powder.
2. Pour 1 L of sterile tissue culture grade water into a sterile 2 L Erlenmeyer flask (*see* **Note 9**).
3. Place a heavy gauge stir bar into the Erlenmeyer flask and start stirring the water at 350–500 rpm.
4. Slowly add the MCC powder into the water and, once all of the powder has been added, continue stirring for an additional 30 min–1 h until it is a smooth opaque white suspension, and no clumps can be seen (*see* **Note 10**).
5. Once properly combined, autoclave the 2.4% MCC in water suspension for at least 20 min on a liquid autoclave cycle. Leave the stir bar in the flask during autoclaving.
6. After autoclaving and while still hot, stir the 2.4% MCC in water at 200–400 rpm for another 30 min–1 h, to ensure a completely homogenous opaque white suspension.

7. Allow to cool to room temperature overnight.
8. Once cooled, add one packet of powdered DMEM media (Gibco 12800–017) to 1 L of 2.4% MCC in water in a biosafety cabinet (*see* **Note 11**). Adjust the pH by adding sterile cell culture grade sodium bicarbonate solution to a concentration of at least 1.3 g/L. This will bring the pH of the 2.4% MCC into workable range and the pH does not need to be measured or adjusted further. Add antibiotic–antimycotic to 1× concentration.
9. Store at 4°C while not in use. The prepared DMEM supplemented 2.4% MCC stock suspension should be used within 6 months of its preparation.

3.3.2 SARS-CoV-2 Plaque Assay

1. Plate 1–2 × 10^5 Vero E6 cells in DMEM/ 10% FBS into each well of a 24-well plate the day before the plaque assay.
2. All subsequent steps in this protocol should be performed within a biosafety level 3 (BSL3) laboratory unless otherwise indicated.
3. Prepare ten-fold serial dilutions of SARS-CoV-2 samples in DMEM/ 2% FBS (Fig. 1a; *see* **Note 12**).
4. Gently aspirate the growth media from the Vero E6 cells and plate 100 μL of each desired dilution of virus (Fig. 1b; *see* **Note 13**).
5. Gently rock the plates to ensure even coverage of the monolayer.
6. Move the 24-well plates to a cell culture incubator at 37 °C/5% CO_2 and gently rock the plates every 10–15 min for 1 h to allow for virus adsorption.
7. During virus adsorption, prepare 0.6% MCC overlay by adding 1 part DMEM supplemented 2.4% MCC overlay stock (prepared in 3.3.1) to 3 parts DMEM 2% FBS media. For 100 mL of 0.6% MCC overlay add 25 mL of DMEM supplemented 2.4% MCC overlay stock to 75 mL of DMEM 2% FBS media (*see* **Note 14**).
8. After the adsorption period is complete, shake the 0.6% MCC overlay for 15 s to ensure it is homogenous (*see* **Note 15**) and add 1 mL of 0.6% MCC overlay to each well of the 24 well plate, gently rocking the plate to ensure even coverage (*see* **Note 16**).
9. Incubate plaque assays at 37 °C/5% CO_2 for 72 h. It is very important that the plates are not moved during this incubation. The MCC overlay will remain semiliquid and moving the plates can result in "comet-like" streaks as opposed to distinct plaques.

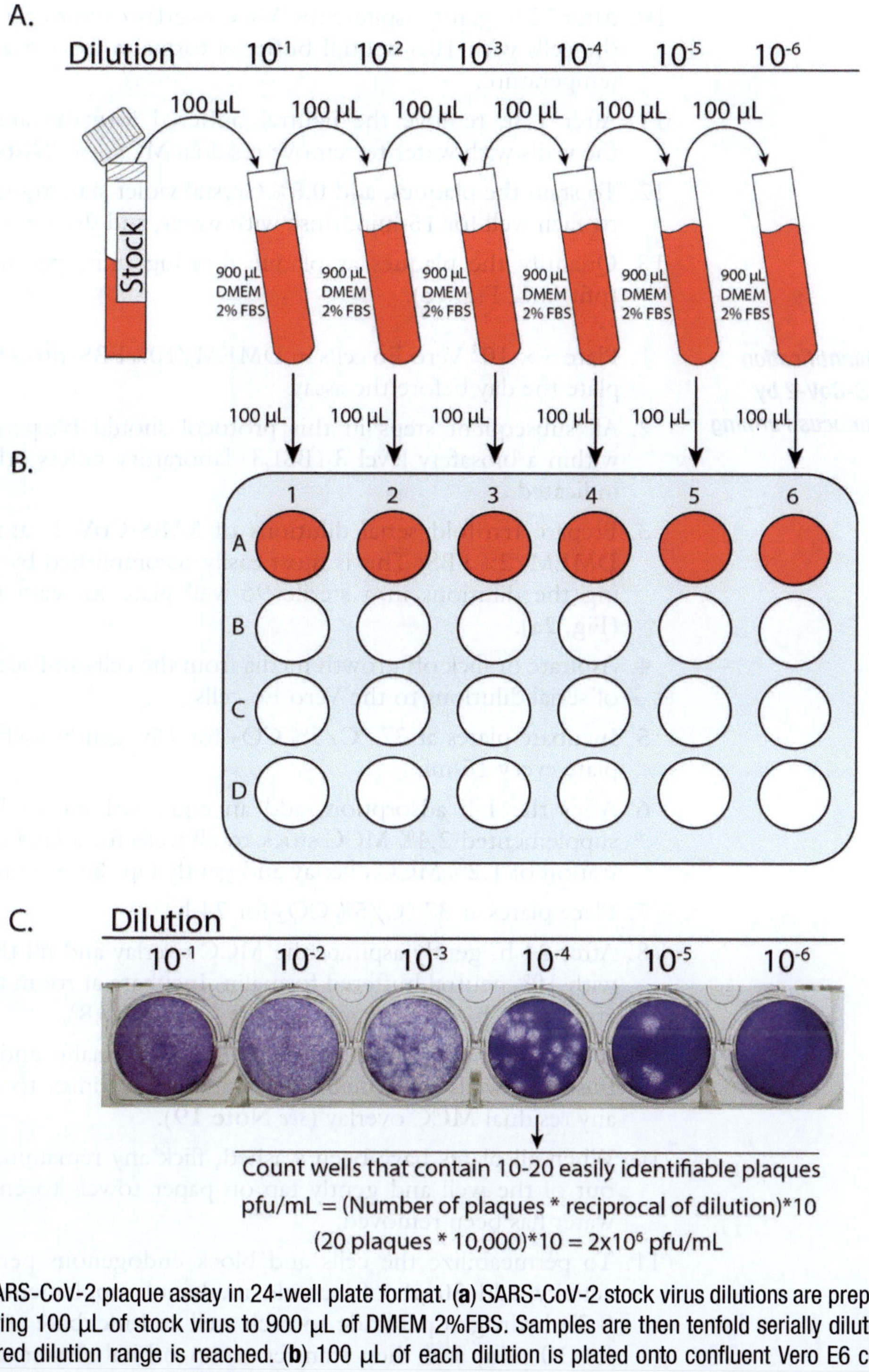

Fig. 1 SARS-CoV-2 plaque assay in 24-well plate format. (**a**) SARS-CoV-2 stock virus dilutions are prepared by transferring 100 μL of stock virus to 900 μL of DMEM 2%FBS. Samples are then tenfold serially diluted until the desired dilution range is reached. (**b**) 100 μL of each dilution is plated onto confluent Vero E6 cells. (**c**) Representative image of a SARS-CoV-2 plaque assay in 24-well plate format. It is important to count wells that contain 10–20 plaques to ensure the accuracy of the final results. The result of (number of plaques* reciprocal of dilution) should be multiplied by 10 due to only plating one-tenth of the original dilution (100 μL). This will equal the number of plaque forming units per mL (pfu/mL)

10. After 72 h, gently aspirate the MCC overlay off the cells and fix the cells with 10% neutral buffered formalin for 1 h at room temperature.
11. After 1 h, remove the neutral buffered formalin and wash the wells with water to remove residual MCC (*see* **Note 17**).
12. To stain the plaques, add 0.5% Crystal violet staining solution to each well for 15 min, rinse with water, and dry the plates.
13. Quantify the plaques as plaque forming units per milliliter (pfu/mL; Fig. 1c).

3.4 Quantification of SARS-CoV-2 by Immunofocus Forming Assay

1. Plate 5 × 10^4 Vero E6 cells in DMEM/10% FBS into a 96-well plate the day before the assay.
2. All subsequent steps in this protocol should be performed within a biosafety level 3 (BSL3) laboratory unless otherwise indicated.
3. Prepare ten-fold serial dilutions of SARS-CoV-2 samples in DMEM/2% FBS. This is most easily accomplished by preparing the dilutions in a sterile 96-well plate for easy transfer (Fig. 2a).
4. Aspirate or flick off growth media from the cells and add 50 μL of serial dilutions to the Vero E6 cells.
5. Incubate plates at 37 °C/5% CO_2 for 1 h, gently rocking the plate every 15 min.
6. After the 1-h adsorption, add an equal volume of DMEM supplemented 2.4% MCC stock to all wells for a final concentration of 1.2% MCC overlay and gently tap the plate to mix.
7. Place plates at 37 °C/5% CO_2 for 24 h.
8. After 24 h, gently aspirate the MCC overlay and fill the wells with 10% neutral buffered formalin. Incubate at room temperature for 1 h to fix the monolayers (*see* **Note 18**).
9. After 1 h, remove the neutral buffered formalin and gently rinse the wells with deionized water three times to remove any residual MCC overlay (*see* **Note 19**).
10. When all plates have been washed, flick any remaining water out of the well and gently tap on paper towels to ensure all water has been removed.
11. To permeabilize the cells and block endogenous peroxidase activity, add 30 μL of ice-cold methanol supplemented with 0.3% hydrogen peroxide to each well and incubate at −20°C for 10 min, and then transfer to an orbital shaker at room temperature for another 20 min.
12. After incubation in methanol–hydrogen peroxide, flick off the methanol into an appropriate waste container and gently wash all plates once with deionized water.

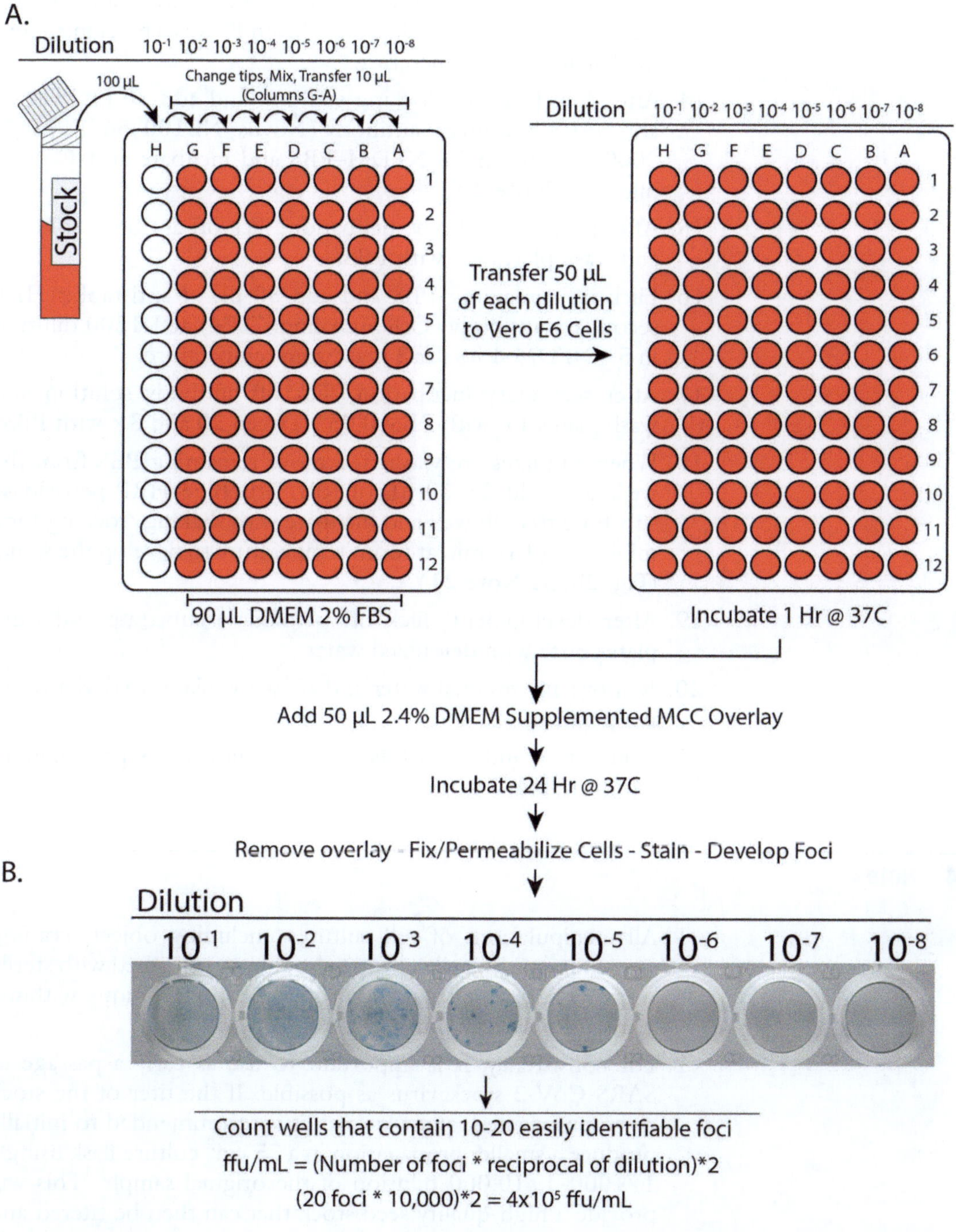

Fig. 2 SARS-CoV-2 immunofocus forming assay in 96-well format. (**a**) SARS-CoV-2 stock virus dilutions are prepared by transferring 100 μL of stock virus to the first well of a 96-well plate. Samples are then ten-fold serially diluted across all columns of the 96 well plate. 50 μL of each dilution is plated onto confluent Vero E6 cells in a 96-well plate. (**b**) Representative image of a SARS-CoV-2 immunofocus forming assay in 96-well plate format after staining with rabbit anti-SARS-CoV-2 N antibody and anti-rabbit HRP secondary antibody and developing using the KPL HRP peroxidase TrueBlue substrate. It is important to count wells that contain

13. Once washed, remove the water from the wells and add 30 μL/well of 5% nonfat dry milk (NFDM) in PBS as a blocking agent for 1 h at room temperature.
14. After 1 h, flick off blocking solution and add 30 μL/well of anti-SARS N primary antibody (Novus; NB100–56576) at 1:2000 dilution in 5% NFDM–PBS and incubate at 4°C overnight (*see* **Note 20**).
15. After primary antibody incubation, flick off antibody solution and wash plates 4× with PBS.
16. Flick off remaining PBS and add 30 μL of anti-rabbit HRP secondary antibody (Cell Signaling; 7074) at 1:1500 dilution in 5% NFDM–PBS for 2 h at room temperature.
17. After secondary incubation, flick off antibody solution and wash plates 1× with PBS–0.05% Tween 20 and 3× with PBS.
18. When all plates are washed, remove remaining PBS from the wells and add 25–30 μL of KPL TrueBlue HRP peroxidase substrate to each well and incubate with shaking/rocking for a minimum of 10 min at room temperature to develop the signal (Fig. 2b; *see* **Note 21**).
19. After development, flick off peroxidase substrate and rinse plates once with deionized water.
20. Remove any residual water and allow the plates to dry at room temperature.
21. Count foci and record as focus forming units per milliliter (ffu/mL; Fig. 2b).

4 Notes

1. All manipulations of cell cultures, including objects coming into contact with cell cultures, should be performed with sterile reagents and supplies using proper aseptic technique within a Class II biosafety cabinet.
2. For consistency, it is important to use as early a passage of SARS-CoV-2 stock virus as possible. If the titer of the stock to be propagated is not known, it is recommended to initially produce a smaller propagation in a 75 cm^2 culture flask using a 1/1000–1/10,000 dilution of the original sample. This will provide a high-quality seed stock that can then be titered and used to produce more stocks in the future.

Fig. 2 (continued) 10–20 foci to ensure the accuracy of the final results. The result of (number of foci* reciprocal of dilution) should be multiplied by two due to only plating half (50 μL) of the original dilution (100 μL). This will equal the number of focus forming units per mL (ffu/mL)

3. We have found that either 2% FBS or 0.3% BSA work equally well as culture media additives and have noticed no difference in overall cell health during culture or in final virus yield. However, if a virus stock propagation is intended for the purification of inactivated whole virions by ultracentrifugation, we recommend 0.3% BSA as we have noticed the purified virus pellets are significantly cleaner than with 2% FBS. DMEM supplemented with 2% FBS or 0.3% BSA can also be used to successfully propagate SARS-CoV-2 stock virus in Vero E6 cells to high titers, but we have found that Opti-MEM tends to result in more reliable titers from batch to batch.
4. We have found this primary antibody to consistently work well in the SARS-CoV-2 immunofocus forming assay. It is possible that other antibodies could work as well as the one referenced here, although potential troubleshooting may be required. Other studies have shown that anti-spike antibodies work as well in similar assays [12].
5. We have found that TrueBlue performs very well and results in minimal background staining, even with long development times (>30 min). It is likely that other colorimetric HRP peroxidase substrates will also work in this assay.
6. These are the splitting regimes and sub-cultivation ratios that we have found to work best for each cell line. For Caco-2 and Calu3 cells it is recommended to have several flasks of each since these cells grow slowly and the sub-cultivation ratios required for optimal growth do not result in large yields for experiments.
7. It is generally advisable to use low multiplicities of infection (MOI) when preparing virus stocks. Repeated passage at high MOI can result in accumulation of defective-interfering virus particles in the stock. Based on the titer of the stock virus, calculate the volume of virus needed to achieve the desired multiplicity of infection (infectious units/cell). Dilute this in the appropriate volume for inoculation.

$$\text{Virus stock } (\mu\text{L}) = \left(\frac{(\text{Cell number} \div \text{inoculum volume (mL)}) \times \text{MOI}}{\text{Virus Titer } \left(\frac{\text{pfu}}{\text{mL}}\right)}\right) \times 1000$$

8. It is important when performing titrations to quantify virus stocks to ensure that all stock aliquots have been treated equally. Because the virus aliquots are stored at −80 °C, they will be subjected to a thawing step at the time of use. Therefore, we recommend that the titer of the virus stock be determined after a single freeze-thaw cycle. To ensure the accuracy

of the MOI being used, only use virus stock that has been frozen once.

9. The empty tissue culture grade water bottle can be saved and used to store the 2.4% MCC overlay stock once it is prepared.
10. MCC will initially clump when added to water and does not disperse well. This is normal and following stirring for 30 min–1 h it should be a very smooth opaque white suspension, resembling a vanilla milkshake. While stirring the MCC, the mixture will become quite thick and adjustments to the stirring speed may be required, but stir speeds should be maintained as fast as possible for your stir plate/stir bar.
11. Alternatively, if 2× DMEM is available, it is not necessary to add powdered DMEM or antibiotics to the 2.4% MCC stock. 2× DMEM can be mixed directly with 2.4% MCC stock and sterile water at a 2:1:1 ratio respectively to reach the working 0.6% MCC concentration. Unsupplemented 2.4% MCC in water is indefinitely shelf stable at room temperature as long as it is kept sterile. It is also not necessary to add serum or additional L-glutamine to the DMEM supplemented 2.4% MCC stock.
12. When planning dilutions for plaque assays, it is important that the first dilution takes into account the plating volume. For example, if the first dilution is 1/10 and only 0.1 mL of that dilution is plated, the final effective dilution is 1/100 or 10^{-2}. This will allow for accurate calculation of the virus titers which are typically recorded as plaque forming units per milliliter (pfu/mL).
13. With SARS-CoV-2, we have found that the most useful range of dilutions to plate for unknown samples is 10^{-2}–10^{-7}. If titering samples from cell lines that do not propagate SARS-CoV-2 to high titers, such as early timepoints in CaCo2 infections (≤24 h post-infection), we recommend plating 10^{-1} through 10^{-6} dilutions.
14. As mentioned in **Note 10**, if 2× DMEM media is available for the purposes of making the working 0.6% MCC overlay there is no need to supplement the MCC with powdered DMEM media. With 2× DMEM media, to make 100 mL of the 0.6% MCC overlay, add 25 mL of 2.4% MCC in water to 25 mL of sterile tissue culture water and 50 mL of 2× DMEM 4% FBS. This will result in a 0.6% MCC overlay in 1× DMEM 2% FBS.
15. After preparing the 0.6% MCC overlay, the MCC may begin to separate out of the media. It is important to shake the 0.6% MCC overlay suspension well to ensure it is homogenous before adding to the cells.

16. It is not necessary to remove the inoculum from the cells. As long as all plaque assays are performed the same way in a given experiment, we have observed no ill-effects from not removing the inoculum.
17. It is important to wash the plates with water as the methanol in the 0.5% crystal violet stain will cause any residual cellulose to melt onto the monolayer.
18. 10% neutral buffered formalin effectively inactivates SARS-CoV-2 [9, 15, 16]. If inactivation by this method has been validated and approved per institutional guidelines, the remaining steps of this protocol can be accomplished at BSL2.
19. With all wash steps in this protocol, if you are processing multiple plates, it is recommended to leave the last wash in the plate until all plates are processed. Once all of the plates have been washed the requisite number of times, the last wash can be removed from each plate as you move it to the next step to ensure the monolayer does not dry out between steps.
20. While not strictly necessary, an additional 1–3 h incubation at room temperature on an orbital shaker after the overnight 4 °C incubation increases signal during development of the assay.
21. Small blue foci will begin to form within the wells after about 10–15 min. Holding a white piece of paper behind the plate assists in seeing the foci clearly. Once visible, incubate for an additional 10–15 min before continuing to the next step. If using the KPL TrueBlue peroxidase substrate, incubation time is not critical as this substrate does not produce strong background staining. Incubate the plates as long as necessary for the foci to develop.

References

1. Holshue ML, DeBolt C, Lindquist S, Lofy KH, Wiesman J, Bruce H, Spitters C, Ericson K, Wilkerson S, Tural A, Diaz G, Cohn A, Fox L, Patel A, Gerber SI, Kim L, Tong S, Lu X, Lindstrom S, Pallansch MA, Weldon WC, Biggs HM, Uyeki TM, Pillai SK, Washington State -nCo VCIT (2020) First case of 2019 novel coronavirus in the United States. N Engl J Med 382(10):929–936. https://doi.org/10.1056/NEJMoa2001191
2. Zhu N, Zhang D, Wang W, Li X, Yang B, Song J, Zhao X, Huang B, Shi W, Lu R, Niu P, Zhan F, Ma X, Wang D, Xu W, Wu G, Gao GF, Tan W, China Novel Coronavirus I, Research T (2020) A novel coronavirus from patients with pneumonia in China, 2019. N Engl J Med 382(8):727–733. https://doi.org/10.1056/NEJMoa2001017
3. Weekly Epidemiological Update (2021, 5 January) World Health Organization. Switzerland, Geneva
4. Hoffmann M, Kleine-Weber H, Schroeder S, Kruger N, Herrler T, Erichsen S, Schiergens TS, Herrler G, Wu NH, Nitsche A, Muller MA, Drosten C, Pohlmann S (2020) SARS-CoV-2 cell entry depends on ACE2 and TMPRSS2 and is blocked by a clinically proven protease inhibitor. Cell 181(2):271–280. e278. https://doi.org/10.1016/j.cell.2020.02.052
5. Shang J, Ye G, Shi K, Wan Y, Luo C, Aihara H, Geng Q, Auerbach A, Li F (2020) Structural basis of receptor recognition by SARS-CoV-2. Nature 581(7807):221–224. https://doi.org/10.1038/s41586-020-2179-y

6. Wang Q, Zhang Y, Wu L, Niu S, Song C, Zhang Z, Lu G, Qiao C, Hu Y, Yuen KY, Wang Q, Zhou H, Yan J, Qi J (2020) Structural and functional basis of SARS-CoV-2 entry by using human ACE2. Cell 181(4):894–904. e899. https://doi.org/10.1016/j.cell.2020.03.045
7. Yan R, Zhang Y, Li Y, Xia L, Guo Y, Zhou Q (2020) Structural basis for the recognition of SARS-CoV-2 by full-length human ACE2. Science 367(6485):1444–1448. https://doi.org/10.1126/science.abb2762
8. Chu H, Chan JF, Yuen TT, Shuai H, Yuan S, Wang Y, Hu B, Yip CC, Tsang JO, Huang X, Chai Y, Yang D, Hou Y, Chik KK, Zhang X, Fung AY, Tsoi HW, Cai JP, Chan WM, Ip JD, Chu AW, Zhou J, Lung DC, Kok KH, To KK, Tsang OT, Chan KH, Yuen KY (2020) Comparative tropism, replication kinetics, and cell damage profiling of SARS-CoV-2 and SARS-CoV with implications for clinical manifestations, transmissibility, and laboratory studies of COVID-19: an observational study. Lancet Microbe 1(1):e14–e23. https://doi.org/10.1016/S2666-5247(20)30004-5
9. Jureka AS, Silvas JA, Basler CF (2020) Propagation, inactivation, and safety testing of SARS-CoV-2. Viruses 12(6). https://doi.org/10.3390/v12060622
10. Harcourt J, Tamin A, Lu X, Kamili S, Sakthivel SK, Murray J, Queen K, Tao Y, Paden CR, Zhang J, Li Y, Uehara A, Wang H, Goldsmith C, Bullock HA, Wang L, Whitaker B, Lynch B, Gautam R, Schindewolf C, Lokugamage KG, Scharton D, Plante JA, Mirchandani D, Widen SG, Narayanan K, Makino S, Ksiazek TG, Plante KS, Weaver SC, Lindstrom S, Tong S, Menachery VD, Thornburg NJ (2020) Isolation and characterization of SARS-CoV-2 from the first US COVID-19 patient. bioRxiv. https://doi.org/10.1101/2020.03.02.972935
11. Case JB, Bailey AL, Kim AS, Chen RE, Diamond MS (2020) Growth, detection, quantification, and inactivation of SARS-CoV-2. Virology 548:39–48. https://doi.org/10.1016/j.virol.2020.05.015
12. Case JB, Rothlauf PW, Chen RE, Liu Z, Zhao H, Kim AS, Bloyet LM, Zeng Q, Tahan S, Droit L, Ilagan MXG, Tartell MA, Amarasinghe G, Henderson JP, Miersch S, Ustav M, Sidhu S, Virgin HW, Wang D, Ding S, Corti D, Theel ES, Fremont DH, Diamond MS, Whelan SPJ (2020) Neutralizing antibody and soluble ACE2 inhibition of a replication-competent VSV-SARS-CoV-2 and a clinical isolate of SARS-CoV-2. Cell Host Microbe 28(3):475–485. e475. https://doi.org/10.1016/j.chom.2020.06.021
13. Gordon DE, Hiatt J, Bouhaddou M, Rezelj VV, Ulferts S, Braberg H, Jureka AS, Obernier K, Guo JZ, Batra J, Kaake RM, Weckstein AR, Owens TW, Gupta M, Pourmal S, Titus EW, Cakir M, Soucheray M, McGregor M, Cakir Z, Jang G, O'Meara MJ, Tummino TA, Zhang Z, Foussard H, Rojc A, Zhou Y, Kuchenov D, Huttenhain R, Xu J, Eckhardt M, Swaney DL, Fabius JM, Ummadi M, Tutuncuoglu B, Rathore U, Modak M, Haas P, Haas KM, ZZC N, Pulido EH, Shi Y, Barrio-Hernandez I, Memon D, Petsalaki E, Dunham A, Marrero MC, Burke D, Koh C, Vallet T, Silvas JA, Azumaya CM, Billesbolle C, Brilot AF, Campbell MG, Diallo A, Dickinson MS, Diwanji D, Herrera N, Hoppe N, Kratochvil HT, Liu Y, Merz GE, Moritz M, Nguyen HC, Nowotny C, Puchades C, Rizo AN, Schulze-Gahmen U, Smith AM, Sun M, Young ID, Zhao J, Asarnow D, Biel J, Bowen A, Braxton JR, Chen J, Chio CM, Chio US, Deshpande I, Doan L, Faust B, Flores S, Jin M, Kim K, Lam VL, Li F, Li J, Li YL, Li Y, Liu X, Lo M, Lopez KE, Melo AA, Moss FR 3rd, Nguyen P, Paulino J, Pawar KI, Peters JK, Pospiech TH Jr, Safari M, Sangwan S, Schaefer K, Thomas PV, Thwin AC, Trenker R, Tse E, TKM T, Wang F, Whitis N, Yu Z, Zhang K, Zhang Y, Zhou F, Saltzberg D, QSB C, Hodder AJ, Shun-Shion AS, Williams DM, White KM, Rosales R, Kehrer T, Miorin L, Moreno E, Patel AH, Rihn S, Khalid MM, Vallejo-Gracia-A, Fozouni P, Simoneau CR, Roth TL, Wu D, Karim MA, Ghoussaini M, Dunham I, Berardi F, Weigang S, Chazal M, Park J, Logue J, McGrath M, Weston S, Haupt R, Hastie CJ, Elliott M, Brown F, Burness KA, Reid E, Dorward M, Johnson C, Wilkinson SG, Geyer A, Giesel DM, Baillie C, Raggett S, Leech H, Toth R, Goodman N, Keough KC, Lind AL, Zoonomia C, Klesh RJ, Hemphill KR, Carlson-Stevermer J, Oki J, Holden K, Maures T, Pollard KS, Sali A, Agard DA, Cheng Y, Fraser JS, Frost A, Jura N, Kortemme T, Manglik A, Southworth DR, Stroud RM, Alessi DR, Davies P, Frieman MB, Ideker T, Abate C, Jouvenet N, Kochs G, Shoichet B, Ott M, Palmarini M, Shokat KM, Garcia-Sastre A, Rassen JA, Grosse R, Rosenberg OS, Verba KA, Basler CF, Vignuzzi M, Peden AA, Beltrao P, Krogan NJ (2020) Comparative host-coronavirus protein interaction networks reveal pan-viral disease mechanisms. Science 370(6521). https://doi.org/10.1126/science.abe9403

14. Pinto D, Park YJ, Beltramello M, Walls AC, Tortorici MA, Bianchi S, Jaconi S, Culap K, Zatta F, De Marco A, Peter A, Guarino B, Spreafico R, Cameroni E, Case JB, Chen RE, Havenar-Daughton C, Snell G, Telenti A, Virgin HW, Lanzavecchia A, Diamond MS, Fink K, Veesler D, Corti D (2020) Cross-neutralization of SARS-CoV-2 by a human monoclonal SARS-CoV antibody. Nature 583(7815):290–295. https://doi.org/10.1038/s41586-020-2349-y
15. Patterson EI, Prince T, Anderson ER, Casas-Sanchez A, Smith SL, Cansado-Utrilla C, Solomon T, Griffiths MJ, Acosta-Serrano A, Turtle L, Hughes GL (2020) Methods of inactivation of SARS-CoV-2 for downstream biological assays. J Infect Dis 222(9): 1462–1467. https://doi.org/10.1093/infdis/jiaa507
16. Welch SR, Davies KA, Buczkowski H, Hettiarachchi N, Green N, Arnold U, Jones M, Hannah MJ, Evans R, Burton C, Burton JE, Guiver M, Cane PA, Woodford N, Bruce CB, Roberts ADG, Killip MJ (2020) Analysis of inactivation of SARS-CoV-2 by specimen transport media, nucleic acid extraction reagents, detergents, and fixatives. J Clin Microbiol 58(11). https://doi.org/10.1128/JCM.01713-20

Chapter 9

Quantification of Infectious SARS-CoV-2 by the 50% Tissue Culture Infectious Dose Endpoint Dilution Assay

C. Korin Bullen, Stephanie L. Davis, and Monika M. Looney

Abstract

A number of viral quantification methods are used to measure the concentration of infectious severe acute respiratory syndrome coronavirus 2 (SARS-CoV-2). While the traditional plaque-based assay allows for direct enumeration of replication competent lytic virions and remains the gold standard for the quantification of infectious virus, the 50% tissue culture infectious dose ($TCID_{50}$) endpoint dilution assay allows for a more rapid, large-scale analysis of experimental samples. In this chapter, we describe a well-established $TCID_{50}$ assay protocol to measure the SARS-CoV-2 infectious titer in viral stocks, in vitro cell or organoid models, and animal tissue. We also present alternative assays for scoring the cytopathic effect of SARS-CoV-2 in cell culture and comparable methods to calculate the 50% endpoint by serial dilution.

Key words SARS-CoV-2, Virus quantification, $TCID_{50}$, Cytopathic effects (CPE), Cell culture, Cell viability assay, Reed and Muench, Trimmed Spearman, Karber

1 Introduction

Virus quantification assays of infectious SARS-CoV-2 particles are essential for comparative tropism, replication kinetics and therapeutic studies in cell culture and animal models [1, 2]. As SARS-CoV-2 preclinical and basic research continues to surge, validated and consistent protocols are necessary to ensure reproducibility and accuracy. This is particularly true when quantifying infectious viral particles, where individual cell culture and dilution techniques may result in varying values. While quantitative real-time polymerase chain reaction (RT-qPCR) is a sensitive and rapid method for quantifying viral genome copies, RT-qPCR also amplifies and detects defective virions and particle-free nucleic acids, resulting in higher values than the actual number of replication competent virus particles [3–7]. Two common methods for reliable quantification of infectious virions are the plaque-forming assay (PFA), which relies on visualization and direct counting of small discrete

Justin Jang Hann Chu et al. (eds.), *SARS-CoV-2: Methods and Protocols*,
Methods in Molecular Biology, vol. 2452, https://doi.org/10.1007/978-1-0716-2111-0_9,

areas of cell death, plaques, that form upon spread of virus progeny from one cell infected by a single virion [8], and the 50% tissue culture infectious dose ($TCID_{50}$) endpoint dilution assay, which utilizes assessment of cytopathic effects (CPE) in host cells infected in vitro with varying dilutions of virus to back calculate the viral concentration or titer in a given sample [6]. While these methods are restricted to viruses that induces visible cell damage or death, SARS-CoV-2 infection induces substantial cell death in the nonhuman primate kidney VeroE6 cell line, and thus the PFA and $TCID_{50}$ assay are well suited to reliably quantify infectious SARS-CoV-2 titers [1, 2, 9, 10]. Although traditionally low throughput, these assays are a key tool to study new and emerging viral pathogens as they do not require virus-specific reagents such as antibodies or target probes, both reducing cost and enabling viral quantification when no such reagents are available.

The viral plaque assay is one of the most widely used techniques to isolate a clonal population of virus or to determine infectious viral titers by counting plaques formed in a monolayer of host cells [11, 12]. This assay requires a densely confluent monolayer of cells, typically plated in 6- or 12-well plates, to ensure a sufficient surface area for plaques to form. The low-throughput format greatly prolongs the time required for liquid handling and dilution preparation and increases the number of plates required to evaluate large sample sets, significantly reducing sample throughput. In addition, monolayer immobilization with agar and manual counting of visual plaques are both laborious and subjected to variability [13]. However, a major advantage of the plaque assay method is the ability to generate a quantitative and precise measurement of replication competent virions.

The $TCID_{50}$ assay offers another reliable, yet higher-throughput, cell-based approach to quantify infectious virus particles. This method provides a qualitative measurement of the ratio between uninfected and infected cells at varying dilutions of virus [14, 15]. In this assay, seven serial dilutions of a SARS-CoV-2 sample are added to Vero-E6 cells in six replicates on a 96-well plate format (Fig. 1). After a period of 2–5 days, CPE or cell death is assessed by crystal violet staining or cell viability/cytotoxicity microplate assays, enabling each well to be scored as "nonviable" or "viable," representing infected or uninfected cells respectively. Virus quantity, often expressed as $TCID_{50}$/mL, represents the concentration of virus required to induce CPE, or death in 50% of inoculated host cell.

Although the $TCID_{50}$ assay is not a direct quantitative measurement of the virus, this method offers critical advantages over the plaque assay, in that it allows more rapid, large-scale analysis of SARS-CoV-2 infected experimental samples. The 96-well plate format not only allows fewer plates per sample but also enables the use of automated multichannel pipettes for more efficient cell plating, serial dilutions preparation, washing, and inoculum

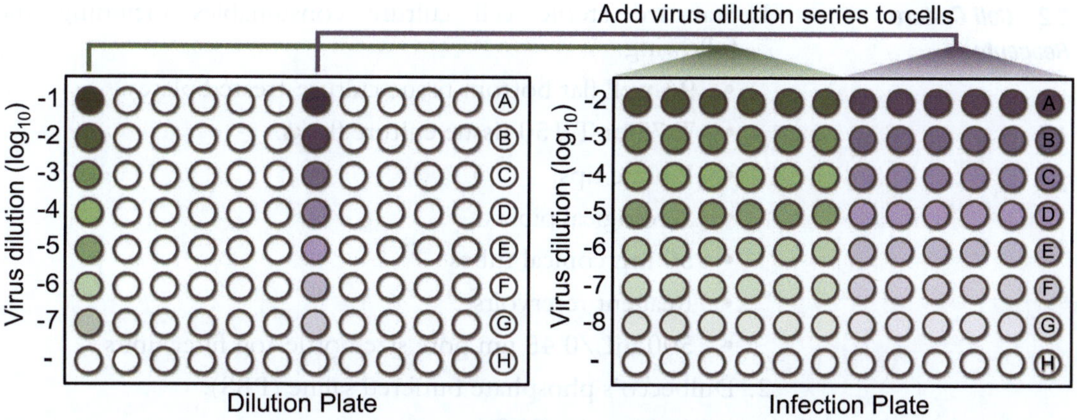

Fig. 1 TCID$_{50}$ end point dilution assay for infectious SARS-CoV-2 titer quantification. Schematic diagram of serial dilution of and subsequent inoculation of cells with virus from two separate sample replicates (green and purple). Each sample containing SARS-Cov-2 is first serially diluted down one column of a 96-well "dilution plate" from row A to G (tenfold dilution series for this example). One set of serially diluted sample is then added in six technical replicates to Vero-E6 cells in a separate 96-well "infection plate" prepared the day prior, resulting in an additional tenfold-dilution of each sample as indicated on the Y axis above. No virus is added to row H for an uninfected control

addition. The TCID$_{50}$ assay has been shown to be more sensitive than the plaque assay [6], and TCID$_{50}$ shares a theoretical relationship with a plaque forming unit (PFU) based on the Poisson distribution, where 1 PFU = ~(0.7)*TCID$_{50}$ [16]. This relationship is derived by applying the Poisson distribution, P(0) = e(−m), where P(0) is the proportion of negative wells, and m is the mean number of plaque-forming units per volume (PFU/mL). Since P (0) = 0.5 when 50% of host cells or animals killed by the virus, e (−m) = 0.5, which can be rearranged to m = −ln 0.5 = ~0.7. An approximate PFU can therefore be calculated from TCID$_{50}$, and vice versa.

Here we describe a TCID$_{50}$ assay protocol for titering infec tious SARS-CoV-2 stocks and experimental samples, including infected animal tissue specimens. In addition, we present the two main mathematical methods for calculating TCID$_{50}$ and a comparison of two methods of scoring CPE.

2 Materials

2.1 Cell Lines

1. Tissue culture cell lines that are both susceptible to SARS-CoV-2 infection and show viral cytopathic effects such as Huh-7.5 human hepatoma cells or Vero African green monkey kidney cells (Vero-E6 or Vero CCL-81 ATCC) [2, 10]. The TMPRSS2-expressing VeroE6 cell line is also highly susceptible to SARS-CoV-2 infection and shows a greater number of SARS-CoV-2–infected cells compared the parental VeroE6 cells [17], resulting in slightly higher TCID^{50} values.

2.2 Cell Culture Reagents

1. Standard sterile cell culture consumables including the following.
   - 96-well flat bottom tissue culture-treated plates.
   - T-75 or T-150 tissue culture flasks.
   - Pipette tips.
   - Serological pipets.
   - 50 mL conical tubes
   - Reagent reservoirs.
   - 500 mL/0.45 μm pore size bottle-top filter units
2. Dulbecco's phosphate buffered saline (PBS).
3. 0.25% Trypsin–EDTA.
4. Complete growth media for Vero-E6 cells.
   - Dulbecco's Modified Eagle Medium with low glucose (1 g/L).
   - 10% heat-inactivated fetal bovine serum.
   - 2 mM L-glutamine.
   - 1 mM sodium pyruvate.
   - 100 U/mL penicillin and 100 μg/mL streptomycin.
5. Low-serum infection media for Vero-E6 cells.
   - Dulbecco's Modified Eagle Medium with low glucose (1 g/L).
   - 2% heat-inactivated fetal bovine serum.
   - 2 mM L-glutamine.
   - 1 mM sodium pyruvate.
   - 100 U/mL penicillin and 100 μg/mL streptomycin.
6. Hemocytometer or automated cell counter.
7. Incubator for cell culture set at 37 °C and 5% CO_2.

2.3 Virus Isolation From Infected Animal Tissue

1. Sterile 2 mL Screw-Cap Micro Tube with O-Ring Seal filled with sterile ceramic 1.4 or 2.8 mm beads.
2. Bead beating tissue homogenizer instrument such as Precellys Evolution or FastPrep-24.
3. Virus Titer Buffer.
   - Low-serum infection media (see Subheading 2.2, **item 5**).
   - 200 U/mL penicillin and 200 μg/mL streptomycin
   - 100 μg/mL gentamicin
   - 0.5 μg/mL amphotericin B.

2.4 Virus Sample Dilution and Vero-E6 Infection

1. 96-well round bottom non–tissue culture-treated plates for serial dilution of samples.
2. Low-serum infection media for Vero-E6 cells (see Subheading 2.2, **item 5**).

2.5 Cell Viability Assays

2.5.1 Crystal Violet Staining

1. Fixation Buffer: 4% paraformaldehyde solution in PBS.
2. Crystal violet solution: crystal violet (0.05% w/v) and 20% methanol in deionized water.
3. Deionized water for washing.
4. Optional: white-light transilluminator (light box) to aid in visualizing stained cells.

2.5.2 Cell Viability/Cytotoxicity Microplate Assay

1. Commercially available colorimetric, fluorescence, or bioluminescence-based cell viability or cytotoxicity microplate assay kits, such as the CellTiter-Glo® Luminescent Cell Viability Assay (Promega, Cat # G9241), as exemplified in this protocol.
2. 96-well opaque-walled plate specified by the microplate assay manufacture protocol.
3. Microplate shaker (required for the CellTiter-Glo® Luminescent Cell Viability Assay).
4. Microplate reader capable of UV-visible absorbance, fluorescence and/or luminescence detection like the BMG FLUOstar® Omega or equivalent.

2.6 Calculating $TCID_{50}$

1. Reed and Meunch or Spearman and Kaerber $TCID_{50}$ calculator programs using widely available software (e.g., Microsoft Excel) [18, 19].

3 Methods

The World Health Organization biosafety guidance recommends SARS-CoV-2 isolation, propagation, and experimental inoculation of cells be conducted in a Biosafety Level 3 (BSL-3) laboratory using BSL-3 practices [20]. Working with SARS-CoV-2 in laboratory animal models requires an Animal BSL-3 facility. Follow institution and facility specific Standard Operating Procedures (SOP) when handling high concentrations of live virus and inoculating host cells and animals.

A schematic diagram of serial dilution of virus and subsequent infection of cells is shown in Fig. 1.

3.1 Seeding and Preparation Cells for SARS-CoV-2 Inoculation

1. Maintain Vero-E6 cells in T-175 cell culture flasks using complete growth media in a cell culture incubator set at 37 °C, 5% CO_2. See Current Protocols in Cell Biology for more detailed basic cell culture methods [21].
2. One day prior to virus inoculation, aspirate media from a confluent T175 flask of Vero-E6 cells and wash once with 10 mL of PBS (*see* **Note 1**).
3. Aspirate the PBS and add 2–3 mL of 0.25% trypsin–EDTA and incubate for 3–4 min at 37 °C, 5% CO_2 to detach the adherent cells.
4. Add 10 mL of complete growth media to each flask to neutralize trypsin. Resuspend cells by pipetting up and down to break up cell clumps and transfer cell solution to a 50 mL conical tube (*see* **Note 2**).
5. Centrifuge cells at 200 × *g* for 5 min at 25 °C.
6. Remove supernatant and resuspend cells in 10 mL of complete growth media.
7. Count cells using a hemocytometer or automated cell counter. Prepare a cell suspension to a concentration of 1×10^5 cells/mL in complete growth media.
8. For each virus stock or experimental sample to be titered, seed cells on a 96-well flat bottom tissue culture-treated plate, herein termed "infection plate." Plate cells in 6 columns × 8 rows (one half of the infection plate) for each replicate dilution series of virus sample.
9. Incubate the Vero-E6 cell infection plates at 37 °C, 5% CO_2 for 18–24 h.
10. Examine the cells by light microscopy 18–24 h after seeding. Proceed with the SARS-CoV-2 infection when the cells have reached a minimum of 80% confluence. The optimal density of Vero-E6 cells for this protocol is 90% confluence.
11. Remove the media from the Vero-E6 cells seeded the previous day and wash once with 100–200 μL of PBS per well. Pipet gently to ensure the cells do not lift off of the plate.
12. Remove the PBS wash and add 180 μL of viral infection media to each well (*see* Subheading 2.2, **item 5**).
13. Incubate the cells at 37 °C, 5% CO_2 while preparing the dilution series for each virus stock or experimental sample to be titered (see Subheading 3.3).

3.2 Preparing Virus-Containing Homogenate from Animal Tissue

The $TCID_{50}$ assay can also be employed to quantify the amount of infectious SARS-CoV-2 present in tissue from experimental animal models. This application uses the same methodology as described herein for measuring SARS-CoV-2 in viral stocks or cell culture specimens; however, a suspension of virus must first be extracted from the animal tissue or organ without diminishing infectivity. Our method for preparing infectious virus-containing homogenate from animal tissue is described below.

1. Weigh a sterile 2 mL tube, preferably with an O-ring, containing ceramic 2.8 mm beads. Place an animal tissue section of interest inside the tube. Weigh the tube again and subtract the new weight from the original weight to calculate the weight of the tissue.
2. Snap freeze the weighed tissue on dry ice and store at −80 °C (*see* **Note 3**).
3. When ready to process the tissue for viral quantification, remove the samples from the freezer and place on ice (*see* **Note 4**).
4. Immediately add ice-cold virus titer buffer at a 10% weight to volume ratio to each tube (*see* Subheading 2.3, **item 3** and **Note 5**).
5. Homogenize the tissue for 2 cycles of 20 s at a speed setting of 5000 rpm using a Precelley's Evolution homogenizer. This setting instrument dependent and should be converted to an equivalent speed according to the manufacture.
6. Incubate on ice for 2 min between homogenization cycles.
7. Centrifuge tubes containing the tissue homogenate at 10,000 × *g* for 1 min at 4 °C to pellet the beads and tissue debris.
8. Transfer the supernatant to a new 2 mL microcentrifuge O-ring tube.
9. Store samples on ice and proceed immediately to Subheading 3.3 to avoid decreases in virus infectivity.

3.3 Preparing Dilution Series of Virus Samples

1. In a 96-well round bottom plate, herein termed "dilution plate," add 180 μL of infection media to one column (rows A-H) per virus stock or experimental sample to be titered. This plate will be used to generate serial dilutions (Fig. 1).
2. Thaw samples to be titered on ice (*see* **Note 6**).
3. Thoroughly mix each sample by pipetting up and down multiple times. Do not vortex.
4. For a ten-fold dilution series, add 20 μL of neat sample to row A of one column on the dilution plate to create a 10^{-1} dilution of the sample. To generate technical replicates of each dilution series, add 20 μL of the same neat sample to row A of two or more columns on the dilution plate (Fig. 1; *see* **Note 7**).

5. Serially dilute each sample starting in row A (10^{-1} dilution) down to row G (10^{-7}dilution) within the same column of wells (*see* **Note 8**).
6. Proceed immediately to Subheading 3.4 to avoid decreases in virus infectivity.

3.4 Host Cell Inoculation with Dilution Series of Virus samples

1. Using a multichannel pipet, transfer 20 μL from one column of serially diluted sample on the dilution plate to the corresponding columns of VeroE6 cells on the infection plate as prepared above in Subheading 3.3 (Fig. 1).
2. For each dilution series in one column on the dilution plate, repeat **step 1** above such that the Vero-E6 host cells are inoculated with each dilution (10^{-2} to 10^{-8}) in six replicates.
3. Row H of the infection plate will serve as a no infection control.
4. Incubate plates for 3–5 days at 37 °C, 5% CO_2 (*see* **Note 9**).

3.5 Scoring Host Cells for Viral Induced Cytopathic Effects

A number of methods for determining viral induced CPE, both quantitative and qualitative, have been evaluated for reliability, sensitivity, and reproducibility. Traditional methods include microscopic observations of CPE and cell death through various cell-staining methods such as naphthol blue-black, neutral red, or crystal violet reagents. Colorimetric, fluorescence, or bioluminescence cell-based microplate assays have also been applied to rapidly measure cell viability or cytotoxicity as an indicator of viral infection [16, 22–25]. The rapid measurement of light output from these assays using a microplate reader great reduces the time-consuming and subjective manual assessment of cell health. While colorimetric and luminescent cell viability assays show increased sensitivity compared to conventional manual-based $TCID_{50}$ methods, they have been well optimized and validated for signal endpoint detection of CPE for a variety of viruses and have been reported to be within range of within close range of titers obtained by the plaque assay [16, 22–27].

The CellTiter-Glo® Luminescent Cell Viability Assay (Promega), which produces a luminescent signal directly proportional to the amount of adenosine triphosphate (ATP) and thus the number of cells present in a culture, was shown to be a superior to Neutral-Red cell viability staining for high-throughput quantification of CPE in Vero-E6 cells infected with the original SARS coronavirus (SARS-CoV) [22]. Importantly, this assay requires no washing, minimal pipetting steps, and allows for the processing of multiple plates in under 1 h, which greatly reduces the materials and time in a BSL-3 environment. However, cell-staining methods for microscopic enumeration of CPE are considerably less expensive and do not require a microplate reader instrument in a BSL-3 facility. In effort to determine the correlation between the results

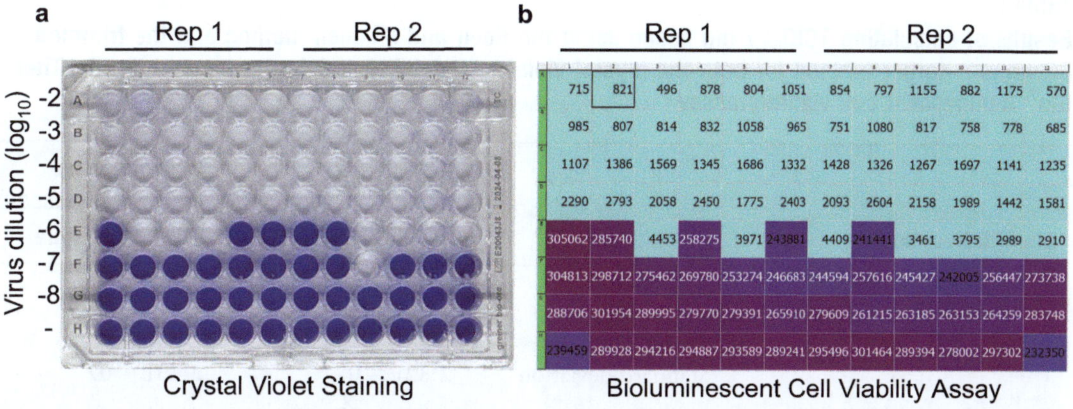

Fig. 2 Scoring host cells for viral-induced cytopathic effects. CPE enumeration of Vero-E6 cells inoculated for 5 days with serial dilutions of a lung homogenate sample from a Syrian Golden Hamster infected with SARS-Cov-2 by (**a**) crystal violet staining or (**b**) CellTiter-Glo® Luminescent Cell Viability Assay (Promega). The plates are scored for the number of wells per dilution with (+) or without (–) viral induced cytopathic effects as visualized by well color: (+) clear or (–) purple **(a)** or as determined by raw luminescence values: (+) < mean – 5 times standard deviation of noninfected controls or (–) > mean – 5 times standard deviation of noninfected controls **(b)**

of each method for measuring CPE in Vero-E6 cells, we performed crystal violet cell viability staining and the CellTiter-Glo® Luminescent Cell Viability Assay in parallel using the same lung homogenate sample from a Syrian Golden Hamster infected with SARS-CoV-2 (Fig. 2a, b). The luminescent cell-based assay was more time-efficient, but slightly more variable than crystal violet viability staining. Both assays produced comparable numerical results as determined by both the Reed and Muench and the Spearman-Karber methods for enumerating TCID$_{50}$ (Table 1).

3.5.1 Crystal Violet Cell Viability Staining Assay

1. Five days after virus inoculation (*see* Subheading 3.4), remove the media from the Vero-E6 cells and add 100 μL of 4% paraformaldehyde solution (PFA) per well.
2. Incubate for 20 min at 20–25 °C.
3. Remove the PFA and add 100 μL of crystal violet solution per well.
4. Incubate for 30 min at 20 to 25 °C.
5. Remove the crystal violet solution and wash cells twice with 100–200 μL of deionized water per well.
6. Remove all of the water from each well, invert the plate and pat dry on a paper towel.
7. Score each plate for the number of wells per dilution that are positive or negative for SARS-CoV-2 induced cell death as visualized by well color: clear (+) or purple (–) (Fig. 2a; *see* **Note 10**).

Table 1
Results of calculating $TCID_{50}$ / mg tissue using the Reed and Muench method and the trimmed Spearman–Karber method for both the crystal violet cell viability staining assay and the CellTiter-Glo® luminescent cell viability assay

Quantitative method	Sample	Cell viability staining assay TCID50 / g tissue	Luminescence cell viability assay TCID50/g tissue
Reed and Muench	Replicate 1	5.00E + 07	2.81E + 07
	Replicate 2	1.16E + 08	1.26E + 08
	Mean	8.28E + 07	7.69E + 07
	Standard deviation	2.32E + 07	3.45E + 07
Trimmed Spearman–Karber	Replicate 1	5.00E + 07	3.41E + 07
	Replicate 2	1.08E + 08	1.08E + 08
	Mean	7.89E + 07	7.09E + 07
	Standard deviation	2.04E + 07	2.60E + 07

Datasets used for calculations are shown in Fig. 2a

3.5.2 Luminescence Cell Viability Assay

1. Five days after virus inoculation, (*see* Subheading 3.4), remove 100 μL of media from the Vero-E6 cells, leaving approximately 100 μL of media remaining in each well.
2. Follow the manufactures instructions for the CellTiter-Glo® Luminescent Cell Viability Assay (Promega, Cat # G9241) or an alternate cell viability microplate assay of choice.
3. Record raw luminescence values using a microplate reader in luminescent detection mode (*see* **Note 11**).
4. Score each plate for the number of wells per dilution that are positive or negative for SARS-CoV-2 induced cell death as determined by raw luminescence values (Fig. 2b; *see* **Note 12**).

3.6 Methods for Calculating the $TCID_{50}$

The two most common methods for calculating $TCID_{50}$ are the Reed and Muench method and the Spearman–Karber method [15, 28–31]. The Reed and Muench method has a slightly lower limit of quantification, namely the lowest dilution that shows CPE in >50% of the wells, while the Spearman–Karber method's limit of quantification is the lowest viral dilution at which all the wells show CPE. In Table 1, we show both calculation methods produce comparable results when using the same CPE scoring data. Both approaches can be calculated either by hand or using widely available software (e.g., Microsoft Excel) found online [18, 19].

3.6.1 Reed and Muench Method

The Reed and Muench method uses the proportionate distance, also known as the interpolated value, to calculate a difference in logarithms so that [31]:

Table 2
Table setup for quantifying TCID$_{50}$ using the Reed and Muench Method

	Well score		Cumulative totals			
Log$_{10}$ virus dilution	CPE (+)	Viable (-)	CPE (+)	Viable (-)	Total	Percent CPE+ (%)
−2	6	0	29	0	29	100.0
−3	6	0	23	0	23	100.0
−4	6	0	17	0	17	100.0
−5	6	0	11	0	11	100.0
−6	4	2	5	2	7	71.43
−7	1	5	1	7	8	12.50
−8	0	6	0	13	13	0.0

Dataset used for calculations are shown in Fig. 2a, replicate 2

$D = [(X_G)-\%]/[(X_G)-(X_L)]$, where.
$D =$ the difference of logarithms;
$X_G =$ mortality at dilution next above %; and.
$X_L =$ mortality at dilution next below %.
TCID$_{50}$ is then calculated as follows:
Log$_{10}$% end point dilution = log$_{10}$ of X−(D*F), where.
$X =$ dilution showing a mortality next above %; and.
$F =$ log$_{10}$ of dilution factor.
$TCID_{50} = 10^{(-\log 10\% \text{ end point dilution})}$

End point dilution CPE data from a lung homogenate sample from a Syrian Golden Hamster infected with SARS-Cov-2 is used as an example in the following section (Fig. 2a, Table 2).

1. Calculate the cumulative total number of wells that scored positive for CPE (dead cells) for each dilution starting at 10^{-2} downward to 10^{-8}.
2. Calculate the cumulative total number of wells that scored negative for CPE (viable cells) for each dilution starting at 10^{-8} upward to 10^{-2}.
3. Add the cumulative total number of CPE (+) and Viable (−) wells for each row of the table to give the final cumulative total for each dilution.
4. Calculate the percent of wells that scored positive for CPE for each dilution.

   % Mortality = $100(d/t)$, where
   $d =$ the value in the "died" column; and.
   $t =$ the value in the "total" column.

5. Calculate the difference of logarithms.
   $D = (X_G-\%)/(X_G-X_L) = (71.43-50)/(71.43-12.) = 0.36$.
6. Calculate the log10% end point dilution using the difference of logarithms.
   $Log_{10}\%$ end point dilution = log_{10} of X−(D*F) = −6−(0.36*1) = −6.36.
7. Calculate the $TCID_{50}$.
   $TCID_{50} = 10^{(-\log 10\% \text{ end point dilution})} = 10^{-(-6.36)} = 2.31 \times 10^6$.
8. $TCID_{50}$ is often presented as the count of the $TCID_{50}$ per mL. If needed, convert using the formula:
   $TCID_{50}/mL = TCID_{50}/m$, where.
   m = final volume of media per well of the Vero-E6 cell infection plate (mL).
   $TCID_{50}/mL = (2.31 \times 10^6)/0.2 = 1.16 \times 10^7$.
9. For $TCID_{50}$ quantification of samples from animal tissue, convert $TCID_{50}$ to $TCID_{50}$ per gram of tissue input using the formula:
   $TCID_{50}/g = TCID_{50}/mL*(l/w)$, where.
   l = the volume of virus titer buffer added to the tissue sample (mL); and.
   w = the weight of the tissue sample (g).

   $TCID_{50}/g = (1.16 \times 10^7)*(1.6/0.16) = 1.16 \times 10^8$.

3.6.2 Trimmed Spearman–Karber Method

The Trimmed Spearman–Karber method calculates $TCID_{50}$ as follows [15, 31]:

$Log_{10}\%$ end point dilution = $-(X_0-d/2 + d\,\Sigma\, ri/ni)$, where.

X_0 = log_{10} of the reciprocal of the highest dilution at which all cell culture wells are positive;

d = log10 of the dilution factor.

ni = number of wells used in each individual dilution.

ri = number of positive cell culture wells (out of ni).

And the summation is started at dilution ×0.

$TCID_{50} = 10^{(-\log 10\% \text{ end point dilution})}$.

End point dilution CPE data from a lung homogenate sample from a Syrian Golden Hamster infected with SARS-Cov-2 is used as an example in the following section (Fig. 2a, Table 3).

1. To apply the trimmed Spearman–Karber method to enumerate the $TCID_{50}$ value of each viral stock or experimental sample, determine X_0 and calculate the log_{10}% end point dilution.
   $Log_{10}\%$ end point dilution = $-(X_0-d/2 + d\,\Sigma\, ri/ni) = -(5-(½) + 1(11/6)) = -6.33$.

Table 3
Table setup for quantifying $TCID_{50}$ using the trimmed Spearman–Karber method

Log_{10} virus dilution	Well score CPE (+)	Total wells inoculated
−2	6	6
−3	6	6
−4	6	6
−5	6	6
−6	4	6
−7	1	6
−8	0	6

Dataset used for calculations is shown in Fig. 2a, replicate 2

2. Calculate the $TCID_{50}$.
   $TCID_{50} = 10^{(-\log 10\% \text{ end point dilution})} = 10^{-(-6.33)} = 2.15 \times 10^6$.
3. Convert the $TCID_{50}$ value to $TCID_{50}$ / mL (see Subheading 3.6.1, **step 8**).
   $TCID_{50}/mL = (2.15 \times 10^6)/0.2 = 1.08 \times 10^7$.
4. Convert $TCID_{50}$ to $TCID_{50}$ per gram of tissue (see Subheading 3.6.1, **step 9**).
   $TCID_{50}/g = (1.08 \times 10^7)*(1.6/0.1) = 1.08 \times 10^8$.

4 Notes

1. One confluent T175 flask yields approximately 1.2×10^7–1.5×10^7 Vero-E6 cells. A total of 0.96×10^6 Vero-E6 cells are seeded per 96-well infection plate to measure $TCID_{50}$ of one virus sample in technical duplicates. Therefore, one confluent T175 flask of Vero-E6 cells will provide enough cells to analyze approximately 12–15 viral stocks or experimental samples.
2. Two to three volumes of prewarmed complete growth media are sufficient to neutralize the trypsin.
3. Avoid freeze-thawing of tissue specimens, virus stocks and experimental samples containing virus to prevent decreases in SARS-CoV-2 infectivity. Samples can be stored long term at −80 °C before moving on to the next step. If no −70 °C is available, store samples at −20 °C.
4. Only process frozen tissue samples to be used in the endpoint dilution assay on the same day to avoid freeze-thaw cycles.

5. For example, if the tissue weights 50 mg then add 0.5 mL of virus titer buffer.
6. If measuring $TCID_{50}$ from infected animal tissue, use homogenized tissue supernatant prepared in Subheading 3.2 at this point rather than virus stocks.
7. Dilutions can be made on any scale. Some optimization may be required to identify the range needed so that the $TCID_{50}$ value lies in the middle rows of the plate.
8. To ensure samples are homogenous during serial dilution, thoroughly mix row A before transferring 20 μL from row A to row B. Change pipet tips, then thoroughly mix row B before transferring 20 μL from row B to row C. Repeat through row G.
9. This step requires optimization to determine the amount of time required for the virus to spread and induce measurable CPE in the specific cell line and dilutions being used.
10. Each well should be scored as positive or negative for the presence of SARS-CoV-2, that is, SARS-CoV-2 has killed all of the cells, so there is minimal crystal violet stain or the cells have all survived, so the well is fully stained violet, respectively. The crystal violet stain should be visible to the naked eye, but a light box is useful to discriminate wells that have moderate staining.
11. Instrument settings depend on the manufacturer. An integration time of 0.25–1 s per well is the recommended range.
12. Each well should be scored as positive or negative for the presence of SARS-CoV-2. A well is considered positive for SARS-CoV-2 infection when the luminescence signal is less than the mean minus five times the standard deviation of the noninfected control wells [26].

Acknowledgments

The authors would like to thank Dr. Andrew S. Pekosz and the members of his laboratory at The Johns Hopkins School of Public Health for guidance, materials, and training on quantification of SARS-CoV-2 and for providing their $TCID_{50}$ protocol from which we further developed the methods described herein. We are grateful to Kirsten Littlefield who took the time out of her busy schedule during an emerging SARS pandemic to educate us on CPE morphology in Vero-E6 cells and $TCID_{50}$ calculation methods. We would also like to thank Dr. William R. Bishai for his valuable scientific discussions and logistical support. The Mercatus Center Fast Grant Award #2167 funded this work.

References

1. Mendoza EJ, Manguiat K, Wood H, Drebot M (2020) Two Detailed Plaque Assay Protocols for the Quantification of Infectious SARS-CoV-2. Curr Protoc Microbiol 57 (1): ecpmc105. https://doi.org/10.1002/cpmc.105
2. Chu H, Chan JF, Yuen TT, Shuai H, Yuan S, Wang Y, Hu B, Yip CC, Tsang JO, Huang X, Chai Y, Yang D, Hou Y, Chik KK, Zhang X, Fung AY, Tsoi HW, Cai JP, Chan WM, Ip JD, Chu AW, Zhou J, Lung DC, Kok KH, To KK, Tsang OT, Chan KH, Yuen KY (2020) Comparative tropism, replication kinetics, and cell damage profiling of SARS-CoV-2 and SARS-CoV with implications for clinical manifestations, transmissibility, and laboratory studies of COVID-19: an observational study. Lancet Microbe 1(1):e14–e23. https://doi.org/10.1016/S2666-5247(20)30004-5
3. Keiser PT, Anantpadma M, Staples H, Carrion R, Davey RA (2021) Automation of infectious focus assay for determination of filovirus titers and direct comparison to plaque and TCID50 assays. Microorganisms 9(1):156
4. Chu DKW, Pan Y, Cheng SMS, Hui KPY, Krishnan P, Liu Y, Ng DYM, Wan CKC, Yang P, Wang Q, Peiris M, Poon LLM (2020) Molecular diagnosis of a novel coronavirus (2019-nCoV) causing an outbreak of pneumonia. Clin Chem 66(4):549–555. https://doi.org/10.1093/clinchem/hvaa029
5. Corman VM, Landt O, Kaiser M, Molenkamp R, Meijer A, Chu DK, Bleicker T, Brunink S, Schneider J, Schmidt ML, Mulders DG, Haagmans BL, van der Veer B, van den Brink S, Wijsman L, Goderski G, Romette JL, Ellis J, Zambon M, Peiris M, Goossens H, Reusken C, Koopmans MP, Drosten C (2020) Detection of 2019 novel coronavirus (2019-nCoV) by real-time RT-PCR. Euro Surveill 25(3). https://doi.org/10.2807/1560-7917.ES.2020.25.3.2000045
6. Smither SJ, Lear-Rooney C, Biggins J, Pettitt J, Lever MS, Olinger GG Jr (2013) Comparison of the plaque assay and 50% tissue culture infectious dose assay as methods for measuring filovirus infectivity. J Virol Methods 193(2): 565–571. https://doi.org/10.1016/j.jviromet.2013.05.015
7. Jonsson N, Gullberg M, Lindberg AM (2009) Real-time polymerase chain reaction as a rapid and efficient alternative to estimation of picornavirus titers by tissue culture infectious dose 50% or plaque forming units. Microbiol Immunol 53(3):149–154. https://doi.org/10.1111/j.1348-0421.2009.00107.x
8. Baer A, Kehn-Hall K (2014) Viral concentration determination through plaque assays: using traditional and novel overlay systems. J Vis Exp 93:e52065. https://doi.org/10.3791/52065
9. van Doremalen N, Bushmaker T, Morris DH, Holbrook MG, Gamble A, Williamson BN, Tamin A, Harcourt JL, Thornburg NJ, Gerber SI, Lloyd-Smith JO, de Wit E, Munster VJ (2020) Aerosol and surface stability of SARS-CoV-2 as compared with SARS-CoV-1. N Engl J Med 382(16):1564–1567. https://doi.org/10.1056/NEJMc2004973
10. Harcourt J, Tamin A, Lu X, Kamili S, Sakthivel SK, Murray J, Queen K, Tao Y, Paden CR, Zhang J, Li Y, Uehara A, Wang H, Goldsmith C, Bullock HA, Wang L, Whitaker B, Lynch B, Gautam R, Schindewolf C, Lokugamage KG, Scharton D, Plante JA, Mirchandani D, Widen SG, Narayanan K, Makino S, Ksiazek TG, Plante KS, Weaver SC, Lindstrom S, Tong S, Menachery VD, Thornburg NJ (2020) Severe acute respiratory syndrome coronavirus 2 from patient with coronavirus disease, United States. Emerg Infect Dis 26(6):1266–1273. https://doi.org/10.3201/eid2606.200516
11. Cooper PD (1961) The plaque assay of animal viruses. Adv Virus Res 8:319–378. https://doi.org/10.1016/s0065-3527(08)60689-2
12. Juarez D, Long KC, Aguilar P, Kochel TJ, Halsey ES (2013) Assessment of plaque assay methods for alphaviruses. J Virol Methods 187(1):185–189. https://doi.org/10.1016/j.jviromet.2012.09.026
13. Masci AL, Menesale EB, Chen WC, Co C, Lu X, Bergelson S (2019) Integration of fluorescence detection and image-based automated counting increases speed, sensitivity, and robustness of plaque assays. Mol Ther Methods Clin Dev 14:270–274. https://doi.org/10.1016/j.omtm.2019.07.00714
14. Flint S, Enquist W, Racaniello V, Skalka A (2009) Virological methods, Principles of virology. ASM Press, In
15. Ramakrishnan MA (2016) Determination of 50% endpoint titer using a simple formula. World J Virol 5(2):85–86. https://doi.org/10.5501/wjv.v5.i2.85
16. Pourianfar HR, Javadi A, Grollo L (2012) A colorimetric-based accurate method for the determination of enterovirus 71 titer. Indian J Virol 23(3):303–310. https://doi.org/10.1007/s13337-012-0105-0

17. Matsuyama S, Nao N, Shirato K, Kawase M, Saito S, Takayama I, Nagata N, Sekizuka T, Katoh H, Kato F, Sakata M, Tahara M, Kutsuna S, Ohmagari N, Kuroda M, Suzuki T, Kageyama T, Takeda M (2020) Enhanced isolation of SARS-CoV-2 by TMPRSS2-expressing cells. Proc Natl Acad Sci USA 117(13):7001–7003. https://doi.org/10.1073/pnas.2002589117
18. Lindenbach BD (2009) Measuring HCV infectivity produced in cell culture and in vivo. Methods Mol Biol 510:329–336. https://doi.org/10.1007/978-1-59745-394-3_24
19. Hierholzer JC, Killington RA (1996) Virus isolation and quantitation. Virology Methods Manual:25–46. https://doi.org/10.1016/B978-012465330-6/50003-8
20. World Health Organization. (2020, February 12). Laboratory biosafety guidance related to coronavirus disease 2019 (COVID-19): interim guidance. https://apps.who.int/iris/handle/10665/331138. License: CC BY-NC-SA 3.0 IGO
21. Phelan K, May KM (2015) Basic techniques in mammalian cell tissue culture. Curr Protoc Cell Biol 66:111–112. https://doi.org/10.1002/0471143030.cb0101s66
22. Severson WE, Shindo N, Sosa M, Fletcher T 3rd, White EL, Ananthan S, Jonsson CB (2007) Development and validation of a high-throughput screen for inhibitors of SARS CoV and its application in screening of a 100,000-compound library. J Biomol Screen 12(1): 33–40. https://doi.org/10.1177/1087057106296688
23. Heldt CL, Hernandez R, Mudiganti U, Gurgel PV, Brown DT, Carbonell RG (2006) A colorimetric assay for viral agents that produce cytopathic effects. J Virol Methods 135(1):56–65. https://doi.org/10.1016/j.jviromet.2006.01.022
24. Niles A Noah J Rasmussen L Lazar D (2013) Determine viral-induced cytopathic effect using a luminescent assay Promega. https://www.promega.com/resources/pubhub/determine-viral-induced-cytopathic-effect-using-a-luminescent-assay/.. Accessed May 1, 2021
25. Noah JW, Severson W, Noah DL, Rasmussen L, White EL, Jonsson CB (2007) A cell-based luminescence assay is effective for high-throughput screening of potential influenza antivirals. Antivir Res 73(1):50–59. https://doi.org/10.1016/j.antiviral.2006.07.006
26. Chung D-H, Moore BP, Matharu DS, Golden JE, Maddox C, Rasmussen L, Sosa MI, Ananthan S, White EL, Jia F, Jonsson CB, Severson WE (2013) A cell based high-throughput screening approach for the discovery of new inhibitors of respiratory syncytial virus. Virol J 10(1):19. https://doi.org/10.1186/1743-422X-10-19
27. Phillips T, Jenkinson L, McCrae C, Thong B, Unitt J (2011) Development of a high-throughput human rhinovirus infectivity cell-based assay for identifying antiviral compounds. J Virol Methods 173(2):182–188. https://doi.org/10.1016/j.jviromet.2011.02.002
28. Reed LJ, Muench H (1938) A simple method of estimating fifty per cent endpoints12. Am J Epidemiol 27(3):493–497. https://doi.org/10.1093/oxfordjournals.aje.a118408
29. Kärber G (1931) Beitrag zur kollektiven Behandlung pharmakologischer Reihenversuche. Naunyn Schmiedebergs Arch Exp Pathol Pharmakol 162(4):480–483. https://doi.org/10.1007/BF01863914
30. Hamilton MA, Russo RC, Thurston RV (1977) Trimmed Spearman-Karber method for estimating median lethal concentrations in toxicity bioassays. Environ Sci Technol 11(7): 714–719. https://doi.org/10.1021/es60130a004
31. Lei C, Yang J, Hu J, Sun X (2021) On the calculation of TCID50 for quantitation of virus infectivity. Virol Sin 36(1):141–144. https://doi.org/10.1007/s12250-020-00230-5

Chapter 10

One-Step Reverse Transcription Droplet Digital PCR Protocols for SARS-CoV-2 Detection and Quantification

Raphael Nyaruaba, Xiohong Li, Caroline Mwaliko, Faith Ogolla, Changchang Li, Lu Zhao, Hang Yang, Junping Yu, and Honping Wei

Abstract

Droplet digital polymerase chain reaction (ddPCR) is a third generation of PCR that was recently developed to overcome the limitation of direct quantification observed in real-time quantification PCR (qPCR). Recent studies have shown that ddPCR is more sensitive than the gold standard reverse transcription real-time quantitative polymerase chain reaction (RT-qPCR) in detecting severe acute respiratory syndrome coronavirus 2 (SARS-CoV-2) samples. In combination with multiplexing, multiple RT-ddPCR assays can be developed to directly quantify different SARS-CoV-2 nucleic acid targets within a single sample, significantly saving on cost and time. Since ddPCR is tolerant to a number of inhibitors unlike qPCR, it can be used to detect and quantify samples from complex environments like wastewater. Here we present three one-step RT-ddPCR protocols on how to develop simplex (one target), duplex (two targets), and triplex probe mix (three targets) assays for SARS-CoV-2 detection and quantification. The assays can be used for diagnosis or other research-related SARS-CoV-2 applications.

Key words Droplet digital, Polymerase chain reaction, qPCR, SARS-CoV-2, COVID-19, RT-ddPCR, Quantification, Detection

1 Introduction

SARS-CoV-2, the causative agent of coronavirus disease 2019 (COVID-19) recently emerged from Wuhan, China, in December 2019 to spread across the globe causing a global public health crisis [1, 2]. The current gold standard method for isolation and treatment of COVID-19 patients remains to be RT-qPCR [3, 4]. However effective this method may be, current and ongoing research has shown that it is prone to false positives and negatives when testing samples with low abundant SARS-CoV-2 targets [5–7]. Therefore, it is crucial to explore alternative strategies that can overcome the limits faced by RT-qPCR.

Justin Jang Hann Chu et al. (eds.), *SARS-CoV-2: Methods and Protocols*,
Methods in Molecular Biology, vol. 2452, https://doi.org/10.1007/978-1-0716-2111-0_10,

Among these strategies includes droplet digital PCR (ddPCR) which has recently been shown to be superior to RT-qPCR in detecting low abundant SARS-CoV-2 targets [1, 4–9]. ddPCR is a third generation of PCR that was recently developed to overcome the limitation of direct quantification observed in real time quantitative PCR (qPCR) [10, 11]. Using ddPCR, nucleic acid targets are distributed into thousands of droplets that will later be counted as positive or negative. Using Poisson statistics, the exact amount of nucleic acid targets in copies/μL will be estimated directly without the need for a standard curve. Since its advent, it has been used to study a wide range of microorganisms. In most of these studies, ddPCR has been said to be far superior to qPCR in terms of sensitivity, accuracy, limits of detection, and reproducibility [10, 12–14].

ddPCR also has a unique advantage over qPCR in multiplexing. Unlike qPCR where nucleic acid targets have to be detected in discrete channels (using single fluorophores), ddPCR has the ability to detect more than one target in the same channel (using the same fluorophore). This technique is known as higher order multiplexing [1, 15]. Using this approach, one can either use amplitude-based multiplexing or probe mix multiplexing to detect three or more nucleic acid targets in two discrete channels [1, 11]. Briefly, when using amplitude-based multiplexing, one target is added in low concentration (0.5×) when compared to other targets in the same channel (1×), hence the targets will be located at different amplitudes. However, for probe mix multiplexing, one target's probe is conjugated (two probe dyes mixed together in half concentrations (0.5×)) to constitute the final probe concentration (1×) while the other target probes are mixed to their final concentrations (1×) in discrete channels using only a single dye. Using this approach, Nyaruaba et al. demonstrated that the method can be used to detect SARS-CoV-2 samples with improved sensitivity compared to RT-qPCR [1]. Additionally, multiplexing will improve the number of targets that can be sensitively detected within a sample, saving on costs and time needed to perform simplex assays.

In this current work, using a one-step RT-ddPCR kit and already tested China CDC primer and probe sets [1, 4, 5, 16], we aim to demonstrate steps on how to develop RT-ddPCR SARS-CoV-2 assays including high order multiplex assays for a two-color ddPCR system. These assays can detect one (simplex), two (duplex), or three (triplex probe mix) nucleic acid targets sensitively within a single sample. The assays can be used for a wide range of applications including diagnosis and other research related activities.

2 Materials

2.1 Sample Inactivation

1. Water bath.

2.2 RNA Extraction

1. FineMag Rapid Bead Virus DNA/RNA Extraction Kit (Genefine Biotech).
2. 32-channel fully automatic nucleic acid extractor Purifier 32 (Genefine Biotech).

2.3 Mastermix Components

1. 2× One-Step RT-ddPCR Advanced Kit for Probes (Bio-Rad).
2. Primer-probe mix (PP mix) to a final concentration of 900 nm and 270 nm respectively.
3. Nuclease free water.

2.4 Automated Droplet Generation

1. QX200 Automated Droplet Generator (Bio-Rad).
2. ddPCR 96-Well Plates (Bio-Rad).
3. AutoDG Oil for Probes (Bio-Rad).
4. DG32 AutoDG Cartridges (Bio-Rad).
5. Cooling block (Bio-Rad).
6. Pipet Tips for the AutoDG (Bio-Rad).
7. Pipet Tip Waste Bins for the AutoDG (Bio-Rad).

2.5 PCR

1. PX1 PCR Plate Sealer (Bio-Rad).
2. Pierceable Foil Heat Seals (Bio-Rad).
3. T100 Thermal Cycler (Bio-Rad) or equivalent (*see* **Note 1**).

2.6 Droplet Reading

1. Droplet Reader Oil (Bio-Rad).
2. QX100/QX200 Droplet Reader (Bio-Rad).

2.7 Data Analysis

1. QuantaSoft 1.7 Software (Bio-Rad).
2. QuantaSoft Analysis Pro 1.0 (Bio-Rad).

3 Methods

WARNING: Owing to the infectious nature and lethality of SARS-CoV-2 samples, all samples should be treated as if they can transmit infectious agents even when using safe laboratory procedures. Live sample processing steps should be done in designated hospitals or biosafety level 2/3 (BSL-2/3) laboratories following strict hospital and/or BSL-2/3 rules including wearing of appropriate PPE.

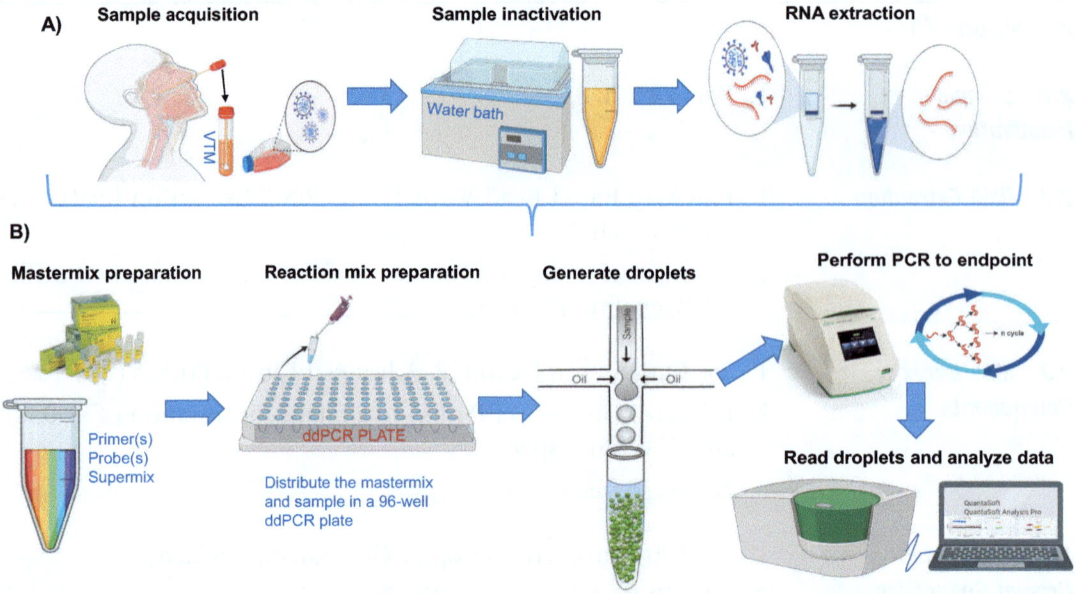

Fig. 1 The general SARS-CoV-2 detection and quantification workflow using ddPCR. (**a**) Sample handling workflow. (**b**) General ddPCR workflow

The protocols developed can be used for quantifying SARS-CoV-2 viral RNA from both clinical and research samples. The general workflow is summarized in Fig. 1.

3.1 Sample Acquisition

Sample acquisition is not described in this protocol as SARS-CoV-2 samples can come from different sources including hospitals and research facilities. However, for showing the process of ddPCR assay development in this work, we used SARS-CoV-2 samples cultured in Vero E6 cells. Pooled human oral swabs from healthy volunteers were used for obtaining the human gene.

3.2 Sample Categories

Four sample categories were developed to be used for the location of SARS-CoV-2 target droplets. Since the assays can be used in research or diagnostics, the four sample categories were also used to demonstrate representative results when performing diagnostic tests or research tests (*see* **Note 2**).

1. SARS-CoV-2 + IC – SARS-CoV-2 virus from Vero E6 cells spiked in a background of pooled oral swabs of healthy volunteers to represent positive clinical samples (*see* **Note 3**).
2. SARS-CoV-2 – SARS-CoV-2 only sample obtained from cultured virus in Vero E6 cells to represent research work.
3. IC - pooled oral swabs of healthy volunteers to represent negative clinical samples.
4. No-template control (NTC) – nuclease free water with no targets to detect any environmental or reagent contaminations.

3.3 Sample Inactivation

1. Take SARS-CoV-2 samples to a biosafety laboratory.
2. Heat-inactivate samples by heating at 65 °C for 30 min (*see* **Note 4**).
3. Take samples to a biosafety cabinet (BSC) and let them stand for 10 min to allow potential aerosols to settle.
4. Store samples in 4 °C for up to 24 h. If not processed immediately.

3.4 RNA Extraction

Many extraction kits and instruments exist for RNA extraction and one can use any locally available RNA extraction kit/instrument(s). In this work we use GenFine's FineMag Rapid Magnetic Bead Extraction Kit for viral RNA extraction using the 32-channel fully automated nucleic acid extractor Purifier 32 extraction machine according to the manufacturer's instructions.

1. Take out the prefilled 96-well plate, mix it upside-down gently to resuspend the magnetic beads, and gently shake the plate to concentrate reagent and magnetic beads at the plate's bottom.
2. Carefully tear off the aluminum foil sealing film to avoid vibration of the plate and liquid spillage.
3. Add 200 μL of sample to rows 2 and 8 of the 96-well plate.
4. Insert the magnetic bar sleeve into the card slot of the 32-channel automatic nucleic acid extraction instrument.
5. Select the program "GF-FM502T5-TR"1"GF-FM502T5YH" (quick version) and run it.
6. After extraction completion, take out the nucleic acid samples in line 6 and 12 (about 100 μL/sample).
7. Immediately use or store the samples at 4 °C for up to 24 h till use or at −20 °C for a longer time.

3.5 Reaction Mix Preparation

Primer and probe stock solutions should be stored in high concentrations of 100 μM, in −20 °C. To reduce repeated freeze-thawing cycles, smaller volumes of working solutions can be stored in lower concentrations of 20–4 0 μM (diluted in nuclease free water) and stored at −20 °C for up to 3 months. In this work, all primer working solutions were stored at 40 μM, while all the probe working solutions were stored at 20 μM.

1. Based on the number of samples, calculate the amount of primers-probe mix (PP mix) and volume of reagents that will be needed to develop the different master mix assays as follows (*see* **Note 5**):

(a) For one well the simplex assay contains:
- 5.5 μL supermix (1× final concentration).
- 2.2 μL reverse transcriptase (20 U/μL final concentration).
- 1.1 μL 300 mM DTT (15 mM final concentration).
- 7.5 μL nuclease free water.
- 1.3 μL of either PP mix code 1, 2, 3, or 4 as seen in Table 1.

(b) For one well the duplex assay contains:
- 5.5 μL supermix (1× final concentration).
- 2.2 μL reverse transcriptase (20 U/ μL final concentration).
- 1.1 μL 300 mM DTT (15 mM final concentration).
- 6.2 μL nuclease free water.
- 2.6 μL of either PP mix code 1, 2, or 3 as seen in Table 2 (*see* **Note 6**).

(c) For one well the triplex probe mix assay contains:
- 5.5 μL supermix (1× final concentration).
- 2.2 μL reverse transcriptase (20 U/ μL final concentration).
- 1.1 μL 300 mM DTT (15 mM final concentration).
- 4.9 μL nuclease free water.
- 3.9 μL of either PP mix code 1, 2 or 3 as seen in Table 3 (*see* **Note 6**).

2. Distribute 17.6 μL of the above master mix into the wells of a nuclease-free 96-well ddPCR plate based on the number of samples (*see* **Note 7**).
3. Add 4.4 μL RNA sample to each well with master mix (*see* **Note 8**).
4. Mix by pipetting up and down or vortex briefly (15–30 s), and centrifuge briefly (10–15 s) to settle contents at the bottom of the plate.
5. Proceed directly to droplet generation.

3.6 Automated Droplet Generation

Different instruments may be used to generate droplets. In this study we used Bio-Rad's QX200 Automated Droplet Generator (AutoDG) as seen in Fig. S1. To avoid amplicon contamination, ensure the droplet generator and readers have dedicated space in separate areas.

1. Gather all consumables needed to setup the QX200 AutoDG (*see* **Note 9**).
2. On the AutoDG touch screen, touch "Configure Sample Plate" and select columns where samples are located on sample

Table 1
Primers and probes used in the simplex assays

PP mix code	Target and ref	Name	Sequence of primer and probes (5′ to 3′)	Final concentration in PCR (nmol/L)	Volume added per PCR well (μL)
1	RBD2 [3]	F-RBD2	CTCAAGTGTCTGTGGATCACG	900	0.5
		R-RBD2	CCTGTGCCTGTTAAACCATTG	900	0.5
		P-RBD2	FAM-ACAGCATCAGTAGTGTCAGCAA TGTCTC-BHQ1	270	0.3
2	ORF1ab [17]	F-ORF1ab	CCCTGTGGGTTTTACACTTAA	900	0.5
		R-ORF1ab	ACGATTGTGCATCAGCTGA	900	0.5
		P-ORF1ab	FAM/HEX-CCGTCTGCGGTATGTGGA AAGGTTATGG-BHQ1	270	0.3
3	N [17]	F-N	GGGGAACTTCTCCTGCTAGAAT	900	0.5
		R-N	CAGACATTTTGCTCTCAAGCTG	900	0.5
		P-N	FAM/HEX-TTGCTGCTGCTTGAC AGATT-BHQ1	270	0.3
4	IC [16]	F-IC	AGTGCATGCTTATCTCTGACAG	900	0.5
		R-IC	GCAGGGCTATAGACAAGTTCA	900	0.5
		P-IC	HEX-TTTCCTGTGAAGGCG ATTGA CCGA-BHQ1	270	0.3

All primer solutions working concentrations are 40 μM while all probe working concentrations are 20 μM

Table 2
Primers and probes used in the duplex assays

PP mix code	Target and ref	Name	Sequence of primer and probes (5′ to 3′)	Final concentration in PCR (nmol/L)	Volume added per PCR well (μL)
1	RBD2 [3] and ORF1ab [17]	F-RBD2	CTCAAGTGTCTGTGGATCACG	900	0.5
		R-RBD2	CCTGTGCCTGTTAAACCATTG	900	0.5
		P-RBD2	FAM- ACAGCATCAGTAGTGTCAGCAATGTCTC -BHQ1	270	0.3
		F-ORF1ab	CCCTGTGGGTTTTACACTTAA	900	0.5
		R-ORF1ab	ACGATTGTGCATCAGCTGA	900	0.5
		P-ORF1ab	HEX-CCGTCTGCGGTATGTGGAAAGGTTATGG-BHQ1	270	0.3
2	N and IC [16] [17]	F-N	GGGGAACTTCTCCTGCTAGAAT	900	0.5
		R-N	CAGACATTTTGCTCTCAAGCTG	900	0.5
		P-N	FAM-TTGCTGCTGCTTGACAGATT-BHQ1	270	0.3
		F-IC	AGTGCATGCTTATCTCTGACAG	900	0.5
		R-IC	GCAGGGCTATAGACAAGTTCA	900	0.5
		P-IC	HEX-TTTCCTGTGAAGGCG ATTGACCGA-BHQ1	270	0.3
3	ORF1ab and IC [16] [17]	F-ORF1ab	CCCTGTGGGTTTTACACTTAA	900	0.5
		R-ORF1ab	ACGATTGTGCATCAGCTGA	900	0.5
		P-ORF1ab	FAM-CCGTCTGCGGTATGTGGAAAGGTTATGG-BHQ1	270	0.3
		F-IC	AGTGCATGCTTATCTCTGACAG	900	0.5
		R-IC	GCAGGGCTATAGACAAGTTCA	900	0.5
		P-IC	HEX-TTTCCTGTGAAGGCG ATTGACCGA-BHQ1	270	0.3

All primer solutions working concentrations are 40 μM while all probe working concentrations are 20 μM

Table 3
Primers and probes used in the triplex probe mix assays

PP mix code	Target and ref.	Name	Sequence of primer and probes (5′–3′)	Final concentration in PCR (nmol/l)	Volume added per PCR well (μL)
1	ORF1ab, N [17], IC [16]	F-ORF1ab	CCCTGTGGGTTTTACACTTAA	900	0.5
		R-ORF1ab	ACGATTGTGCATCAGCTGA	900	0.5
		P-ORF1ab	FAM-CCGTCTGCGGTATGTGGAAAGGT TATGG-BHQ1	270	0.3
		F-N	GGGGAACTTCTCCTGCTAGAAT	900	0.5
		R-N	CAGACATTTTGCTCTCAAGCTG	900	0.5
		P-N	FAM-TTGCTGCTGCTTGACAGATT-BHQ1	135	0.15
		P-N	HEX-TTGCTGCTGCTTGACAGATT-BHQ1	135	0.15
		F-IC	AGTGCATGCTTATCTCTGACAG	900	0.5
		R-IC	GCAGGGCTATAGACAAGTTCA	900	0.5
		P-IC	HEX-TTTCCTGTGAAGGCG ATTGACCGA-BHQ1	270	0.3
2	N, ORF1ab [17], IC [16]	F-N	GGGGAACTTCTCCTGCTAGAAT	900	0.5
		R-N	CAGACATTTTGCTCTCAAGCTG	900	0.5
		P-N	FAM-TTGCTGCTGCTTGACAGATT-BHQ1	270	0.3
		F-ORF1ab	CCCTGTGGGTTTTACACTTAA	900	0.5
		R-ORF1ab	ACGATTGTGCATCAGCTGA	900	0.5
		P-ORF1ab	FAM-CCGTCTGCGGTATGTGGAAAGGT TATGG-BHQ1	135	0.15
		P-ORF1ab	HEX-CCGTCTGCGGTATGTGGAAAGGT TATGG-BHQ1	135	0.15
		F-IC	AGTGCATGCTTATCTCTGACAG	900	0.5
		R-IC	GCAGGGCTATAGACAAGTTCA	900	0.5
		P-IC	HEX-TTTCCTGTGAAGGCG ATTGACCGA-BHQ1	270	0.3
3	RBD2 [3], ORF1ab [17], IC [16]	F-RBD2	CTCAAGTGTCTGTGGATCACG	900	0.5
		R-RBD2	CCTGTGCCTGTTAAACCATTG	900	0.5
		P-RBD2	FAM- ACAGCATCAGTAGTGTCAG CAATGTCTC -BHQ1	270	0.3
		F-ORF1ab	CCCTGTGGGTTTTACACTTAA	900	0.5

(continued)

Table 3 (continued)

PP mix code	Target and ref.	Name	Sequence of primer and probes (5′–3′)	Final concentration in PCR (nmol/l)	Volume added per PCR well (μL)
		R-ORF1ab	ACGATTGTGCATCAGCTGA	900	0.5
		P-ORF1ab	FAM-CCGTCTGCGGTATGTGGAAAGGT TATGG-BHQ1	135	0.15
		P-ORF1ab	HEX-CCGTCTGCGGTATGTGGAAAGGT TATGG-BHQ1	135	0.15
		F-IC	AGTGCATGCTTATCTCTGACAG	900	0.5
		R-IC	GCAGGGCTATAGACAAGTTCA	900	0.5
		P-IC	HEX-TTTCCTGTGAAGGCG ATTGACCGA-BHQ1	270	0.3

All primer solutions working concentrations are 40 μM while all probe working concentrations are 20 μM

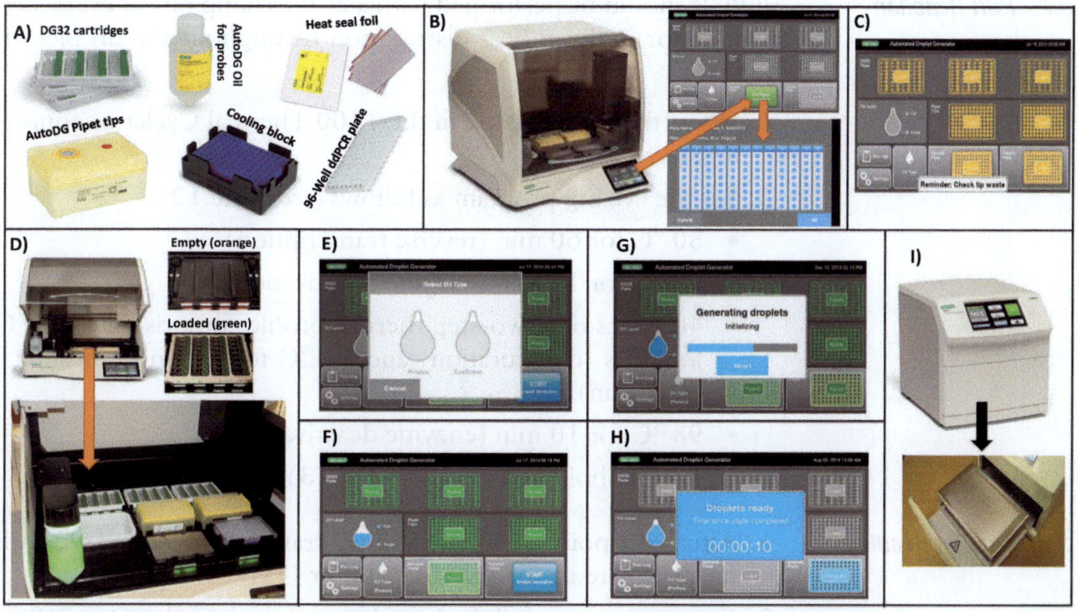

Fig. S1 Automated droplet generation using Bio-Rad's QX200 AutoDG Droplet Generator. (**a**) Consumables needed to setup the AutoDG Automated Droplet Generator. (**b**) On the AutoDG touch screen, touch Configure Sample Plate and select columns where samples are located on sample plate and press OK. (**c**) Once selected, the screen turns yellow indicating where consumables should be loaded. (**d**) Open the AutoDG door and load consumables in respective places. Load consumables from the back working toward the front. Make sure each time a consumable is added the light turns from yellow to green at that location. (**e**) Ensure you use oil for probes type. (**f**) Close the AutoDG door and ensure all reagents are set in place by checking if the AutoDG touch screen is green. (**g**) Press START Droplet Generation to generate droplets. (**h**) After droplets generation is complete, open the AutoDG and remove the sample plate. (**i**) Seal the droplets using a pierceable foil heat seal

plate and press "OK". The screen will turn yellow indicating LOAD where consumables should be loaded.

3. Open the AutoDG door and load consumables in their respective places (*see* **Note 10**).
4. Load the sample plate and droplet generation plate (*see* **Note 11**).
5. Close the AutoDG door, ensure the screen is green (consumables loaded), and Oil for Probes is selected before pressing "START Droplet Generation".
6. Wait for droplets to be generated to completion. Open the door once the screen shows "Droplets ready" and remove the plate containing droplets.
7. Seal the plate with a pierceable foil seal using the PX1 Plate Sealer set to run at 185 °C for 5 s.
8. Proceed to PCR amplification within 30 min after droplet generation.

3.7 PCR Reaction

All PCR should be performed with a 2 °C/s ramp rate at every step to allow for uniform heat distribution throughout the droplet oil mixture.

1. Insert the droplet plate in the T100 Thermal Cycler or equivalent (*see* **Note 1**).
2. Set the cycling program as follows (*see* **Note 12**).
   - 50 °C for 60 min (reverse transcription).
   - 95 °C for 10 min (enzyme deactivation).
   - 40 cycles of a two-step thermal profile comprising of 95 °C for 30 s (denaturation) and 57 °C for 1 min (annealing/ extension).
   - 98 °C for 10 min (enzyme deactivation)
   - Infinite hold at 4 °C (*see* **Note 13**).

3.8 Droplet Reading

1. After end-point thermal cycling, transfer the sealed 96-well droplet plate to the QX100/200 or QXDx Droplet reader.
2. From a computer linked to the reader open the QuantaSoft Software.
3. In the "Setup" mode select "Template", then "New".
4. Double click one well to open the Well Editor dialog box. Select the wells to be read and complete the required information as follows (*see* **Note 14**):
   - Experiment: ABS.
   - Supermix: One-Step RT-ddPCR Kit for Probes.
   - Target 1 Type: Ch1 Unknown.
   - Target 2 Type: Ch2 Unknown.
5. Designate sample names per well based on the plate layout, select "Ok" and save the created experiment.
6. Select "Run" to start reading (*see* **Note 15**).
7. When prompted, choose the appropriate color option (FAM/HEX).
8. Let the droplets to be read to completion and save the experiment data before removing the plate from the machine (*see* **Note 16**).

3.9 Data Analysis

Figure 2 is a schematic guide of how the data output should look like in simplex, duplex, and triplex probe mix assays. This scheme can also help you to know how to assign targets. In all assays, wells will be accepted for data analysis if only the total droplet count is >8000. A well will be considered positive if three or more droplets show fluorescence signals above the threshold values in all assays (*see* **Note 2**).

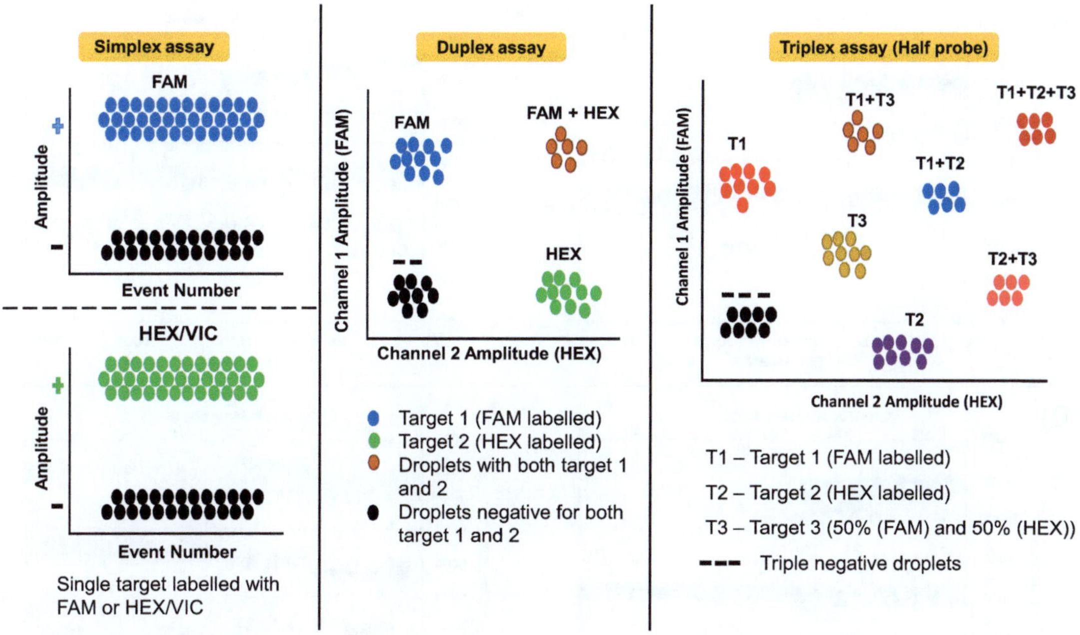

Fig. 2 Schematic representation of how the expected results will look like during analysis of simplex, duplex, and triplex probe mix assays. This scheme is generally for a two-color detection system where target droplets are detected in two discrete channels, for example, FAM/HEX as used in this work. In simplex assays; only two droplet clusters are expected, for duplex assays; four droplet clusters are expected, and for triplex probe mix assays; eight droplet clusters are expected

3.9.1 Simplex and Duplex Assays

Both the QuantaSoft and QuantaSoft Analysis Pro Software's may be used to analyze simplex and duplex experiments. However, for ease, we recommend using the QuantaSoft Software.

1. For QuantaSoft, double click on the .qlp file to be analyzed to open it and select "Analyze" (*see* **Note 17**).
2. Use the threshold tools in the 1D Amplitude and 2D Amplitude to distinguish between positive and negative droplets for each well in the correct channel. The NTC and positive control samples can be used as a guidance for threshold setting.
3. Figures 3 and 4 are examples of droplet readout of simplex and duplex assays respectively using the different sample categories highlighted in Subheading 3.2 (FAM results are seen in Channel 1 while HEX/VIC in Channel 2).
4. Record the results on copy number directly or export results as .csv file for further analysis by Excel or equivalent software.

3.9.2 Triplex Probe Mix Assay

Before analyzing the triplex probe mix assay results, download the QuantaSoft Analysis Pro Software from https://www.bio-rad.com/en-us/product/qx200-droplet-digital-pcr-system

Refer to the minimal system requirements before installing.

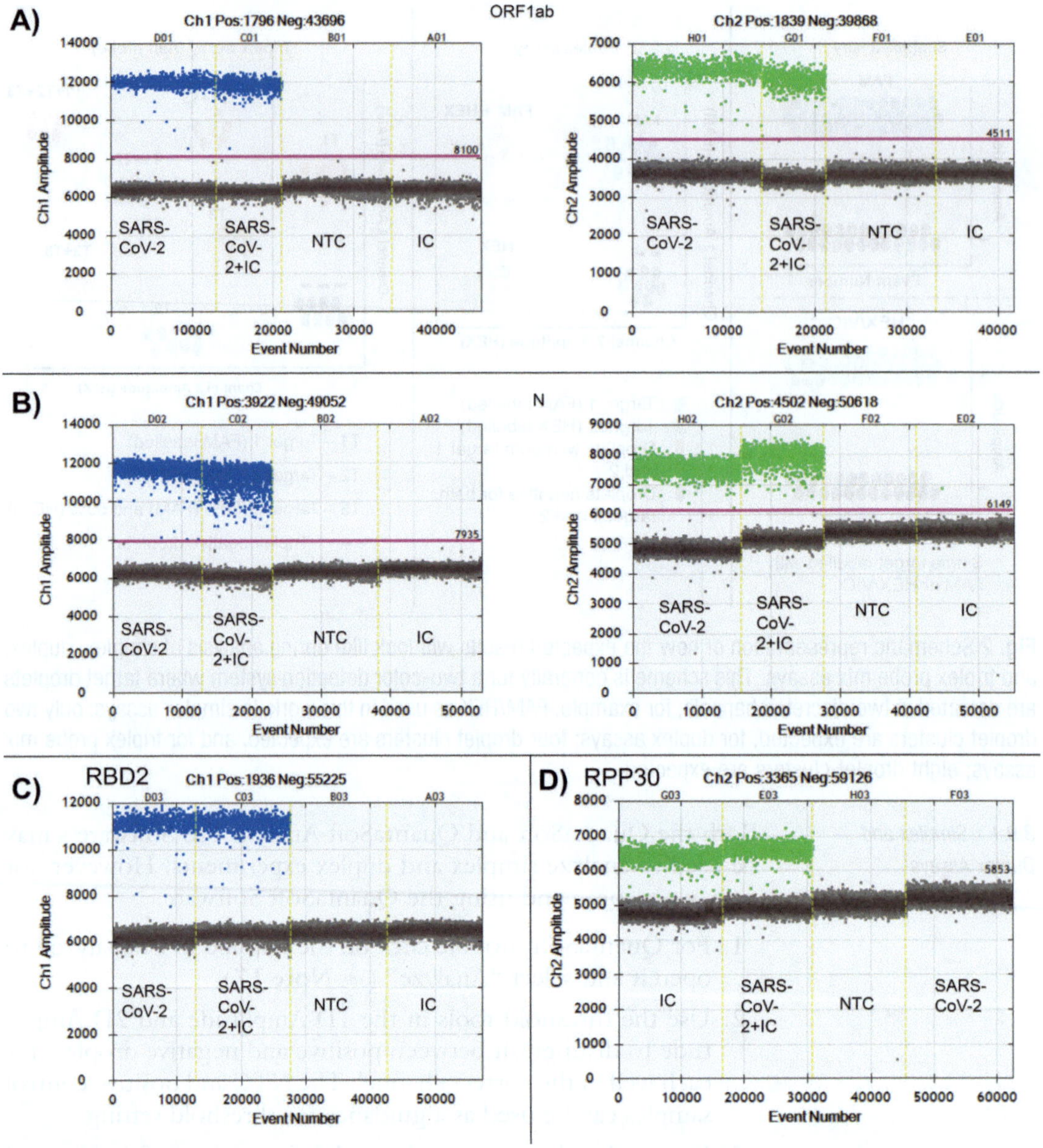

Fig. 3 Example of one-step RT-ddPCR simplex assay results using different targets and sample categories. Results were read in FAM (blue droplets) and HEX (green droplets) channels using single targets ORF1ab (**a**), N (**b**), RBD2 (**c**), and RPP30 (**d**)

1. Right click the .qlp file and select "Open With QuantaSoftAnalysisPro" (*see* **Note 17**).
2. In the plate editor tab, select wells to be analyzed and on the right side of the tab complete the required information as follows:

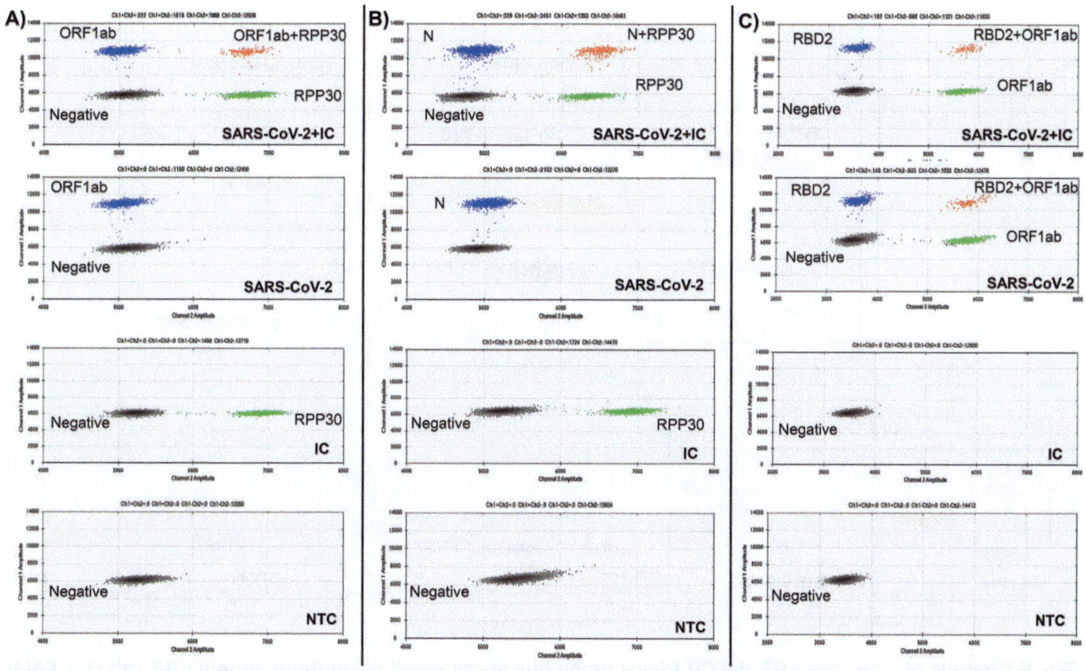

Fig. 4 Example of one-step RT-ddPCR duplex assay results using different targets and sample categories. (**a**) Representative results when ORF1ab is FAM labelled and RPP30 is HEX labelled. (**b**) Representative results when N is FAM labelled and RPP30 is HEX labelled. (**c**) Representative results when RBD2 is FAM labelled and ORF1ab is HEX labelled. For C, four clusters of droplets are seen in both the SARS-CoV-2 + IC and SARS-CoV-2 only sample categories because no IC gene was used here to demonstrate how one can maximize on the number of targets to be detected when performing research work that does not need the human gene

- Experiment Type: Direct Quantification (DQ).
- Assay Information: Probe Mix Triplex (and enter target names accordingly).
- Click "Apply".

3. On the left side of the 2D Amplitude tab, use the "Graph Tools" to set threshold clusters and assign specific colors to different target clusters following the "select to assign cluster" window pop up suggestions (*see* **Note 18**).
4. Figures 5, S2 and S3 are examples of droplet readout of triplex probe mix assays using the different sample categories highlighted in Subheading 3.2.
5. Once target colors are assigned, quantification data can be read in the "Well Data Window" on the lower right.
6. Export data to Excel/csv for further analysis using the triple-bar icon on the upper right hand of the "Well Data" table.

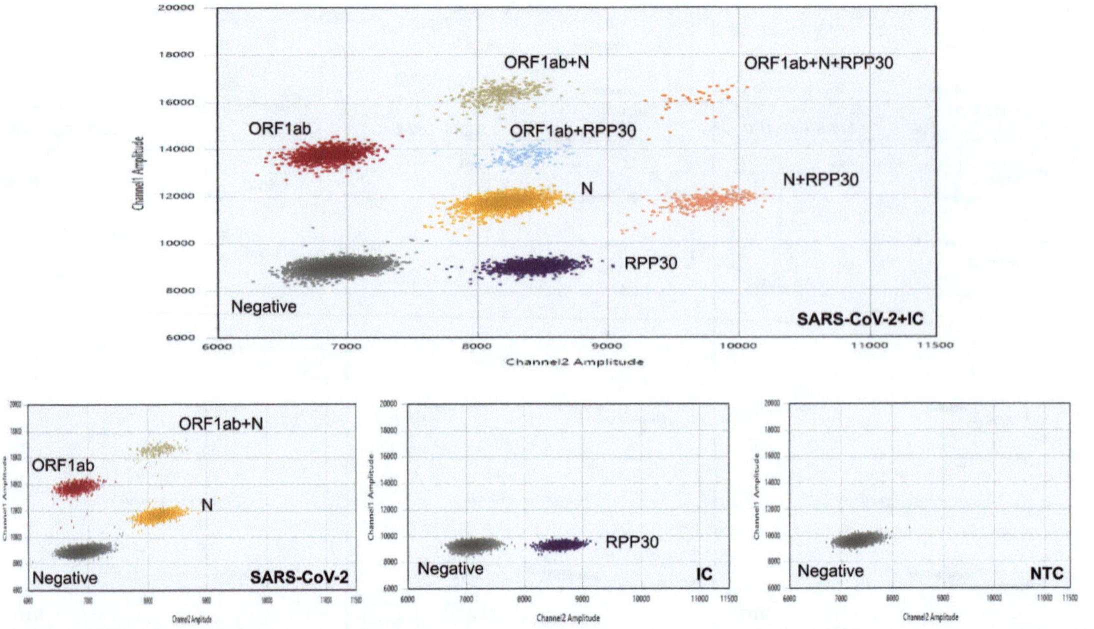

Fig. 5 Example of a one-step RT-ddPCR triplex probe mix assay result using three targets (ORF1ab (1× FAM), N (0.5× FAM and 0.5× HEX), and RPP30 (1× HEX)) and different sample categories. Eight droplet clusters were observed in the SARS-CoV-2 + IC sample category which reduced when differed sample categories (SARS-CoV-2 and IC) with less targets were used

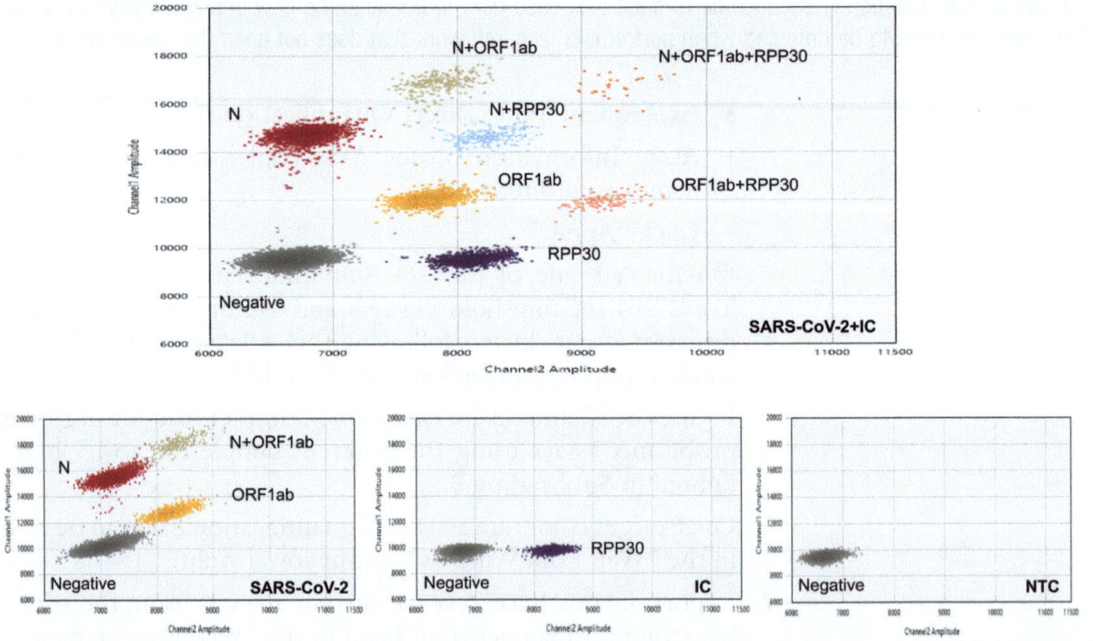

Fig. S2 Example of a one-step RT-ddPCR triplex probe mix assay result using three targets (N (FAM), ORF1ab (50% FAM and 50% HEX), and RPP30 (HEX)) and different sample categories. Eight droplet clusters were observed in the SARS-CoV-2 + IC sample category which reduced when differed sample categories with less targets were used

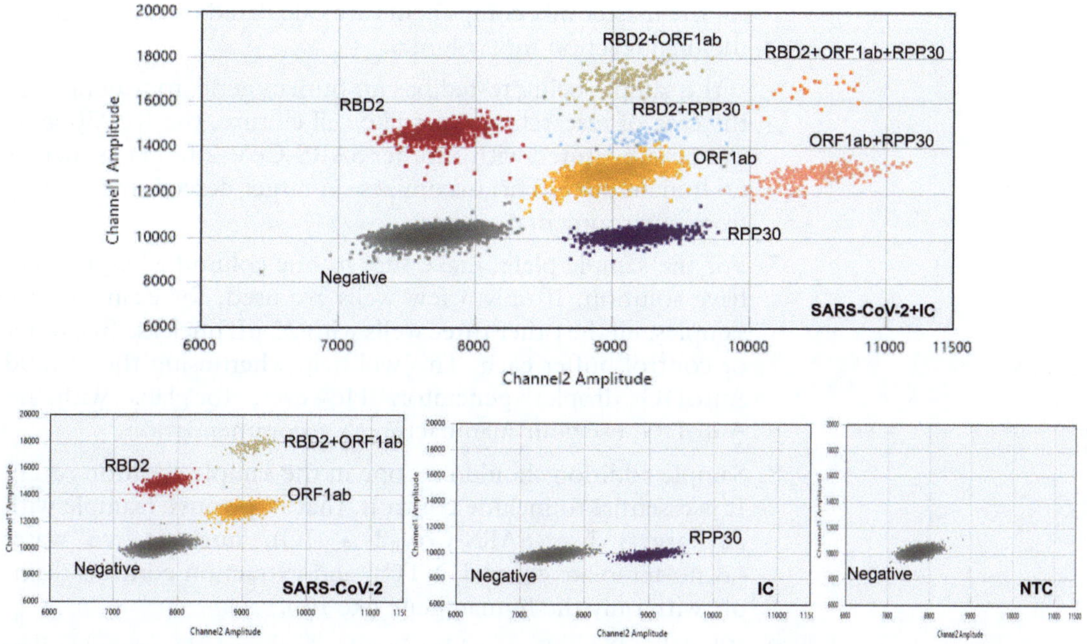

Fig. S3 Example of a one-step RT-ddPCR triplex probe mix assay result using three targets (RBD2 (FAM), ORF1ab (50% FAM and 50% HEX), and RPP30 (HEX)) and different sample categories. Eight droplet clusters were observed in the SARS-CoV-2 + IC sample category which reduced when differed sample categories with less targets were used

4 Notes

1. It is recommended to use a thermal cycler with deep wells. This will be highly compatible with Bio-Rad's ddPCR 96-well plates ensuring optimal distribution of heat throughout the droplets.
2. When reading droplets, it is essential to add controls including positive, negative, and no-template controls (NTC). These will help in setting thresholds for different target fluorophores especially using the positive control. In our sample categories, SARS-CoV-2 + IC may be seen as the positive control and IC as a negative/extraction control.
3. IC here means internal control but in reality, refers to the endogenous human reference gene RPP30.
4. Surfactants and lysis buffers may also be used for direct inactivation. The researcher can use their preferred sample inactivation methods.
5. Ensure to include extra volume to account for pipetting errors. For our work, we multiply each volume by × 1.05. To help in subsequent automated droplet generation, we used a total reaction mix volume of 22 μL from the recommended 20 μL.

All the master mix components are calculated in adjustment to the final reaction mix volume.

6. If the sample being tested has no human gene, for example, in the case of research samples like cell culture, the RPP30 gene can be substituted with another SARS-CoV-2 target as there is no human gene. This maximizes on target detection especially from environmental samples.
7. For the sample plate, make sure in one column all eight wells have solution. If only a few wells are used, for example, five samples, fill the other three wells with 22 μL nuclease free water or control buffer each. This will help when using the QX200 AutoDG droplet generator. However, for labs with no AutoDG, use your manufacturer's recommendation.
8. Sample addition should be done in the sample addition room. It is essential to include controls, that is, positive (sample with all targets like SARS-CoV-2 + IC), nuclease-free water (as no-template control, NTC), and extraction controls (sample with only the human gene like IC).
9. When loading consumables, it is recommended to start loading from the back of the instrument working toward the front to avoid moving your hands over consumables as it may be a potential source of cross-contamination.
10. The respective places where consumables should be loaded will have a yellow indicator that will later turn green if the consumables are loaded properly. This is similar to the touchscreen. Note that for each sample, two tips are used. Hence, in one row of 8 samples, one will need 16 AutoDG pipette tips.
11. The droplet generation plate should be placed on the cooling block. Ensure the cooling block is purple (ready to use) to avoid sample evaporation during droplet generation and not pink (should be kept in a freezer for about 30 min).
12. A thermal gradient (ranging from 65 to 50 °C) at the annealing step should be performed to achieve better separation and optimal annealing temperature of new primer/probe sets.
13. Let the amplified droplet plate stand for at least 30 min at 4 °C post PCR to allow for droplet stabilization. This improves the droplet reading results.
14. When selecting Target Type 1 or 2, one can also select Positive/Negative/Blank/NTC if they used controls. Ch1/Ch2 Unknown are for unknown sample types.
15. It is recommended to "Prime" the system first before running the first experiment of the day or after the instrument has stayed for long without operation. Additionally, check the oil levels and take note of any color changes on the status of the

droplet reader. Change the oil type or adjust the plate position in case of any problems as stated in the droplet reader guide.

16. While acquiring data, the software will display preliminary quality of acquired data and expected completion time. This data can be used to judge samples positive or negative before the whole experiment is completed for further analysis. The first wells can be the controls and one can see if the reading is OK and let the rest of the plate to run.
17. Other forms also exist on how to locate and open the files. Read the SOP on how to use the different software to explore the other options if necessary.
18. The preferred threshold cluster mode is circular however one can use the other options based on their data output.

References

1. Nyaruaba R, Li C, Mwaliko C, Mwau M, Odiwuor N, Muturi E, Muema C, Xiong J, Li J, Yu J, Wei H (2021) Developing multiplex ddPCR assays for SARS-CoV-2 detection based on probe mix and amplitude based multiplexing. Expert Rev Mol Diagn 21(1): 119–129. https://doi.org/10.1080/14737159.2021.1865807
2. Zhu N, Zhang D, Wang W, Li X, Yang B, Song J, Zhao X, Huang B, Shi W, Lu R, Niu P, Zhan F, Ma X, Wang D, Xu W, Wu G, Gao GF, Tan W, China Novel Coronavirus I, Research T (2020) A novel coronavirus from patients with pneumonia in china, 2019. N Engl J Med 382(8):727–733. https://doi.org/10.1056/NEJMoa2001017
3. Nyaruaba R, Zhang B, Muema C, Muturi E, Oyejobi GK, Xiong J, Li B, Shi Z, Mwaliko C, Yu J, Li X, Wei H (2020) Development of a field-deployable RT-qPCR Workflow for COVID-19 detection. https://doi.org/10.20944/preprints202004.0216.v1
4. Suo T, Liu X, Feng J, Guo M, Hu W, Guo D, Ullah H, Yang Y, Zhang Q, Wang X, Sajid M, Huang Z, Deng L, Chen T, Liu F, Xu K, Liu Y, Zhang Q, Liu Y, Xiong Y, Chen G, Lan K, Chen Y (2020) ddPCR: a more accurate tool for SARS-CoV-2 detection in low viral load specimens. Emerg Microbes Infect 9(1): 1259–1268. https://doi.org/10.1080/22221751.2020.1772678
5. Liu X, Feng J, Zhang Q, Guo D, Zhang L, Suo T, Hu W, Guo M, Wang X, Huang Z, Xiong Y, Chen G, Chen Y, Lan K (2020) Analytical comparisons of SARS-COV-2 detection by qRT-PCR and ddPCR with multiple primer/probe sets. Emerg Microbes Infect 9(1):1175–1179. https://doi.org/10.1080/22221751.2020.1772679
6. Lv J, Yang J, Xue J, Zhu P, Liu L, Li S (2020) Detection of SARS-CoV-2 RNA residue on object surfaces in nucleic acid testing laboratory using droplet digital PCR. Sci Total Environ 742:140370. https://doi.org/10.1016/j.scitotenv.2020.140370
7. Nyaruaba R, Li C, Mwaliko C, Mwau M, Odiwour N, Muturi E, Muema C, Xiong J, Li J, Yu J, Wei H (2020) Developing multiplex ddPCR assays for SARS-CoV-2 detection based on probe mix and amplitude based multiplexing. medRxiv.:2020.2010.2005.20207506. https://doi.org/10.1101/2020.10.05.20207506
8. Yu F, Yan L, Wang N, Yang S, Wang L, Tang Y, Gao G, Wang S, Ma C, Xie R, Wang F, Tan C, Zhu L, Guo Y, Zhang F (2020) Quantitative detection and viral load analysis of SARS-CoV-2 in infected patients. Clin Infect Dis 71(15): 793–798. https://doi.org/10.1093/cid/ciaa345
9. Liu Y, Ning Z, Chen Y, Guo M, Liu Y, Gali NK, Sun L, Duan Y, Cai J, Westerdahl D, Liu X, Xu K, Ho KF, Kan H, Fu Q, Lan K (2020) Aerodynamic analysis of SARS-CoV-2 in two Wuhan hospitals. Nature 582(7813): 557–560. https://doi.org/10.1038/s41586-020-2271-3
10. Nyaruaba R, Mwaliko C, Kering KK, Wei H (2019) Droplet digital PCR applications in the tuberculosis world. Tuberculosis (Edinb) 117: 85–92. https://doi.org/10.1016/j.tube.2019.07.001
11. Whale AS, Huggett JF, Tzonev S (2016) Fundamentals of multiplexing with digital PCR.

Biomol Detect Quantif 10:15–23. https://doi.org/10.1016/j.bdq.2016.05.002

12. Nyaruaba R, Xiong J, Mwaliko C, Wang N, Kibii BJ, Yu J, Wei H (2020) Development and evaluation of a single dye duplex droplet digital PCR assay for the rapid detection and quantification of mycobacterium tuberculosis. Microorgan 8(5). https://doi.org/10.3390/microorganisms8050701
13. Demeke T, Dobnik D (2018) Critical assessment of digital PCR for the detection and quantification of genetically modified organisms. Anal Bioanal Chem 410(17): 4039–4050. https://doi.org/10.1007/s00216-018-1010-1
14. Li H, Bai R, Zhao Z, Tao L, Ma M, Ji Z, Jian M, Ding Z, Dai X, Bao F, Liu A (2018) Application of droplet digital PCR to detect the pathogens of infectious diseases. Biosci Rep 38(6). https://doi.org/10.1042/BSR20181170
15. Nyaruaba R, Li X, Mwaliko C, Li C, Mwau M, Odiwour N, Muturi E, Muema C, Li J, Yu J, Wei H (2021) Two-step reverse transcription droplet digital PCR Protocols for SARS-CoV-2 detection and quantification. J Vis Exp 169. https://doi.org/10.3791/62295
16. Lu R, Wang J, Li M, Wang Y, Dong J, Cai W (2020) SARS-CoV-2 detection using digital PCR for COVID-19 diagnosis, treatment monitoring and criteria for discharge. medRxiv.:2020.2003.2024.20042689. https://doi.org/10.1101/2020.03.24.20042689
17. Viral Disease Control Institute n.d. http://www.chinaivdc.cn/kyjz/202001/t20200121_211337.html

Chapter 11

Profiling SARS-CoV-2 Infection by High-Throughput Shotgun Proteomics

Lucia Grenga, Duarte Gouveia, and Jean Armengaud

Abstract

A comprehensive cartography of viral and host proteins expressed during the different stages of SARS-CoV-2 infection is key to decipher the molecular mechanisms of pathogenesis. For the most detailed analysis, proteins should be first purified and then proteolyzed with trypsin in the presence of detergents. The resulting peptide mixtures are resolved by reverse phase ultrahigh pressure liquid chromatography and then identified by a high-resolution tandem mass spectrometer. The thousands of spectra acquired for each fraction can then be assigned to peptide sequences using a relevant protein sequence database, comprising viral and host proteins and potential contaminants from the growth medium or from the operator. The peptides are evidencing proteins and their intensities are used to infer the abundance of their corresponding proteins. Data analysis allows for highlighting the viral and host proteins dynamics. Here, we describe the sample preparation method adapted to profile SARS-CoV-2 -infected cell models, the shotgun proteomics pipeline to record experimental data, and the workflow for data interpretation to analyze infection-induced proteomic changes in a time-resolved manner.

Key words High-throughput proteomics, Tandem mass spectrometry, Protein extraction, Peptide preparation, Quantitative proteomics, Data analysis, Virus, Coronavirus

1 Introduction

The COVID-19 pandemic, instigated by the SARS-CoV-2 betacoronavirus, is an ongoing pneumonia which raises a lot of concerns worldwide and causes terrific casualties since its appearance in Wuhan, China and its global spread in a few weeks [1, 2]. The current number of SARS-CoV-2 positive patients is above 25 million and the death toll is superior to 851 thousand (2020/08/31). Before SARS-CoV-2, six coronavirus strains were known to cause diseases in humans, with four (229E, OC43, NL63, and HKU1) associated with mild, common cold symptoms, and two (SARS-CoV and MERS-CoV) responsible of fatal outbreaks in 2002/2003 and 2012, respectively [3]. SARS-CoV-2 belongs to the sarbecovirus lineage [4] and is the seventh coronavirus known to

Justin Jang Hann Chu et al. (eds.), *SARS-CoV-2: Methods and Protocols*,
Methods in Molecular Biology, vol. 2452, https://doi.org/10.1007/978-1-0716-2111-0_11,

infect humans. Despite the limited knowledge available a semester ago, once its genome was sequenced, molecular protocols were quickly established to detect the virus either based on quantitative reverse transcription PCR [5], mass spectrometry [6–8], immunoassays, or other alternatives. Besides these, SARS-CoV-2 infected host cells and hijacked cellular processes were profiled using shotgun proteomics as the main tool [9–12].

SARS-CoV-2 infected cells are manipulated in BSL3 facilities and should be inactivated prior to protein extraction and tandem mass spectrometry analysis. While many protocols have been conceived for proteomics analysis of biological samples, the present protocol is especially adapted for hazardous samples that can be thermo-inactivated in the presence of detergents. Specific precautions should be taken with BSL3 biological agents and therefore full inactivation of the samples should be checked before transferring the processed samples to the mass spectrometry facility. The method outlined here combines electrophoresis, chromatography, and high-resolution mass spectrometry to reveal the most abundant proteins in the sample. The depth of analysis depends directly on the gradient length applied for the chromatography and the power of the mass spectrometer used. Furthermore, the label-free method for quantifying proteins performs well in comparison with other methods relying on isotopic or chemical labeling which are time-consuming and required extra costs [13]. The main advantage of the protocol is its simplicity and robustness, making possible its application to numerous samples and providing a quick answer regarding the proteins modulated during SARS-CoV-2 infection. Using the method described below, we explored the production of SARS-CoV-2 viral proteins and simultaneously the modulation of host proteins following the infection of Vero E6 cells in a time-dependent manner [12]. We could detect with a high degree of confidence (below 1% peptide and protein false discovery rate) the presence of more than 5000 host proteins and 21 of the annotated proteolytically mature proteins encoded by SARS-CoV-2. The viral proteins were three out of the four structural proteins (N, S, and M), 15 of the 16 nonstructural proteins encoded by the Orf1ab and three of the nine accessory factors (ORF3a, ORF9b, and ORF7a).

2 Materials

Prepare all solutions using ultrapure water. Reagents used for mass spectrometry must be analytical grade. We recommend working with small aliquots and preparing fresh reagents regularly to avoid contamination.

2.1 Harvesting Cell Cultures Infected with SARS-CoV-2

1. Vero E6 cells (ATCC, CLR-1586) infected with SARS-CoV-2 strain 2019-nCoV/Italy-INMI1 (GenBank MT066156) (*see* **Note 1**).
2. Centrifuge. Typically, any centrifuge for 2 mL and 50 mL tubes operated at high speed will be appropriate.
3. Wash buffer: 10× Phosphate Buffered Saline (PBS). Store at 4 °C.
4. Lithium dodecyl sulfate (LDS) buffer (4×) (Invitrogen) which after dilution to 1× contains 10% glycerol (w/v), 2% LDS (w/v), 1 M triethanolamine-Cl pH 7.6, 1.4 M Tris base, 0.16 mM phenol red, 0.025 mM SERVA Blue G-250, 0.5 mM EDTA disodium, buffered pH 8.5. The LDS 4× solution is commercially available and is supplied concentrated in order to be incorporated into any sample without a strong diluting of the sample.
5. Heating block.
6. Ultrasonic probe.

2.2 SDS PAGE

1. Protein gel sample buffer: Dilute 1 mL of LDS sample buffer (4×) in 2.85 mL of water. Add 150 μL of β-mercaptoethanol. Store aliquots at −20 °C. Allow to reach room temperature before use.
2. Polyacrylamide gels: for example, 4%–12% Bis-Tris gradient 10-well gels. Store at 4 °C.
3. Electrophoresis system.
4. PAGE running buffer: Dilute 100 mL of 20× MES SDS Running Buffer (2-(N-morpholino)ethanesulfonic acid Buffered Saline) in 1.9 L of water. Store at 4 °C. The final solution MES 1× contains 50 mM MES, 50 mM Tris Base, 0.1% SDS, and 1 mM EDTA, and is buffered at pH 7.3. The 20× MES SDS solution is commercially available.
5. Coomassie blue stain.
6. Heating block.
7. Loading tips or syringe.
8. Ultrapure water to rinse gel.

2.3 Reduction, Alkylation, and Digestion of Proteins

1. Clean scalpel or razor blade.
2. Orbital Shaker.
3. Dehydration solution: 50% (v/v) acetonitrile solution in 50 mM ammonium bicarbonate.
4. SpeedVac.
5. Deionized water.

6. Reduction solution: 25 mM dithiothreitol in 50 mM ammonium bicarbonate.
7. Alkylation solution: 55 mM iodoacetamide in 50 mM ammonium bicarbonate. Store in the dark.
8. Enzyme solution: Reconstitute lyophilized sequencing-grade trypsin to a final concentration of 0.1 μg/μL in 0.01% trifluoroacetic acid. Reconstituted enzyme can be aliquoted and stored at −20 °C for several months or at 4 °C for up to one week (*see* **Note 2**).
9. Trypsin enhancer solution: Reconstitute 1 mg lyophilized ProteaseMax (Promega) with 100 μL of 50 mM ammonium bicarbonate (*see* **Note 3**). The resulting solution contains 1% ProteaseMax and should be aliquoted and stored at −20 °C.
10. Digestion solution: 16 μL of 50 mM ammonium bicarbonate, 2 μL of 0.1% ProteaseMax, and 2 μL of 0.1 μg/μL trypsin (20 μL total volume) per gel piece. Keep on ice until use.
11. Ice.
12. Heating block—37 °C, 56 °C.
13. Trifluoroacetic acid: 5% stock solution in water. Prepare 0.5% and 0.1% solutions in water.

2.4 NanoLC-MS/MS

1. Mass spectrometer: Q-Exactive HF (ThermoFisher) coupled to an UltiMate 3000 nano-LC system (Dionex).
2. Reverse-phase Acclaim PepMap100 C18 μ-precolumn (5 μm, 100 Å, 300 μm i.d. × 5 mm, ThermoFisher).
3. Nanoscale Acclaim PepMap100 C18 nano-LC column (3 μm, 100 Å, 75 μm i.d. × 50 cm, ThermoFisher).
4. LC solvent A: 0.1% formic acid in water.
5. LC solvent B: 0.1% formic acid, 80% acetonitrile in water.

2.5 Identification and Quantitation

1. Mascot Daemon software (version 2.6.0, Matrix Science).
2. MaxQuant software (version 1.6.6.0).
3. R software (version 4.0.2).
4. Perseus software (version 1.6.14.0).

3 Methods

Carry out all procedures at room temperature unless otherwise specified. Be sure to carry out all the steps involving manipulation of samples infected with SARS-CoV-2 in a Biosafety level 3 laboratory.

3.1 Harvesting Infected Cell Cultures

1. Following a gentle cell scraping directly in the culture media, recover the suspension in a 15 mL centrifuge tube and centrifuge 5 min at 2800 × *g* at room temperature (*see* **Note 4**).
2. Wash 2× with 5 mL of 1× PBS (*see* **Note 4**). Remove residual washing liquid after brief centrifugation.
3. Resuspend the cell pellet directly in 200 μL of 1× LDS buffer (*see* **Note 5**).
4. Heat at 95 °C for 10 min in a heating block.
5. Sonicate briefly (approximately 20 s at 60% amplitude).

At this step the samples may be processed outside of the Biosafety level 3 laboratory. We recommend to check for appropriate virus full inactivation by plaque assay. Adequate precautions (use gloves, clean surfaces with ethanol) must be taken to avoid keratin contamination during sample preparation for mass spectrometry (Fig. 1).

3.2 SDS-PAGE

1. Dilute the sample 1:2 in 1× LDS buffer (*see* **Note 6**).
2. Heat at 95 °C for 5 min in a heating block.
3. Load 20 μL onto a polyacrylamide gel (*see* **Note** 7).
4. Run at 200 V for approximately 5 min or until all of the sample has entered into the gel but has not progressed more than 10 mm.
5. Rinse the gel three times with water.
6. Stain with Coomassie blue for 30–60 min.
7. Rinse well with water.

Figure 1 shows the different steps for the sample preparation for mass spectrometry and illustrates the polyacrylamide gel after 5 min electrophoresis and Coomassie blue staining.

3.3 Reduction, Alkylation, and Digestion

1. Excise bands of interest using a clean scalpel or razor blade and transfer to microcentrifuge tube (*see* **Note 8**).
2. Dehydrate with 200 μL of dehydration solution, shake for 5 min at 600 rpm, and discard the fluid.
3. Dehydrate with 200 μL of pure acetonitrile, shake for 1 min at 600 rpm, and discard the fluid.
4. Dry in a SpeedVac for 2–5 min.
5. Rehydrate the gel pieces with 100 μL of reduction solution; incubate for 10–20 min at 56 °C, shaking at 500 rpm. Discard the fluid.
6. Add 100 μL of alkylation solution; incubate for 10–20 min at room temperature in the dark. Discard the fluid.

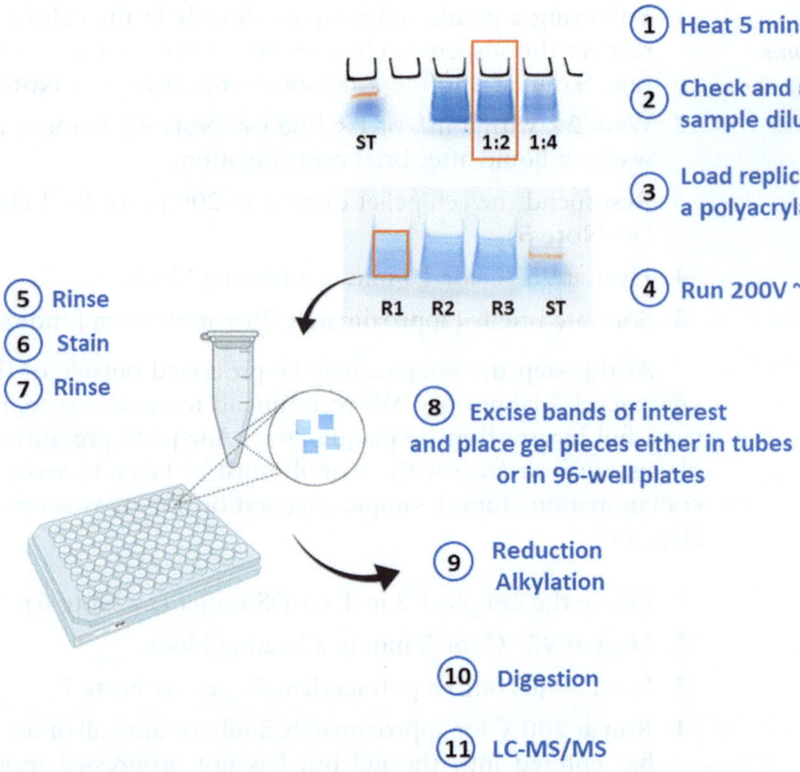

Fig. 1 Main steps of the procedures involving the use of SDS-PAGE gel for shotgun proteomic analysis of SARS-CoV-2 infected cells. After adjusting and checking sample concentration (e.g., by densitometry), replicates (here: R1, R2, and R3) and a standard (ST, here: See Blue Plus2 marker from Invitrogen) are loaded and run until they enter the gel. Bands of interest are excised and, depending on the number of sample, placed either in 1.5 mL tubes or in 96-well plates for in-gel trypsin proteolysis

7. Wash with 400 μL of deionized water; shake for 1 min at 600 rpm; discard the fluid. Repeat this step once (*see* **Note 9**).
8. Dehydrate with 200 μL of dehydration solution, shake for 5 min 600 rpm, and discard the fluid.
9. Dehydrate with 200 μL of pure acetonitrile, shake for 1 min at 600 rpm, and discard the fluid.
10. Dry in a SpeedVac for 2–5 min.
11. Rehydrate the gel piece(s) with 20 μL of enzyme solution, and incubate for 15 min on ice. Remove excess liquid.
12. Add 50 μL of 0.01% ProteaseMax; shake briefly.
13. Incubate for 1 h at 50 °C.
14. Transfer the solution to a clean tube (or well) and add 5 μL of 5% trifluoroacetic acid. If the recovered volume is less than 50 μL, which is often the case for larger gel pieces, add 0.1% trifluoroacetic acid equivalent to the lost volume, shake for

5 min at 500 rpm, and pool the solution with the previously recovered volume (*see* **Note 10**).

As illustrated in Fig. 1, the whole procedure may be performed in 96-well plate if numerous samples have to be treated in parallel.

3.4 NanoLC-MS/MS

Settings and conditions are described for the Q-Exactive HF (ThermoFisher) coupled to an UltiMate 3000 nano-LC system (ThermoFisher) equipped with a reverse-phase Acclaim PepMap100 C18 μ-precolumn (5 μm, 100 Å, 300 μm i.d. × 5 mm, ThermoFisher) followed by a nanoscale Acclaim PepMap100 C18 capillary column (3 μm, 100 Å, 75 μm i.d. × 15 cm, ThermoFisher).

1. Load 1–10 μL (maximum volume allowed by the system) of the acidified peptide mixture and resolve over a 100 min linear gradient from 4% to 25% solvent B and 20 min linear gradient from 25% to 40% solvent B using a flow rate of 0.2 μL/min. Adjust the loading volume as a function of the total current measured by the mass spectrometer to avoid saturating the detector.
2. Collect full-scan mass spectra over the 350–1500 *m/z* range and MS/MS on the 20 most abundant precursor ions (minimum signal required set at $5e^4$, possible charge states: 2+ and 3+), apply a 10 s dynamic exclusion window (*see* **Note 11**). Isolate the secondary ions within a 2.0 *m/z* window.

3.5 Identification

Independently of the search engine of choice for peptide and protein identification, an appropriate database should be created to search the MS/MS spectra. The database should contain the *Chlorocebus sabaeus* (Vero cells) and the SARS-CoV-2 protein sequences. An additional database should be created to remove contaminant spectra and it should include the sequence of the contaminants classically observed in proteomics such as keratin and trypsin and *Bos taurus* sequences corresponding to the 23 most abundant proteins from fetal calf serum (FCS) as defined previously [14] and present in the cell culture medium (*see* **Note 12**).

3.5.1 Identification Using the Mascot Deamon Software (version 2.6.0, Matrix Science)

1. Generate peak lists using the Mascot Daemon software (version 2.6.0, Matrix Science). Data import filter options should be as follows: 400 (minimum mass), 5000 (maximum mass), 0 (grouping tolerance), 0 (intermediate scans), 10 (minimum peaks), 2 (extract MSn), and 1000 (threshold) (*see* **Note 13**).
2. Search MS/MS spectra against an appropriate database using the following parameters: 2 (maximum number of missed cleavages), 5 ppm (mass tolerance for the parent ion), 0.02 Da (mass tolerance for the product ions), carbamidomethylated

cysteine residues (fixed modification), and oxidized methionine residues (variable modification).

3.5.2 Identification Using the MaxQuant Software (version 1.6.6.0)

1. Perform the MaxQuant searches using the built-in search engine Andromeda with the default settings. Select carbamidomethylated cysteine as fixed modification and oxidation of methionine as variable modification. Specify trypsin as the proteolytic enzyme (*see* **Note 14**) and allow maximum two missed cleavages. Set peptide and protein FDRs at 1%. If time course experiments are performed, enable the match between runs option and the FTMS recalibration to align the different runs.

3.6 Quantitation

Different features can be considered to assess protein abundance in label free experiments. Here, quantitation using spectral counts (SCs) and extracted ion chromatograms (XICs) is presented. Depending on the experimental settings and the biological questions, one approach could be preferred over the other.

3.6.1 Quantitation Based on Spectral Counts

1. Parse the MASCOT DAT files using the ms_peptidesummary function of msparser (version 2.6.0, Matrix Science).
2. Using MASCOT's homology threshold option, validate the peptide-to-spectrum matches (PSMs) with expectation values corresponding to a 1% False Discovery Rate (FDR). Allow multiple PSMs for MS/MS spectra if ion scores is higher than 98% of the top ion score.
3. Group proteins if they share at least one peptide, base the label-free quantification of each group on PSM counts for each protein following the principle of parsimony. Retain the proteins identified by one or more specific peptides for the analysis (protein FDR 1%).

3.6.2 Quantitation Based On Extracted Ion Chromatograms

In case of robust workflow, peptide-ion intensities could provide more accurate quantification than spectral counts due to the higher dynamic range of the foremost. For label-free quantification via XICs, MaxQuant uses the MaxLFQ algorithm [15]. A profile of LFQ intensities is calculated for each protein as a proxy for absolute protein abundance and the results of the MaxQuant analysis are returned as txt tables.

3.7 Data Preprocessing

Typically, proteomic data is presented in the form of a matrix table with each sample being described by the abundance values of an extensive list of proteins. Before performing the statistical analysis, these data should be preprocessed. Preprocessing proteomics data involves filtering out low abundant proteins detected in a limited number of samples, handling missing/zero values, and normalizing to correct for interrun variability. Different normalization strategies can sometimes be applied depending on the statistical analysis being performed. The approaches presented here are one example among a large panel of options that should be carefully selected, depending on the experimental design and the biological questions, before performing these operations.

3.7.1 Spectral Count data

1. In Excel or in other data treatment software, normalize the spectral count abundances of each protein to the total number of spectral counts in the sample (*see* **Note 15**).

3.7.2 XIC Data

1. Load the Peptides.txt or ProteinGroups.txt file from the MaxQuant search into the Perseus software and remove contaminants, reverse identifications and proteins identified only by a modification site (only from the ProteinGroups.txt file) from further data analysis.
2. Apply a log2-transformation to the LFQ abundances.
3. Impute missing values for each replicate individually by sampling values from a normal distribution calculated from the original data distribution (width $= 0.3 \times$ s.d., downshift $= -1.8 \times$ s.d.).
4. Remove protein groups that do not contain a minimum of two nonimputed values for tested condition.

3.8 Data Analysis

Typically, the statistical analysis of proteomics data is based on: (i) nonsupervised multivariate methods to explore data variability in a nonsupervised manner; (ii) feature selection based on multivariate models, univariate pairwise comparisons, or coexpression analysis, and (iii) biological analysis of the features of interest (Fig. 2). Here, using the data detailed in Grenga et al. [12], we describe data reduction and exploration with principal component analysis (PCA), coexpression analysis of host proteins during infection, and process and pathway enrichment analysis.

3.8.1 Exploratory Nonsupervised Analysis (Principal Components Analysis)

1. Import the normalized protein abundance table into the R software. The table should be in the following format: one sample per line (first column); the second column contains a label corresponding to the experimental condition (optional); every other column contains the proteins and their abundances in each sample.

```
prot_data <- read.delim("quant.txt")
row.names(prot_data)<-prot_data$Sample
prot_data <- prot_data[,2:1403]
```

2. Install and load the R package "factoextra" (*see* **Note 16**).

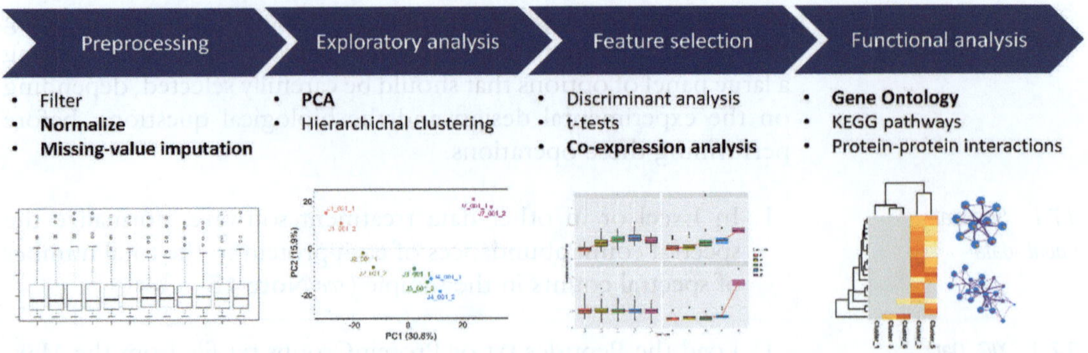

Fig. 2 Summary of a typical data analysis pipeline for shotgun proteomics datasets, starting from a protein abundance list. Text in bold corresponds to the approaches selected in the protocol described herein

```
library(factoextra)
```

3. Calculate the principal components for data reduction using the function "prcomp".

```
res.pca<-prcomp(prot_data[,2:1402], scale=TRUE).

fviz_eig(res.pca)
```

4. Visualize the scores plot with the function "fviz_pca_ind", analyze sample and replicate clustering on the first two components and look for outliers. See Fig. 2, Exploratory analysis panel.

```
groups <- as.factor(SC$Group)

fviz_pca_ind(res.pca, geom = c("point", "text"), labelsize =
4, pointsize = 3, pointshape = 16, habillage = "none", mean.
point = FALSE, axes.linetype = "blank", col.ind = groups,
palette = c("#F8766D", "#A3A500", "#07AD73", "#00F0F6",
"#E76BF3"), addEllipses = F, ellipse.type = "confidence",
legend.title = "", repel = F, legend = "none", title = "") +
theme(text = element_text(size = 15), axis.title = element_-
text(size = 15), axis.text = element_text(size = 15), panel.
border = element_rect(colour = "black", linetype = "solid",
size = 0.5, fill = NA), panel.background = element_blank(),
panel.grid.major = element_blank(), panel.grid.minor = ele-
ment_blank())
```

Nonsupervised multivariate analyses provides a straightforward way of checking the variability present in our dataset (given by the percentage of variance in the first components), replicate reproducibility, and predetermine which samples/conditions seem to differ in their proteomic profiles.

3.8.2 Coexpression Analysis for Temporal Data

1. Import the protein abundance table into the R software (samples in lines, proteins in columns).

```
prot_data2  <  - read.csv("quant_coseq.txt", header  =  TRUE,
sep  =  "\t",row.names  =  1)
```

```
conds <- c(rep("J1", 2), rep("J2", 2), rep("J3", 2), rep("J4",
2), rep("J7", 2))
```

2. Install and load the Bioconductor R package "coseq" (here we used version 1.5.2), and the packages "Biobase" and "corrplot."

```
library(coseq)
```

```
library(Biobase)
library(corrplot)
```

3. Use the "coseq" function to perform the clustering of proteins based on their coexpression profiles along a time-series experiment. Here we performed coseq with the k-means algorithm on log-transformed data.

```
set.seed(12345)
KM <- coseq(prot_data2, K=2:20, model="kmeans", transforma-
tion="logclr")
summary(KM)
```

```
## ************************************************
## K-means algorithm
## Transformation: logclr
## ************************************************
# #                Clusters                fit:
2,3,4,5,6,7,8,9,10,11,12,13,14,15,16,17,18,19,20
## Clusters with errors: ---
## Selected number of clusters via capushe: 4
## ************************************************
## Number of clusters = 4
## ************************************************
## Cluster sizes:
```

```
## Cluster 1 Cluster 2 Cluster 3 Cluster 4
##       295       222       867        17
```

4. Plot the clustered profiles with the "plot" function on the coseq result, with either a line graph (graphs = "profiles") or boxplots (graphs = "boxplots"). See Fig. 2, Feature selection panel.

```
plot(KM, graphs="boxplots", conds = conds, collapse_reps =
"average")
```

5. Export the list of proteins coexpressed in each cluster.

```
Elem <- assay(KM)
write.table(elem, file ="kmeans_clusters.txt", quote = TRUE,
sep = "\t",eol = "\n")
```

In this example with data from Grenga et al. [12], clusters two (222 coexpressed proteins) and four (17 proteins) present interesting profiles regarding the infection process. The 339 proteins from these clusters are therefore of utmost interest and can be further explored via pathway and network analysis to determine their biological roles within the cells.

3.8.3 Process and Pathway Enrichment Analysis

Numerous options are available to retrieve statistically enriched terms starting from a list of proteins of interest. Among them, web-based portal Metascape incorporates a core set of default ontologies for enrichment analysis, including GO processes, KEGG pathways, Reactome gene sets, canonical pathways, and CORUM complexes [16]. The most widely used analysis options and best analysis practices are encapsulate in the Express Analysis but a certain degree of freedom on the settings is available through the Custom Analysis workflow. The output of the analysis could be visualized using several R packages including ggplot2 and could, for instance, contribute to the evaluation of the biological significance of the infection-modulated host proteins. See Fig. 2, Functional analysis panel.

4 Notes

1. Different host cell models and SARS-CoV-2 strains could be selected to profile SARS-CoV-2 infection by shotgun proteomics. The present protocol has been designed using Vero E6 cells (ATCC, CLR-1586) and the SARS-CoV-2 strain 2019-nCoV/Italy-INMI1 (GenBank MT066156) but can be

adopted without major changes to profile SARS-CoV-2 infection under different experimental conditions.

2. Passing reconstituted trypsin through multiple freeze-thaw cycles is not recommended. Solutions containing enzyme should be kept cold at all times.
3. ProteaseMax is a specific reagent made by Promega. Other detergents could be used, but this protocol was designed using ProteaseMax, and others may not work exactly the same. The ratio of detergent to digest solution is an important parameter for digestion efficiency and later LC separation, and it may not be the same for other products.
4. The volumes described in this protocol are adjusted for cells seeded in 25 cm^2 flasks in 2–5 mL of growth medium. Different culture volumes could require centrifuge tubes of different size and adjustment of the volume of 1× PBS used during the wash steps.
5. At this point, cell pellets can be frozen for later analysis.
6. For less or more concentrated samples omit or adjust this dilution step.
7. We use 4%–12% Bis-Tris gels. However, for such short run times, the actual composition of the gel is not important.
8. Larger gel sections can be further cut into pieces approximately 4 mm^3 to increase the surface area in contact with solution. Also, depending on the number of samples, it may be advantageous to place gel pieces either in microcentrifuge tubes or in 96-well plates.
9. If using 96-well plates, wash with 200 μL of water.
10. Peptide solutions that will not be analyzed immediately should be frozen and stored at −20 °C or colder [17]. Peptides in solution are generally stable for 1–2 weeks at +4 °C, for 3–4 months at −20 °C and for 1 year at −80 °C. For optimal long-term storage, peptide solutions should be lyophilized in low-adsorption tubes which prevent peptides to stick to the tube walls.
11. Dynamic exclusion parameters should be adjusted depending on the amount and diversity of peptides expected in the sample [18].
12. Customized databases can be generated by exporting the appropriate references from the National Center for Biotechnology Information database. Alternatively, the NCBI or SwissProt databases, or subsets thereof, can be searched. Databases can also be constructed from in-house sequencing results; however, the appropriate parse rules will need to be

determined empirically. Mass tolerances can be adjusted based on system performance.

13. ThermoFisher supplies the "extract_msn.exe data" import filter as part of its Xcalibur FT package (version 2.0.7). The exact mechanism for transferring raw data will depend on the system used.
14. If additional proteolytic enzymes are used, they should be selected.
15. Optional: if you wish to avoid zero values, add one pseudo spectral count (+1) per protein and per sample, before normalization, as recommended in Carvalho et al. [19].
16. Many packages and functions that can apply and visualize PCA in R.

Acknowledgments

This work was supported by the Commissariat à l'Energie Atomique et aux Energies Alternatives, and the Agence Nationale de la Recherche (ANR-12-BSV6-0012-01). We thank our colleagues from CEA-Li2D for stimulating discussions.

References

1. Armengaud J, Delaunay-Moisan A, Thuret JY, van Anken E, Acosta-Alvear D, Aragon T, Arias C, Blondel M, Braakman I, Collet JF, Courcol R, Danchin A, Deleuze JF, Lavigne JP, Lucas S, Michiels T, Moore ERB, Nixon-Abell J, Rossello-Mora R, Shi ZL, Siccardi AG, Sitia R, Tillett D, Timmis KN, Toledano MB, van der Sluijs P, Vicenzi E (2020) The importance of naturally attenuated SARS-CoV-2in the fight against COVID-19. Environ Microbiol 22(6):1997–2000. https://doi.org/10.1111/1462-2920.15039
2. Zhou P, Yang XL, Wang XG, Hu B, Zhang L, Zhang W, Si HR, Zhu Y, Li B, Huang CL, Chen HD, Chen J, Luo Y, Guo H, Jiang RD, Liu MQ, Chen Y, Shen XR, Wang X, Zheng XS, Zhao K, Chen QJ, Deng F, Liu LL, Yan B, Zhan FX, Wang YY, Xiao GF, Shi ZL (2020) A pneumonia outbreak associated with a new coronavirus of probable bat origin. Nature 579(7798):270–273. https://doi.org/10.1038/s41586-020-2012-710.1038/s41586-020-2012-7. [pii]
3. Zheng J (2020) SARS-CoV-2: an emerging coronavirus that causes a global threat. Int J Biol Sci 16(10):1678–1685. https://doi.org/10.7150/ijbs.45053ijbsv16p1678. [pii]
4. Boni MF, Lemey P, Jiang X, Lam TT, Perry BW, Castoe TA, Rambaut A, Robertson DL (2020) Evolutionary origins of the SARS-CoV-2 sarbecovirus lineage responsible for the COVID-19 pandemic. Nat Microbiol 5 (11):1408–1417. https://doi.org/10.1038/s41564-020-0771-410.1038/s41564-020-0771-4. [pii]
5. Kubina R, Dziedzic A (2020) Molecular and serological tests for COVID-19 a comparative review of SARS-CoV-2 coronavirus laboratory and point-of-care diagnostics. Diagnostics (Basel) 10(6):E434. https://doi.org/10.3390/diagnostics10060434. [pii]
6. Gouveia D, Grenga L, Gaillard JC, Gallais F, Bellanger L, Pible O, Armengaud J (2020) Shortlisting SARS-CoV-2 peptides for targeted studies from experimental data-dependent acquisition tandem mass spectrometry data. Proteomics 20(14):e2000107. https://doi.org/10.1002/pmic.202000107
7. Gouveia D, Miotello G, Gallais F, Gaillard JC, Debroas S, Bellanger L, Lavigne JP, Sotto A, Grenga L, Pible O, Armengaud J (2020) Proteotyping SARS-CoV-2 virus from nasopharyngeal swabs: a proof-of-concept focused on a 3 min mass spectrometry window. J

Proteome Res 19(11):4407–4416. https://doi.org/10.1021/acs.jproteome.0c00535

8. Nachtigall FM, Pereira A, Trofymchuk OS, Santos LS (2020) Detection of SARS-CoV-2 in nasal swabs using MALDI-MS. Nat Biotechnol 38(10):1168–1173. https://doi.org/10.1038/s41587-020-0644-710.1038/s41587-020-0644-7. [pii]

9. Bojkova D, Klann K, Koch B, Widera M, Krause D, Ciesek S, Cinatl J, Munch C (2020) Proteomics of SARS-CoV-2-infected host cells reveals therapy targets. Nature 583(7816):469–472. https://doi.org/10.1038/s41586-020-2332-710.1038/s41586-020-2332-7. [pii]

10. Bouhaddou M, Memon D, Meyer B, White KM, Rezelj VV, Correa Marrero M, Polacco BJ, Melnyk JE, Ulferts S, Kaake RM, Batra J, Richards AL, Stevenson E, Gordon DE, Rojc A, Obernier K, Fabius JM, Soucheray M, Miorin L, Moreno E, Koh C, Tran QD, Hardy A, Robinot R, Vallet T, Nilsson-Payant BE, Hernandez-Armenta C, Dunham A, Weigang S, Knerr J, Modak M, Quintero D, Zhou Y, Dugourd A, Valdeolivas A, Patil T, Li Q, Huttenhain R, Cakir M, Muralidharan M, Kim M, Jang G, Tutuncuoglu B, Hiatt J, Guo JZ, Xu J, Bouhaddou S, CJP M, Gaulton A, Manners EJ, Felix E, Shi Y, Goff M, Lim JK, McBride T, O'Neal MC, Cai Y, JCJ C, Broadhurst DJ, Klippsten S, De Wit E, Leach AR, Kortemme T, Shoichet B, Ott M, Saez-Rodriguez J, tenOever BR, Mullins RD, Fischer ER, Kochs G, Grosse R, Garcia-Sastre-A, Vignuzzi M, Johnson JR, Shokat KM, Swaney DL, Beltrao P, Krogan NJ (2020) The global phosphorylation landscape of SARS-CoV-2 infection. Cell 182(3):685–712 e619. S0092-8674(20)30811-4 [pii]. https://doi.org/10.1016/j.cell.2020.06.034

11. Gordon DE, Jang GM, Bouhaddou M, Xu J, Obernier K, White KM, O'Meara MJ, Rezelj VV, Guo JZ, Swaney DL, Tummino TA, Huttenhain R, Kaake RM, Richards AL, Tutuncuoglu B, Foussard H, Batra J, Haas K, Modak M, Kim M, Haas P, Polacco BJ, Braberg H, Fabius JM, Eckhardt M, Soucheray M, Bennett MJ, Cakir M, McGregor MJ, Li Q, Meyer B, Roesch F, Vallet T, Mac Kain A, Miorin L, Moreno E, Naing ZZC, Zhou Y, Peng S, Shi Y, Zhang Z, Shen W, Kirby IT, Melnyk JE, Chorba JS, Lou K, Dai SA, Barrio-Hernandez I, Memon D, Hernandez-Armenta C, Lyu J, Mathy CJP, Perica T, Pilla KB, Ganesan SJ, Saltzberg DJ, Rakesh R, Liu X, Rosenthal SB, Calviello L, Venkataramanan S, Liboy-Lugo J, Lin Y, Huang XP, Liu Y, Wankowicz SA, Bohn M, Safari M, Ugur FS, Koh C, Savar NS, Tran QD, Shengjuler D, Fletcher SJ, O'Neal MC, Cai Y, Chang JCJ, Broadhurst DJ, Klippsten S, Sharp PP, Wenzell NA, Kuzuoglu-Ozturk D, Wang HY, Trenker R, Young JM, Cavero DA, Hiatt J, Roth TL, Rathore U, Subramanian A, Noack J, Hubert M, Stroud RM, Frankel AD, Rosenberg OS, Verba KA, Agard DA, Ott M, Emerman M, Jura N, von Zastrow M, Verdin E, Ashworth A, Schwartz O, d'Enfert C, Mukherjee S, Jacobson M, Malik HS, Fujimori DG, Ideker T, Craik CS, Floor SN, Fraser JS, Gross JD, Sali A, Roth BL, Ruggero D, Taunton J, Kortemme T, Beltrao P, Vignuzzi M, Garcia-Sastre A, Shokat KM, Shoichet BK, Krogan NJ (2020) A SARS-CoV-2 protein interaction map reveals targets for drug repurposing. Nature 583(7816): 459–468. https://doi.org/10.1038/s41586-020-2286-910.1038/s41586-020-2286-9. [pii]

12. Grenga L, Gallais F, Pible O, Gaillard JC, Gouveia D, Batina H, Bazaline N, Ruat S, Culotta K, Miotello G, Debroas S, Roncato MA, Steinmetz G, Foissard C, Desplan A, Alpha-Bazin B, Almunia C, Gas F, Bellanger L, Armengaud J (2020) Shotgun proteomics analysis of SARS-CoV-2-infected cells and how it can optimize whole viral particle antigen production for vaccines. Emerg Microbes Infect 9(1):1712–1721. https://doi.org/10.1080/22221751.2020.1791737

13. Gouveia D, Grenga L, Pible O, Armengaud J (2020) Quick microbial molecular phenotyping by differential shotgun proteomics. Environ Microbiol 22(8):2996–3004. https://doi.org/10.1111/1462-2920.14975

14. Pereira S, Malard V, Ravanat J-L, Davin A-H, Armengaud J, Foray N, Adam-Guillermin C (2014) Low doses of gamma-irradiation induce an early bystander effect in zebrafish cells which is sufficient to radioprotect cells. PLoS One 9(3):e92974

15. Cox J, Hein MY, Luber CA, Paron I, Nagaraj N, Mann M (2014) Accurate proteome-wide label-free quantification by delayed normalization and maximal peptide ratio extraction, termed MaxLFQ. Mol Cell Proteomics 13(9):2513–2526. https://doi.org/10.1074/mcp.M113.031591M113.031591. [pii]

16. Zhou Y, Zhou B, Pache L, Chang M, Khodabakhshi AH, Tanaseichuk O, Benner C, Chanda SK (2019) Metascape provides a biologist-oriented resource for the analysis of systems-level datasets. Nat Commun 10(1): 1523. https://doi.org/10.1038/s41467-

019-09234-610.1038/s41467-019-09234-6. [pii]

17. Rosenling T, Slim CL, Christin C, Coulier L, Shi S, Stoop MP, Bosman J, Suits F, Horvatovich PL, Stockhofe-Zurwieden N, Vreeken R, Hankemeier T, van Gool AJ, Luider TM, Bischoff R (2009) The effect of preanalytical factors on stability of the proteome and selected metabolites in cerebrospinal fluid (CSF). J Proteome Res 8(12):5511–5522. https://doi.org/10.1021/pr9005876
18. Hartmann EM, Allain F, Gaillard JC, Pible O, Armengaud J (2014) Taking the shortcut for high-throughput shotgun proteomic analysis of bacteria. Methods Mol Biol 1197:275–285. https://doi.org/10.1007/978-1-4939-1261-2_16
19. Carvalho PC, Fischer JS, Chen EI, Yates JR 3rd, Barbosa VC (2008) PatternLab for proteomics: a tool for differential shotgun proteomics. BMC Bioinformatics 9:316. https://doi.org/10.1186/1471-2105-9-3161471-2105-9-316. [pii]

Chapter 12

Detection of SARS-CoV-2 by Mass Spectrometry

Petra Wandernoth, Katharina Kriegsmann, Jörg Kriegsmann, and Mark Kriegsmann

Abstract

Amplification of viral ribonucleic acid by real-time reverse transcriptase polymerase chain reaction is the gold standard to detect severe acute respiratory syndrome coronavirus 2 (SARS-CoV-2). However, alternative reliable, fast, and cost-effective methods for the detection of SARS-CoV-2 are still needed. In this chapter, the mass spectrometry–based detection of amplified polymerase chain reaction products of SARS-CoV-2 genes from oral or nasopharyngeal swabs is described. The respective SARS-CoV-2 test has previously been shown to meet standard quality criteria and was therefore approved by the authorities in Europe and the USA.

Key words Mass Spectrometry, MassARRAY, SARS-CoV-2, COVID-19

1 Introduction

Severe acute respiratory syndrome coronavirus 2 (SARS-CoV-2) is a positive-sense single-stranded ribonucleic acid (RNA) virus and has been identified as the causative agent of coronavirus disease 2019 (COVID-19) disease [1]. To control the pandemic and to avoid uncontrolled viral spread, high-throughput testing and subsequent isolation of infected individuals were advised and performed worldwide [1]. Thus, there was a need for fast, reliable, and cost-effective testing modalities. Despite the development of effective vaccines, the need to test for SARS-CoV-2 is expected to stay high, at least until immunity of a substantial proportion of the population is reached [2, 3].

Real-time reverse transcriptase polymerase chain reaction (rRT-PCR) is the current gold standard for the detection of SARS-CoV-2 [4]. However, as a result of the large demand for commercially available test kits, there was a shortage in materials and other methods were needed to complement the diagnostic arsenal [5, 6].

Justin Jang Hann Chu et al. (eds.), *SARS-CoV-2: Methods and Protocols*,
Methods in Molecular Biology, vol. 2452, https://doi.org/10.1007/978-1-0716-2111-0_12,

Mass spectrometric (MS) techniques have complemented or replaced traditional methods in laboratory medicine, toxicology, microbiology as well as molecular pathology and are suitable for reliable, cost-effective, and rapid detection of amplified PCR products [7–11]. The method is based on matrix assisted laser desorption/ionization time-of-flight and has previously been shown suitable to detect SARS-CoV-2 on oral or nasopharyngeal swabs [12]. The test described herein was recently certified (CE-IVD) and was approved by the Food and Drug Administration [13].

The procedure can be divided into five steps, which will be covered in detail in this chapter.

1. RNA extraction.
2. RT-PCR amplification.
3. Shrimp alkaline phosphatase (SAP) reaction.
4. iPLEX Pro extension reaction.
5. Data analysis.

2 Materials

2.1 RNA Extraction

- Chemagic 360 Nucleic Acid Extractor instrument (PerkinElmer LAS, Germany).
- Chemagic 96 Rod Head Set and software (PerkinElmer LAS, Germany).
- Chemagic Viral DNA/RNA 300 Kit H96 (PerkinElmer LAS, Germany) (see Table 1).
- Nuclease free water.

2.2 SARS-CoV-2 Detection with the MassARRAY System

- MassARRAY System with Chip Prep Module (CPM) 96 (Agena Bioscience, Germany).
- Typer software v.5.0.1 or greater (Agena Bioscience, Germany).
- RT-Workstation v4.1 or greater (Agena Bioscience, Germany).
- Chip Prep Controller v.2.2 or greater (Agena Bioscience, Germany).
- SARS-CoV-2 Report v1.1 or greater (Agena Bioscience, Germany).
- MassARRAY® SARS-CoV-2 Panel Kit, PCR Reagent Set with dUTP, iPLEX® Pro Reagent Set, and SpectroCHIP CPM Array (Agena Bioscience, Germany) (see Table 2).

Table 1
Content of the chemagic viral DNA/RNA 300 kit H96

Chemagic viral DNA/RNA 300 kit H96
• Magnetic beads
• Poly(A)-RNA
• Poly-RNA buffer
• Proteinase K
• Lysis buffer
• Binding buffer
• Wash buffer 1, 2, 3
• Elution buffer
• Deep-well-plate
• Low-well-plate
• Rack with disposable Tips

3 Methods

3.1 RNA Extraction

Patient samples are collected according to appropriate clinical guidelines and transferred to a S2 laboratory in viral transport medium. RNA isolation and preparation are performed using the Chemagic Viral DNA/RNA 300 Kit H96 on 96-well plates at room temperature.

1. Initialize the Chemagic 360 instrument.
2. Start Chemagic software and follow the instructions.
3. Check and confirm the volumes of the connected Wash Buffer 1, 2, 3 and the binding Buffer containers.
4. Check and empty the liquid waste if necessary.
5. Run the "check manifolds 1-6" protocol in the Chemagic software. A small volume of buffer will be dispensed by each manifold.
6. Select the isolation protocol "Chemagic Viral300 360 H96 drying prefilling VD200309.che" in the Chemagic software.
7. Design sample ID layout for the 96-well plate format in the software.
8. Place the disposable tips rack in position 1 on the tracking system.
9. Prefill a low-well-plate with 150 μl resuspended magnetic beads per well and place the plate in position 2 on the tracking system.

Table 2
Content of the MassARRAY® SARS-CoV-2 Panel Kit, PCR Reagent Set with dUTP, iPLEX® Pro Reagent Set, and SpectroCHIP CPM Array

MassARRAY SARS-CoV-2 panel kit
1. MassARRAY SARS-CoV-2 panel PCR primers
2. MassARRAY SARS-CoV-2 panel extend primers
3. UNG (heat labile)
4. MMLV Enzyme
5. RNase Inhibitor
6. MS2 Control
PCR reagent set with dUTP
1. MgCl2, 25 mM
2. 10X PCR Buffer
3. dUTP/dNTP Mix
4. PCR Enzyme
iPLEX pro reagent set
1. 3-Point Calibrant
2. iPLEX Termination Mix
3. iPLEX Buffer Plus, GPR
4. iPLEX Pro Enzyme
5. SAP Buffer
6. SAP
SpectroCHIP CPM Arrays
1. HPLC-grade water (sensitivity >18.2 MΩ)
2. Nuclease-free water
3. Synthetic SARS-CoV-2 RNA Control (Twist Bioscience, USA)
4. Clean Resin (Agena Bioscience, Germany)
5. 96-well non- or semiskirted microtiter plates (ThermoScientific, USA)

CoV Coronavirus, *dNTP* Nucleotide triphosphate, *dUTP* Deoxyuridine triphosphate, *MMLV* Moloney Murine Leukemia Virus, *MS2* MS2 RNA bacteriophage, *PCR* Polymerase chain reaction, *UNG* Uracil-DNA Glycosylase, *SAP* Shrimp Alkaline Phosphatase, *SARS* Severe respiratory distress syndrome

10. Place empty deep-well-plates in position 4, 5, and 6 on the tracking system.
11. Prefill a deep-well-plate with 80 μl elution buffer per well and place the plate in position 7 on the tracking system.
12. Prefill a deep-well plate each with 300 μl Sample.

Table 3
Lysis cocktail

Reagent	Per reaction (μl)
Lysis Buffer	300
Dissolved Poly(A)-RNA	4
Dissolved Proteinase K	10
MS2 Extraction Control (SARS-CoV-2 Panel Kit content)	10

CoV Corona Virus, *MS2* MS2 RNA bacteriophage, *RNA* Ribonucleic acid, *SARS* Severe respiratory distress syndrome

13. Reconstitute lyophilized Poly(A)-RNA by adding 440 μl of Poly(A)-RNA buffer to the Poly(A)-RNA tube and mix thoroughly.
14. Dissolve lyophilized Proteinase K in 11 ml nuclease-free water.
15. Premix the Lysis cocktail as shown in Table 3 in a 5 ml tube.
16. Add 300 μl lysis cocktail to each sample and place the plate in position 3 on the tracking system.
17. Check all plates for accurate orientation and fitting.
18. Close the door and start the selected isolation process.
19. After finishing RNA isolation, unload and seal eluate plate in position 7 for further application.
20. Discard all used plates from position 1 to 6.
21. Once per week clean the Chemagic dispenser by starting the "regular cleaning procedure.che" protocol.

3.2 Mass Spectrometric Analyses

3.2.1 RT-PCR Amplification, SAP Reaction, and iPLEX Pro Extension Reaction

1. Prepare positive Twist SARS-CoV-2 RNA control dilution on ice using nuclease-free water.
   - Prepare a 100-fold dilution by pipetting 990 μl nuclease-free water into a microfuge tube, then adding 10 μl of Twist SARS-CoV-2 RNA stock (final concentration 10,000 copies/μl). Mix well, then centrifuge briefly.
   - Prepare a ten-fold dilution by pipetting 90 μl nuclease-free water into a microfuge tube, then adding 10 μl of the dilution from a) (final concentration 1000 copies/μl). Mix well, then centrifuge briefly.
   - Prepare a ten-fold dilution by pipetting 90 μl nuclease-free water into a microfuge tube, then adding 10 μl of the dilution from b) (final concentration 100 copies/μl). Mix well, then centrifuge briefly. Store aliquots at −80 °C until use.
   - Pipette 50 μl of nuclease-free water into a microfuge tube, then add 50 μl of the dilution from c) (final concentration

Table 4
RT-PCR reaction

Reagent	Per reaction (μl)
10× PCR buffer	0.500
$MgCl_2$, 25 mM	0.400
dUTP/dNTP Mix	0.100
UNG (heat labile)	0.050
RNase inhibitor	0.125
PCR enzyme	0.200
MMLV enzyme	0.125
SARS-CoV-2 PCR primer	0.500
RT-PCR Cocktail final volume	**2.000**
Sample RNA/positive control/negative control	3.000
RT-PCR Reaction final volume	**5.000**

CoV Corona Virus, *dNTP* Nucleotide triphosphate, *dUTP* Deoxyuridine triphosphate, *MMLV* Moloney Murine Leukemia Virus, *PCR* Polymerase chain reaction, *RT* reverse transcriptase, *SARS* Severe acute respiratory syndrome

50 copies/μl). Mix well, then centrifuge briefly. Store aliquots at −80 °C until use.

2. Prepare the RT-PCR cocktail as shown in Table 4 in a 0.5 ml tube placed on ice.
3. Pulse vortex and centrifuge the cocktail tube briefly.
4. Dispense 2 μl of RT-PCR cocktail into each well of a new 96 well plate on ice.
5. Add 3 μl of either sample RNA, positive Twist SARS-CoV-2 RNA control final concentration 100 copies/μl and negative H_2O control (Table 4) to each well of the reaction plate.
6. Seal the RT-PCR reaction plate and centrifuge at 1000 × *g* for 30 s.
7. Thermocycle the RT-PCR reaction plate with the following conditions.
   (a) 1 cycle 25 °C 5 min.
   (b) 1 cycle 55 °C 10 min.
   (c) 1 cycle 95 °C 2 min.
   (d) 40 cycles 95 °C 30 s, 60 °C 30 s, 72 °C 30 s.
   (e) 1 cycle 72 °C 5 min and finally 4 °C.
8. Centrifuge RT-PCR reaction plate at 1000 × *g* for 30 s and keep the plate on a cold block until further access.
9. Prepare the SAP cocktail (Table 5) in a 0.5 ml tube on ice.

Table 5
SAP reaction

Reagent	Per reaction (μl)
HPLC-grade water	1.53
SAP Buffer	0.17
SAP	0.30
SAP cocktail final volume	**2.00**

HPLC high performance liquid chromatography, *SAP* Shrimp Alkaline Phosphatase

Table 6
iPLEX pro extension reaction

Reagent	Per reaction (μl)
HPLC-grade water	0.62
iPLEX Buffer Plus, GPR	0.20
iPLEX Termination Mix	0.20
iPLEX Pro Enzyme	0.04
SARS-CoV-2 Extend Primer	0.94
Extension cocktail final volume	**2.00**

CoV Corona Virus, *HPLC* high performance liquid chromatography, *SARS* Severe acute respiratory syndrome

10. Pulse vortex and centrifuge the cocktail tube briefly.
11. Dispense 2 μl of SAP cocktail into each well of the RT-PCR reaction plate on ice.
12. Seal the reaction plate and centrifuge at 1000 × *g* for 30 s.
13. Thermocycle the reaction plate by the following conditions:
    (a) 1 cycle 37 °C 10 min.
    (b) 1 cycle 85 °C 5 min and 4 °C.
14. Centrifuge reaction plate at 1000 × *g* for 30 s and keep the plate on a cold block until further access.
15. Prepare iPLEX Pro Extension reaction cocktail as shown in Table 6 in a 0.5 ml tube placed on ice.
16. Pulse vortex and centrifuge the cocktail tube briefly.
17. Dispense 2 μl of extension reaction cocktail into each well of the reaction plate on ice.
18. Seal the reaction plate and centrifuge at 1000 × *g* for 30 s.

19. Thermocycle the reaction plate by the following conditions.
    (a) 1 cycle 95 °C 30 s.
    (b) 40 cycles 95 °C 5 s, 52 °C 15 s, 80 °C 5 s.
    (c) 1 cycle 72 °C 3 min and finally 4 °C.
20. Centrifuge reaction plate at 1000 × *g* for 30 s.
21. Add 41 μl HPLC-grade water to each well of the reaction plate.
22. Seal the plate and centrifuge at 1000 × *g* for 1 min.

3.2.2 Data Acquisition

The MassARRAY system and the Chip Prep Module are configured at installation for the SARS-CoV-2 Panel.

1. Create a sample ID text file for the 96-well plate.
2. Open the MassARRAY Typer Plate Editor software and design a new sample plate in the directory database with a unique plate ID.
3. Load "SARS-CoV-2_ADF.tsv" assay, add the created sample ID text file and save the sample plate.
4. Open the MassARRAY Typer Chip linker software and connect the designed sample plate to a new, unique SpectroCHIP Array.
5. Select the designed sample plate in the directory tree, iPLEX as the terminator chemistry, "Genotype and Area" as the processing method and "Nanodispenser 96 to 96" as the dispenser method.
6. Enter an experiment name.
7. Enter the SpectroCHIP Array barcode and add SpectroCHIP Array to create an input XML file.
8. Open the MassARRAY Typer SpectroACQUIRE window.
9. Check the green "okay" status of the waste tank, system fluid, and resin level in the status section in the run setup tab.
10. Extend the deck by the button "Chip prep module Deck In/Out."
11. Transfer the linked SpectroCHIP Array into the chip holder position 1.
12. Pipette 75 μl 3 Point Calibrant into the calibrant vial and place the calibrant vial holder on the deck.
13. Load the extracted/analyte sample microtiter plate in plate position 1 on the deck.
14. Retract the full deck by the button "Chip prep module Deck In/Out."
15. Select the XML input file created earlier in the experiment setup section in the Run Setup tab MPT1.

16. Select "Automatic" in the "wells to process" field.
17. Check "Experiment," "Chip Prep Module," and "Analyzer Setup settings."
18. Start chip prep module run.
19. After the chip prep module has finished, discharge the analyzed SpectroCHIP Array from the MassARRAY Analyzer using the "Run Setup" tab.
20. Extend the deck by the button "Chip prep module Deck In/Out" and remove the analyzed SpectroCHIP Array, calibrant and extracted/analyte sample microtiter plate.

3.2.3 Data Analysis

1. Open the MassARRAY typer analyzer software.
2. Select the experiment in the project explorer and the SpectroCHIP Array of interest for "Chip loading and opening."
3. Select the report file "SARS-CoV-2 Report-v1" in the MassARRAY Typer Analyzer menu bar.
4. SARS-CoV-2 Report Details.csv will automatically open when the report is complete and results will be made available in the respective SARS-CoV-2 report folder.
5. Interpretation of the results is performed by the Agena SARS-CoV-2 report software. Patient sample test results must be performed after the positive and negative controls (including on each plate) have been examined and determined to be valid and acceptable (QC Status: "PASS"). Results are interpreted according to the criteria described in Table 7. A SARS-CoV-2 negative and positive example is shown in Fig. 1.

Table 7
Patient sample results for MassARRAY® SARS-CoV-2 Panel (CE-IVD (Europe) Report v.1.1 CE-IVD)

5 SARS-CoV-2 targets (N1,N2, N3, ORF1, ORF1ab)	QC extraction control	QC Status	Report output
$\geq$ 2 SARS CoV-2 targets detected	Detected	PASS	SARS-CoV-2 detected
$\geq$ 2 SARS CoV-2 targets detected	Not detected	WARNING	SARS-CoV-2 detected
< 2 SARS CoV-2 targets detected	Detected	PASS	SARS-CoV-2 Not detected
< 2 SARS CoV-2 targets detected	Not detected	FAIL	Invalid

CoV Corona Virus, *SARS* Severe acute respiratory syndrome, QC quality control

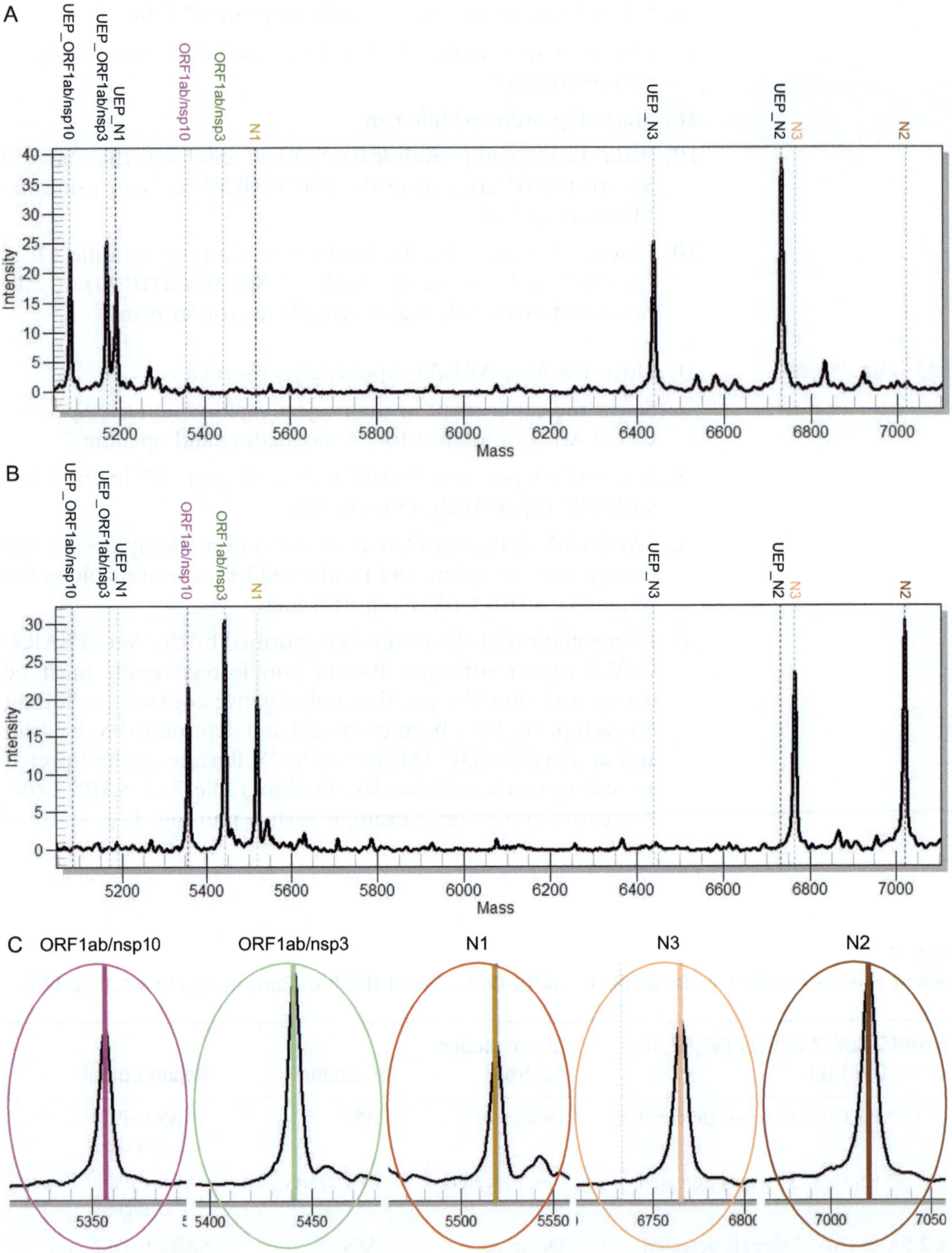

Fig. 1 Examples for a positive and negative result of the commercial MS assay ([15]). In the sum spectra (**a, b**) the expected mass peaks for the respective amplicons are highlighted with a dashed line. In the negative case (**a**) peaks of unextended primers are observed (UEP). In the positive case (**b**) peaks for amplicons of the ORF1ab/nap3, ORF1ab/nap10, N1, N2 and N3 can be identified. Zoomed spectra of these amplicons are depicted in **c**

4 Notes

4.1 RNA Extraction

Isolation and preparation of RNA should be in a separate lab area to prevent any contamination [14].

Storage temperature of all reagents should be 15 °C to 25 °C. Additionally, lysis buffer and poly(A)-RNA buffer should be stored in the dark. Reconstituted lyophilized Poly(A)-RNA in Poly(A)-RNA Buffer is stable for 4 weeks at 4 °C and reconstituted Proteinase K is stable for 2 weeks at 4 °C. After finishing RNA isolation, start as fast as possible with the MassARRAY Protocol, as storing at 4 °C is only possible for a short period of time.

Deep-well plates and low well plates are delivered optionally. It is possible to adapt the Chemagic™ instrument protocol to your own plates.

When you check the wash and binding buffer containers, make sure that the liquid level for all buffers is above 125 ml for 96 isolations. Otherwise, fill up the buffer supply containers or exchange against new ones. When performing several runs per day, it is only necessary to check the manifolds once at the beginning.

The isolation protocol runs for about 60 min. It is also possible to run a shorter protocol of 35 min, depending on the components of the transport media, which includes only two instead of three washing steps. When prefilling the plates with magnetic beads and elution buffer, use a multichannel pipette to decrease manual work time.

Proteinase K activity will decrease after incubation >10 min in Lysis buffer. Be sure that all samples are mixed within this time. The use of the guanidinium based lysis buffer is the standard isolation procedure. Depended on the viral transport medium, it is possible to replace the lysis buffer by nuclease free water to prevent sample precipitation in the next steps.

In some cases, you may find traces of magnetic beads left in the eluate after RNA isolation which can be separated using an appropriate Chemagic™ magnetic stand [6].

When discarding all used plates, check accurate manifold performance as a decrease in volume might indicate pump blockage or failure. Also, control the uniform volume of the wash plates (500 μl per well) and lysis plate (1600 μl per well).

4.2 Mass Spectrometric Analyses

Take care of the well depth of the plates as usage without reconfiguration can damage the MassARRAY instrument or compromise assay performance.

Shipping of the mass spectrometric reagents are on dry ice and storage temperature is −25 °C to −10 °C, except for SpectroChip CPM Arrays which can be stored at 15 °C to 25 °C.

4.2.1 RT-PCR Amplification, SAP Reaction, and iPLEX Pro Extension Reaction

For the RT-PCR, SAP and iPLEX Pro extension reaction cocktail, prepare a 10-percentage overhang of the RT-PCR cocktail to overcome typical pipetting variation. Keep enzymes, UNG (heat labile) and RNA inhibitor on ice during use. For all further reagents make sure they are thawed completely and homogenized at room temperature.

To maintain the sample ID layout of the RNA eluate plate, use a multichannel pipette to transfer sample RNA accurate and equal to the RT-PCR reaction plate. Also, use a multichannel pipette to transfer HPLC-grade water to each well of the reaction plate during the iPLEX Pro extension reaction.

For the RT-PCR, SAP cocktail and iPLEX Pro extension reaction seal the plate very well to avoid evaporation during thermal cycling and check the total volume condition by visually inspecting the 96 wells from the bottom of the reaction plate before and after each thermal cycling step to confirm that uniform and adequate reaction volume is present.

After the RT-PCR reaction and centrifugation, the reaction plate can be stored at 4 °C for less than 24 h. Likewise, the prepared SAP mix and the iPLEX Pro extension reaction plate can be stored at 4 °C for less than 24 h.

4.2.2 Data Acquisition and Reporting

Use the designed sample ID layout from the Chemagic™ software to avoid sample confusion. Define "PC" as positive control samples and "NC" as negative samples to enable the software to recognize the sample IDs.

When requested to enter the barcode from the SpectroCHIP Array, remember that it is located on the right corner next to the logo. Note that the created input XML file will be selected in the automatic run set up for data acquisition and report generation.

When the SpectroACQUIRE window displays a red status symbol, perform the respective indicated maintenance.

When you transfer the SpectroCHIP Array into the chip holder, make sure to wear gloves and to use a tweezer for the transfer, as the matrix coated chip is very sensitive. Never touch the chip surface. Orient the SpectroCHIP Array such that the Agena logo and barcode are at the bottom. Make sure that the SpectroCHIP Array is properly seated and that the chip holder and SpectroCHIP Array surface are flush.

When the extracted/analyte sample microtiter plate is loaded in plate position 1 on the deck, make sure that the A1 well is in the front left corner. Note that plate position 1 corresponds to chip position 1.

The Analyzer Setup settings are configurated during the installation by the manufacturer.

References

1. Guan WJ, Ni ZY, Hu Y, Liang WH, Ou CQ, He JX, Liu L, Shan H, Lei CL, DSC H, Du B, Li LJ, Zeng G, Yuen KY, Chen RC, Tang CL, Wang T, Chen PY, Xiang J, Li SY, Wang JL, Liang ZJ, Peng YX, Wei L, Liu Y, Hu YH, Peng P, Wang JM, Liu JY, Chen Z, Li G, Zheng ZJ, Qiu SQ, Luo J, Ye CJ, Zhu SY, Zhong NS, China Medical Treatment Expert Group for C (2020) Clinical characteristics of coronavirus disease 2019 in China. N Engl J Med 382(18):1708–1720. https://doi.org/10.1056/NEJMoa2002032
2. Baden LR, El Sahly HM, Essink B, Kotloff K, Frey S, Novak R, Diemert D, Spector SA, Rouphael N, Creech CB, McGettigan J, Kehtan S, Segall N, Solis J, Brosz A, Fierro C, Schwartz H, Neuzil K, Corey L, Gilbert P, Janes H, Follmann D, Marovich M, Mascola J, Polakowski L, Ledgerwood J, Graham BS, Bennett H, Pajon R, Knightly C, Leav B, Deng W, Zhou H, Han S, Ivarsson M, Miller J, Zaks T, Group CS (2020) Efficacy and safety of the mRNA-1273 SARS-CoV-2 vaccine. N Engl J Med. https://doi.org/10.1056/NEJMoa2035389
3. Polack FP, Thomas SJ, Kitchin N, Absalon J, Gurtman A, Lockhart S, Perez JL, Perez Marc G, Moreira ED, Zerbini C, Bailey R, Swanson KA, Roychoudhury S, Koury K, Li P, Kalina WV, Cooper D, Frenck RW Jr, Hammitt LL, Tureci O, Nell H, Schaefer A, Unal S, Tresnan DB, Mather S, Dormitzer PR, Sahin U, Jansen KU, Gruber WC, Group CCT (2020) Safety and efficacy of the BNT162b2 mRNA Covid-19 vaccine. N Engl J Med 383(27):2603–2615. https://doi.org/10.1056/NEJMoa2034577
4. Corman VM, Landt O, Kaiser M, Molenkamp R, Meijer A, Chu DK, Bleicker T, Brunink S, Schneider J, Schmidt ML, Mulders DG, Haagmans BL, van der Veer B, van den Brink S, Wijsman L, Goderski G, Romette JL, Ellis J, Zambon M, Peiris M, Goossens H, Reusken C, Koopmans MP, Drosten C (2020) Detection of 2019 novel coronavirus (2019-nCoV) by real-time RT-PCR. Euro Surveill 25(3). https://doi.org/10.2807/1560-7917.ES.2020.25.3.2000045
5. Mascuch SJ, Fakhretaha-Aval S, Bowman JC, Ma MTH, Thomas G, Bommarius B, Ito C, Zhao L, Newnam GP, Matange KR, Thapa HR, Barlow B, Donegan RK, Nguyen NA, Saccuzzo EG, Obianyor CT, Karunakaran SC, Pollet P, Rothschild-Mancinelli B, Mestre-Fos S, Guth-Metzler R, Bryksin AV, Petrov AS, Hazell M, Ibberson CB, Penev PI, Mannino RG, Lam WA, Garcia AJ, Kubanek J, Agarwal V, Hud NV, Glass JB, Williams LD, Lieberman RL (2020) A blueprint for academic laboratories to produce SARS-CoV-2 quantitative RT-PCR test kits. J Biol Chem 295(46):15438–15453. https://doi.org/10.1074/jbc.RA120.015434
6. Klein S, Muller TG, Khalid D, Sonntag-Buck V, Heuser AM, Glass B, Meurer M, Morales I, Schillak A, Freistaedter A, Ambiel I, Winter SL, Zimmermann L, Naumoska T, Bubeck F, Kirrmaier D, Ullrich S, Barreto Miranda I, Anders S, Grimm D, Schnitzler P, Knop M, Krausslich HG, Dao Thi VL, Borner K, Chlanda P (2020) SARS-CoV-2 RNA extraction using magnetic beads for rapid large-scale testing by RT-qPCR and RT-LAMP. Viruses 12(8). https://doi.org/10.3390/v12080863
7. Seger C, Salzmann L (2020) After another decade: LC-MS/MS became routine in clinical diagnostics. Clin Biochem. https://doi.org/10.1016/j.clinbiochem.2020.03.004
8. Borden SA, Palaty J, Termopoli V, Famiglini G, Cappiello A, Gill CG, Palma P (2020) Mass spectrometry analysis of drugs of abuse: challenges and emerging strategies. Mass Spectrom Rev. https://doi.org/10.1002/mas.21624
9. Fung AWS, Sugumar V, Ren AH, Kulasingam V (2020) Emerging role of clinical mass spectrometry in pathology. J Clin Pathol 73(2):61–69. https://doi.org/10.1136/jclinpath-2019-206269
10. Angeletti S, Ciccozzi M (2019) Matrix-assisted laser desorption ionization time-of-flight mass spectrometry in clinical microbiology: an updating review. Infect Genet Evol 76:104063. https://doi.org/10.1016/j.meegid.2019.104063
11. De Roock W, Claes B, Bernasconi D, De Schutter J, Biesmans B, Fountzilas G, Kalogeras KT, Kotoula V, Papamichael D, Laurent-Puig P, Penault-Llorca F, Rougier P, Vincenzi B, Santini D, Tonini G, Cappuzzo F, Frattini M, Molinari F, Saletti P, De Dosso S, Martini M, Bardelli A, Siena S, Sartore-Bianchi A, Tabernero J, Macarulla T, Di Fiore F, Gangloff AO, Ciardiello F, Pfeiffer P, Qvortrup C, Hansen TP, Van Cutsem E, Piessevaux H, Lambrechts D, Delorenzi M, Tejpar S (2010) Effects of KRAS, BRAF, NRAS, and PIK3CA mutations on the efficacy of cetuximab plus chemotherapy in chemotherapy-refractory metastatic colorectal cancer: a retrospective consortium analysis.

Lancet Oncol 11(8):753–762. https://doi.org/10.1016/S1470-2045(10)70130-3

12. Wandernoth P, Kriegsmann K, Groh-Mohanu-C, Daeumer M, Gohl P, Harzer O, Kriegsmann M, Kriegsmann J (2020) Detection of severe acute respiratory syndrome coronavirus 2 (SARS-CoV-2) by mass spectrometry. Viruses 12(8). https://doi.org/10.3390/v12080849
13. Administration FaD (2020) Fact sheet for healthcare providers: massarray SARS-CoV-2 panel
14. Mifflin TE (2007) Setting up a PCR laboratory. CSH Protoc 2007:pdb top14. https://doi.org/10.1101/pdb.top14
15. Kriegsmann M, Wandernoth P, Lisenko K, Casadonte R, Longuespee R, Arens N, Kriegsmann J (2017) Detection of HPV subtypes by mass spectrometry in FFPE tissue specimens: a reliable tool for routine diagnostics. J Clin Pathol 70(5):417–423. https://doi.org/10.1136/jclinpath-2016-204017

Chapter 13

Bioinformatics-Based Approaches to Study Virus–Host Interactions During SARS-CoV-2 Infection

Muhammad Saad Khan, Qudsia Yousafi, Shabana Bibi, Muhammad Azhar, and Awais Ihsan

Abstract

As the knowledge of biomolecules is increasing from the last decades, it is helping the researchers to understand the unsolved issues regarding virology. Recent technologies in high-throughput sequencing are providing the swift generation of SARS-CoV-2 genomic data with the basic inside of viral infection. Owing to various virus–host protein interactions, high-throughput technologies are unable to provide complete details of viral pathogenesis. Identifying the virus–host protein interactions using bioinformatics approaches can assist in understanding the mechanism of SARS-CoV-2 infection and pathogenesis. In this chapter, recent integrative bioinformatics approaches are discussed to help the virologists and computational biologists in the identification of structurally similar proteins of human and SARS-CoV-2 virus, and to predict the potential of virus–host interactions. Considering experimental and time limitations for effective viral drug development, computational aided drug design (CADD) can reduce the gap between drug prediction and development. More research with respect to evolutionary solutions could be helpful to make a new pipeline for virus–host protein–protein interactions and provide more understanding to disclose the cases of host switch, and also expand the virulence of the pathogen and host range in developing viral infections.

Key words SARS-CoV-2 virus, Protein domain interaction, Protein interaction networks, Structural bioinformatics, Virus–host protein–protein interactions

1 Introduction

Coronavirus (CoV) is the most serious disease caused by single-stranded RNA viruses, which belong to the family Coronaviridae [1, 2]. CoVs have been reported in several hosts which includes camels, dogs, bats, and masked palm civets, which could easily affect the people with low immune system [3, 4]. While in Guangdong China, severe acute respiratory syndrome (SARS) in 2002 was initiated via a new beta-CoV type and about 8100 patients were diagnosed with a mortality rate of 9%, while that was about 50% in patients of 60 years or above [5, 6]. It was claimed that its origin

Justin Jang Hann Chu et al. (eds.), *SARS-CoV-2: Methods and Protocols*,
Methods in Molecular Biology, vol. 2452, https://doi.org/10.1007/978-1-0716-2111-0_13,

was infected bats and CoV infection was transmitted by direct contact with infected bats or infected patients. A distinct group of corona virus was associated with an outbreak in Saudi Arabia in 2012 and this kind of infection was recognized as Middle East Respiratory Syndrome (MERS). However, it was responsible for an initial 50% mortality rate, but later did not spread substantially out of that region [7].

While in December 2019, serious pneumonia alarmed the people in Wuhan, China and now it has become a health care challenge to the world. It was due to a major respiratory infection caused by SARS-CoV, Corona Virus Disease 2019 (COVID-19) proved to be a disastrous infection compared to SARS-CoV discovered before. Hence, this novel strain of CoV identified in 2019 was recognized as SARS-CoV-2 or COVID-19 [8]. COVID-19 is a very serious infectious disease which has become a pandemic. According to world health organization (WHO), 59,481,313 cases of COVID-19 has been reported from December 2019 to 26 November 2020, along with the serious number of deaths which accounts 1,404,542 [9].

The SARS-CoV-2 virus is linked with the angiotensin-converting enzymes 2 (ACE2) activity as the central cellular receptor with the membrane serine protease. TMPRSS2, stand-in as an adjunct protein for stabilization cell entry and cleavage of the spike protein, ensuing viral fusion with the cell membrane [10, 11]. The virus route into the host and replicates inside the cytoplasm, initiating translation of the replicase gene and assembling a viral replicase complex [12]. The complex and nonstructural genes accomplish blocking of host cell translation while endorsing host mRNA degradation and aggregate the viral RNA replication [12, 13]. The process linked with the generation of genomic and sub-genomic RNA by the negative-strand intermediates and the spike envelope and membrane structural proteins that transfer it into the endoplasmic reticulum linked Golgi intermediate compartment, where the virus is surrounded by the N protein shell [13, 14]. After assembly of virions, the virus is transported to the cell surface and released by exocytosis. While in few cases of coronaviruses, additional S protein could arbitrate the cell fusion with the adjacent cells, which could permit the fast transmission of CoV without detecting via host humoral immune response [13, 15].

SARS-CoV-2 has endangered the population worldwide since unfortunately, there is no effective vaccine or drug to treat the infected patients. Early supportive measures are in consideration to treat the patients, such as oxygen therapy, extra-corporeal membrane oxygenation etc [16, 17]. Scientists investigated the potential of seven selected antiviral drugs: chloroquine, favipiravir (T-750), penciclovir, nitazoxanide, nafamostat, remdesivir (GS-5734), and ribavirin to treat SARS-CoV-2 on Vero E6 cells. In an in-vitro study, Chinese researchers found that the treatments of Remdesivir and chloroquine could effectively the inhibit the SARS-CoV-2 [18]. These medicines are already in use for the treatment of several viral infectious diseases, so it could also be used to check their potential effectiveness against SARS-CoV-2 infection, this technique is known as drug repurposing. It is the requisite approach during current pandemic until the FDA approved medicine can come into the market for SARS-CoV-2. In contrast, several researchers are using a computer-aided drug design (CADD) approaches to conclude with novel drugs that might be highly significant against SARS-CoV-2 infection [19].

Bioinformatics tools, databases and several other online platforms could be most valuable in revealing molecular bases of host-virus interaction for SARS-CoV-2 by utilizing the available data properly to extract the information embedded in them [20]. Such as functional genomic information of SARS-CoV-2 with associated clinical and epidemiological information to the researchers all around the globe has been provided by "Global Initiative on Sharing All Influenza Data" (GISAD) (http://www.gisaid.org). An open-source tool named Nextstain (http://nextstrain.org) has provided the researcher with updated information on SAR-CoV-2 genome. It also provided the current trend of new cases in SARS-CoV-2 patients of the world [21]. Bioinformatics study performed by Huang and coworkers have highlighted the ACE2 expression correlated significantly with immune cell infiltration and SARS-CoV-2 in cancer patients [22].

In this chapter, our bioinformatics based approaches highlight the peculiarities of SARS-CoV-2 biological characteristics regarding applications of predictive models for virus–host PPIs. Our results could help to develop new pipelines for PPI inhibition and offer a new vision to understand the cases of host switch, also expands the pathogen virulence and host range in emergent viral diseases.

2 Methods

2.1 Identification of Structurally Similar Proteins Between Human and Virus

Obtain the structurally similar human and SARS-CoV-2 (*see* **Note 1**) counterparts of SARS-CoV-2 proteins by submitting the viral spike glycoprotein protein structures (PDB ID:5XLR) [23] to Daliserver [24]. Select the Human and SARS-CoV-2 protein as structurally similar when the Z-score≥2 (*see* **Note 1**).

2.2 Prediction of Subcellular Localization Prediction

Predict subcellular localization of SRS-CoV-2 proteins and human ACE2 proteins, while two proteins theoretically must share at least one cellular compartment for direct interaction among them. Obtain the cellular compartment and type of proteins, that is, soluble or membrane is predicted by an online available server DeepLoc-1.0 [25] (*see* **Note 2**).

2.3 Protein Interaction Networks Underlying Viral Processes

Molecular interactions should be formed with their specific hosts for pathogens the highjack the cellular machinery and replicate themselves. Such type of interactions are commonly exemplified with the help of networks, where proteins demonstrate nodes the and edges make contacts between them denoting direct physical interactions [26, 27]. Numerous characteristics of respective proteins can be considered from its network topology, such as, centrality and connectivity (degree) while constructing a Protein interaction network [28]. It has been advocated that viral proteins incline to target highly connected and more central host proteins during virus–host interactions [29, 30].

The virus–host PPI data that existed so far let us to carry out detailed comparative studies to better understanding the basic of cellular functions targeted by viruses belonging from different viral families, as well as the evolution as high confidence data are still rare [31, 32].

2.4 Protein–Protein interaction

Having a comprehensive understanding of virus–host PPIs and their interfaces, is critical for the progress of development of a new antiviral treatment, such as the design of a small drug-like entity capable of binding and blocking essential interactions of viral processes[27, 33]. Identification of human interacting protein partners with SARS-CoV-2 spike protein was made using online available servers and databases to understand host-virus PPI.BIOGRID: Biological General Repository for Interaction Datasets [34, 35] and STRING [36] databases were used to explore interacting human partners with SARS-CoV-2 spike protein. All the literature curated interactions established through in vitro and/or in vivo methods among human proteins are enlisted in these datasets (*see* **Note 3**).

2.5 Virus–Host Interactions with respect to Protein Domains

Based on viral classification SARS-CoV-2 is Positive-sense single-stranded RNA viruses as revealed in Table 1 [37].

Domains are fundamental protein structures that evolve independently from each other and have specific biological functions [38, 39]. From the level of DDIs, PPIs can be better understood [40, 41]. From DDIs perspective, RNA viruses are considered to be better in case of the encoded domains and functional activity they possess. Submitted the 3D domain structure of both ACE2 and SARS-CoV-2 spike protein on online available servers Patchdock [42] or Zdock [43]. The 2D and 3D resulting docked complex file is visualized through LigPlot (DimPlot) [44] (Wallace et al., 1995) and MOE [45], PyMol [46] respectively to find out the interacting residues of the protein.

Table 1
Viral classification system

Group	Members	Acronym
I	Double-stranded DNA viruses	dsDNA
II	Single-stranded DNA viruses	ssDNA
III	Double-stranded RNA viruses	dsRNA
IV	**Positive-sense single-stranded RNA viruses**	ssRNA(+)
V	Negative-sense single-stranded RNA viruses	ssRNA(-)
VI	RNA reverse-transcribing viruses	ssRNA-RT
VII	DNA reverse-transcribing viruses	dsDNA-RT

2.6 Interfaces of Protein Interactions

Respective binding sites of host and viral protein necessary to be in direct physical contact, moreover in transient or in stable mode so that proteins can interact with each other [47]. Three-dimensional organizations designed by sets of amino acid residues directly accountable for the recognition of binding partners Such binding sites are called "interfaces" [31, 48]. With the help of different bioinformatics tools like PyMol, MOE, and Discovery Studio, interfaces can be determined and visualized as presented in Fig. 1.

As demonstrated in the case of virus–host PPIs; exogenous interfaces facilitate interactions among proteins of distinct proteomes [48].

2.7 Sources of Biomolecular Data for Virus–Host Interaction Studies

For a deep understanding of the evolutionary characteristics of virus–host PINs, integrative bioinformatics approaches are essential. For different biological dimensions different types of data like gene ontology terms, protein sequence, domain composition, DDIs, PPIs, and taxonomic data come. In Table 2, several databases providing such biological information are listed.

Multiple databases provide Viral and host PPI, but for virus–host interactions, VirHostNet [49] and VirusMentha [50] are entirely dedicated viral databases. Almost all records come from semi-independent and external sources, such as IntAct [51], DIP [52], MINT [53], and BIOGRID [54], which predominantly store literature-derived and experimental data.

To understand the evolution of viral PPIs through the integration of bioinformatics techniques, data of three different kinds are of specific interest: sequence diversity, protein structures and binary protein interactions. It is possible to assess the level of data availability for several viral families from data collected from NCBI (Viral Genomes), VirHostNet (PPIs), and PDB (Structures) [31].

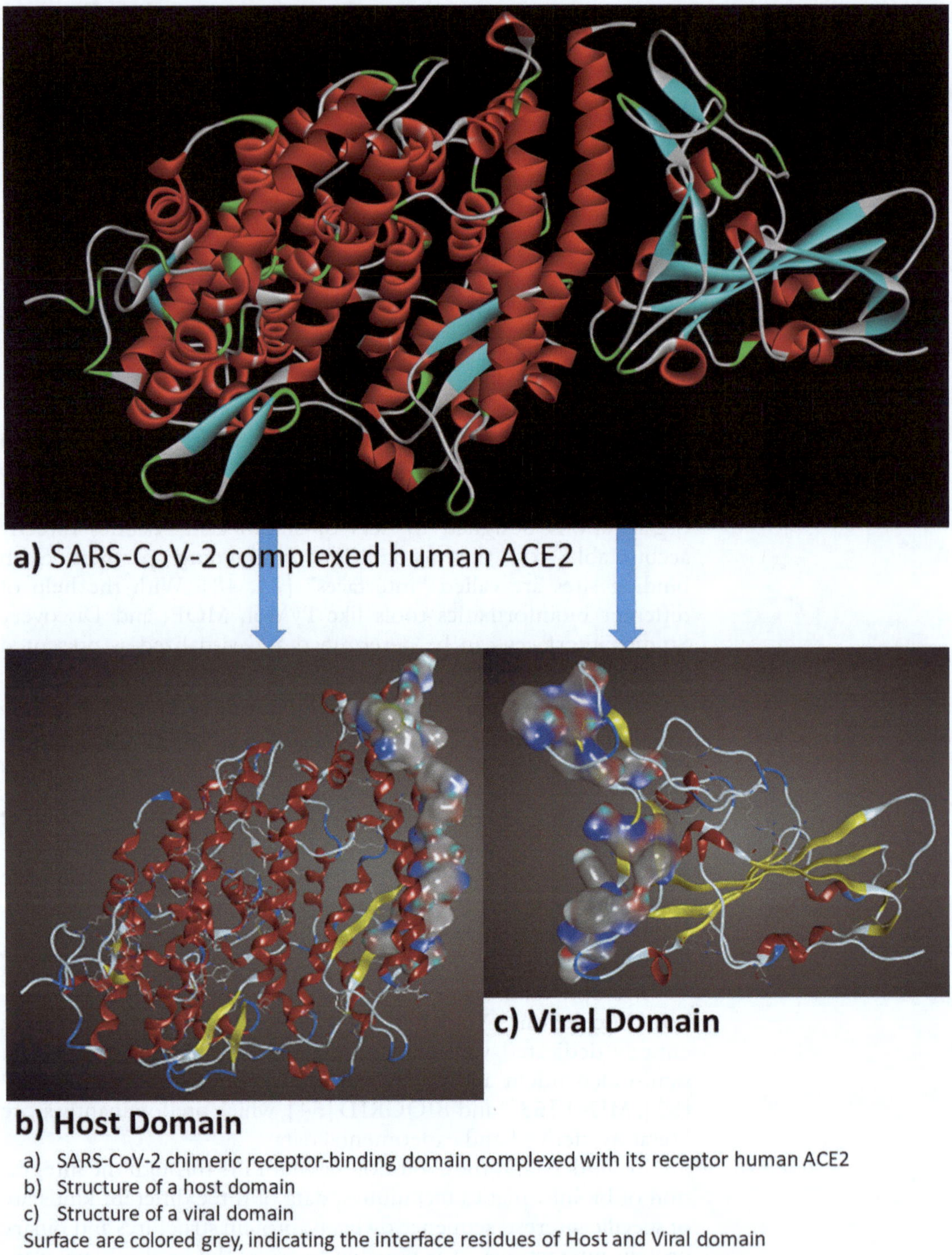

Fig. 1 (**a**) SARS-CoV-2 chimeric receptor-binding domain complexed with its human receptor ACE2; (**b**) Structure of a host domain; (**c**) Structure of a viral domain; Surface are colored gray, indicating the interface residues of Host and Viral domain

Table 2
Bioinformatics Sources to investigate SAR-CoV-2 virus–host interactions

Data types and databases	Genomic sequence	GO terms	Protein sequence	Protein classification	Domain interaction	Protein Interaction	Protein structure	PSICQUIC member?	Data source[a]
GENERIF	–	–	–	–	–	•	–	No	[E] [T]
3DID	–	o	–	–	•	•	o	No	[I]
MINT	–	o	o	o	–	•	o	Yes	[E] [T]
BIOGRID	–	–	–	–	–	•	–	Yes	[E] [T]
BIND	–	–	–	o	–	•	–	Yes	[E]
HPIDb	–	–	–	–	–	•	–	Yes	[I] [P]
InnateDB	–	o	–	–	–	•	–	Yes	[E] [T]
Pfam	–	o	•	•	•	•	o	No	[I] [P]
PDB	–	o	•	o	•	•	•	No	[E]
iPfam	–	–	–	•	•	•	o	No	[I]
HPRD	–	o	o	o	–	•	o	Yes	[E]
MatrixDB	–	o	o	o	–	•	o	Yes	[E] [P]
IntAct	–	o	o	•	–	•	–	Yes	[E] [T]
PQS/PISA	–	–	–	–	•	•	•	No	[A] [I]
SCOP	–	–	•	•	–	–	o	–	[A] [I]
Reactome	–	o	–	•	–	•	–	Yes	[I] [T]
UniProt	•	o	•	o	–	o	o	Yes	[I]
InterPro	–	o	o	•	–	o	o	–	[E] [I]
VirHostNet	–	–	–	–	–	•	–	Yes	[I] [T]
Viral Genomes NCBI	•	–	o	–	–	–	–	–	[I]
VirusMentha	–	–	–	–	–	•	–	No	[I] [T]

[a] *E* Experimental data, *A* Annotation, • primary data type, *P* Prediction, *I* Integration of multiple databases, *T* Text mining, ° data retrieved from secondary sources

3 Results Interpretation

By exploring the protein structure information with accession ID: 5XLR (Spike glycoprotein), which is an example as input, the computer-assisted technique highlighted in this chapter will predict 98 human proteins that have structurally similar Domains with the highest Z score 26 for 6WV1 (Human ACE2). The result of STRING Virus will show seven human protein interaction partners of SARS-CoV-2spike protein, as demonstrated in Fig. 2.

ACE2 will not appear as a direct interaction partner; rather it interacted through another human protein CLEC4M (C-type lectin domain family 4, member M). BIOGRID results will present an interaction network in which COV Spike protein network will show 621 human interaction partners with 621 nodes and 1054 edges. SARS COV spike protein will show only ten direct interaction partners when the minimum evidence threshold value was set to three, that is, ACE2 (Thickest edge reflects strong evidence) and others are demonstrated in Fig. 3.

Another important indirect human protein indirect interaction partner is the epidermal growth factor receptor (EGFR). The DeepLoc 1.0 results will reflect the membrane localization for three SARS-CoV-2 proteins, spike, envelope, and membrane, and for human ACE2 with a strong probability of more than 0.9 as demonstrated in Fig. 4.

The DimPlot (LigPlot), as displayed in Fig. 5, demonstrated the 2D results in the form of Hydrogen bonding between host and viral docked proteins.

4 Conclusion

With the increase of knowledge and biomolecules research, new researchers could solve the issues related to virus–host interactions through data integration and understand the biological complexity. Integrative bioinformatics approached studied in this chapter will only helpful for the virologist but also for computational biologist to understand Human-SAR-CoV-2 PPIs mechanism and peculiarities of viral biology to design, and develop new predictive models. Similar studies on evolutionary perspectives might be significant to design new pipelines for PPI inhibition and aid the researchers by adding information to describe cases of host switch, in addition to the expansion of pathogen virulence, and host range in emerging viral diseases.

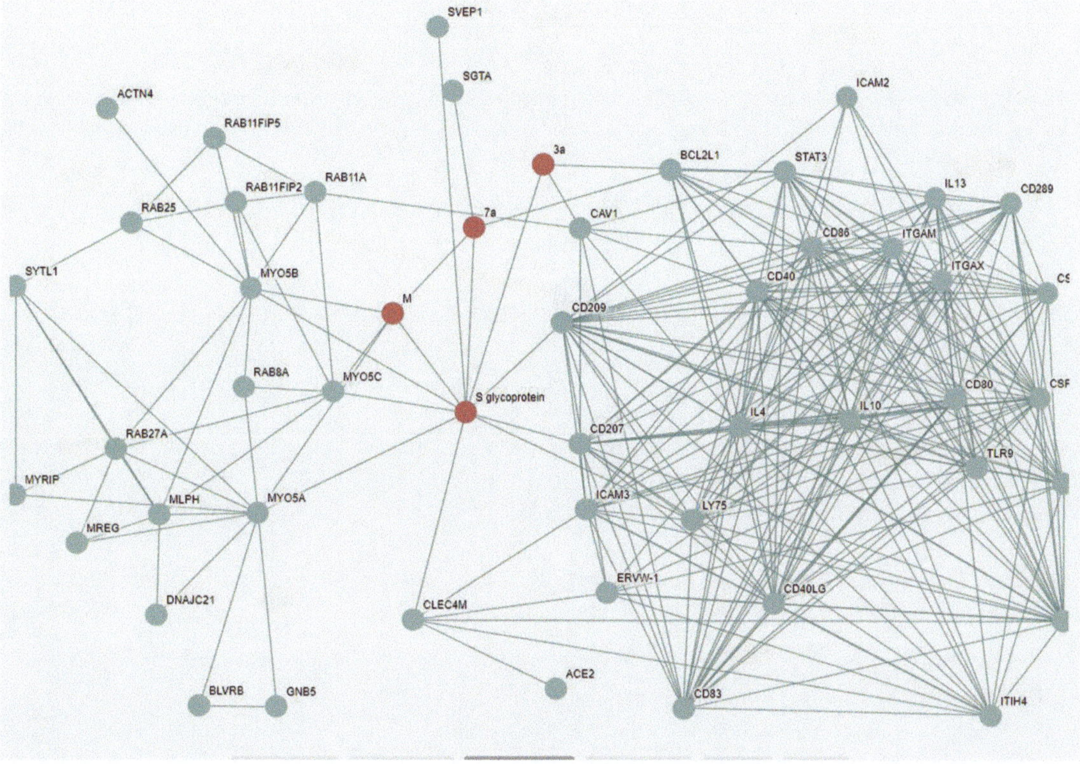

Fig. 2 Protein–Protein interaction network of viral spike protein with human proteins generated by string database

5 Future Perspective

In the modern era of advanced science and technology, no vaccine has been developed to aid human immune system and to protect against corona virus infection. However, certain viral medicines are tested as a proposed treatment for COVID-19 patients but still have limitations. Information related to patients has been used to apply "Machine learning and Data mining" methods to predict, diagnose, prognoses, and selection of most suitable drug to target certain patient groups [55]. Scientists in many countries like in "Mexico" used classical ensemble methods to predict the positive, negative and susceptive patients classified in various groups based on age, gender, and connectivity. Also, these patients were traced back according to their medical condition such as (a) Recovered, (b) Deaths, and (c) Assets [56]. Muhammad and his coworkers compared the several machine learning methods such as Logistic Regression (LT), Support Vector Machine (SVM), Decision Tree (DT), Naive Bayes (NB), Random Forest (RF), and K-Nearest Neighbor (K-NN) to predict COVID-19 infected patients in South Korea. According to prediction results, DT was established

Fig. 3 SARS-CoV-2 virus spike protein interaction network generated by BoiGRID

to be the effectual presenting the maximum accuracy in terms of percentages (99.85%), followed by RF (99.60%), SVM (98.85%), K-NN (98.06%), NB (97.52%), and LR (97.49%) exactly [57]. Several studies have revealed the importance of machine learning to elucidate normal and toxicological responses using nanomaterials in biological systems [58–60]. Hence, machine learning methods have implications ranging from diagnosis, prognosis, prediction models, patient history, and nanoparticles to drug prediction.

6 Notes

1. DaliLite v.3 Web server uses a weighted sum-of-similarity (sum of-pairs method) for comparing the intra-molecular distances between amino acids. The comparison correlates with the expert classification as the structures of homologues proteins get higher scores for similarity relative to the structures of evolutionary unrelated proteins. DALI or distance alignment matrix method server strategically screens all the structures available in PDB to ensure that no significant similarity is

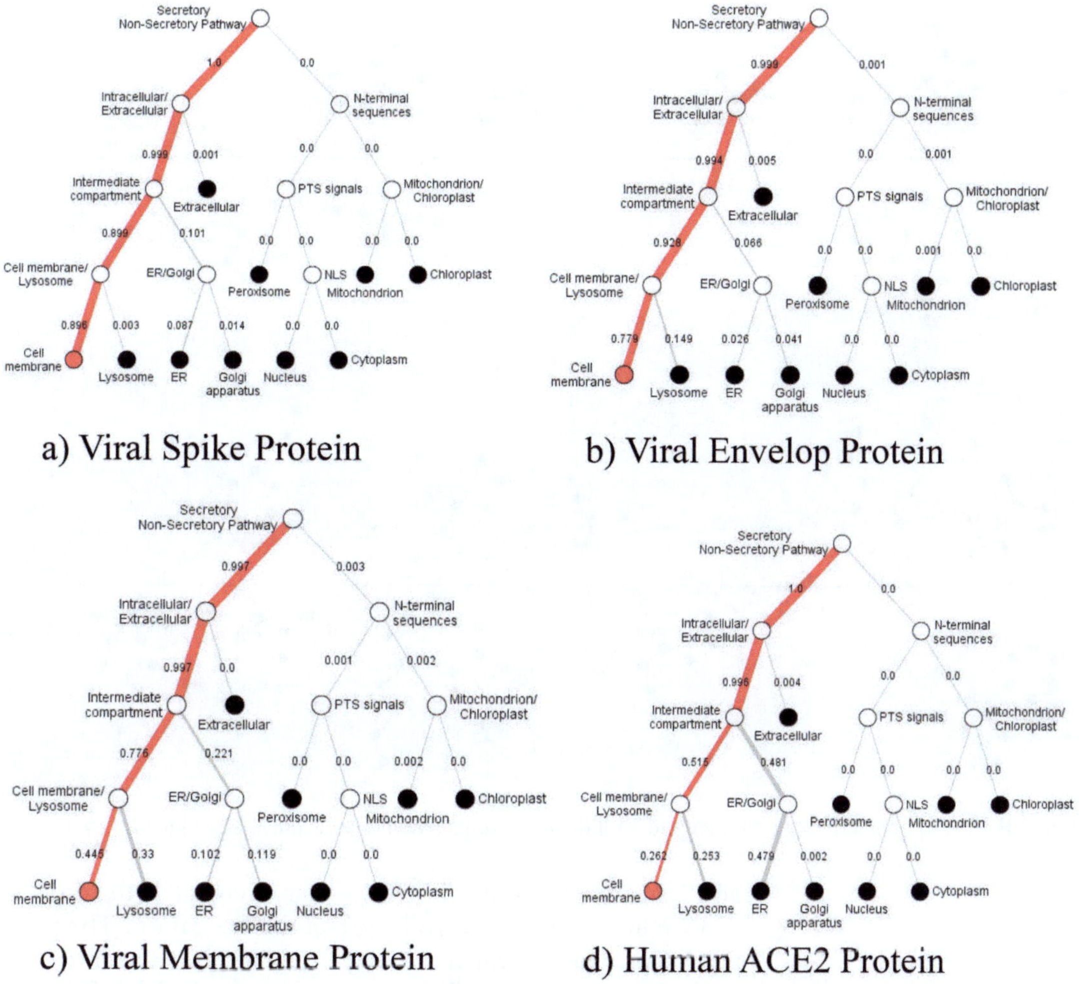

Fig. 4 Subcellular localization of SRS-CoV-2 proteins and human ACE2 proteins predicted by Deeploc 0.1

missed. This server divides the input structures into hexapeptide fragments and calculates a distance matrix by evaluating the contact patterns between successive fragments. While determining the similarity between two proteins, the 3D structural coordinates of these proteins can be compared by alignment of alpha carbon distance matrices. When two proteins distance matrices share the same or similar features in approximately the same positions, they can be said to have similar folds with similar-length loops connecting their secondary structure elements.

2. DeepLoc-1.0 predicts the subcellular localization of eukaryotic proteins. .It predicted the localization within eight different sub cellular locations as well as the extracellular parts of the protein, that is, endoplasmic reticulum, nucleus, cytoplasm, mitochondrion, cell membrane, peroxisome, Golgi apparatus,

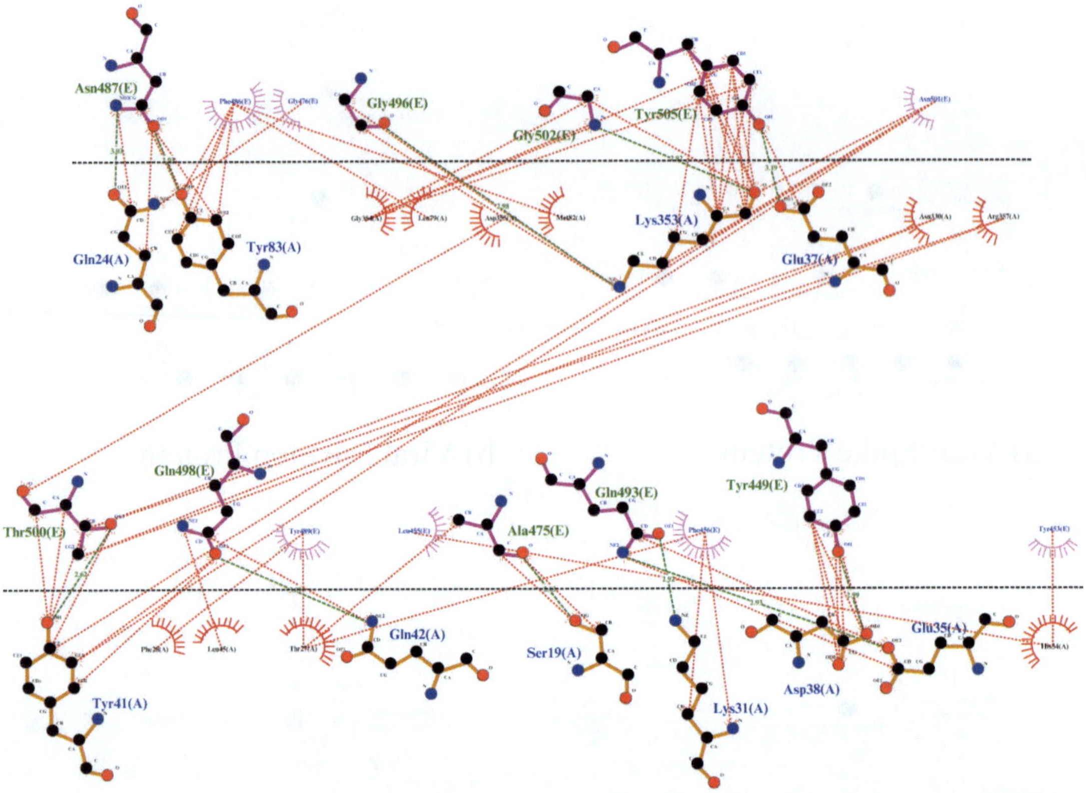

Fig. 5 Hydrogen bonding between host and viral docked proteins complex displayed by LigPlot

and lysosome/vacuole. The results are generated in terms of probability of likelihood to be present in a location. The pathway of protein is defined by first bifurcation of secretory and nonsecretory pathway. Then further scoring is done for compartmentalization. The joint probability of decisions, on the whole, is used to calculate a likelihood score, that is, 0–1.

3. The Biological General Repository for Interaction Datasets (BioGRID) is a public database that archives and disseminates genetic and protein interaction data from model organisms and humans (thebiogrid.org). BioGRID currently holds over 1,740,000 interactions curated from both high-throughput datasets and individual focused studies, as derived from over 70,000+ publications in the primary literature. Complete coverage of the entire literature is maintained for budding yeast (*S. cerevisiae*), fission yeast (*S. pombe*), and thale cress (*A. thaliana*), and efforts to expand curation across multiple metazoan species are underway. Current curation drives are focused on particular areas of biology to enable insights into conserved networks and pathways that are relevant to human health. The BioGRID 3.5 web interface contains new search and display features that enable rapid queries across multiple data types and sources.

References

1. Su S, Wong G, Shi W, Liu J, Lai ACK, Zhou J, Liu W, Bi Y, Gao GF (2016) Epidemiology, genetic recombination, and pathogenesis of coronaviruses. Trends Microbiol 24(6): 490–502. https://doi.org/10.1016/j.tim.2016.03.003
2. Cascella M, Rajnik M, Aleem A, Dulebohn SC, Di Napoli R (2021) Features, evaluation, and treatment of coronavirus (COVID-19). StatPearls Publishing LLC, Copyright © 2021, Treasure Island (FL)
3. Zhong NS, Zheng BJ, Li YM, Poon XZH, Chan KH, Li PH, Tan SY, Chang Q, Xie JP, Liu XQ, Xu J, Li DX, Yuen KY, Peiris GY (2003) Epidemiology and cause of severe acute respiratory syndrome (SARS) in Guangdong, people's Republic of China, in February, 2003. Lancet 362(9393):1353–1358. https://doi.org/10.1016/s0140-6736(03)14630-2
4. Bradburne AF, Bynoe ML, Tyrrell DA (1967) Effects of a "new" human respiratory virus in volunteers. Br Med J 3(5568):767–769. https://doi.org/10.1136/bmj.3.5568.767
5. Peiris JS, Yuen KY, Osterhaus AD, Stohr K (2003) The severe acute respiratory syndrome. N Engl J Med 349(25):2431–2441. https://doi.org/10.1056/NEJMra032498
6. Rehman SU, Shafique L, Ihsan A, Liu Q (2020) Evolutionary trajectory for the emergence of novel coronavirus SARS-CoV-2. Pathogens 9(3). https://doi.org/10.3390/pathogens9030240
7. van Boheemen S, de Graaf M, Lauber C, Bestebroer TM, Raj VS, Zaki AM, Osterhaus AD, Haagmans BL, Gorbalenya AE, Snijder EJ, Fouchier RA (2012) Genomic characterization of a newly discovered coronavirus associated with acute respiratory distress syndrome in humans. MBio 3(6). https://doi.org/10.1128/mBio.00473-12
8. Guan WJ, Ni ZY, Hu Y, Liang WH, Ou CQ, He JX, Liu L, Shan H, Lei CL, DSC H, Du B, Li LJ, Zeng G, Yuen KY, Chen RC, Tang CL, Wang T, Chen PY, Xiang J, Li SY, Wang JL, Liang ZJ, Peng YX, Wei L, Liu Y, Hu YH, Peng P, Wang JM, Liu JY, Chen Z, Li G, Zheng ZJ, Qiu SQ, Luo J, Ye CJ, Zhu SY, Zhong NS, China Medical Treatment Expert Group for C (2020) Clinical characteristics of coronavirus disease 2019 in China. N Engl J Med 382(18):1708–1720. https://doi.org/10.1056/NEJMoa2002032
9. World Health Organization (WHO), (2020) WHO- Coronavirus disease (COVID-19) pandemic report https://www.who.int/emergencies/diseases/novel-coronavirus-2019
10. Hofmann H, Pyrc K, van der Hoek L, Geier M, Berkhout B, Pohlmann S (2005) Human coronavirus NL63 employs the severe acute respiratory syndrome coronavirus receptor for cellular entry. Proc Natl Acad Sci U S A 102(22): 7988–7993. https://doi.org/10.1073/pnas.0409465102
11. Belouzard S, Chu VC, Whittaker GR (2009) Activation of the SARS coronavirus spike protein via sequential proteolytic cleavage at two distinct sites. Proc Natl Acad Sci U S A 106(14):5871–5876. https://doi.org/10.1073/pnas.0809524106
12. Kamitani W, Narayanan K, Huang C, Lokugamage K, Ikegami T, Ito N, Kubo H, Makino S (2006) Severe acute respiratory syndrome coronavirus nsp1 protein suppresses host gene expression by promoting host mRNA degradation. Proc Natl Acad Sci U S A 103(34):12885–12890. https://doi.org/10.1073/pnas.0603144103
13. Kim J, Zhang J, Cha Y, Kolitz S, Funt J, Escalante Chong R, Barrett S, Kusko R, Zeskind B, Kaufman H (2020) Advanced bioinformatics rapidly identifies existing therapeutics for patients with coronavirus disease-2019 (COVID-19). J Transl Med 18(1):257. https://doi.org/10.1186/s12967-020-02430-9
14. Ujike M, Taguchi F (2015) Incorporation of spike and membrane glycoproteins into coronavirus virions. Viruses 7(4):1700–1725. https://doi.org/10.3390/v7041700
15. Glowacka I, Bertram S, Muller MA, Allen P, Soilleux E, Pfefferle S, Steffen I, Tsegaye TS, He Y, Gnirss K, Niemeyer D, Schneider H, Drosten C, Pohlmann S (2011) Evidence that TMPRSS2 activates the severe acute respiratory syndrome coronavirus spike protein for membrane fusion and reduces viral control by the humoral immune response. J Virol 85(9): 4122–4134. https://doi.org/10.1128/JVI.02232-10
16. Chen N, Zhou M, Dong X, Qu J, Gong F, Han Y, Qiu Y, Wang J, Liu Y, Wei Y, Xia J, Yu T, Zhang X, Zhang L (2020) Epidemiological and clinical characteristics of 99 cases of 2019 novel coronavirus pneumonia in Wuhan, China: a descriptive study. Lancet 395(10223):507–513. https://doi.org/10.1016/S0140-6736(20)30211-7
17. Holshue ML, DeBolt C, Lindquist S, Lofy KH, Wiesman J, Bruce H, Spitters C, Ericson K, Wilkerson S, Tural A, Diaz G, Cohn A,

Fox L, Patel A, Gerber SI, Kim L, Tong S, Lu X, Lindstrom S, Pallansch MA, Weldon WC, Biggs HM, Uyeki TM, Pillai SK, Washington State -nCo VCIT (2020) First case of 2019 novel coronavirus in the United States. N Engl J Med 382 (10):929-936. https://doi.org/10.1056/NEJMoa2001191

18. Wang M, Cao R, Zhang L, Yang X, Liu J, Xu M, Shi Z, Hu Z, Zhong W, Xiao G (2020) Remdesivir and chloroquine effectively inhibit the recently emerged novel coronavirus (2019-nCoV) in vitro. Cell Res 30(3):269–271. https://doi.org/10.1038/s41422-020-0282-0
19. Low ZY, Farouk IA, Lal SK (2020) Drug repositioning: new approaches and future prospects for life-debilitating diseases and the COVID-19 pandemic outbreak. Viruses 12(9). https://doi.org/10.3390/v12091058
20. Hufsky F, Lamkiewicz K, Almeida A, Aouacheria A, Arighi C, Bateman A, Baumbach J, Beerenwinkel N, Brandt C, Cacciabue M, Chuguransky S, Drechsel O, Finn RD, Fritz A, Fuchs S, Hattab G, Hauschild AC, Heider D, Hoffmann M, Holzer M, Hoops S, Kaderali L, Kalvari I, von Kleist M, Kmiecinski R, Kuhnert D, Lasso G, Libin P, List M, Lochel HF, Martin MJ, Martin R, Matschinske J, McHardy AC, Mendes P, Mistry J, Navratil V, Nawrocki EP, O'Toole AN, Ontiveros-Palacios N, Petrov AI, Rangel-Pineros G, Redaschi N, Reimering S, Reinert K, Reyes A, Richardson L, Robertson DL, Sadegh S, Singer JB, Theys K, Upton C, Welzel M, Williams L, Marz M (2021) Computational strategies to combat COVID-19: useful tools to accelerate SARS-CoV-2 and coronavirus research. Brief Bioinform 22(2): 642–663. https://doi.org/10.1093/bib/bbaa232
21. Moore JH, Barnett I, Boland MR, Chen Y, Demiris G, Gonzalez-Hernandez G, Herman DS, Himes BE, Hubbard RA, Kim D, Morris JS, Mowery DL, Ritchie MD, Shen L, Urbanowicz R, Holmes JH (2020) Ideas for how informaticians can get involved with COVID-19 research. BioData Mining 13(1): 3. https://doi.org/10.1186/s13040-020-00213-y
22. Huang X, He C, Hua X, Kan A, Sun S, Wang J, Li S (2020) Bioinformatic analysis of correlation between immune infiltration and COVID-19 in cancer patients. Int J Biol Sci 16(13): 2464–2476. https://doi.org/10.7150/ijbs.48639
23. Gui M, Song W, Zhou H, Xu J, Chen S, Xiang Y, Wang X (2017) Cryo-electron microscopy structures of the SARS-CoV spike glycoprotein reveal a prerequisite conformational state for receptor binding. Cell Res 27(1):119–129. https://doi.org/10.1038/cr.2016.152
24. Holm L, Laakso LM (2016) Dali server update. Nucleic Acids Res 44(W1): W351–W355. https://doi.org/10.1093/nar/gkw357
25. JAlmagro Armenteros JJ, Sonderby CK, Sonderby SK, Nielsen H, Winther O (2017) DeepLoc: prediction of protein subcellular localization using deep learning. Bioinformatics 33(21):3387–3395. https://doi.org/10.1093/bioinformatics/btx431
26. Dyer MD, Murali TM, Sobral BW (2008) The landscape of human proteins interacting with viruses and other pathogens. PLoS Pathog 4(2):e32. https://doi.org/10.1371/journal.ppat.0040032
27. Bailer SM, Haas J (2009) Connecting viral with cellular interactomes. Curr Opin Microbiol 12(4):453–459. https://doi.org/10.1016/j.mib.2009.06.004
28. Gursoy A, Keskin O, Nussinov R (2008) Topological properties of protein interaction networks from a structural perspective. Biochem Soc Trans 36(Pt 6):1398–1403. https://doi.org/10.1042/BST0361398
29. Zheng LL, Li C, Ping J, Zhou Y, Li Y, Hao P (2014) The domain landscape of virus-host interactomes. Biomed Res Int 2014:867235. https://doi.org/10.1155/2014/867235
30. Dyer MD, Murali TM, Sobral BW (2011) Supervised learning and prediction of physical interactions between human and HIV proteins. Infect Genet Evol 11(5):917–923. https://doi.org/10.1016/j.meegid.2011.02.022
31. Brito AF, Pinney JW (2017) Protein-protein interactions in virus-host systems. Front Microbiol 8:1557. https://doi.org/10.3389/fmicb.2017.01557
32. Navratil V, de Chassey B, Meyniel L, Delmotte S, Gautier C, Andre P, Lotteau V, Rabourdin-Combe C (2009) VirHostNet: a knowledge base for the management and the analysis of proteome-wide virus-host interaction networks. Nucleic Acids Res 37(Database issue):D661–D668. https://doi.org/10.1093/nar/gkn794
33. Gardner MR, Fellinger CH, Kattenhorn LM, Davis-Gardner ME, Weber JA, Alfant B, Zhou AS, Prasad NR, Kondur HR, Newton WA, Weisgrau KL, Rakasz EG, Lifson JD, Gao G, Schultz-Darken N, Farzan M (2019) AAV-delivered eCD4-Ig protects rhesus macaques from high-dose SIVmac239 challenges.

Sci Transl Med 11(502). https://doi.org/10.1126/scitranslmed.aau5409

34. Oughtred R, Stark C, Breitkreutz BJ, Rust J, Boucher L, Chang C, Kolas N, O'Donnell L, Leung G, McAdam R, Zhang F, Dolma S, Willems A, Coulombe-Huntington J, Chatr-Aryamontri A, Dolinski K, Tyers M (2019) The BioGRID interaction database: 2019 update. Nucleic Acids Res 47(D1): D529–D541. https://doi.org/10.1093/nar/gky1079
35. Stark C, Breitkreutz BJ, Reguly T, Boucher L, Breitkreutz A, Tyers M (2006) BioGRID: a general repository for interaction datasets. Nucleic Acids Res 34(Database issue): D535–D539. https://doi.org/10.1093/nar/gkj109
36. Szklarczyk D, Franceschini A, Kuhn M, Simonovic M, Roth A, Minguez P, Doerks T, Stark M, Muller J, Bork P, Jensen LJ, von Mering C (2011) The STRING database in 2011: functional interaction networks of proteins, globally integrated and scored. Nucleic Acids Res 39(Database issue):D561–D568. https://doi.org/10.1093/nar/gkq973
37. Baltimore D (1971) Expression of animal virus genomes. Bacteriol Rev 35(3):235–241. https://doi.org/10.1128/br.35.3.235-241.1971
38. Vogel C, Bashton M, Kerrison ND, Chothia C, Teichmann SA (2004) Structure, function and evolution of multidomain proteins. Curr Opin Struct Biol 14(2):208–216. https://doi.org/10.1016/j.sbi.2004.03.011
39. Apic G, Gough J, Teichmann SA (2001) An insight into domain combinations. Bioinformatics 17(Suppl 1):S83–S89. https://doi.org/10.1093/bioinformatics/17.suppl_1.s83
40. Lee H, Deng M, Sun F, Chen T (2006) An integrated approach to the prediction of domain-domain interactions. BMC Bioinformatics 7:269. https://doi.org/10.1186/1471-2105-7-269
41. Yellaboina S, Tasneem A, Zaykin DV, Raghavachari B, Jothi R (2011) DOMINE: a comprehensive collection of known and predicted domain-domain interactions. Nucleic Acids Res 39(Database issue):D730–D735. https://doi.org/10.1093/nar/gkq1229
42. Schneidman-Duhovny D, Inbar Y, Nussinov R, Wolfson HJ (2005) PatchDock and SymmDock: servers for rigid and symmetric docking. Nucleic Acids Res 33(Web Server issue): W363–W367. https://doi.org/10.1093/nar/gki481
43. Chen R, Li L, Weng Z (2003) ZDOCK: an initial-stage protein-docking algorithm. Proteins 52(1):80–87. https://doi.org/10.1002/prot.10389
44. Wallace AC, Laskowski RA, Thornton JM (1995) LIGPLOT: a program to generate schematic diagrams of protein-ligand interactions. Protein Eng 8(2):127–134. https://doi.org/10.1093/protein/8.2.127
45. Vilar S, Cozza G, Moro S (2008) Medicinal chemistry and the molecular operating environment (MOE): application of QSAR and molecular docking to drug discovery. Curr Top Med Chem 8(18):1555–1572. https://doi.org/10.2174/156802608786786624
46. W.L. Delano, The PyMOL molecular graphics system, schrodinger. (2002). http://www.pymol.sourceforge.net
47. Byrum S, Smart SK, Larson S, Tackett AJ (2012) Analysis of stable and transient protein-protein interactions. Methods Mol Biol 833:143–152. https://doi.org/10.1007/978-1-61779-477-3_10
48. Franzosa EA, Xia Y (2011) Structural principles within the human-virus protein-protein interaction network. Proc Natl Acad Sci U S A 108(26):10538–10543. https://doi.org/10.1073/pnas.1101440108\
49. Guirimand T, Delmotte S, Navratil V (2015) VirHostNet 2.0: surfing on the web of virus/host molecular interactions data. Nucleic Acids Res 43(Database issue):D583–D587. https://doi.org/10.1093/nar/gku1121
50. Calderone A, Licata L, Cesareni G (2015) VirusMentha: a new resource for virus-host protein interactions. Nucleic Acids Res 43 (Database issue):D588–D592. https://doi.org/10.1093/nar/gku830
51. Hermjakob H, Montecchi-Palazzi L, Lewington C, Mudali S, Kerrien S, Orchard S, Vingron M, Roechert B, Roepstorff P, Valencia A, Margalit H, Armstrong J, Bairoch A, Cesareni G, Sherman D, Apweiler R (2004) IntAct: an open source molecular interaction database. Nucleic Acids Res 32(Database issue): D452–D455. https://doi.org/10.1093/nar/gkh052
52. Xenarios I, Salwinski L, Duan XJ, Higney P, Kim SM, Eisenberg D (2002) DIP, the database of interacting proteins: a research tool for studying cellular networks of protein interactions. Nucleic Acids Res 30(1):303–305. https://doi.org/10.1093/nar/30.1.303
53. Licata L, Briganti L, Peluso D, Perfetto L, Iannuccelli M, Galeota E, Sacco F, Palma A, Nardozza AP, Santonico E, Castagnoli L, Cesareni G (2012) MINT, the molecular interaction database: 2012 update. Nucleic Acids

Res 40(Database issue):D857–D861. https://doi.org/10.1093/nar/gkr930

54. Chatr-Aryamontri A, Oughtred R, Boucher L, Rust J, Chang C, Kolas NK, O'Donnell L, Oster S, Theesfeld C, Sellam A, Stark C, Breitkreutz BJ, Dolinski K, Tyers M (2017) The BioGRID interaction database: 2017 update. Nucleic Acids Res 45(D1):D369–D379. https://doi.org/10.1093/nar/gkw1102
55. Khanday A, Rabani ST, Khan QR, Rouf N Mohi Ud Din M (2020) Machine learning based approaches for detecting COVID-19 using clinical text data. Int J Inf Technol:1–9. https://doi.org/10.1007/s41870-020-00495-9
56. C Medel-Ramirez, H Medel-Lopez (2020), Data mining for the study of the epidemic (SARS-CoV-2) COVID-19: algorithm for the identification of patients (SARS-CoV-2) COVID 19 in Mexico, Available SSRN 3619549
57. Muhammad LJ, Islam MM, Usman SS, Ayon SI (2020) Predictive data mining models for novel coronavirus (COVID-19) infected patients' recovery. SN Comput Sci 1(4):206. https://doi.org/10.1007/s42979-020-00216-w
58. Duan Y, Coreas R, Liu Y, Bitounis D, Zhang Z, Parviz D, Strano M, Demokritou P, Zhong W (2020) Prediction of protein corona on nanomaterials by machine learning using novel descriptors. Nano Impact 17. https://doi.org/10.1016/j.impact.2020.100207
59. Ban Z, Yuan P, Yu F, Peng T, Zhou Q, Hu X (2020) Machine learning predicts the functional composition of the protein corona and the cellular recognition of nanoparticles. Proc Natl Acad Sci U S A 117(19):10492–10499. https://doi.org/10.1073/pnas.1919755117
60. Papa E, Doucet JP, Sangion A, Doucet-Panaye A (2016) Investigation of the influence of protein corona composition on gold nanoparticle bioactivity using machine learning approaches. SAR QSAR Environ Res 27(7):521–538. https://doi.org/10.1080/1062936X.2016.1197310

Chapter 14

Human Nasal Epithelial Cells (hNECs) Generated by Air-Liquid Interface (ALI) Culture as a Model System for Studying the Pathogenesis of SARS-CoV-2

Kai Sen Tan, Akshamal M. Gamage, and Jing Liu

Abstract

The nasal epithelium lining the human upper airway is the primary portal of entry for several respiratory pathogens, including the recently emerged SARS-CoV-2 virus responsible for the ongoing COVID-19 pandemic. Here, we describe in detail methods for in vitro ALI differentiation of primary cells collected from human donors, to obtain differentiated hNECs. This can serve as a physiologically relevant model to investigate various aspects of host–pathogen responses to SARS-CoV-2 and other emerging respiratory viruses.

Key words Human nasal epithelial cells, Air–liquid interface, SARS-CoV-2, Viral kinetics, Host–pathogen interaction

1 Introduction

The emergence of SARS-CoV-2 in late-2019 and its remarkably rapid global spread emphasizes the urgent need for suitable human model systems to study respiratory viral infection [1, 2]. For SARS-CoV-2, the high transmissibility, and the ability to cause severe complications from a novel coronavirus species with no prior anti-viral therapeutics or vaccines resulted in a serious healthcare and economic crisis [3, 4]. Future responses to novel respiratory pathogens can be greatly helped by the presence of such model systems to enable quick assessment of various aspects of the host–pathogen interaction. The system should provide a close mimic of the airway and readily allow assessments of viral replicative capacity and kinetics, the magnitude of the host response elicited and viral antagonism of the responses, within-host viral evolution, entry factors and host susceptibility factors, the impact of variants of concern during the course of the pandemic, and efficacy of novel and repurposed antiviral therapeutics [1, 5].

Justin Jang Hann Chu et al. (eds.), *SARS-CoV-2: Methods and Protocols*,
Methods in Molecular Biology, vol. 2452, https://doi.org/10.1007/978-1-0716-2111-0_14,

The human upper airway (nasal) epithelium is the primary portal of entry for respiratory viruses and serves as the first line of defense against invading respiratory viral pathogens by providing a mechanical barrier and producing early immune response cytokines [5–7]. During SARS-CoV-2 infection, the nasal epithelium expresses the ACE2 and TMPRSS2 viral entry factors and mounts an immune response against the virus [8]. Studies have demonstrated that hNECs are readily susceptible to SARS-CoV-2 infection [6, 9, 10]. These findings highlight the importance of hNECs in basic and clinical investigations of SARS-CoV-2 infection and warrant the establishment of standardized protocols for hNEC growth and infection to improve their usability worldwide, particularly due to the relatively laborious nature of preparing this model [11, 12]. The protocol aims to provide step-by-step guidance for differentiating hNECs from tissue biopsies or nasal brushes, for infection with SARS-CoV-2 and related downstream analysis. These protocols can be readily extended and adapted for studying current and emerging respiratory viruses.

2 Materials

Preprepare all cell culture reagents prior to commencing each culture of hNECs. Prepared cell culture reagents should be stored for no more than 3 months before use. Prepare and store all reagents at 4 °C, unless indicated otherwise. Preparation of reagents should be performed under aseptic conditions to avoid downstream contamination during cell culture. The cell culture reagents listed here are examples of media that can support the expansion of human nasal epithelial stem/progenitor cells (hNESPCs); and media that allow the hNESPC differentiation into hNECs under ALI culture conditions. Other iteration of media with specifications that allows expansion of hNESPCs and differentiation of hNECs can be used as well. All viruses are stored at −80 °C and are thawed immediately before use. Avoid repeated freeze–thaw cycles for virus stocks.

2.1 Cell Culture Mediums

1. Complete Dulbecco's Minimum Essential Media (DMEM)-High Glucose: DMEM containing 10% Fetal Bovine Serum (FBS) and 1× antibiotic–antimycotic (anti-anti) for growing 3T3 fibroblast feeder cells. For making 500 mL of complete DMEM, mix 445 mL of DMEM with 50 mL of heat-inactivated FBS and 5 mL of anti-anti (100×).
2. Medium 3: Medium for expansion and enrichment of hNESPCs. In 500 mL of DMEM/Nutrient Mixture F-12 medium, add 5 mL of N-2 supplement (final concentration

10 μL/mL) and 5 mL of anti-anti (100×) and the following reagents to achieve their respective final concentration: human epithelial growth factor 5 ng/mL, insulin 2.5 μg/mL, cholera toxin 0.1 nM, hydrocortisone 0.5 μg/mL, and 3,3′,5-triiodo-l-thyronine 2 nM.

3. Mitomycin C-containing Medium 3: Medium for treating 3T3 fibroblast to inhibit their proliferative capability prior to seeding hNESPCs. In complete Medium 3, add 500 μL of 200 μg/mL mitomycin C stock into 10 mL of medium 3 to achieve a final mitomycin C concentration of 10 μg/mL.

4. Differentiation media (STEMCELL Technologies): Medium for differentiating hNESPCs into hNECs in ALI Transwell culture. For 500 mL of media, mix 441 mL of PneumaCult-ALI Basal Medium, 50 mL of PneumaCult-ALI 10× supplement, 2.5 mL of hydrocortisone solution (200×), 1 mL of 0.2% (2 mg/mL; 1000 IU/mL) heparin sodium salt in phosphate-buffered saline, and 5 mL of anti-anti (100×). Aliquot the medium into 50 mL aliquots. Prior to use, take a 50 mL aliquot and add 500 μL of PneumaCult-ALI Maintenance Supplement (100×) to obtain a complete differentiation media.

2.2 ALI culture Transwell Inserts

1. Transwell inserts for differentiation of hNESPCs into hNECs: 0.4 μm pore size clear polyester (PET) membrane Transwell insert is required for optimum growing and differentiation into hNECs. Transwell size of 6.5 mm (for 24-well plate) and 12 mm (for 12-well plate) is recommended.

2.3 Transepithelial Electrical Resistance (TEER)

1. TEER measurement for ensuring epithelial integrity is performed using an EVOM voltohmmeter device with STX2 electrodes.

3 Methods

Preparation of cell culture should be done under BSL-2 conditions and handling of inactivated viral samples from BSL-3 should be done under BSL-2 conditions with appropriate precautions. Handling of active viral samples should only be done in BSL-3 conditions following the designated BSL-3 standard operating procedure of each institute. Any infected samples to be used for downstream analysis outside of BSL-3 must be inactivated accordingly and validated to be noninfectious prior.

3.1 Preparation of 3T3 Feeder Layer for Expansion and Selection of hNESPCs from Tissue Biopsies or Nasal Brushes

1. Thaw cryovial containing NIH/3T3 fibroblast from liquid nitrogen storage and add 2 mL of complete DMEM dropwise.
2. Centrifuge cell suspension at 300 × *g* for 5 min at room temperature/pressure (RTP) and aspirate supernatant. Resuspend cells in complete DMEM. Grow cells for 2–3 passages prior to treatment for generating feeder layer cells.
3. Count cells with trypan blue exclusion staining and seed cells at a density of 1×10^4 cells/cm^2 in an appropriate culture dish (e.g., T25 flask). Incubate cells for 3 days in a tissue culture incubator at 37 °C with 5% CO_2.
4. After an approximately 3-day incubation, the culture should be at 60–80% confluency (*see* **Note 1**). Remove medium and wash cells with 1× Phosphate-Buffered Saline (PBS).
5. Add mitomycin C containing medium 3 to the cells and incubate at 37 °C, 5% CO_2 for 3.5 h. After incubation, remove the mitomycin C containing media and wash the cells 2× with 1× PBS.
6. Incubate cells with 3 mL of 1× trypsin–EDTA (for T25 flask) for 3–5 min. Collect the dissociated cells and centrifuge at 300 × *g* for 5 min RTP. Aspirate the supernatant and resuspend cells in complete DMEM.
7. Count cells with trypan blue exclusion staining and seed cells in a 6-well plate at 7.5×10^5–2.5×10^6 cells/well. Incubate the plate overnight at 37 °C, 5% CO_2 to establish the feeder layer.

3.2 Expansion of hNESPCs and Establishment of Differentiated hNECs in ALI Culture (see Note 2)

1. Obtain fresh nasal tissue biopsies or brush from donors (*see* **Note 3**) and wash tissue in 1× Dulbecco's PBS (dPBS) (*see* **Note 4**) containing 1× anti-anti. Submerge tissue in the dPBS-anti-anti mixture, cut tissue into small fragments with a scissor and treat tissue with 10 mg/mL of a neutral protease (e.g., Dispase II) overnight at 4 °C with shaking. If a nasal brush was used for obtaining primary cells, immerse brush in dPBS-anti-anti mixture and proceed directly to **step 3**.
2. Centrifuge cells at 200 × *g* for 5 min RTP and aspirate supernatant. Incubate the resulting pellet with 1–2 mL of 1× trypsin–EDTA at 37 °C for 15 min. After incubation, quench by adding FBS to a final concentration of 10% of the tube contents.
3. Mechanically dissociate the cells from the digested tissue by pipetting repeatedly with a 1 mL micropipette. If processing from nasal brush, skip step 4.
4. Pass the resulting suspension through a 70 μm cell strainer. Gently push contents through the strainer with the plunger of a 5 mL syringe.
5. Resuspend in complete DMEM (*see* **Note 5**).

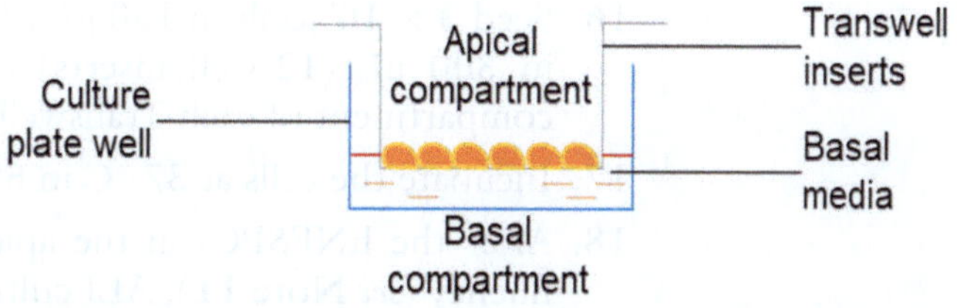

Fig. 1 Schematic of a Transwell ALI culture. Transwell inserts placed within cell culture plates containing culture medium. The medium compartment forms the basal compartment, while the top of the Transwell where cells are seeded make up the apical compartment

6. Centrifuge at 300 × *g* for 5 min RTP, aspirate supernatant and resuspend with 3–5 mL of medium 3.
7. Count cells via trypan blue exclusion staining. Seed 1×10^6 cells in 2 mL of medium 3 onto the 3T3 feeder layer in a 6-well plate. Incubate cells at 37 °C, 5% CO_2 (*see* **Note 6**).
8. Change medium 3 every 2–4 days (*see* **Note 7**).
9. After the hNESPCs is cultured for 7–10 days, islands of squamous-shaped cells (hNESPCs) should be observed to be expanding and dispersing the elongated 3T3 feeder cells. When these hNESPCs "islands" comprise >70% of the well surface, they can be dissociated and transferred onto Transwell inserts for ALI culture.
10. To transfer the hNESPCs onto Transwell inserts, remove medium 3 gently and flush each well with 500 μL of 1× dPBS three times using a micropipette to remove the 3T3 feeder layer (*see* **Note 8**). Visually observe under a microscope to ensure that most of the feeder cells has been dissociated and only the hNESPCs remain adherent in the wells.
11. Add 500 μL of Accutase into each well and incubate at 37 °C with 5% CO_2 for 5–10 mins. Ensure that all hNESPCs are detached from the well. If not, lengthen the digestion time, and pipette up and down with a 1 mL micropipette to aid dissociation.
12. Collect hNESPCs suspension and centrifuge the cells at 300 × *g* for 5 min rtp. Aspirate supernatant.
13. Resuspend cell pellet in medium 3. Count cells with trypan blue exclusion staining.
14. Prepare cell culture plates for Transwell inserts by adding 350 μL (24-well plate) or 700 μL (12-well plate) of medium 3 into the basal compartment of each well (Fig. 1).
15. Lay Transwell inserts onto the preadded media, ensure that the bottom side of the Transwell membrane is in contact with the medium in the basal compartment.

16. Seed 3×10^4 cells in 150 μL (24-well inserts) or 1×10^5 cells in 300 μL (12-well inserts) of medium 3 onto the apical compartment of each Transwell (*see* **Note 9**).
17. Incubate the cells at 37 °C in 5% CO_2 (*see* **Note 10**).
18. After the hNESPCs in the apical compartment reaches confluency (*see* **Note 11**), ALI culture is initiated by changing the basal media into differentiation media and removing the media from the apical compartment. Continue to replace the differentiation media in the basal compartment every 2–3 days, until hNECs are fully differentiated and ready for the experiment at around 21–28 days post-ALI.

3.3 Preparing Differentiated hNECs for Infection (Removal of Excess Mucus and Debris, Cell Counting and TEER Measurements)

1. Upon maturation of hNECs, start washing the hNECs one week prior to the infection experiment to remove excess debris and mucus. Before replacing the basal medium, add 50 μL (24-well) or 150 μL (12-well) of 1× dPBS onto the apical compartment of the Transwell and incubate cells at 37 °C for 10 min. After incubation, remove the 1× dPBS and change basal media as per normal.
2. Select one well for cell counting to obtain a representative cell number for multiplicity of infection (MOI) calculation during infection. Trypsinize hNECs by adding 150 μL and 350 μL (24-well); or 300 μL and 700 μL (12-well) of trypsin–EDTA to the apical and basal compartment, respectively, and incubate at 37 °C for 10–30 min until all cells have detached from the apical compartment.
3. Collect the digested cells from the apical compartment, then flush the apical compartment using the remaining trypsin–EDTA from the basal compartment. Repeat until all trypsin–EDTA added to both apical and basal compartment is collected.
4. Quench trypsin–EDTA action by adding 20% total cell suspension volume with complete DMEM.
5. Count cells using trypan blue exclusion staining and multiply cell count per mL with total volume collected (including the volume added for quenching) to obtain cell count per well.
6. For TEER measurements (*see* **Note 12**), wash cells similarly to **step 1** of this section prior to measuring.
7. Rinse the STX2 electrodes with 70% ethanol and sterilize under ultraviolet light for 15 min, followed by rinsing with 1× dPBS prior to using the electrodes.
8. Equilibrate the electrodes by placing them in prewarmed (37 °C) differentiation media while the voltohmmeter is switched on.

9. Prepare two empty wells and add 1 mL (24-well) or 3 mL (12-well) of prewarmed (37 °C) differentiation media to the basal compartment. In one well, put an empty Transwell insert into the well to serve as a blank reading.
10. In the other well, transfer a Transwell insert with hNECs and fill both blank and hNECs Transwell inserts with 200 μL (24-well) and 600 μL (12-well) prewarmed differentiation media, respectively.
11. Position electrodes in the blank with one electrode submerged in the apical compartment and the other in the basal compartment (*see* **Note 13**). Record the reading when a constant reading is observed over a period of at least 5 min. Repeat measurement at two different positions in the Transwell for 3 readings.
12. Repeat measurement for the hNECs well.
13. Wash electrode with 1× dPBS and repeat **steps 11–12** with three other Transwell inserts of the same hNEC sample to obtain an average of 4 readings of TEER.
14. Calculate epithelial integrity of each sample by subtracting the blank measurement and multiplying the net TEER measurement (Ω) with the area of Transwell insert (cm^2) to obtain readings in Ωcm^2 (*see* **Note 14**).
15. Once these steps are completed, the cells can be transferred into the BSL-3 for infection experiment using SARS-CoV-2.

3.4 Infection of hNECs with SARS-CoV-2 Virus

1. Wash the hNECs with 1× dPBS as detailed in Subheading 3.3, **step 1**, and replace the basal media with fresh differentiation media prior to virus inoculation.
2. Calculate the required MOI based on the cell numbers of hNECs per well obtained. Inoculate 30 μL (24-well) or 100 μL (12-well) of the virus onto the apical compartment and incubate at 37 °C in 5% CO_2 for 1 h.
3. After incubation, remove inoculum from the apical compartment (*see* **Note 15**), and wash Transwell by adding 100 μL (24-well) or 300 μL (12-well) of 1× dPBS. Incubate at 37 °C in 5% CO_2 for 10 min and remove the 1× dPBS (*see* **Note 16**).
4. Replace basal media (350 μL for 24-well; 700 μL for 12-well) with fresh differentiation media.
5. Incubate the infected hNECs in 37 °C, 5% CO_2 up to the desired time point.
6. To collect the infected supernatant for viral quantification at each time point, add 50 μL (24-well) or 150 μL (12-well) of 1× dPBS onto the apical compartment of the Transwell and incubate cells at 37 °C for 10 min. After incubation, collect the

supernatant which can then be used for plaque assay [6, 12, 13] or PCR [14] to quantify the virus present, or cytokine analysis [13, 14] for detecting apically secreted cytokines (*see* **Note 17**).

7. For cellular RNA or protein analysis, add appropriate lysis buffer to collect the cell lysate and extract the respective material from the infected hNECs [13, 15]. To fix cells for further immunofluorescence staining, wash cells twice with 1x dPBS for 5 min (37 °C in 5% CO_2), then dissociate cells with trypsin–EDTA as per Subheading 3.3, **steps 2–4**, and centrifuge cells at 300 × *g* for 5 mins RTP and resuspend cells in 4% PFA for Cytospin preparation; or add 4% PFA directly to the apical and basal compartment of the ALI culture (> 15 min) for Transwell cross section staining (*see* **Note 18**).
8. Perform validation experiments to ensure that SARS-CoV-2 has been inactivated by each specific treatment (PFA fixation, RNA lysis buffer, etc.), before transferring materials out of the BSL3 for further assays.

3.5 Immunofluorescence Staining of Transwell Cross Section to Visualize SARS-CoV-2 Infection in the Nasal Epithelium

1. The step follows the cross section staining portion of **step 7** from Subheading 3.4.
2. After PFA fixation of the Transwell, remove 4% PFA from both apical and basal compartment and wash membrane inserts by adding 1× PBS (or 1× dPBS) to both apical and basal compartments to submerge the fixed hNECs.
3. Discard PBS and start dehydration of the hNECs in the inserts by adding 70% ethanol to the Transwell two times, 10 min each.
4. Following the 70% ethanol treatment, carefully cut the Transwell membrane (*see* **Note 19**) out from the Transwell insert and place the membrane into a cassette (*see* **Note 20**) for subsequent dehydration steps.
5. Continue dehydration by immersing cassette in increasing concentrations of ethanol of 95%, two times; 100%, three times, each time immersing for 10 min.
6. Immerse cassette in xylene two times, 10 min each.
7. Immerse cassette in liquid paraffin (*see* **Note 21**) two times, 30 min each.
8. Cut the solidified membrane into two and embed in a paraffin boat by standing the membrane upright on the base of the semicircle. Note the side of the membrane with cells on it (*see* **Note 22**).
9. Section embedded Transwell cross section with a microtome to obtain 4–5 μm thickness section.

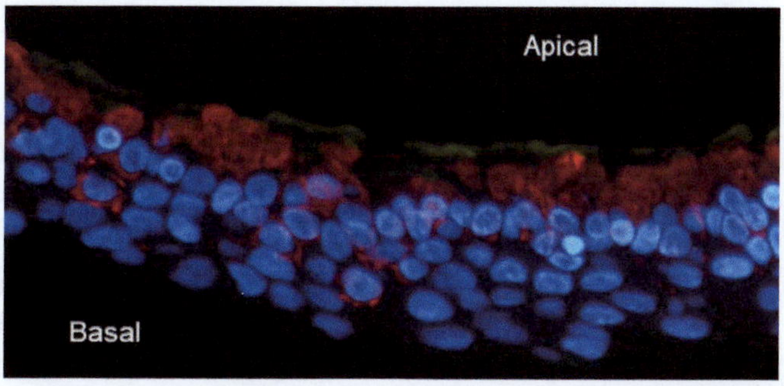

Fig. 2 Representative Transwell cross section immunofluorescence image of SARS-CoV-2 infected hNECs. Confocal microscopy image showing immunofluorescence staining of infected hNECs in a cross section detailing the multicellular, multilayered morphology of differentiated hNECs in ALI culture. The apical side (with ciliated cells, stained green) of the hNECs is the primary localization region of SARS-CoV-2 virus. Green—βIV-Tubulin; Red—SARS-CoV-2 nucleocapsid; Blue—DAPI

10. Deparaffinize, antigen-retrieve, permeabilize, stain, and mount section on slides for viewing under fluorescence microscope [6]. A representative result of the IF staining from the resulting cross section staining is shown in Fig. 2.

4 Notes

1. It is important that the 3T3 cells are not fully confluent otherwise it will affect the effectiveness of mitomycin C treatment.
2. The protocol here outlines the steps for feeder-based expansion of hNESPCs. For feeder-free expansion of hNESPCs (using lower starting material, or for other purposes, e.g., gene editing), collagen coating and the use of specific small molecule inhibitors can be considered to ensure the proper expansion of hNESPCs [16].
3. Donor tissue should be obtained from patients free from upper respiratory tract infection.
4. dPBS—Mg^{2+} and Ca^{2+} free PBS.
5. After completing these steps, the resulting cell suspension is a mixture of differentiated cells and hNESPCs. The subsequent steps serve to select and enrich hNESPCs that exhibit proliferative capacity under the stated culture conditions.
6. Take care not to agitate the resulting culture from this step as it will affect hNESPCs attachment.
7. The timing to replace medium 3 depends on the growth dynamics. The interval of change is determined by the acidity

of the culture media, and the media should be replaced when it turns yellow.

8. 3T3 cells are unable to grow after mitomycin C treatment and medium 3 does not sustain them. Therefore, their weakened adhesion enables them to be selectively flushed off the wells while the hNESPCs remain adherent.
9. During the initial expansion stage of the hNESPCs on the Transwell membrane, medium 3 is kept on the apical compartment in order to ensure the hNESPCs grow to confluency before initiating ALI.
10. Do not disturb the medium in the apical compartment after seeding to allow attachment of the hNESPCs to the Transwell membrane.
11. The hNESPCs should reach 100% confluency for proper differentiation to occur.
12. TEER measurements are performed to ensure that the epithelial layer is intact and healthy for infection experiment.
13. Make sure the longer (external) electrode is in contact with the bottom of the plate to prevent the shorter (internal electrode) from touching the bottom of the Transwell membrane.
14. TEER values for viral infection experiments should be >1000 Ωcm^2.
15. The infected hNECs remain in ALI conditions after removal of viral inoculum.
16. The washing step is to remove unbound virus particles to prevent confounding of the kinetics readout at subsequent time-points.
17. Basal media can also be collected for similar analysis of viral quantification and cytokine secretion into the basal compartment [6].
18. As the cells cannot be dissociated from the Transwell prior to the different collection for downstream analysis, an individual Transwell of the same experimental setting for each collected component (RNA, protein, fixing) should be prepared if multiple downstream analysis is planned.
19. The Transwell membrane can be removed by carefully running a scalpel blade through the circumference of the membrane on the inserts.
20. Use cassettes for tissue/biopsy processing.
21. Preheat paraffin to 65 °C to obtain liquid paraffin.
22. The orientation of the Transwell during the embedding is to allow a cross section of the Transwell to be cut during sectioning. If embedded with the right orientation, the section cut should have the Transwell appearing as a single line (the membrane) on the section.

Acknowledgments

This work was supported by the National Medical Research Council, Singapore No. MOH-COVID19RF2-0001 and COVID19RF-003. We thank the staff and surgeon team in the Department of Otolaryngology, Heck and Neck surgery, NUH, Singapore for help with acquiring human biopsy samples; and the operations team at the Duke-NUS Medical School ABSL3 facility for logistics management and technical assistance.

References

1. Guo YR, Cao QD, Hong ZS, Tan YY, Chen SD, Jin HJ, Tan KS, Wang DY, Yan Y (2020) The origin, transmission and clinical therapies on coronavirus disease 2019 (COVID-19) outbreak - an update on the status. Mil Med Res 7(1):11. https://doi.org/10.1186/s40779-020-00240-0
2. Zhu N, Zhang D, Wang W, Li X, Yang B, Song J, Zhao X, Huang B, Shi W, Lu R, Niu P, Zhan F, Ma X, Wang D, Xu W, Wu G, Gao GF, Tan W, China Novel Coronavirus I, Research T (2020) A novel coronavirus from patients with pneumonia in China, 2019. N Engl J Med 382:727–733. https://doi.org/10.1056/NEJMoa2001017
3. Guan WJ, Ni ZY, Hu Y, Liang WH, Ou CQ, He JX, Liu L, Shan H, Lei CL, DSC H, Du B, Li LJ, Zeng G, Yuen KY, Chen RC, Tang CL, Wang T, Chen PY, Xiang J, Li SY, Wang JL, Liang ZJ, Peng YX, Wei L, Liu Y, Hu YH, Peng P, Wang JM, Liu JY, Chen Z, Li G, Zheng ZJ, Qiu SQ, Luo J, Ye CJ, Zhu SY, Zhong NS, China Medical Treatment Expert Group for C (2020) Clinical characteristics of coronavirus disease 2019 in China. N Engl J Med 382:1708–1720. https://doi.org/10.1056/NEJMoa2002032
4. Hu Y, Sun J, Dai Z, Deng H, Li X, Huang Q, Wu Y, Sun L, Xu Y (2020) Prevalence and severity of corona virus disease 2019 (COVID-19): a systematic review and meta-analysis. J Clin Virol 127:104371. https://doi.org/10.1016/j.jcv.2020.104371
5. Tan KS, Lim RL, Liu J, Ong HH, Tan VJ, Lim HF, Chung KF, Adcock IM, Chow VT, Wang Y (2020) Respiratory viral infections in exacerbation of chronic airway inflammatory diseases: novel mechanisms and insights from the upper airway epithelium. Front Cell Dev Biol 8:99. https://doi.org/10.3389/fcell.2020.00099
6. Gamage AM, Tan KS, Chan WOY, Liu J, Tan CW, Ong YK, Thong M, Andiappan AK, Anderson DE, Wang Y, Wang LF (2020) Infection of human Nasal Epithelial Cells with SARS-CoV-2 and a 382-nt deletion isolate lacking ORF8 reveals similar viral kinetics and host transcriptional profiles. PLoS Pathog 16(12):e1009130. https://doi.org/10.1371/journal.ppat.1009130
7. Vareille M, Kieninger E, Edwards MR, Regamey N (2011) The airway epithelium: soldier in the fight against respiratory viruses. Clin Microbiol Rev 24(1):210–229. https://doi.org/10.1128/CMR.00014-10
8. Sungnak W, Huang N, Becavin C, Berg M, Queen R, Litvinukova M, Talavera-Lopez C, Maatz H, Reichart D, Sampaziotis F, Worlock KB, Yoshida M, Barnes JL, Network HCALB (2020) SARS-CoV-2 entry factors are highly expressed in nasal epithelial cells together with innate immune genes. Nat Med 26(5):681–687. https://doi.org/10.1038/s41591-020-0868-6
9. Liu X, Mondal AM (2020) Conditional cell reprogramming for modeling host-virus interactions and human viral diseases. J Med Virol 92(11):2440–2452. https://doi.org/10.1002/jmv.26093
10. Hatton CF, Botting RA, Duenas ME, Haq IJ, Verdon B, Thompson BJ, Spegarova JS, Gothe F, Stephenson E, Gardner AI, Murphy S, Scott J, Garnett JP, Carrie S, Hussain R, Coxhead J, Davey T, Simpson AJ, Haniffa M, Hambleton S, Brodlie M, Ward C, Trost M, Reynolds G, Duncan CJA (2021) Delayed induction of type I and III interferons and nasal epithelial cell permissiveness to SARS-CoV-2. BioRxiv:431591. https://doi.org/10.1101/2021.1102.1117.431591
11. Zhao XN, Yu FG, Li CW, Li YY, Chao SS, Loh WS, Pan XL, Shi L, Wang D, Y. (2012) The use of nasal epithelial stem/progenitor cells to produce functioning ciliated cells in vitro. Am J Rhinol Allergy 26(5):345–350
12. Lew ZZR, Liu J, Ong HH, Tan VJ, Luukkainen A, Ong YK, Thong M, Puan KJ,

Chow VT, Tan KS, Wang DY (2021) Contact-free co-culture model for the study of innate immune cell activation during respiratory virus infection. J Vis Exp 168:e62115

13. Yan Y, Tan KS, Li C, Tran T, Chao SS, Sugrue RJ, Shi L, Chow V, Wang de Y (2016) Human nasal epithelial cells derived from multiple individuals exhibit differential responses to H3N2 influenza virus infection in vitro. J Allergy Clin Immunol 138(1):276–281

14. Tan KS, Ong HH, Yan Y, Liu J, Li C, Ong YK, Thong KT, Choi HW, Wang DY, Chow VT (2018) In vitro model of fully differentiated human nasal epithelial cells infected with rhinovirus reveals epithelium-initiated immune responses. J Infect Dis 217(6):906–915. https://doi.org/10.1093/infdis/jix640

15. Tan KS, Andiappan AK, Lee B, Yan Y, Liu J, Tang SA, Lum J, He TT, Ong YK, Thong M, Lim HF, Choi HW, Rotzschke O, Chow VT, Wang Y (2019) RNA sequencing of H3N2 influenza virus-infected human nasal epithelial cells from multiple subjects reveals molecular pathways associated with tissue injury and complications. Cell 8(9):986. https://doi.org/10.3390/cells8090986

16. Zhang C, Lee HJ, Shrivastava A, Wang R, McQuiston TJ, Challberg SS, Pollok BA, Wang T (2018) Long-term in vitro expansion of epithelial stem cells enabled by pharmacological inhibition of PAK1-ROCK-myosin II and TGF-beta signaling. Cell Rep 25(3): 598–610. e595. https://doi.org/10.1016/j.celrep.2018.09.072

Part III

Immunology and Animal Models

Chapter 15

Animal Models of COVID-19: Nonhuman Primates

Dhiraj K. Singh, Journey Cole, Ruby A. Escobedo, Kendra J. Alfson, Bindu Singh, Tae-Hyung Lee, Xavier Alvarez, Shashank R. Ganatra, Ricardo Carrion, Jr, and Deepak Kaushal

Abstract

With the advent of the novel SARS-CoV-2, the entire world has been thrown into chaos with severe disruptions from a normal life. While the entire world was going chaotic, the researchers throughout the world were struggling to contribute to the best of their capabilities to advance the understanding of this new pandemic and fast track the development of novel therapeutics and vaccines. While various animal models have helped a lot to understand the basic physiology, nonhman primates have been promising and much more successful in modelling human diseases compared to other available clinical models. Here we describe the different aspects of modelling the SARS-CoV-2 infection in NHPs along with the associated methods used in NHP immunology.

Key words Animal models, NHP, Rhesus macaques, COVID-19, SARS-CoV-2, Nonhuman primates, Computed tomography, X-rays, Cone beam, In vivo imaging, Flow cytometry, Confocal microscopy, IHC, BAL, FACS

1 Introduction

Human COVID-19 disease presents a broad clinical spectrum ranging from asymptomatic to mild and severe cases. Patients with COVID-19 pneumonia exhibit high-grade pyrexia, fatigue, dyspnea, and dry cough accompanied by a rapidly progressing pneumonia. Individuals with immunocompromised conditions and comorbidities are at the highest risk of worse outcomes of COVID-19. Nonhuman primate (NHP) models of infectious diseases have been proven to be useful for both investigating the pathogenesis of infection [1, 2] and testing therapeutic [3] and vaccine candidates [4]. During the SARS and MERS outbreaks, NHP models were developed with a moderate degree of success [5]. Expedited studies have clearly demonstrated utility of NHPs for SARS-CoV-2 infection, and for evaluating vaccine candidates

Justin Jang Hann Chu et al. (eds.), *SARS-CoV-2: Methods and Protocols*,
Methods in Molecular Biology, vol. 2452, https://doi.org/10.1007/978-1-0716-2111-0_15,

[6–9]. They have proved to be a seminal tool in not only deciphering the basic biology of the disease but also in addressing the translational aspects of the disease ranging from testing of vaccine candidates to evaluating the potential of new drug candidates. Nonhuman primate models of SARS-CoV-2 range from tiny marmosets to huge Baboons covering a wide spectrum of physiology and pathogenesis. These experimental models have provided a significantly deeper understanding of COVID-19 which would not have been as fast and accurate by clinical observations. However, while each model replicates some aspects of the disease, none of the models can be an exact replica of the disease in humans. Therefore, care must be taken in drawing out inferences from the observations made out from these models. This book chapter describes the common methods currently being utilized in SARS-CoV-2 research in NHPs which has been successfully validated in the Indian origin Rhesus macaques and the African origin Baboon model of COVID-19.

2 The Virus

2.1 Generation of Virus Stock

SARS-CoV-2 isolate USA-WA1/2020 (BEI resources NR-52281, GenBank accession number MN985325.1) is used to generate the animal exposure stock used at Texas Biomedical Research Institute. Vero E6 cells (VERO C1008, catalog number NR-596, BEI resources)—grown in Dulbecco's modified essential media (DMEM) with 10% heat-inactivated fetal bovine serum (FBS) at 37 °C with 5% CO_2—are used. This method is intended to be used as a guidance for the production of live SARS-CoV-2 stocks used as challenge material for vaccine and therapeutic animal studies.

2.1.1 Materials and Equipment

- Sterile Dulbecco's Modified Eagle's Medium (DMEM) (supplemented with 10% fetal bovine serum [FBS]).
- Fetal bovine serum (FBS), certified, heat-inactivated.
- Serological pipettes, sterile, various sizes.
- Pipette tips.
- 50 mL conical tubes.
- Cryovials.
- Tissue culture flasks.
- Refrigerated high-speed centrifuge capable of 10,000 × *g* for 10 min.
- Fixed angle rotor appropriate for centrifuge and volume of product prepared.
- Incubator set to 37 °C ± 2 °C with 5% CO_2 (±2%), and 95% humidity (±10%).

- Water bath (37 °C ± 2 °C).
- Inverted microscope.

2.1.2 Method

1. On the day before infection, seed the appropriate number of tissue culture flasks for infection. Include at least one flask that will serve as a negative control. Use either T-225 tissue culture flasks seeded with 1.2×10^7cells or T-150 tissue culture flasks with 8–9 × 10^6 cells.
2. On the day of infection, cells must look healthy and be ≥85% confluent before use.
3. Prepare the sterile viral growth medium using DMEM +2% HI FBS. Mix well by swirling the bottle. Antibiotics are not recommended. The media can be stored for up to 30 days at 4 °C until required. Media should be warmed to approximately 37 °C prior to use.
4. If the titer of the starting virus is known, calculate the appropriate volume of virus needed to achieve desired multiplicity of Infection (MOI). Example: Plaque forming units (PFU) needed for Infection = MOI × total number of cells to be infected. Amount of Virus (PFU) needed for infection / Viral titer of the seed stock (PFU/mL) = mL of stock needed for infection. If titer is not known, estimate 1 × 10^6 PFU/mL.
   (a) An MOI of approximately 0.01 or 0.001 can be used to generate a small master stock, depending on titer and volume of the progenitor virus. The larger working stock should then be generated using an MOI of approximately 0.01.
5. Aseptically prepare diluted virus in viral growth media. Thaw virus and add an appropriate amount of virus to warmed media.
6. Set up a minimum of one mock-infected flask for use as a negative control. Aseptically aspirate the media from the flask, transfer 20 mL (T-150) or 30 mL (T-225) of warmed media to the negative control flask.
7. For infected flasks, aseptically aspirate the media from each flask, and transfer 20 mL (T-150) or 30 mL (T-225) of the virus suspension to each of the flasks.
8. Place all flasks into a 37 °C, 5% CO_2 incubator.
9. Periodically observe for cytopathic effect (CPE). Compare the infected cell monolayer to that of the control which should remain healthy and intact. If possible, take picture(s) of both infected and control cell monolayers.
10. Once the infected cell monolayers exhibit 2+ (50% of cells showing signs of infection) to 3+ (75% of cells showing signs

of infection) CPE, proceed with the harvest. This should take 2–4 days, depending on the isolate used.

11. Remove medium from the infected flasks and deposit into clean 50 mL conical tubes. Clarify the supernatant by centrifuging at approximately 3000 rpm for 10 ± 2 min at 4 ± 2 °C in a fixed angle rotor.
12. Carefully remove the supernatant from all tubes into a single, clean container. Be careful so as to not disrupt the cell pellet.
13. Add additional HI FBS to achieve the desired final concentration of 10%.
14. Aseptically dispense the supernatant into single use vials, periodically mixing to ensure homogeneity of filled vials. Place vials into boxes and store the vials at ≤ -70 °C.
15. Harvest control flask in a similar manner either before or after the infected flasks. This media can be used subsequently as a negative control for conducting characterization assays.

2.2 Quantification of Infectious Virus Via Plaque Assay

This method describes the process for performance of plaque assay for SARS-CoV-2 to determine infectious viral load.

2.2.1 Materials and Equipment

- Sterile serological pipettes.
- Micropipettes.
- Cell culture media, 1× DMEM.
- Cell culture media, 2× MEM.
- Heat inactivated FBS.
- Prepared methylcellulose.
- Sterile plates or tubes suitable for dilution preparation (at least 1 mL volume).
- 5 mm steel beads
- 10% neutral buffered formalin
- Crystal violet stain.
- Calibrated scale.
- Tissue homogenizer.
- Humidified CO_2 incubator at 5% ±2% CO_2, and 37° ± 2 °C.
- Plate rocker.

2.2.2 Method

1. Prepare twelve-well tissue culture plates with Vero E6 cells. Either seed wells with 2.0×10^5 cells/well no more than 32 h prior to use or seed wells with 1.0×10^5 cells/well no more than 48 h prior to use. Cells should be confirmed to be ≥75% on the day of the assay.

2. Initial sample processing:
   (a) Dry tissue samples (not frozen in media or RNAlater) should first be weighed. An empty tube identical to the sample tube can be used to tare the scale before each sample. The weight of each sample should be documented. After weighing, homogenize the samples with a sterile 5 mm steel bead and 1 mL of refrigerated or room temperature DMEM cell culture media, at 30 Hz for 2 min. If any samples contain large pieces of tissue after homogenization, they should be rehomogenized one additional time at 30 Hz for 2 min. Do not exceed three homogenization cycles. Following homogenization, centrifuge samples for 1 min at 845 × *g*. After centrifugation, the supernatant can be used to perform the plaque assay. Care should be taken to avoid any larger pieces of tissue that remain in the supernatant.
   (b) Liquid samples do not require weighing or homogenization.
3. Prepare serial five or ten fold dilutions of test samples in DMEM containing 2% HI FBS (DMEM-2). If a sufficient volume of the test sample is available, a 1:1 dilution can also be prepared to lower the limit of detection. For example, 150 μL of sample mixed with 150 μL of DMEM-2. Ensure each dilution is mixed sufficiently by pipetting up and down several times after the addition of the test sample to be diluted. A new pipette tip should be used to prepare each dilution.
4. Apply 100 μL of test sample or diluent (as negative control) to each well. Plating should be performed in triplicate.
5. Incubate the cells with the test samples for 60 ± 10 min at 37 °C ± 2 °C and 5% CO_2 with slow rocking or hand rocking every 15 min, to allow for the virus to adsorb.
6. At the end of the incubation/adsorption period, remove the test sample from each well and add overlay media.
   (a) Slowly add 2 mL DMEM-2 containing approximately 30% methylcellulose (DMEM-2/methylcellulose) (temperature between approximately 25–40 °C) to each well. Modifying the percentage of methylcellulose will impact the plaque size and morphology. Less methylcellulose will result in larger plaques, but can lead to amorphous and ill-defined plaques if the incubator or plates are disturbed. A higher concentration will result in smaller plaques that may require enumeration via microscope but the risk of ill-defined plaques is decreased.
   (b) Alternatively, slowly add 1 mL of primary overlay media consisting of 2× MEM with 4% HI FBS (final

concentration will be 2%) mixed 1:1 with 1% agarose (final concentration will be 0.5%).

7. Incubate the assay plates in a humidified incubator at 37 °C ± 2 °C and 5% CO_2. The assay plates should not be moved or disturbed during the post-infection incubation period.
8. For plates overlaid with DMEM-2/methylcellulose, incubate for 4 ± 1 days. At the end of this incubation, remove the DMEM-2/Methylcellulose overlay and add ~1 mL of room temperature 10% neutral buffered formalin as a fixative.
   (a) Allow the formalin fixation to proceed for at least 10 min. Dispose the formalin in an appropriate hazardous waste container. Stain each well with room temperature crystal violet at approximately 0.5 mL per well. After 10 min, wash the crystal violet from each plate by repeatedly dunking in fresh tap water until no more dye runs from the monolayer. Allow the plates to dry prior to counting the plaques per well.
9. For plates overlaid with primary overlay containing agarose, incubate the plates for 2 days. On the morning of day 2 post infection, slowly add 1 mL of secondary overlay media consisting of 2× MEM with 4% HI FBS (final concentration will be 2%) and 8% (final concentration will be 4%) aqueous neutral red stain mixed 1:1 with 1% agarose (final concentration will be 0.5%).
   (a) Once the overlay has solidified, return the plates to the incubator for approximately 6–8 h. After 4 h of incubation, the plates can be monitored for visible plaques. Disturbing the plates will not impact plaque morphology due to the solid overlay. If the plates are allowed to incubate overnight, the stain may diffuse out and plaques will no longer be visible.
10. If plaques are visible by eye, the plates can be scanned at 800 dpi on a flatbed scanner.
11. Enumerate the plaques from either the plates (by eye or via microscope) or from the scanned images. Wells with uninfected control cells should have formed confluent monolayers whereas plaques will have formed in wells containing infected monolayers.
12. Only wells with an intact monolayer and plaque numbers which are less than too numerous to count (TNTC), that is, plaques do not merge or overlap such that counting efficiency is compromised, should be used to calculate a reportable titer. The mean of the replicates for each dilution of a sample should be used to calculate a titer. If only one replicate for a given

dilution is less than TNTC, that value can be used to calculate the titer whereas the value from the other replicates will be omitted.

13. Contingencies for Plaque Counting and Titer Calculation.
    (a) If the number of plaques present for a sample is TNTC for all dilutions, the assay must be repeated for that sample with a different dilution series designed to capture plaques at less than TNTC.
    (b) In the event that a well has ≤40% or less monolayer damage and countable plaques are present on the intact monolayer, the total number of plaques in the well can be extrapolated based upon the number of counted plaques and the percent of the monolayer missing as follow:
        - (# of plaques counted / (1−approximate % of monolayer missing)
            - Example: Number of plaques counted = 40 and % of monolayer missing = approximately 20% (1−0.2 = 0.8) + 40 = approximately 50 total plaques.
        - NOTE: If the percentage of monolayer missing is greater than 40%, the well should not be counted.
        - In the event that multiple replicates of the same dilution have monolayer damage, the extrapolated counts can be averaged with the other replicates of that dilution. The overall average will be used for calculating a titer.

2.3 Viral Load Quantification via Quantitative Reverse Transcription Polymerase Chain Reaction

This method provides detailed instructions on the performance of a duplex quantitative reverse transcription polymerase chain reaction (qRT-PCR) in order to determine the viral load in bodily fluids and tissue samples from animals infected with SARS-CoV-2, and simultaneously detect an MS2 bacteriophage internal control.

Two assays can be used to assess SARS-CoV-2 present in the samples. The CDC-developed 2019-nCoV_N1 assay is used to target a region of the N gene. In addition, a secondary qRT-PCR assay can be used to measure subgenomic RNA, targeting a region of E (Envelope).

2.3.1 Materials and Equipment

- NucleoMag Pathogen kit (Macherey-Nagel).
- 1× DPBS
- Steel beads.
- TRIzol LS Isolation Reagent.
- RNaseZap.
- MS2 phage (*Escherichia coli* bacteriophage MS2, ATCC).

- DNA LoBind Tubes.
- RNAsecure Resuspension Solution.
- TaqPath 1-Step RT-qPCR Master Mix.
- Applied Biosystems, MicroAmp Optical 96-well Reaction Plate or Applied Biosystems MicroAmp Optical 384-well Reaction Plate.
- Applied Biosystems MicroAmp Optical Adhesive Film (Applied Biosystems).
- EpMotion M5073c Liquid Handler (Eppendorf).
- Qiagen TissueLyser.
- HulaMixer (Thermo Fisher Scientific Inc.)
- QuantStudio 3 with 96-well plates or QuantStudio 5 with 384 well plates.
- QuantStudio Design and Analysis Software.
- Optical 96-well Reaction Plate Spinner.

2.3.2 Sample Inactivation

1. For biological liquids, such as serum or viral transport media, combine 250 μL of fluid with 750 μL TRIzol LS (ratio of 1:3, following the manufacturer's instructions). Alternatively, if the sample volume is limited, use 100 μL of test sample mixed with 150 μL of 1 × DPBS (Gibco), and add 750 μL TRIzol LS. Shake the tube vigorously by hand for 5 s and incubate for 5 min at room temperature.
   (a) A small amount of TRIzol LS should be added to the O-ring of each cryovial lid to inactivate the threads.
2. For tissues, weigh a piece of ≤100 mg (tissue may need to be cut to achieve appropriate weight). Place this piece into a 1.5-mL tube with a steel bead (Qiagen, Hilden, Germany) and 1 mL TRIzol reagent.
   (a) Using a Qiagen TissueLyser, homogenize the tube for 3 min at 30 Hz. Then rotate the sample block holding the tube to ensure uniform homogenization and homogenize again for 3 min at 30 Hz. After homogenization, the material in the tube should resemble a slurry with no large chunks remaining. If this is not the case, a third round of homogenization (3 min at 30 Hz) should be performed.
   (b) A small amount of TRIzol should be added to the O-ring of each cryovial lid to inactivate the threads.

Notes

- Trizol LS Reagent and Trizol Reagent are commonly used for RNA extraction from samples containing enveloped viruses and have been shown to inactivate high consequence pathogens, such as Ebola virus [10].

- A sample containing no infectious material or RNA should be processed in the same manner as experimental samples, as an inactivation control.

2.3.3 RNA Extraction

1. Prior to RNA extraction, work areas and equipment should be cleaned using RNaseZap.
2. Extraction controls should be prepared with each batch of samples to ensure no cross contamination occurs during extraction.
3. Thaw samples inactivated in Trizol or Trizol LS Reagents, at room temperature.
4. Add 1×10^3 pfu of MS2 phage (Escherichia coli bacteriophage MS2, ATCC) to each biological fluid sample to assess extraction efficiency. MS2 phage addition to tissue samples has not been assessed.
5. Add 10 μg yeast tRNA to each biological fluid sample as carrier. Yeast tRNA is not needed for tissue samples.
6. RNA extraction is performed using the EpMotion M5073c Liquid Handler (Eppendorf) and the NucleoMag Pathogen kit (Macherey-Nagel). (*Alternatively, extraction can be performed manually following the manufacturer instructions for Trizol or Trizol LS Reagents*).
7. Add 200 μL of chloroform to each sample and shake vigorously for approximately 10 s.
8. Incubate for 2–3 min at room temperature.
9. After incubation, centrifuge at $12,000 \times g$ for 15 min at 2–8 °C.
10. Prepare DNA LoBind Tubes (Eppendorf) with 20 μL of NucleoMag B-Beads (NucleoMag Pathogen kit, Macherey-Nagel) and 975 μL of Buffer NPB2 (NucleoMag Pathogen kit, Macherey-Nagel).
11. After centrifugation, the upper aqueous phase of each sample should be transferred to the corresponding new tube containing NucleoMag B-Beads and Buffer NPB2.
12. Mix the samples using a HulaMixer, rotating for 10 min at room temperature.
13. Transfer samples to the sample rack on the EpMotion M5073c Liquid Handler for further processing according to NucleoMag Pathogen kit instructions.
14. After processing, confirm the presence of the eluate (100 μL).
15. Store extracted RNA at −80 °C ± 10 °C.

2.3.4 Determination of Viral Load Via RT-qPCR

Synthetic RNA Standards are produced by Bio-Synthesis Inc., Lewisville, TX. Standards are received from the manufacturer in lyophilized form.

Preparation of Synthetic RNA Standards, Primers, and Probes

1. Spin the tube at 2000 × *g* at room temperature for 1 min to pellet all the powder to the bottom of the tube.
2. Dissolve the synthetic RNA standard in RNAsecure Resuspension Solution (Invitrogen) to a stock concentration of 5 μg/mL.
3. Prepare 5 μL aliquots in 1 mL cryogenic tubes. Store these aliquots in the vapor phase of liquid nitrogen for up to 36 months from the date of preparation.
4. To achieve precise RNA virus standard quantification, thaw one aliquot the next day and measure the RNA concentration using a fluorometer.

   Primers are received from the manufacturer(s) in lyophilized form. Each primer comes with its own unique LOT number.

1. Spin the tubes at 2000 × *g* for 1 min at room temperature to pellet all the powder to the bottom of the tube.
2. Dissolve the primers in nuclease-free water to a stock concentration of 100 μM. The amount of water needed for each batch differs and the required volume is indicated on the specification sheet included with each shipment.
3. Stock primers at 100 μM concentration should be aliquoted at 20 μL per tube and can be stored for up to 24 months in a −20 °C freezer.

Probes are received from the manufacturer as 100 μM solutions.

1. Spin the tubes at 2000 × *g* for 1 min at room temperature to pellet all the liquid to the bottom of the tube.
2. Prepare 20 μL aliquots, which can be stored for up to 24 months in a −20 °C freezer.

Quantitative Reverse Transcription PCR (qRT-PCR)

1. Before starting the procedure:
   (a) Generate a plate layout listing all samples and standard positions.
   (b) Prepare a fresh set of standards:
   - Prepare 10 mL of MS2 RNA standard diluent of 2E+04 molecules /μL. This solution will be then used to dilute the synthetic RNA standards. Thus, the MS2 concentration will be constant throughout the whole standard curve range.

- Prepare serial tenfold dilutions of the RNA synthetic standard from a previously prepared, frozen 5 μL aliquot.

(c) Prepare a 10 μM concentration of SARS-Cov-2 primers and probe mixture, using nuclease-free water as diluent. 10 μM solutions of primers and probes can be frozen at −20 °C but should not be stored for more than 1 month, or 5 freeze-thaw cycles.

2. Prepare the reaction Master Mix using TaqPath 1-Step RT-qPCR Master Mix (Life Technologies).

   (a) Thaw the components on ice and prepare the Master Mix on ice.

   (b) Determine the number of reactions needed and include an extra 5–10% to compensate for pipetting error. Based on the RNA virus and type of sample select the correct master mix composition:

   (c) For one reaction, the Master Mix should include the following materials for a total volume of 15 μL:

   - 5.0 μL 4× TaqPath 1-Step RT-qPCR master mix.
   - 6.4 μL Nuclease Free Water.
   - 1.5 μL SARS-Cov-2 primers-probe mix.
   - 0.8 μL of 10 μM MS2 forward primer (final concentration 400 nM).
   - 0.8 μL of 10 μM MS2 reverse primer (final concentration 400 nM).
   - 0.5 μL of 10 μM MS2 probe (final concentration 250 nM).

3. Aliquot 15 μL of the Master Mix into the appropriate wells of an optical plate on ice.

4. Add 5 μL of RNA sample or synthetic RNA standard to the appropriate wells based on the plate layout generated prior to starting the procedure.

5. Seal the plate with Optical Adhesive Film.

6. Tap the plate from left and right to mix the contents well.

7. Spin the plate to bring the contents to the bottom of the well in a plate spinner for 30 s to 1 min.

8. Start QuantStudio Design and Analysis Software.

9. Verify the current calibration status of the instrument.

10. Turn on the QuantStudio 3 or 5 qRT-PCR machine and load the plate into the machine. Verify that the A1 well is positioned correctly in the top left corner.

11. Begin analysis of the plate using the following cycling parameters: hold stage 2 min at 25 °C; 15 min at 50 °C; 2 min at 95 °C; PCR stage: 45 cycles (N1 assay) or 40 cycles (E assay) of 3 s at 95 °C followed by 30 s at 60 °C (image capture).

Notes

- Two instruments can be used to quantitate the viral load from bodily fluids or tissue samples. Depending on the number of samples, QuantStudio 3 with 96-well plates or QuantStudio 5 with 384 well plates can be used. The software for both machines is the QuantStudio Design and Analysis Software from Applied Biosystems. Applied Biosystems recommends the machines undergo operation qualification/instrument performance verification (OQ\IPV) as follows:
  - After replacement of any part or execution of any service that is critical to the performance of the QuantStudio 3 or 5 PCR System, including:

    Optics Assembly replacement and/or alignment.

    Camera Assembly replacement and/or alignment.

    Lens Assembly replacement.

    Filter Assembly replacement.

    Thermal Cycler replacement.

    Heated Cover replacement.
  - After completion of any procedure or maintenance that may affect instrument performance.
  - After movement or relocation of the instrument.

- The CDC-developed 2019-nCoV_N1 assay targets a region of the N gene. SARS-CoV-2_N1 probe (ACCCCGCAT TACGTTTGGTGGACC) is labelled with 6-FAM fluorescent dye. The forward primer sequence is: GACCCCAAAATCAGC GAAAT, and the reverse primer sequence is: TCTGGTTACTGC CAGTTGAATCTG.
- A secondary qPCR assay to measure subgenomic RNA has also been developed to target a region of E (Envelope) [11]. The probe is also labelled with a 6-FAM fluorescent dye (ACACTAGC CATCCTTACTGCGC TTCG). The forward primer sequence is: CGATCTCTTGTAGATCTGTTCTC , and the reverse primer sequence is: ATATTGCAGCAGTACGCACACA.
- The MS2 probe is labelled with VIC fluorescent dye (CGCTC GAGAACGCAA). The forward primer sequence is: GGGTTT CCGTCTTGCTCGTA, and the reverse primer sequence is: TCGTGCTTTTCGCTGAAGAA.

- If the assay efficiency of the run (Eff%) falls below 85% or above 115%, the run should be considered inefficient and be repeated.
- The average MS2 Ct value plus 5% (AVG MS2 + 5%) value should be calculated by multiplying the AVG MS2 Ct of all the samples on the plate by 1.05. Samples with MS2 Ct values higher than the calculated AVG MS2 + 5% should be considered to have inefficient detection and be repeated.
- If the Ct value of a sample is lower than Ct value of the highest synthetic standard, the sample should be diluted 100× and qRT-PCR repeated.

3 The Host

3.1 Clinical Monitoring

Routine clinical monitoring of immunocytes using peripheral blood when combined with other clinical parameters imparts a strong insight into diseases progression and enables investigators to take well-informed decision during the course of the disease. Here, we provide a simple protocol utilizing IDEXX ProCyte Dx which is an automated hematology analyzer to monitor peripheral blood that evaluates and provides the results of 24 parameters (Fig. 1) for each blood sample in approximately 2 min. The ProCyte Dx analyzer employs three technologies—laser flow cytometry, optical fluorescence, and laminar flow impedance—as well as the SLS-hemoglobin method. With laser flow cytometry, the system performs two separate analyses: Red blood cell optical analyses for mature red blood cells, reticulocytes, and platelets along with White blood cell differential. Laminar Flow Impedance is the fastest method for analyzing the size and number of red blood cells and platelets. Hemoglobin conversion with the SLS-hemoglobin method is fast and does not use poisonous substances like other methods (such as the cyanmethemoglobin method), making it a suitable method for automation. And since it can be used to measure methemoglobin, it can also accurately measure blood containing methemoglobin, as is the case with control samples.

3.1.1 Materials

- IDEXX ProCyte Dx.
- Blood collected in desired EDTA tubes.

3.1.2 Methods

1. Blood is collected in a 0.5 mL EDTA tube by vet technicians.
2. On the ProCyte Dx analyzer, ensure the basic QC is clear and the sample drawer is open. If necessary, press the Open/Close button on the analyzer to open the sample drawer.
3. Ensure the appropriate tube adapter is in the sample position area of the sample drawer.

Parameter	Description
RBC	Total number of erythrocytes
HCT	Hematocrit value: erythrocyte ratio of total blood volume
HGB	Hemoglobin concentration
MCV	Mean erythrocyte volume in total sample
MCH	Mean hemoglobin volume per red blood cell (RBC) count
MCHC	Mean hemoglobin concentration of erythrocytes
RDW	The degree of variation in size of the erythrocyte population
%RETIC	Reticulocyte percent
RETIC	Reticulocyte count
WBC	Total number of leukocytes
%NEU	Neutrophil percent
%LYM	Lymphocyte percent
%MONO	Monocyte percent
%EOS	Eosinophil percent
%BASO	Basophil percent
NEU	Neutrophil count
LYM	Lymphocyte count
MONO	Monocyte count
EOS	Eosinophil count
BASO	Basophil count
PLT	Total number of platelets
MPV	Mean platelet volume
PDW	Platelet distribution width; the degree of variation in size of the platelet population
PCT	Plateletcrit value

Fig. 1 IDEXX ProCyte Dx Panel for peripheral blood monitoring

4. Prepare the sample. **Gently invert the sample 10×** to ensure it is well mixed.
5. Immediately place the sample in the tube adapter.
6. Label the sample to be analyzed with the study number, animal ID and initials of the person who is running the blood.
7. Press the **Start** button on the analyzer. The sample drawer automatically closes and the analyzer begins processing the sample.
8. Print/save the results from this page or if you have abnormal results which the machine will prompt, reanalyze the sample.

Notes

• If you are using a microtube, remove the cap before placing the tube in the tube adapter.

3.2 In Vivo CT Imaging

CT imaging has emerged to be a helpful tool in diagnosing as well as monitoring the wide range of clinical and immunological events manifested during the infection with this respiratory-tropic novel coronavirus [12]. We performed CT imaging of the lung lesions using LFER 150 PET/CT scanner (MEDISO Inc., Budapest,

Hungary) [13]. LFER CT scanner uses a cone beam x-ray source aimed at the region of interest, which rotates around the animal's (subject's) body, producing signals that are processed by the computer to generate a cross sectional-image of the region of interest. Computed tomography (CT) imaging is a noninvasive that enables three dimensional in vivo imaging of live animals providing spatiotemporal information about the evolving lung pathology due to SARS-CoV-2 infection. Due to its noninvasive nature, CT imaging facilitates the longitudinal study of lung pathology. Here, we describe the method for using a cone-beam X-ray for performing breath-hold scans CT imaging in SARS-CoV-2 infected nonhuman primates.

3.2.1 Material

- Induction Anesthesia: Telazol (2–6 mg/kg).
- Maintenance anesthesia: Isoflurane (1%–5%) with supplemental oxygen.
- Bair Hugger blanket.
- Laryngoscope and endotracheal tubes (2.5–3.5F) to intubate macaques and baboons.
- Hallowell 2002 ventilator anesthesia machine with remote breath-holding switch.
- Multiscan LFER150 PET/CT—In vivo dual modality PET/CT imager.

3.2.2 Methods

Animal Subject Preparation

1. The study animals will be identified the previous day and have their food removed 2–6 h, to prevent aspiration during general anesthesia.
2. On the day of the procedure, after performing general evaluation animals will be anaesthetized using Telazol (2–6 mg/kg of body weight) intramuscular injection.
3. NHPs are then intubated using a laryngoscope and an appropriate 2.5–3.5 French endotracheal tube depending on animal size and as determined by a trained veterinarian or veterinary technician.
4. The animals are maintained on the anesthetic plane using inhaled 1%–5% isoflurane in oxygen delivered via an endotracheal tube attached to the ventilator anesthesia system.
5. The animal will be catheterized in the cephalic vein using a 22- or 24-gauge indwelling catheter secured with tape.
6. Sterile ophthalmic ointment/lubricant may be applied to eyes to prevent corneal desiccation.
7. The animal will be transported from the animal procedure area CT scanner suite using a special HEPA filter-fitted biobubble.

8. The animal is identified by the imaging personnel and necessary details entered as the scan ID.

Ventilator-Anesthesia Set-Up

9. Monitor overall animal condition and place it in the supine position on the carbon fibre pallet mounted on the MobileCell and lined with absorbent underpads.
10. The endotracheal tube is connected to the tube attached to the Hallowell 2002 ventilator anesthesia system, providing for 1%–5% isoflurane in oxygen with airflow determined by the animal chest size.
11. A pulse-oximeter probe is applied to the tongue to measure O_2 saturation.
12. The anaesthetized animal is maintained on the bed by using the Bair Hugger blanket supplying a 43 °C airflow from the caudal end. Once the animal is comfortable and has a stable tidal volume being displayed on the bellows of a ventilator, the animal is switched to ventilator mode.

CT Imaging Procedure

13. CT Topogram: For planning a nontrivial 3D reconstruction area, you may need both Side and Top View Topograms. CT Topogram, or Scout view, is a planar x-ray image of the subject from the side (laterolateral) or top view (anteroposterior). This fast snapshot helps to plan the scanning area of a CT scan or a PET acquisition.
14. Scan range is set up by adjusting the blue rectangle in the geometric planner window.
15. Scan parameters are selected by chasing various scan options in the dropdown menu. We performed a Semicircular Single FOV scan, with 520 projections, at 80 kV of tube voltage and 980 μA of current, 62 ms of exposure time and 1:16 binning.
16. After all the parameters have been selected for the scan and making sure that the animal has regular, breathing with no efforts, acquire the scan while performing end-inspiratory breath hold using a remote switch.

Lung Segmentation and Quantitative Analysis of Lung Lesions

17. Image analysis is performed using Vivoquant software (InVicro, Boston, MA).
18. Images are visualized using the axial coronal, sagittal, and 3D views of the thoracic region (Fig. 2a) [14].
19. Lung segmentation is performed using connected thresholding feature in the 3D-ROI tool of Vivoquant software.
20. Lung ROI is defined as the input voxel range from −850 HU (Hounsfield unit) to 500 HU.
21. Smoothening of Lung ROI is performed using filters to reassign the voxel value to the mode value of surrounding voxels.

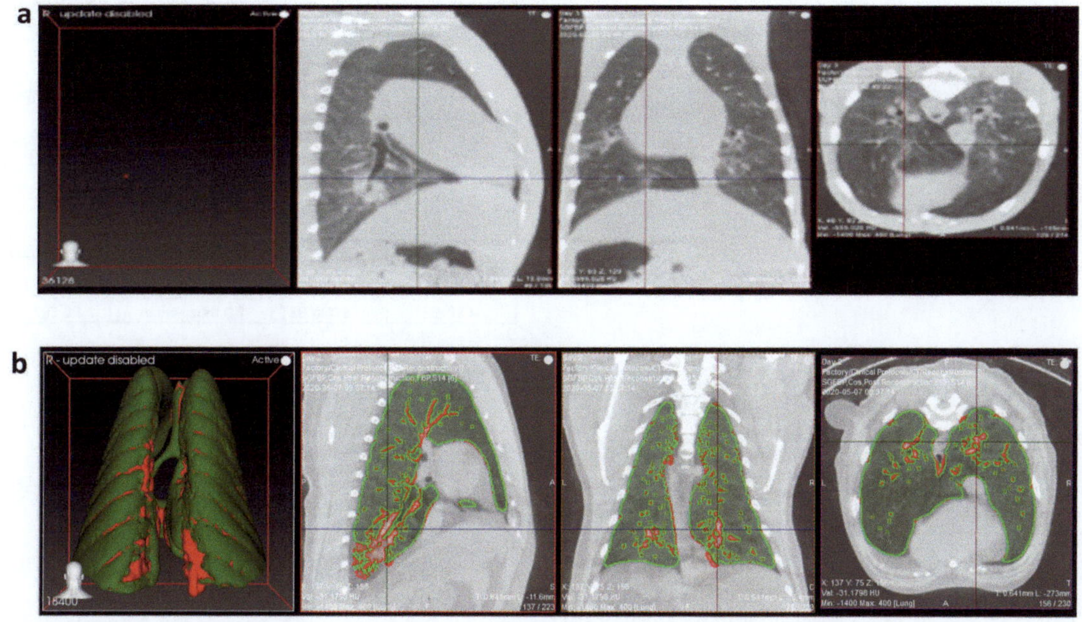

Fig. 2 CT representative images from animal with SARS-CoV-2 infection performed Day 2 after infection, show (Panel A) sagittal, coronal, and axial view in gray scale while (Panel B) represent lung images of lung segmentation with Lung ROI from −850 HU to −500 HU (green) and hyperdensity −490 HU to +500 HU (red)

22. Subsequently, global thresholding is performed on lung ROI, defining the range of −490 HU to +500 HU as hyperdensity (Fig. 2b) [15].
23. The resultant ROI is then rendered into 3D view using VTK feature of #D-ROI tool.

3.3 Immunomonitoring

Immunophenotyping using flow cytometry has been a powerful technique in immunological studies involving infectious diseases, particularly SARS-CoV-2 in elucidating the dynamics of the immunological landscape during the course of infection[2]. The availability of a wide repertoire of existing cross-reactive antibodies and the development of new NHP specific reagents have assisted in accelerating the understanding of immunological dynamics during SARS-CoV-2 infection. Flow cytometry–based immunomonitoring using optimized multicolor immunofluorescence panels (Figs. 3 and 4) has improved our understanding of the basic biology and associated immunological mechanisms of SARS-CoV-2 infection along with accelerating the development of novel vaccines and therapeutic interventions.

3.3.1 Purpose

To stain cells with specific fluorescently tagged antibodies to be used for immunophenotyping and cellular dynamics (Figs. 3 and 4).

Panel T1		Target Cells : T Cells	
Antibody Name	Clone	Vendor	Fluorochrome
CD69	FN50	BD Biosciences	FITC
CD4	L200	BD Biosciences	PerCP-Cy5.5
CD8	RPA-T8	Biolegend	APC
CD3	SP34-2	BD Biosciences	AL 700
CD20	2H7	BD Biosciences	APC-HY7
CD95	DX2	BD Biosciences	BV421
Ki67	B56	BD Biosciences	BV510
CCR7	3D12	BD Biosciences	BV605
CD28	CD28.2	BD Biosciences	PE-Cy7
CCR5	3A9	BD Biosciences	PE

Panel T2		Target Cells : T Cells	
Antibody Name	Clone	Vendor	Fluorochrome
CD69	FN50	BD Biosciences	FITC
CD4	L200	BD Biosciences	PerCP-Cy5.5
CD8	RPA-T8	Biolegend	APC
CD3	SP34-2	BD Biosciences	AL 700
HLA-DR	L243	BD Biosciences	APC-CY7
PD-1	EH12.2H7	Biolegend	BV421
CCR7	3D12	BD Biosciences	BV510
CCR5	3A9	BD Biosciences	BV650
CCR6	11A9	BD Biosciences	BV711
LAG-3	Polyclonal	R&D (Cat No.FAB2319)	PE
CXCR3	1C6/CXCR3	BD Biosciences	PE-Cy7

Panel M1		Target Cells : Myeloids	
Antibody Name	Clone	Vendor	Fluorochrome
CD103	B-Ly7	ThermoFisher	FITC
CD123	7G3	BD Biosciences	PerCP-Cy5.5
CD14	M5E2	BD Biosciences	APC
CD3	SP34-2	BD Biosciences	AL 700
CD20	2H7	BD Biosciences	AL 700
HLA-DR	L243	BD Biosciences	APC-CY7
CD206	19.2	BD Biosciences	BV421
CD16	3G8	BD Biosciences	BV510
CD11c	3.9	Biolegend	BV711
CD163	GHI/61	BD Biosciences	PE
CD40	5C3	BD Biosciences	PE-CY7

Fig. 3 Optimized Flow Cytometry panels for basic lymphoid and myeloid immunomonitoring in BAL of rhesus macaques

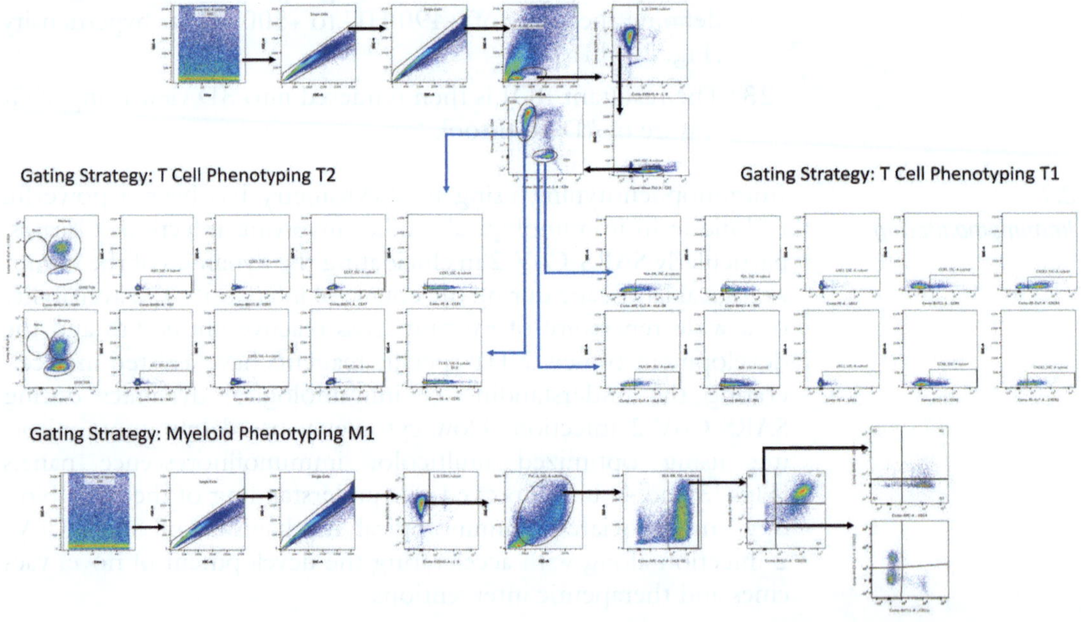

Fig. 4 Representative Flow Cytometry gating strategies for lymphoid and myeloid immunomonitoring in BAL

3.3.2 Safety Procedures

- Be sure all work is performed in a **dark** environment.
- Light will damage the fluorochrome tagged to the antibody.
- All work is to be performed in BSL-3 when dealing with SARS-CoV-2 infected animals.
- Non–SARS-CoV-2 infected samples can be processed at BSL-2 level.
- All work is to be done in a BSC.
- Spray contents in and out of the BSC using the appropriate disinfectant.

3.3.3 Definitions

- Fix/stab—a 3× concentration of fixation/stabilization buffer used to preserve flow staining until analysis can occur.
- Fix/perm—a 1× concentration buffer that creates holes in the cell's membrane allowing for intracellular staining.
- Perm/wash—a 10× concentration buffer used to wash and permeabilize cells.
- Running buffer—autoMACS Running buffer used for cell separation.
- Antibody cocktails—a pool of all antibodies on a panel separated for intra and extracellular staining.

3.3.4 Media and Supplies

- 5 mL round bottom Falcon tubes.
  - "flow tubes"
- 96-well U/V Bottom plates
- 96-well Filter plates
- Various sized pipettes and tips.
- Depends on antibody staining.
- 10 mL serological pipette
- Serological pipette
- Flow racks.
- Aluminum foil.
- Appropriate antibodies.
  - Depends on panel.
- BD 3× Fix/Stab.
- BD 1× Fix/perm.
- BD 10× Perm/wash.
- BD 10× lysing buffer.
- AutoMACS running buffer.

Preparations:

- Make 1:10 dilution of lysing buffer.
- Make 1:10 dilution of perm/wash.
- Make 1:3 dilution of fix/stab.
- Fill out flow cytometry submission sheet and schedule flow samples with the core.
- Label all flow tubes consistent with flow cytometry submission sheet.
- Prepare antibody cocktails.

3.3.5 Protocol: For Staining in Tubes (Low Throughput)

Extracellular Staining

1. Add the required number of cells to a flow tube.
   (a) Aliquot 100 μL of whole blood or 0.2–1 × 10^6 cells of PBMCs, BAL cells or single cells from tissues in 100 μL of RPMI/PBS.
   (b) Concentration of cells can change depending on the panel being run.
2. Add the required amount of antibody cocktail to the sample in the bottom of the flow tube.
3. Pipet to mix thoroughly.
4. Allow the sample to incubate, in the dark, at room temp for 25 min.
5. **Proceed based on sample type**.

   Whole Blood:

6. Once the 25 min incubation is finished, add 1 mL of diluted lysing buffer.
7. Pipet the sample to thoroughly mix.
8. Add an additional 2 mL of diluted lysing buffer to the sample and allow to sit for 10 min.
9. Centrifuge the tube at **400** × *g* at **4 °C** for **5 min** with a **brake of 5**.
10. Dump the supernatant from the flow tube.
    (a) Dab if sample is no longer bloody.
11. If sample remains bloody, add 1 mL of diluted lysing buffer and pipet to mix.
12. Allow sample to sit for 10 min.
13. Centrifuge the tube at **400** × *g* at **4 °C** for **5 min** with a **brake of 5.**
14. Dump the supernatant and dab the tube to rid of any residual lysing buffer.

Bloody BAL and Tissues

15. Once the 25 min incubation is finished, add 1 mL of diluted lysing buffer.
16. Gently vortex the tube, by placing the entire rack on the vortex.
17. Allow the sample to lyse for 10 min.
18. Add 2 mL of running buffer to sample to stop lysis.
19. Gently vortex the tube, by placing the entire rack on the vortex.
20. Centrifuge the tube at **400** × ***g*** at **4 °C** for **5 min** with a **brake of 5**.
21. Dump the supernatant and dab the tube to rid of any residual liquid.

Nonbloody BAL and Tissues

22. Once the 25 min incubation is finished, add 3 mL of running buffer.
23. Gently vortex the tube, by placing the entire rack on the vortex.
24. Centrifuge the tube at **400** × ***g*** at **4 °C** for **5 min** with a **brake of 5**.
25. Dump the supernatant and dab the tube to rid of any residual liquid.

If the sample does not require intracellular staining proceed to fixation.

Intracellular Staining

26. Once the samples have been dabbed, to rid of any residual liquid, resuspend the pellet in 250 μL of Fix/Perm.
27. Gently vortex the tube, by placing the entire rack on the vortex.
28. Allow the sample to sit for 20 min at 4° in the dark.
29. Add 1 mL of diluted perm/wash to the sample.
30. Centrifuge the tube at **400** × ***g*** at **4 °C** for **5 min** with a **brake of 5**.
31. Dump off the supernatant but **DO NOT** dab the sample.
32. Wash the samples again with 1 mL of diluted perm/wash.
33. Centrifuge the tube at **400** × ***g*** at **4 °C** for **5 min** with a **brake of 5**.
34. Dump off the supernatant until a small volume (100 μL) of perm/wash remains.

35. Resuspend the sample by gently vortexing the rack.
36. Add the desired amount of intracellular antibodies to the sample.
37. Allow sample to incubate at room temp for 25 min in the dark.
38. Wash the cells with 3 mL of running buffer.
39. Centrifuge the tube at **400** × *g* at **4 °C** for **5 min** with a **brake of 5.**
40. Dump the supernatant and dab the tube to rid of any residual liquid.

Fixation and Drop Off

41. Resuspend the pellet in 250 μL of diluted fix/stab buffer.
    (a) If you are not using tandem dyes, use 1% paraformaldehyde buffer in PBS instead.
    (b) Filter the samples and transfer into a new tube.
    (c) Protect the samples from light by wrapping them in aluminum foil, surface decontaminate using Wex-Cide.
42. Allow to sit for a minimum of 15 min prior to removing from the BSL3 and store at **4 °C** until they are acquired on a flow cytometer.

3.3.6 Protocol: 96—Well U-Bottom/V-Bottom Plate (High Throughput: For HTS)

(For PBMCs and RBC lysed single cells)

Extracellular Staining

1. Prepare the plate map and add the required amount of cells ($0.2–1 \times 10^6$ cells) to each well following the plate map.
2. Add the required amount of antibody cocktail to the sample in the bottom of the well.
3. Pipet to mix thoroughly.
4. Allow the sample to incubate, in the dark, at room temp for 25 min.
5. Once the 25 min incubation is finished, add 100–150 μL of running buffer and mix.
6. Note: If the volume of antibody cocktail is 100 μL or above, Centrifuge the plate at **400** × *g* at **4 °C** for **5 min** with a **brake of 7** and then add 200 μL of running buffer and mix by pipetting using a multichannel pipette.
7. Centrifuge the plate at **400** × *g* at **4 °C** for **5 min** with a **brake of** 7.
8. Dump the supernatant and resuspend in 200 μL of running buffer.
9. Centrifuge the plate at **400** × *g* at **4 °C** for **5 min** with a **brake of** 7.

10. Proceed based on sample type, If the sample does not require intracellular staining proceed to fixation.

Intracellular Staining

11. Resuspend the pellet in 200 μL of Fix/Perm.
12. Allow the sample to sit for 20 min at 4 °C in the dark.
13. Centrifuge at **400 ×** ***g*** at **4 °C** for **5 min** with a **brake of** 7.
14. Dump off the supernatant.
15. Add 200 μL of diluted perm/wash to the sample.
16. Centrifuge at **400 ×** ***g*** at **4 °C** for **5 min** with a **brake of** 7.
17. Dump off the supernatant.
18. Wash the samples again with 200 μL of diluted perm/wash.
19. Centrifuge at **400 ×** ***g*** at **4 °C** for **5 min** with a **brake of** 7.
20. Dump off the supernatant.
21. Add the desired amount of intracellular antibody cocktail to the sample.
22. Allow sample to incubate at room temp for 25 min in the dark.
23. Centrifuge at **400 ×** ***g*** at **4 °C** for **5 min** with a **brake of** 7.
24. Dump the supernatant and resuspend in 200 μL of running buffer.
25. Centrifuge at **400 ×** ***g*** at **4 °C** for **5 min** with a **brake of** 7.
26. Dump the supernatant.
27. **Proceed to fixation.**

Fixation and Drop Off

28. Resuspend the pellet in 200 μL of diluted fix/stab buffer.
    a. If you are not using tandem dyes, use 1% paraformaldehyde buffer in PBS instead.
    b. Filter the samples using filter plate and transfer into a new plate.
    c. Wrap the plate in aluminium foil and double bag with zip lock.
    d. Surface decontaminate the zip lock with Wex-Cide.
    e. Allow to sit for a minimum of 15 min prior to removing from BSL3.
29. Protect the samples from light by wrapping them in aluminum foil and store at **4 °C** until they are acquired on a flow cytometer.

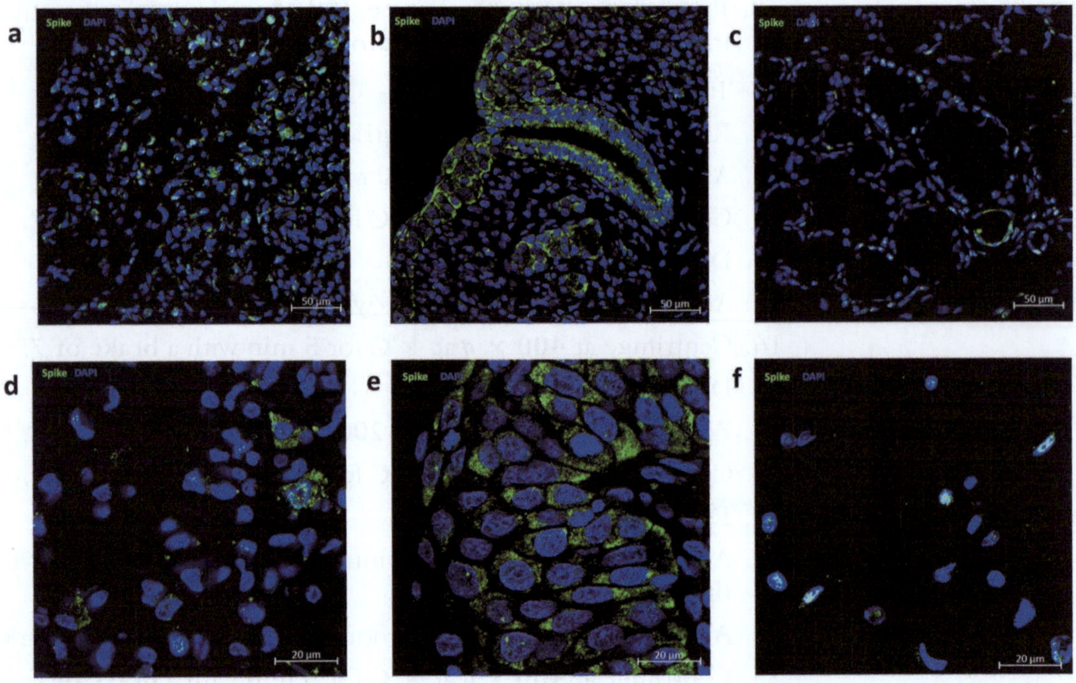

Fig. 5 Confocal images of immunohistofluorescence of SARS-CoV-2 infected (3 dpi) lungs (20×-a, 63×-d, nasal epithelium (20×-b, 63×-e) and tonsil (20×-c,63×-f) with SARS CoV-2 Spike specific antibody (green) and DAPI (blue)

3.4 Immuno-histochemical Detection of SARS CoV-2

Immunohistochemistry has emerged as an effective tool in diagnosis and in-depth study of a wide range of infectious and noninfectious diseases as well as in drug discovery. Here, we present the identification of SARS-CoV-2 virus (Fig. 5) within the host tissues using the method of immunohistochemistry or more precisely immunohistofluorescence. Immunohistochemistry (IHC) is an immunological technique which enables us in detection of proteins expressed in the cells of a tissue section (Fig. 6). This procedure makes use of specificity of antigen–antibody interactions to visualize the distribution and localization of various antigens/proteins in the tissue of interest. IHC is an immensely powerful tool vastly used in diagnosis of various diseases and numerous pathological conditions. Owing to the implication of specific antigen–antibody reactions, multiple antigenic markers can be identified simultaneously. Therefore, IHC has become a crucial technique and is widely used in many medical research laboratories as well as clinical diagnostics [16].

We have used the technique of immunohistochemistry/immunohistofluorescence (IHF) to identify the newly discovered virus, severe acute respiratory syndrome coronavirus-2 (SARS-CoV-2) in nonhuman primate models. SARS-CoV-2 is the causative agent of coronavirus disease-19 (COVID-19), which has created a havoc

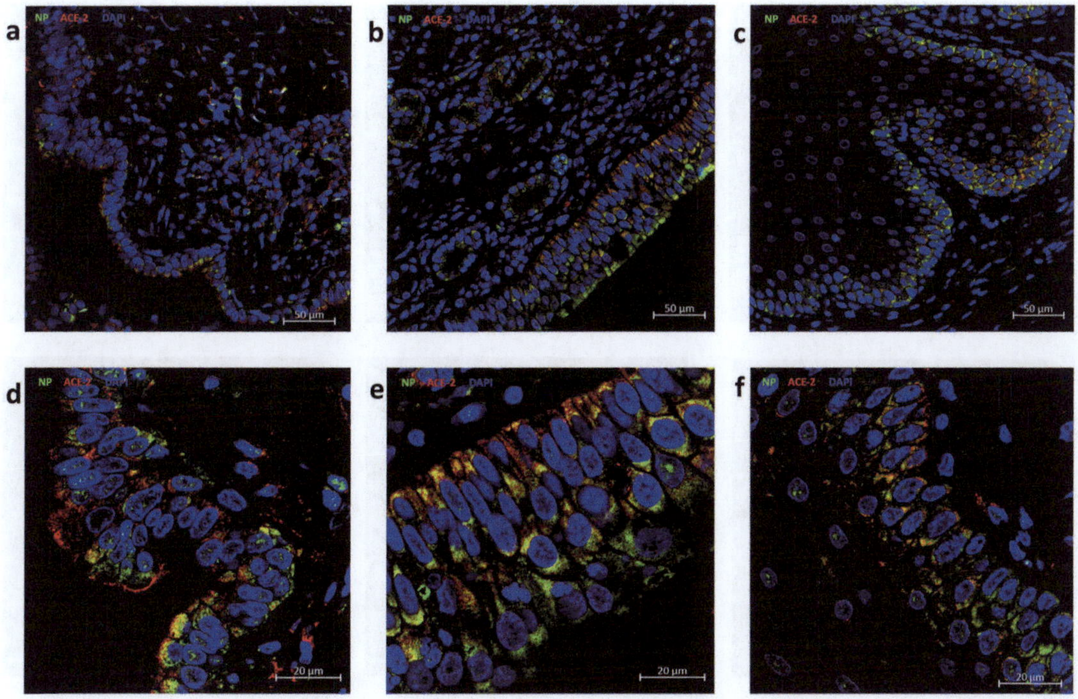

Fig. 6 Confocal images of multilabel immunohistofluorescence of SARS-CoV-2 infected (3 dpi) lungs (20×-a, 63×-d, nasal epithelium (20×-b, 63×-e), and tonsil (20×-c,63×-f) with SARS CoV-2 Nucleocapsid antibody (green), NHP/human ACE-2 antibody (red), and DAPI (blue)

resulting in world-wide health emergency in recent times [17–19]. Histopathologic assessment of tissue obtained from biopsy of affected patients is crucial for diagnostic purposes. Nevertheless, the histologic evaluation of tissues obtained from animal models of COVID-19, is beneficial in gaining deep insight in our understanding of the disease [1, 2, 20].

IHF is defined as a version of IHC where a target antigen in a tissue of interest is detected using an antigen-specific antibody tagged to a fluorophore (direct or indirect) and is subsequently visualized using a fluorescent microscope. Here, in this chapter we present the protocol and reagents used for the detection of the virus and study its distribution among different sites and tissue types. This can be extended to study the infected cell types (Fig. 7) as well as the related immunological changes that occur in host tissues in response to SARS-CoV-2 infection.

3.4.1 Materials

- **Deparaffinization:** xylene.
- **Rehydration:** 100% ethanol, 95% ethanol, 85% ethanol, 70% ethanol, distilled water.
- **Antigen retrieval buffers.**

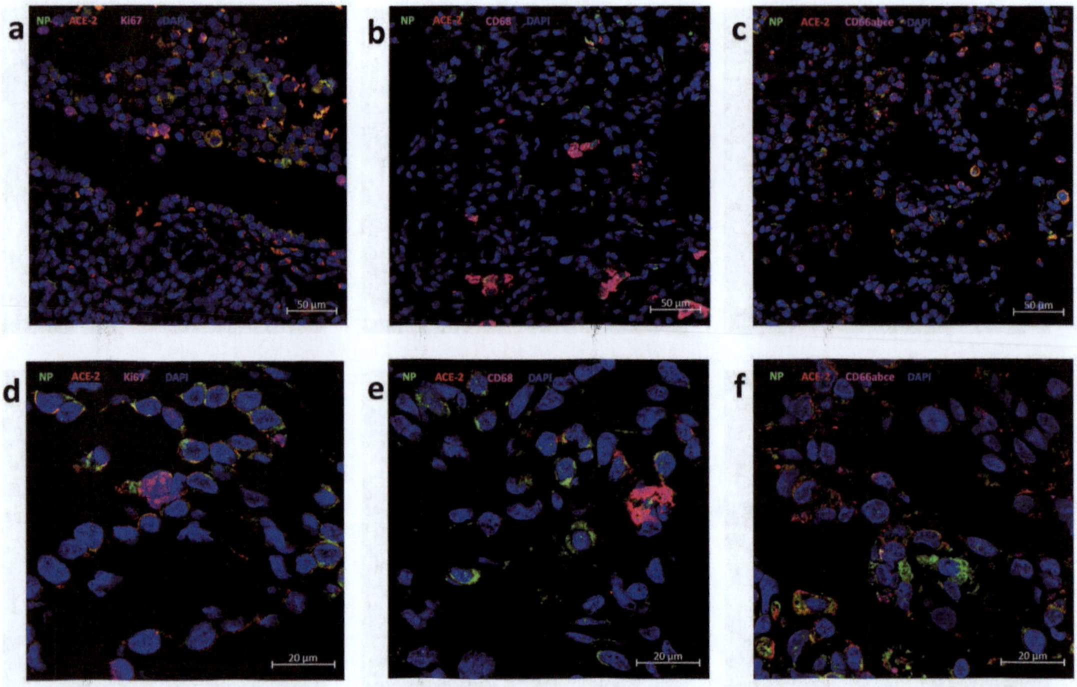

Fig. 7 Multilabel immunohistofluorescence confocal images depicting host cell types expressing ACE-2 receptor and those infected with SARS CoV-2. The images shown are from SARS CoV-2 infected lungs (3dpi) depicting NP (green), ACE-2 (red), and Ki67 (magenta), DAPI (blue)—20×-a, 63×-d; NP (green), ACE-2 (red), and CD68+ macrophages (magenta), DAPI (blue) —20×-b, 63×-e; NP (green), ACE-2 (red), and CD66abce + neutrophils (magenta), DAPI (blue)—20×-c, 63×-f

- **Tris–EDTA buffer:** 10 mM tris base, 1 mM EDTA, 0.05% Tween 20, pH 8.0.
  - Weigh 1.21 g tris base, 0.37 g EDTA and mix in 900 mL of distilled water. Adjust the pH to 9.0 with 1 N NaOH and then add 0.5 mL of Tween-20. Mix the contents using a magnetic stirrer.
- **Citrate buffer:** 10 mM, pH 6.
  - Weigh 1.92 g of citric acid (anhydrous) and dissolve in 900 mL dH_2O. Adjust pH to 6.0 with 1 N NaOH. Make up the volume up to 1 L with dH_2O.
- **Wash buffers.**
  - **Tris buffered saline (TBS, 10×).** Weigh 24.2 g Tris (0.2 M) and 80 g NaCl (1.5 M) and mix with dH_2O. Adjust the pH to 7.4 and make up the volume to 1 L.
  - **TBST.** Add 0.1% Tween-20 to 1 × TBS.
  - **Blocking buffer.** 10% Normal goat serum (Thermo Fisher Scientific, Cat no. 50062Z).

- **Antibody diluent.** 3% BSA in TBST.
  - Weigh 3 g bovine serum albumin and dissolve in 100 mL TBST.
  - Store it at 4 °C.
- **Antibodies.**
  - **Primary antibodies.**
- SARS-CoV-2 (2019-nCoV) Nucleocapsid Antibody, Rabbit PAb (SinoBiological, Cat no. #40588-T62).
- SARS-CoV-2 (2019-nCoV) Spike Antibody, Rabbit PAb (ProSci, Cat no. #3525).
  - **Secondary antibodies.**
- Goat anti-Rabbit IgG (H + L) Cross-Adsorbed Secondary Antibody, Alexa Fluor 488.
- **Other reagents:** 10% neutral buffered formalin, tissue cassettes, paraffin wax, cryotome, oven, water bath, staining boxes, staining trays, coverslips.

3.4.2 Methods

Carry out all procedures at room temperature unless otherwise specified.

Tissue Collection and Fixation

1. Cut the tissues obtained from the necropsied animal into pieces of dimensions 2.5 cm × 2.5 cm × 0.6 cm.
2. Place the tissue samples into plastic tissue cassettes and dip into 10% neutral buffered formalin solution (approx. 4% formaldehyde).
3. The samples need to be kept in 10% NBF for a minimum of 7 days before removal from the ABSL3 lab (*see* **Note 1**).
4. **Embedding and Sectioning** [21].
5. The fixed tissues are then gradually dehydrated by dipping into increasing gradations of ethanol.
6. Subsequently, tissues are placed in xylene to complete the process of dehydration of tissues.
7. Eventually, tissues are subjected to embedment into paraffin wax.
8. Tissue blocks are then sliced into 5 μm thick sections on positively charged slides.

Baking

9. Preheat the oven to 65 °C.
10. Keep the unstained slides in the oven for 30–60 min.

Deparaffinization and Rehydration

11. Incubate the slides in xylene with three changes each for 10 min.
12. Take out the slides from xylene and dip into two changes of absolute ethanol (100%) for 5 min each.
13. Subsequently, hydrate the slides by placing in two changes of 95%, 85%, and 70% ethanol respectively for 5 min each.
14. To complete the rehydration process, rinse the slides in distilled water with two changes of 2 min each.
15. The whole process of de-paraffinization and rehydration is summarized in a tabular from given below.

(a)	Xylene	10 min
(b)	Xylene	10 min
(c)	Xylene	10 min
(d)	100% Ethanol	5 min
(e)	100% Ethanol	5 min
(f)	95% Ethanol	5 min
(g)	95% Ethanol	5 min
(h)	85% Ethanol	5 min
(i)	85% Ethanol	5 min
(j)	70% Ethanol	5 min
(k)	70% Ethanol	5 min
(l)	Distilled water	2 min
(m)	Distilled water	2 min

Antigen Retrieval [22]

We have used double-Heat Induced Epitope Retrieval (HIER) method.

16. Turn on the water bath and set temperature to 95–100 °C (*see* **Note 2**).
17. Fill a staining container with 250 mL of Tris–EDTA buffer and bring it to a boil in a microwave or water bath (*see* **Note 3**).
18. Take the slides out of distilled water and place them in a slide holder and put it in the container filled with preheated Tris–EDTA buffer. Let the slide container stay in the water bath (95–100 °C) for 10 min.
19. Take the staining jar out of the water bath and let it sit.

20. Fill another staining container with 250 mL of Citrate buffer and let it boil in a microwave or water bath (*see* **Note 3**).
21. Rinse the slides in distilled water and place them in the container filled with preheated citrate buffer. Keep the slide container in the water bath (95–100 °C) for 20 min.
22. After 20 min, take out the jar containing slides and let it cool for 20–30 min.
23. Wash the slides in 1 × TBS with 2 changes of 5 min each.

Blocking

24. Mark a boundary around the tissue with hydrophobic Pap Pen. This is to ensure that the reagents are not drained off the tissue while staining process.
25. Put slides in a humidified chamber (*see* **Note 4**).
26. Wash slides twice with 1 × TBST for 5 min each with gentle agitation.
27. Add 10% NGS and incubate for 1 h at RT.
28. Drain slides and carefully wipe around the sections with tissue paper.

Primary Incubation

29. Incubate the sections with SARS CoV-2 Spike (1:200) / Nucleocapsid (1:100) antibody (diluted in antibody diluent) for 2 h at RT in humidified chamber.
30. Rinse slides 3× with TBST with gentle agitation for 5 min each.

Secondary Antibody Incubation

31. Incubate with fluorophore conjugated secondary antibody (1: 400, diluted in antibody diluent) for 1 h at RT in humidified chamber.
32. Wash 3× with TBS for 5 min each.

Nuclear Staining

33. Prepare DAPI solution (1:5000) from 5 mg/mL stock solution of DAPI.
34. Incubate the tissue sections with DAPI for 5 min.
35. Wash 2× with TBST for 5 min each.

Mounting

36. Mount the slides using antifade mounting medium (Thermo Fisher Scientific, Cat no. P36961) and put on the coverslips carefully to avoid the bubbles (*see* **Note 5**).
37. Gently press the coverslips with the thicker end of 1 mL tip to remove the excess mounting media and to get rid of the bubbles if any.

38. Store the slide at 4 °C until ready to visualize under the fluorescence/confocal microscope.

4 Notes

1. As per our institutional protocol, the fixation time for SARS-CoV-2 has been optimized to 7 days.
2. Switch on the water—bath at least 45–60 min earlier so that it attains the temperature of 95—100 °C when we reach the antigen retrieval step.
3. Buffer should be at 95 °C before adding the slides.
4. Incubations need to be carried out in a humidified chamber to avoid drying out of the tissue, which can lead to nonspecific binding and high background staining.
5. Use clean coverslips to prevent background due to dirt. To clean coverslips, dip the coverslips in absolute ethanol and air-dry them beforehand.

References

1. Singh DK, Singh B, Ganatra SR, Gazi M, Cole J, Thippeshappa R, Alfson KJ, Clemmons E, Gonzalez O, Escobedo R, Lee TH, Chatterjee A, Goez-Gazi Y, Sharan R, Gough M, Alvarez C, Blakley A, Ferdin J, Bartley C, Staples H, Parodi L, Callery J, Mannino A, Klaffke B, Escareno P, Platt RN 2nd, Hodara V, Scordo J, Gautam S, Vilanova AG, Olmo-Fontanez A, Schami A, Oyejide A, Ajithdoss DK, Copin R, Baum A, Kyratsous C, Alvarez X, Ahmed M, Rosa B, Goodroe A, Dutton J, Hall-Ursone S, Frost PA, Voges AK, Ross CN, Sayers K, Chen C, Hallam C, Khader SA, Mitreva M, TJC A, Martinez-Sobrido L, Patterson JL, Turner J, Torrelles JB, Dick EJ Jr, Brasky K, Schlesinger LS, Giavedoni LD, Carrion R Jr, Kaushal D (2021) Responses to acute infection with SARS-CoV-2 in the lungs of rhesus macaques, baboons and marmosets. Nat Microbiol 6(1):73–86. https://doi.org/10.1038/s41564-020-00841-4
2. Rosa BA, Ahmed M, Singh DK, Choreno-Parra JA, Cole J, Jimenez-Alvarez LA, Rodriguez-Reyna TS, Singh B, Gonzalez O, Carrion R Jr, Schlesinger LS, Martin J, Zuniga J, Mitreva M, Kaushal D, Khader SA (2021) IFN signaling and neutrophil degranulation transcriptional signatures are induced during SARS-CoV-2 infection. Commun Biol 4(1):290. https://doi.org/10.1038/s42003-021-01829-4
3. Baum A, Ajithdoss D, Copin R, Zhou A, Lanza K, Negron N, Ni M, Wei Y, Mohammadi K, Musser B, Atwal GS, Oyejide A, Goez-Gazi Y, Dutton J, Clemmons E, Staples HM, Bartley C, Klaffke B, Alfson K, Gazi M, Gonzalez O, Dick E Jr, Carrion R Jr, Pessaint L, Porto M, Cook A, Brown R, Ali V, Greenhouse J, Taylor T, Andersen H, Lewis MG, Stahl N, Murphy AJ, Yancopoulos GD, Kyratsous CA (2020) REGN-COV2 antibodies prevent and treat SARS-CoV-2 infection in rhesus macaques and hamsters. Science 370(6520): 1110–1115. https://doi.org/10.1126/science.abe2402
4. Vogel AB, Kanevsky I, Che Y, Swanson KA, Muik A, Vormehr M, Kranz LM, Walzer KC, Hein S, Guler A, Loschko J, Maddur MS, Ota-Setlik A, Tompkins K, Cole J, Lui BG, Ziegenhals T, Plaschke A, Eisel D, Dany SC, Fesser S, Erbar S, Bates F, Schneider D, Jesionek B, Sanger B, Wallisch AK, Feuchter Y, Junginger H, Krumm SA, Heinen AP, Adams-Quack P, Schlereth J, Schille S, Kroner C, de la Caridad Guimil Garcia R, Hiller T, Fischer L, Sellers RS, Choudhary S, Gonzalez O, Vascotto F, Gutman MR, Fontenot JA, Hall-Ursone S, Brasky K, Griffor MC, Han S, Su AAH, Lees JA, Nedoma NL,

Mashalidis EH, Sahasrabudhe PV, Tan CY, Pavliakova D, Singh G, Fontes-Garfias C, Pride M, Scully IL, Ciolino T, Obregon J, Gazi M, Carrion R Jr, Alfson KJ, Kalina WV, Kaushal D, Shi PY, Klamp T, Rosenbaum C, Kuhn AN, Tureci O, Dormitzer PR, Jansen KU, Sahin U (2021) BNT162b vaccines protect rhesus macaques from SARS-CoV-2. Nature 592:283–289. https://doi.org/10.1038/s41586-021-03275-y

5. Gretebeck LM, Subbarao K (2015) Animal models for SARS and MERS coronaviruses. Curr Opin Virol 13:123–129. https://doi.org/10.1016/j.coviro.2015.06.009
6. Rockx B, Kuiken T, Herfst S, Bestebroer T, Lamers MM, Oude Munnink BB, de Meulder D, van Amerongen G, van den Brand J, Okba NMA, Schipper D, van Run P, Leijten L, Sikkema R, Verschoor E, Verstrepen B, Bogers W, Langermans J, Drosten C, Fentener van Vlissingen M, Fouchier R, de Swart R, Koopmans M, Haagmans BL (2020) Comparative pathogenesis of COVID-19, MERS, and SARS in a nonhuman primate model. Science 368(1015):1012. https://doi.org/10.1126/science.abb7314
7. Munster VJ, Feldmann F, Williamson BN, van Doremalen N, Perez-Perez L, Schulz J, Meade-White K, Okumura A, Callison J, Brumbaugh B, Avanzato VA, Rosenke R, Hanley PW, Saturday G, Scott D, Fischer ER, de Wit E (2020) Respiratory disease in rhesus macaques inoculated with SARS-CoV-2. Nature 585:268–272. https://doi.org/10.1038/s41586-020-2324-7
8. Yu J, Tostanoski LH, Peter L, Mercado NB, McMahan K, Mahrokhian SH, Nkolola JP, Liu J, Li Z, Chandrashekar A, Martinez DR, Loos C, Atyeo C, Fischinger S, Burke JS, Slein MD, Chen Y, Zuiani A, FJN L, Travers M, Habibi S, Pessaint L, Van Ry A, Blade K, Brown R, Cook A, Finneyfrock B, Dodson A, Teow E, Velasco J, Zahn R, Wegmann F, Bondzie EA, Dagotto G, Gebre MS, He X, Jacob-Dolan C, Kirilova M, Kordana N, Lin Z, Maxfield LF, Nampanya F, Nityanandam R, Ventura JD, Wan H, Cai Y, Chen B, Schmidt AG, Wesemann DR, Baric RS, Alter G, Andersen H, Lewis MG, Barouch DH (2020) DNA vaccine protection against SARS-CoV-2 in rhesus macaques. Science 369:806–811. https://doi.org/10.1126/science.abc6284
9. Chandrashekar A, Liu J, Martinot AJ, McMahan K, Mercado NB, Peter L, Tostanoski LH, Yu J, Maliga Z, Nekorchuk M, Busman-Sahay K, Terry M, Wrijil LM, Ducat S, Martinez DR, Atyeo C, Fischinger S, Burke JS, Slein MD, Pessaint L, Van Ry A, Greenhouse J, Taylor T, Blade K, Cook A, Finneyfrock B, Brown R, Teow E, Velasco J, Zahn R, Wegmann F, Abbink P, Bondzie EA, Dagotto G, Gebre MS, He X, Jacob-Dolan C, Kordana N, Li Z, Lifton MA, Mahrokhian SH, Maxfield LF, Nityanandam R, Nkolola JP, Schmidt AG, Miller AD, Baric RS, Alter G, Sorger PK, Estes JD, Andersen H, Lewis MG, Barouch DH (2020) SARS-CoV-2 infection protects against rechallenge in rhesus macaques. Science:369, 812–817. https://doi.org/10.1126/science.abc4776
10. Alfson KJ, Griffiths A (2018) Development and testing of a method for validating chemical inactivation of ebola virus. Viruses 10(3). https://doi.org/10.3390/v10030126
11. Wolfel R, Corman VM, Guggemos W, Seilmaier M, Zange S, Muller MA, Niemeyer D, Jones TC, Vollmar P, Rothe C, Hoelscher M, Bleicker T, Brunink S, Schneider J, Ehmann R, Zwirglmaier K, Drosten C, Wendtner C (2020) Virological assessment of hospitalized patients with COVID-2019. Nature 581(7809):465–469. https://doi.org/10.1038/s41586-020-2196-x
12. Sanyaolu A, Okorie C, Marinkovic A, Patidar R, Younis K, Desai P, Hosein Z, Padda I, Mangat J, Altaf M (2020) Comorbidity and its impact on patients with COVID-19. SN Compr Clin Med:1–8. https://doi.org/10.1007/s42399-020-00363-4
13. Sarnyai Z, Nagy K, Patay G, Molnar M, Rosenqvist G, Toth M, Takano A, Gulyas B, Major P, Halldin C, Varrone A (2019) Performance evaluation of a high-resolution nonhuman primate PET/CT system. J Nucl Med 60(12):1818–1824. https://doi.org/10.2967/jnumed.117.206243
14. Salazar-Austin N, Ordonez AA, Hsu AJ, Benson JE, Mahesh M, Menachery E, Razeq JH, Salfinger M, Starke JR, Milstone AM, Parrish N, Nuermberger EL, Jain SK (2015) Extensively drug-resistant tuberculosis in a young child after travel to India. Lancet Infect Dis 15(12):1485–1491. https://doi.org/10.1016/S1473-3099(15)00356-4
15. Cockrell AS, Johnson JC, Moore IN, Liu DX, Bock KW, Douglas MG, Graham RL, Solomon J, Torzewski L, Bartos C, Hart R, Baric RS, Johnson RF (2018) A spike-modified middle east respiratory syndrome coronavirus (MERS-CoV) infectious clone elicits mild respiratory disease in infected rhesus macaques. Sci Rep 8(1):10727. https://doi.org/10.1038/s41598-018-28900-1
16. Duraiyan J, Govindarajan R, Kaliyappan K, Palanisamy M (2012) Applications of

immunohistochemistry. J Pharm Bioallied Sci 4(6):307–309. https://doi.org/10.4103/0975-7406.100281

17. Sohrabi C, Alsafi Z, O'Neill N, Khan M, Kerwan A, Al-Jabir A, Iosifidis C, Agha R (2020) Corrigendum to "World Health Organization declares Global Emergency: a review of the 2019 novel coronavirus (COVID-19)". Int J Surg 77:217. https://doi.org/10.1016/j.ijsu.2020.03.036
18. Sohrabi C, Alsafi Z, O'Neill N, Khan M, Kerwan A, Al-Jabir A, Iosifidis C, Agha R (2020) World Health Organization declares global emergency: a review of the 2019 novel coronavirus (COVID-19). Int J Surg 76: 71–76. https://doi.org/10.1016/j.ijsu.2020.02.034
19. Zhou P, Yang XL, Wang XG, Hu B, Zhang L, Zhang W, Si HR, Zhu Y, Li B, Huang CL, Chen HD, Chen J, Luo Y, Guo H, Jiang RD, Liu MQ, Chen Y, Shen XR, Wang X, Zheng XS, Zhao K, Chen QJ, Deng F, Liu LL, Yan B, Zhan FX, Wang YY, Xiao GF, Shi ZL (2020) A pneumonia outbreak associated with a new coronavirus of probable bat origin. Nature 579(7798):270–273. https://doi.org/10.1038/s41586-020-2012-7
20. Garcia LF (2020) Immune response, inflammation, and the clinical spectrum of COVID-19. Front Immunol 11:1441. https://doi.org/10.3389/fimmu.2020.01441
21. Fischer AH, Jacobson KA, Rose J, Zeller R (2008) Paraffin embedding tissue samples for sectioning. CSH Protoc 2008:pdb prot4989. https://doi.org/10.1101/pdb.prot4989
22. Pileri SA, Roncador G, Ceccarelli C, Piccioli M, Briskomatis A, Sabattini E, Ascani S, Santini D, Piccaluga PP, Leone O, Damiani S, Ercolessi C, Sandri F, Pieri F, Leoncini L, Falini B (1997) Antigen retrieval techniques in immunohistochemistry: comparison of different methods. J Pathol 183(1):116–123. https://doi.org/10.1002/(SICI)1096-9896(199709)183:1<116::AID-PATH1087>3.0.CO;2-2

Chapter 16

Animal Models of COVID-19: Transgenic Mouse Model

Jun-Gyu Park, Paula A. Pino, Anwari Akhter, Xavier Alvarez, Jordi B. Torrelles, and Luis Martinez-Sobrido

Abstract

Severe acute respiratory syndrome coronavirus 2 (SARS-CoV-2), the causative agent of coronavirus disease 2019 (COVID-19), emerged in December 2019 in Wuhan, China, and rapidly spread throughout the world, threatening global public health. An animal model is a valuable and a crucial tool that allows understanding of nature in the pathogenesis of SARS-CoV-2 and its associated COVID-19 disease. Here we introduce detailed protocols of SARS-CoV-2 infection and COVID-19 disease using C57BL/6 (B6) transgenic mice expressing the human angiotensin-converting enzyme 2 (hACE2) from the human cytokeratin 18 promoter (K18 hACE2). To mimic natural SARS-CoV-2 infection, K18 hACE2 transgenic mice are infected intranasally under anesthesia. Upon infection, viral pathogenesis is determined by monitoring changes in body weight (morbidity) and monitoring survival (mortality), cytokine/chemokine responses, gross-lung pathology, histopathology, and viral replication in tissues. The presence of the virus and viral replication is evaluated by immunohistochemistry (IHC) and viral titrations, respectively, from the upper (nasal turbinate) and the lower (lungs) respiratory tracts, and nervous system (brain). Also, the immune response to SARS-CoV-2 infection is measured by cytokine/chemokine enzyme-linked immunosorbent assay (ELISA) from lung, spleen and brain homogenates to characterize the cytokine storm that hallmarks as one of the major causes of death caused by SARS-CoV-2 infection. This small rodent animal model based on the use of K18 hACE2 transgenic mice represents an excellent option to understand the pathogenicity of natural SARS-CoV-2 strains and its recently described Variants of Concern (VoC), and will be applicable to the identification and characterization of prophylactic (vaccine) and therapeutic (antiviral and/or neutralizing monoclonal antibodies) strategies for the prevention or treatment of SARS-CoV-2 infection or its associated COVID-19 disease.

Key words COVID-19, SARS-CoV-2, hACE2, Mouse model, K18 hACE2 transgenic mice

1 Introduction

Severe acute respiratory syndrome coronavirus 2 (SARS-CoV-2) is the causative agent of coronavirus disease 2019 (COVID-19) which is a severe respiratory disease that first emerged in Wuhan,

Jun-Gyu Park and Paula A. Pino contributed equally to this work.

Justin Jang Hann Chu et al. (eds.), *SARS-CoV-2: Methods and Protocols*,
Methods in Molecular Biology, vol. 2452, https://doi.org/10.1007/978-1-0716-2111-0_16,

China, late in 2019 leading to a major threat to public health and socioeconomic activities worldwide of an unprecedented magnitude since the "Spanish flu" pandemic in 1918/1919. According to the World Health Organization (WHO), by the end of 2021, more than 272 million infections and over 5.3 million deaths were identified, with a fatality rate of approximately 1.9 % (http://covid19.who.int). Infected people show clinical manifestations such as fever, cough, shortness of breath and fatigue. Of those, people that succumbed to the infection develop acute respiratory distress syndrome (ARDS), also known as cytokine release syndrome (CRS), or cytokine storm, a major feature of the pathogenesis of COVID-19 [1, 2]. The severe morbidity and mortality of SARS-CoV-2 made it inevitable to develop effective and reliable animal models that enable understanding of the pathogenesis of SARS-CoV-2 and to evaluate prophylactic and therapeutic approaches such as vaccines, antivirals and neutralizing monoclonal antibodies for the efficient treatment of SARS-CoV-2 infection and/or associated COVID-19 disease.

During the previous SARS coronavirus (SARS-CoV) outbreak in 2002–2003, several animal models were developed using wild-type (WT) C57BL/6 (B6) mice [3], young-aged [4] and/or old-aged [5, 6] BALB/C mice, golden Syrian hamsters [7], transgenic mice expressing the human angiotensin converting enzyme 2 (hACE2) from the human cytokeratin 18 promoter (K18 hACE2 transgenic mice) [8, 9], cats [10], and ferrets [10, 11]. However, hamster and cat models did not show mortality [7], and studies using old-aged WT mice or ferrets had reproducibility problems [10, 11]. A mouse-adapted SARS-CoV (MA-15) was also developed but it took a long time for the in vivo passaging in mice, and potential concerns in virus virulence were risen [12]. Efforts have been also directed to establish a SARS-CoV-2 mouse-adapted (MA) strains [12, 13], but concerns also exist that it may not reproduce SARS-CoV-2 infection in humans [12]. Upon the establishment that both SARS-CoV and SARS-CoV-2 use hACE2 as an entry receptor, attention was given to B6.Cg-Tg(K18-ACE2)2Prlmn/J (K18 hACE2) transgenic mice, which previously showed 100% fatality for SARS-CoV [8]. It is now established by our group and others that the K18 hACE2 transgenic mouse model can be infected with SARS-CoV-2, mimicking clinical symptoms to the ones observed in infected people developing severe COVID-19 disease (e.g., weight loss, rapid breathing, hunched posture, and inactivity) [14–17]. Additionally, SARS-CoV-2 infection in this model results in pathology lesions and inflammation in the lungs, as well as with the chemokine/cytokine storm observed in humans infected with SARS-CoV-2, and viral replication in the upper and lower respiratory tract, brain, gut and heart [14]. Thus, the K18 hACE2 transgenic mouse model represents an excellent small animal model to interrogate vaccines and/or antivirals for the prophylactic and/or therapeutic treatment, respectively, of SARS-CoV-2 infection. Moreover, K18 hACE2 transgenic mice represent an excellent option for the high throughput screening of patient convalescent plasma, monoclonal antibodies, and other biologicals for the prevention and/or treatment of SARS-CoV-2 infection, including recently identified Variants of Concern (VoC).

Here we present a detailed protocol defining how to use the K18 hACE2 transgenic mouse model for SARS-CoV-2 infection. Defining parameters such as morbidity (bodyweight loss) and mortality (survival), viral titers, cytokine/chemokine storms, histopathology, and virus tissue tropisms by immunohistochemistry (IHC). In brief, bodyweight of K18 hACE2 transgenic mice is measured and mice are then infected intranasally under isoflurane anesthesia (Fig. 1a). The body weight of infected mice, as well as control mock-infected animals, is recorded daily to determine the morbidity and survival rates (Fig. 1b). At 2 and 4 days postinfection (dpi), mice are euthanized and necropsied to collect nasal turbinate, lungs, spleen and brain (Fig. 1c). Collected lungs are analyzed for gross-pathology, homogenized for viral titration and determining immune responses, and fixed for histopathology and IHC (Fig. 1c). SARS-CoV-2 infected K18 hACE2 transgenic mice decrease their body weight progressively showing 100% fatality between 5 to 6 dpi (Fig. 2a, b, respectively). Gross-lung pathologic changes become severe over time (Fig. 3a, b), where viral titers peak by 2 dpi in the upper and lower respiratory tract (nasal turbinate and lungs, respectively), while the brain shows higher viral titers at 4 and 6 dpi (Fig. 3c). Histopathology analyses demonstrate inflammation associated with SARS-CoV-2 infection. SARS-CoV-2 tropism for specific tissues and locations is confirmed by IHC (Fig. 4). In terms of an immune response, SARS-CoV-2 infection in K18 ACE2 transgenic mice triggers a relatively localized cytokine storm at 2 dpi in the lungs, which is not apparent in the spleen (as an indicator of systemic inflammation). TH1, TH2, and TH17 cytokine responses are increased in the brain at 4 dpi, and these appear to be sex-dependent toward TH2 in females, which may contribute to the delayed viral spread in their nervous system. Similarly, a local and systemic chemokine storm is also observed, mainly driven by an increase in RANTES/CCL5 and IP10/CXCL10 in lungs (at 2 and 4 dpi) and spleen (IP10/CXCL10 at 2 dpi), while the same chemokines are significantly increased in the brain at 4 dpi when compared to mock-infected control mice (Fig. 5).

Nonhuman primates and hamsters are shown to develop mild to moderate COVID-19 clinical symptoms [18–20]. This was also observed in the mouse model expressing hACE2 by other means (e.g., using a mouse promoter, adenoviruses overexpressing mouse ACE2, or expressing hACE2 with other promoters) [8, 21–25]. The ideal SARS-CoV-2 animal model needs to exhibit clear clinical symptoms and characteristics of the severe COVID-19

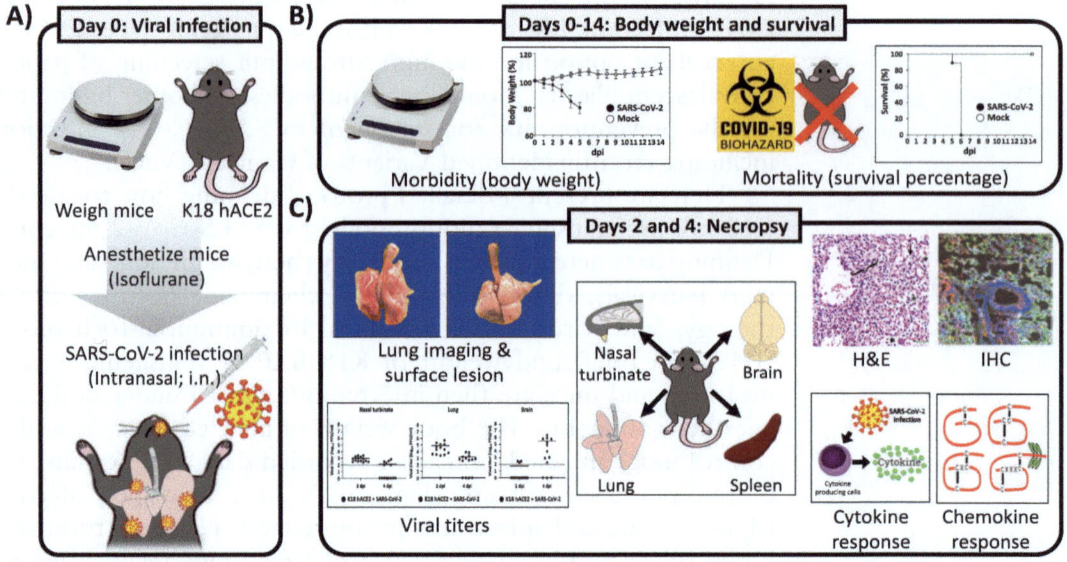

Fig. 1 Schematic experimental procedures for SARS-CoV-2 infection in K18 hACE2 transgenic mice and parameters to measure. (**a**) *Viral infection*: K18 hACE2 transgenic mice are weighed and anesthetized using isoflurane before intranasal (i.n.) infection with 10^5 PFU/ mouse of SARS-CoV-2, WA-1/US strain. (**b**) *Morbidity and mortality*: From 0 to 14 dpi, mice are weighed daily to evaluate changes in body weight loss (morbidity) and mortality (survival percentage). (**c**) *Organ lesions, viral titration, histology and cytokine changes*: At 2 and 4 dpi, SARS-CoV-2 infected K18 hACE2 transgenic mice are sacrificed and internal organs such as nasal turbinate, lung, spleen, and brain are collected for gross pathology changes, viral titers, histopathology and IHC, and cytokine/chemokine analysis. All experiments to characterize SARS-CoV-2 in the K18 hACE2 transgenic mouse model are performed in ABSL3 laboratories. All experiments with SARS-CoV-2 in K18 hACE2 transgenic mice discussed are approved by the IBC and IACUC committees at Texas Biomedical Research Institute. dpi: days postinfection

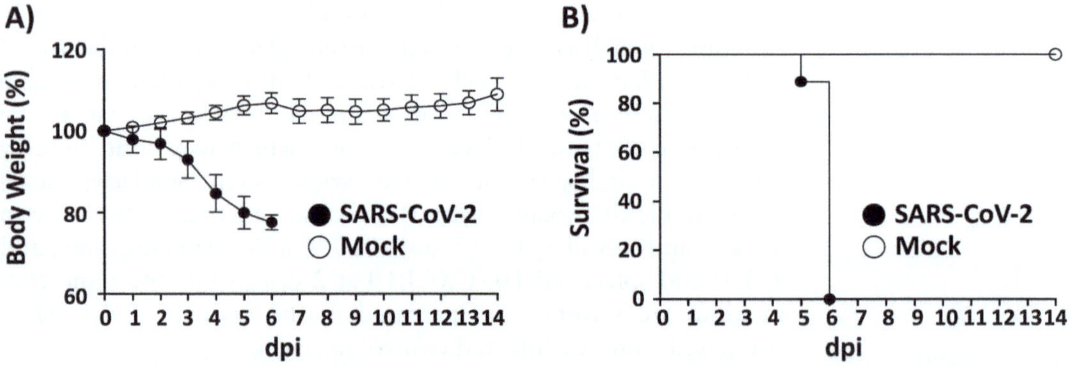

Fig. 2 Morbidity and mortality of SARS-CoV-2 infected K18 hACE2 transgenic mice: K18 hACE2 transgenic mice are mock-infected or infected (10^5 PFU/mouse) with SARS-CoV-2 intranasally ($n = 14$). Changes in body weight (**a**) and survival (**b**) are evaluated daily for 14 days. Mice that loss more than 20% of their initial body weight are humanely euthanized. Error bars represent standard deviations (SD) of the mean for each group of mice. dpi: days postinfection

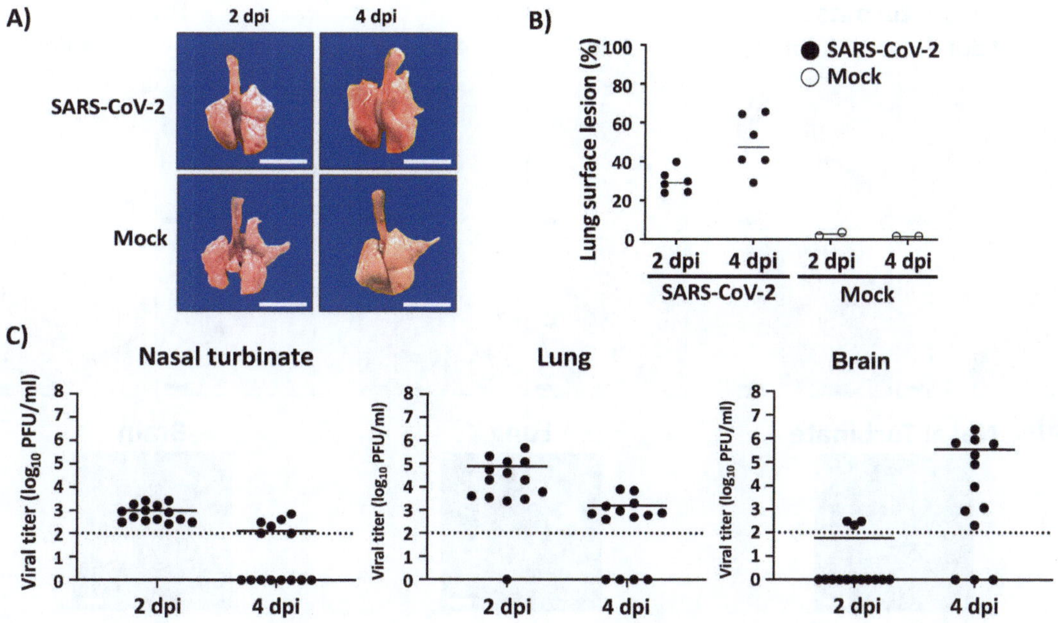

Fig. 3 Gross lung pathology and viral titers. (**a**, **b**) Macroscopic lung pathology lesions: K18 hACE2 transgenic mice are mock-infected or infected with 10^5 PFU/mouse of SARS-CoV-2 and sacrificed at 2 and 4 dpi. Lungs are collected from all sacrificed mice to observe gross pathological changes (**a**). Scale bars = 1 cm. Macroscopic pathology scoring is determined by measuring the distributions of pathological lesions, including consolidation, congestion, and pneumonic lesions using ImageJ software (NIH) (**b**). (**c**) *Viral titers*: Presence of SARS-CoV-2 in the nasal turbinate (left panel), lung (middle panel) and brain (right panel) is determined by plaque assay (PFU/ml) from homogenized tissues at the indicated dpi. Dotted line indicates limit of detection (100 PFU/ml). Each symbol represents an individual animal and bars the geometric means of viral titers. dpi: days postinfection

disease seeing in humans, including mortality rates, viral replication in the lungs and other tissues, histopathological evidence of the disease and dysregulated immunological responses (e.g., cytokine storm). The K18 hACE2 transgenic mouse model shows all of these aspects of the virus-associated disease. Here we introduce a specific SARS-CoV-2 infection method using the K18 hACE2 transgenic mouse model, and subsequent analyses to determine the infectivity and pathogenicity of SARS-CoV-2 as well as some factors associated with COVID-19 disease in vivo.

2 Materials

2.1 Intranasal Infection

1. B6.Cg-Tg(K18-ACE2)2Prlmn/J (K18 hACE2) transgenic mice, The Jackson laboratory.
2. SARS-CoV-2 isolate USA-WA1/2020. Biodefense and Emerging Infections Research Resources Repository, BEI

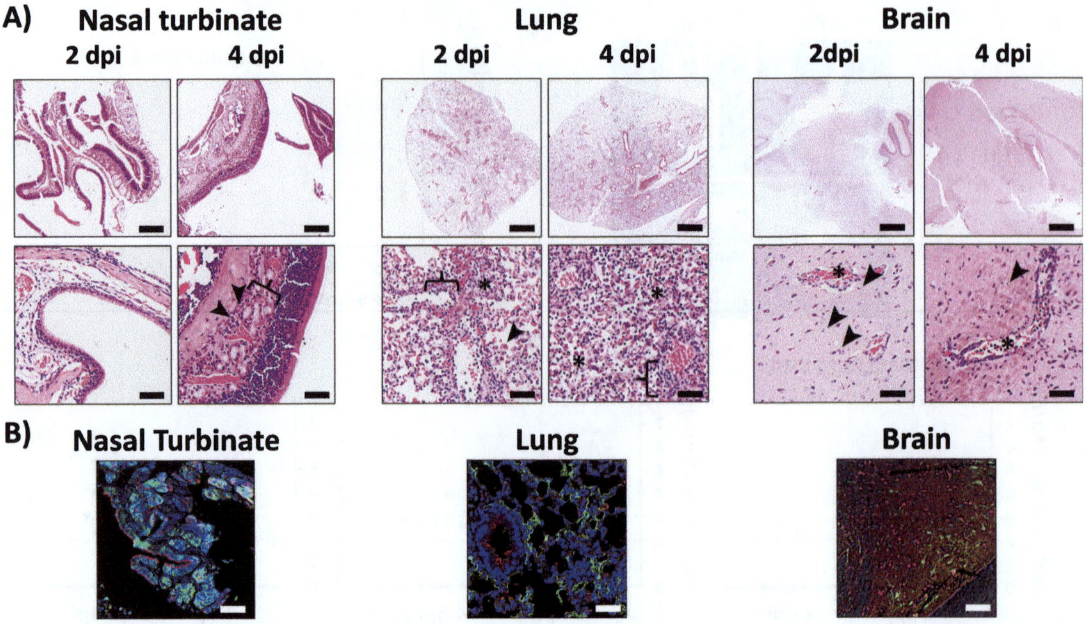

Fig. 4 Histopathologic and IHC examination from SARS-CoV-2 infected K18 hACE2 transgenic mice: Nasal turbinate (left), lung (middle), and brain (right) tissues from K18 hACE2 transgenic mice infected with SARS-CoV-2 are fixed in 10% NBF, embedded in paraffin blocks, and sectioned for hematoxylin and eosin (H&E) (**a**) and confocal microscopy (**b**). *For H&E staining*, nasal turbinate, lung, and brain. Nasal turbinate at 2 dpi are microscopically unremarkable. At 4 dpi there is minimal neutrophilic inflammation (arrowheads) within the submucosa (bracket). Scale bars = 200 μm (upper row) and 50 μm (lower row). Lung at 2 dpi has mild to moderate multifocal interstitial pneumonia associated with inflammatory cell infiltrations (asterisks), mild type II pneumocyte hyperplasia (arrowhead), endothelial cells hyperplasia and vasculitis (bracket). At 4 dpi, it progressed to severe diffuse interstitial pneumonia with inflammatory cell infiltrations in alveolar spaces (asterisks) and interstitium (bracket). Brain at 2 dpi presented multifocal necrotic cellular debris within the brain parenchyma (arrowheads) and vasculitis (inflammation of blood vessels, asterisks). This phenotype extends on the brain over time (4 dpi) causing encephalitis. Scale bars = 1 mm (upper row) and 50 μm (lower row). *For IHC staining*, nasal turbinate, lung and brain tissues from SARS-CoV-2 infected K18 hACE2 transgenic mice at 2 and 4 dpi are immune-stained with hACE2 (red), and SARS-CoV N (green) antibodies. Cell nucleus is stained by DAPI (blue). Note of the tropism of SARS-CoV-2 for pyramidal neurons in the brain. Scale bars = 50 μm. Dpi: days postinfection

Resources, NR-52281: Total 50 μl of inoculum adjusted to 1×10^5 Plaque Forming Units (PFU) of SARS-CoV-2 diluted in phosphate buffer saline (PBS) per mouse.

3. Sterile PBS pH 7.4: 0.155 M NaCl, 0.0010 M KH_2PO_4, 0.0029 M Na_2HPO_4-$7H_2O$ without calcium and magnesium.
4. Anesthesia gas machine (Veterinary Anesthesia System), including small oxygen tank.
5. Isoflurane (Covetrus).
6. Heating pads.
7. Biohazard labels.

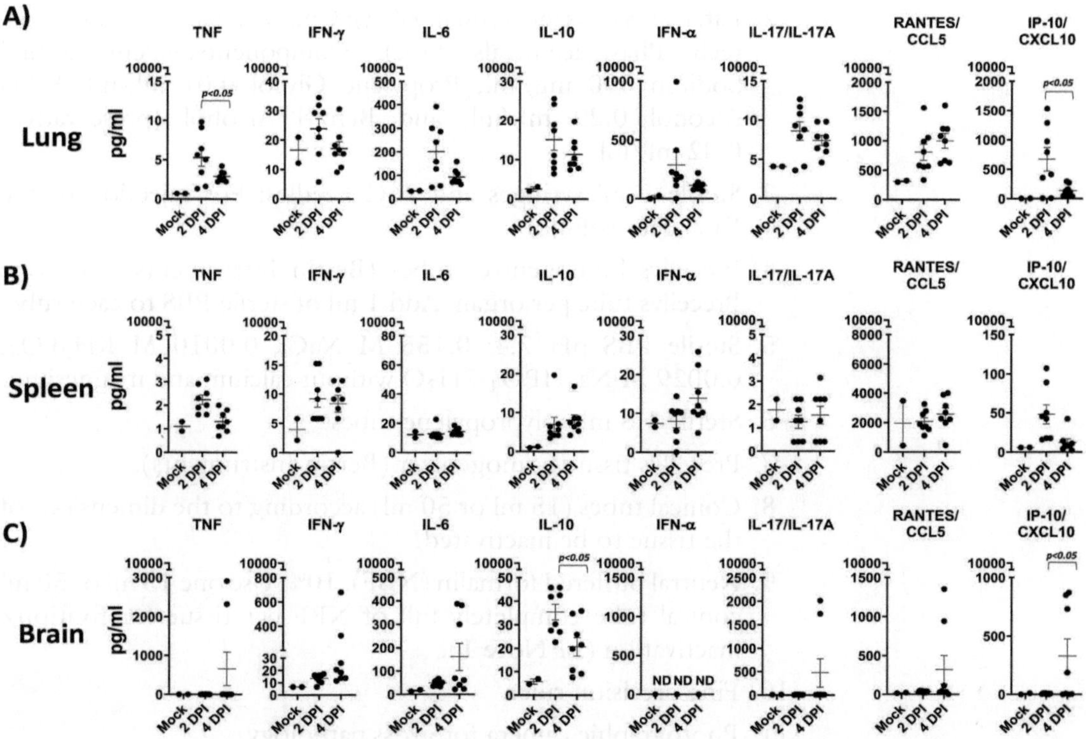

Fig. 5 Cytokine and chemokine responses in SARS-CoV-2 infected K18 hACE2 transgenic mice: Cytokine (TNF, IFN-γ, IL-6, IL-10, IL-17/IL-17A, and IFN-α) (**a**) and chemokine (CCL-5 and CXCL-10) responses in SARS-CoV-2 infected K18 hACE2 transgenic mice at 2 and 4 dpi are measured in lung (**a**), spleen (**b**), and brain (**c**). A 2-WAY ANOVA K18 hACE2 transgenic mice over time. $n = 8$ (per time-point studied, except mock $n = 3$). dpi: days postinfection

8. 2% Wex-Cide (disinfectant, Wexford Labs): Dilute the Wex-Cide in tap water to 2%.
9. 70% ethanol: Mix 7 parts of 100% ethanol and 3 parts of distilled water (dH_2O).

2.2 Measuring Body Weight and Survival

1. Weight scale with three decimal range (0.000).
2. Bowl.
3. 2% Wex-Cide (disinfectant, Wexford Labs): Dilute the Wex-Cide in tap water to 2%.
4. 70% ethanol: Mix 7 parts of 100% ethanol and 3 parts of dH_2O.

2.3 Necropsy

1. Necropsy tools: Necropsy board wrapped with aluminum foil, 18G needles to restrain the mice on the necropsy board, autoclaved surgical tools (including scissors, forceps and a sterile plastic 250 ml beaker to submerge and disinfect the surgical tools in 70% ethanol between animals).

2. Fatal-Plus (dosage >100 mg/kg): Commercially available (Vortech Phasmaceuticals Ltd.). Components: Pentobarbital Sodium 390 mg/ml, Propylene Glycol 0.01 ml/ml, Ethyl Alcohol 0.29 ml/ml, and Benzyl Alcohol (preservative) 0.02 ml/ml.
3. Sterile 1 ml syringes and 26G needles: For injection of the Fatal-Plus solution.
4. Precellys homogenizer tubes (Bertin Instruments): Use one Precellys tube per organ. Add 1 ml of sterile PBS to each tube.
5. Sterile PBS pH 7.4: 0.155 M NaCl, 0.0010 M KH_2PO_4, 0.0029 M Na_2HPO_4-$7H_2O$ without calcium and magnesium.
6. Sterile 1.5 ml polypropylene tubes.
7. Precellys tissue homogenizer (Bertin Instruments).
8. Conical tubes (15 ml or 50 ml) according to the dimensions of the tissue to be inactivated.
9. Neutral buffered formalin (NBF), 10%: Use one 15 ml or 50 ml conical tube completely full of NBF per tissue for fixation/inactivation (*see* **Note 1**).
10. Fine precision ruler.
11. Photographic camera for gross pathology.
12. Dissecting boards to photograph tissues after removal from each animal (Mopec).
13. Paper towels.
14. 2% Wex-Cide (disinfectant, Wexford Labs): Dilute the Wex-Cide in tap water to 2%.
15. 70% ethanol: Mix 7 parts of 100% ethanol and 3 parts of dH_2O.

2.4 Viral Titration

1. Vero E6 cells (American Type Culture Collection, ATCC, CRL-1586).
2. Cell growth media: Dulbecco's modified Eagle medium (DMEM) supplemented with 10% fetal bovine serum (FBS), 1% penicillin–streptomycin, L-glutamine 100× (PSG).
3. Infection media: DMEM supplemented with 1% PSG.
4. Microcrystalline cellulose (Avicel, Sigma-Aldrich), 4%: Weigh 4 g of Avicel in 100 ml of PBS. Autoclave the bottle before use.
5. Postinfection media: 75 ml of DMEM supplemented with 2% FBS and 1% PSG, 25 ml of 4% microcrystalline cellulose solution.
6. Permeabilization solution: 0.5% of Triton X-100 in PBS.
7. Blocking solution: 2.5% bovine serum albumin (BSA) in PBS.

8. Mouse monoclonal antibody anti-SARS nucleocapsid (N) protein 1C7C7 (Leinco): Prepare 1 μg/ml of antibody in 1% BSA.
9. VECTASTAIN® ABC-HRP Kit, Peroxidase, POD (Mouse IgG) (Vector Laboratory).
10. DAB Substrate Kit, POD (HRP), with Nickel (Vector Laboratory).
11. 96-well cell culture plates.
12. Pipettes and pipette tips.
13. Multichannel pipette, manifold dispenser.
14. Vacuum aspirator-multichannel adapter (HandE-Vac Hand Controller Set, Argos Technologies).
15. Sterile 1.5 ml microcentrifuge tubes.
16. CO_2 incubator at 37 °C.
17. CTL ImmunoSpot plate reader and counting software (Cellular Technology Limited).
18. 2% Wex-Cide (disinfectant, Wexford Labs): Dilute the Wex-Cide in tap water to 2%.
19. 70% ethanol: Mix 7 parts of 100% ethanol and 3 parts of dH_2O.

2.5 Histopathology

1. Tissue processing solutions: xylene, 95% alcohol, absolute alcohol.
2. Microscope slides, coverslip.
3. Parapro XLT infiltration and embedding media (StatLab).
4. Hematoxylin stain (StatLab).
5. High Def solution (StatLab).
6. Reserve Bluing Reagent (StatLab).
7. Eosin (StatLab).
8. Tissue Tek VIP tissue processor.
9. Microm HM325 rotary microtome.
10. A flotation water bath set between 46 and 48 °C.
11. Oven at 60 °C.
12. Varistain Gemini AS Automated Slide Stainer (Thermo Scientific).
13. Bx46 microscope (Olympus).
14. CellSense v1.18 Life science Imaging software (Olympus).

2.6 Immunohistochemistry (IHC)

1. Hydration and dehydration solutions: xylene, 100%, 95%, and 80% ethanol (anhydrous denatured, histological grade).

2. 10× Tris Buffered Saline with Triton X-100 (TBST) (Bio-Rad).
3. 1× TBST: add 100 ml 10× TBST to 900 ml dH_2O; add 0.025% Triton X-100 and mix well.
4. Blocking and Antibody diluent: Background punisher (Bio-Care Medical).
5. Antigen Unmasking Solution: Tris Based (Vector Laboratories).
6. Antigen Unmasking Solution: 0.01 M Citrate Acid Based (Vector Laboratories).
7. Recombinant Rabbit monoclonal antibody anti-hACE2, Thermo-Fisher, Cat#SN0754): Prepare 1:50 dilution of antibody in background punisher.
8. Rabbit polyclonal antibody anti-SARS N protein (homemade): Prepare 1:4000 dilution of antibody in background punisher.
9. Detection System: MACH 3 Rabbit or Mouse Alkaline Phosphatase Polymer Detection Kit (BioCare Medical).
10. Substrate: Warp Red Chromogen Kit (BioCare Medical).
11. Mounting Media with 4′,6-diamidino-2-phenylindole, DAPI (Thermo Scientific): Dissolve DAPI in dH_2O to 1 mg/ml as stock solution. Keep the stock solution from light at −20 °C. Dilute the stock solution in PBS to 1 μg/ml for working concentration.
12. PBS pH 7.4: 0.155 M NaCl, 0.0010 M KH_2PO_4, 0.0029 M Na_2HPO_4-$7H_2O$ without calcium and magnesium.
13. PBS–Fish Skin Gelatin (PBS-FSG): 1000 parts PBS to 2 parts FSG (FSG, Sigma-Aldrich).
14. PBS-FSG-Tx100: 1000 parts PBS-FSG to 1-part Triton X-100 (TX-100, Sigma-Aldrich).
15. Normal Goat Serum (GIBCO), diluted 1:10 in PBS-FSG.
16. Beakers.
17. Coplin staining Jars.
18. Slide racks.
19. Slides incubation chamber.
20. Coverslips.
21. Microtiter pipettes and pipette tips.
22. A rocker platform.

2.7 Cytokines/Chemokines Assay

The Multiplex Mouse Cytokines/Chemokines assay (R&D Systems) is a commercial kit performed following manufacturer's instructions.

1. Magnetic Luminex Assay, Mouse Premixed Multi-Analyte Kit (R&D Systems).
2. Hand-held microplate magnet or plate washer with a magnetic platform.
3. Microtiter pipettes and pipette tips.
4. Multichannel pipette, manifold dispenser, or automated dispensing unit.
5. Deionized H_2O or dH_2O.
6. 500 ml graduated cylinder.
7. Horizontal orbital microplate shaker (0.12″ orbit) capable of maintaining a speed of 800 ± 50 revolutions per min (rpm).
8. Microcentrifuge.
9. Polypropylene test tubes for dilution of standards and samples.
10. 96-well, flat bottom, polystyrene plates, Greiner Bio-One 655096, or equivalent.
11. 1% Formaldehyde in PBS: For one 96-well plate, prepare a total volume of 18.5 ml by mixing 18 ml of PBS pH 7.4 (0.137 M NaCl, 0.0027 M KCl, 0.01 M Na_2HPO_4, 0.0018 M KH_2PO_4) and 0.5 ml of 37% formaldehyde (Sigma-Aldrich).
12. Plate sealer.
13. Aluminum foil.
14. 1× sheath fluid (Luminex).
15. Luminex® MAGPIX®, Luminex® 100/200™, Luminex® FLEXMAP 3D®, or Bio-Rad Bio-Plex analyzer with X-Y platform.

2.8 Interferons Assays

The Mouse IFN-α (Type I) enzyme-linked immunosorbent assay (ELISA) (PBL Assay Science) is a commercial kit developed following manufacturer's instructions.

2.8.1 Type I Interferon Alpha (IFN-α) Assay (See Note 2)

1. Mouse IFN-α ELISA Kit (PBL Assay Science).
2. Variable volume microtiter pipettes.
3. Adjustable multichannel pipette (50–300 μl).
4. Reagent reservoirs.
5. Wash bottle or plate washing system.
6. Distilled or deionized H_2O.
7. Serological pipettes (1, 5, 10 or 25 ml).
8. Disposable pipette tips (polypropylene).
9. Plate shaker.
10. Microtiter plate reader capable of reading an optical density (OD) at a wavelength of 450 nm.

2.8.2 Type III Interferon Lambda (IFN-λ) Assay (See ***Note 2****)*

The Mouse IFN-λ (Type III) ELISA (PBL Assay Science) is a commercial kit that is performed according to the instructions provided by the manufacturer.

1. Mouse IFN-λ ELISA Kit (PBL Assay Science).
2. 96-well ELISA plates.
3. PBS pH 7.2–7.4: 0.137 M NaCl, 0.0027 M KCl, 0.0081 M Na_2HPO_4, 0.0015 M KH_2PO_4. Filter sterilize using a 0.2 μm membrane.
4. Wash Buffer: 0.05% Tween 20 in PBS pH 7.2–7.4.
5. Reagent Diluent: 1% BSA in PBS pH 7.2–7.4. Filter sterilize using a 0.2 μm membrane (*see* **Note 3**).
6. Substrate Solution: 1:1 mixture of Color Reagent A (H_2O_2) and Color Reagent B (Tetramethylbenzidine, TMB).
7. Stop Solution: 2 N H_2SO_4.
8. Plate shaker.
9. Squirt bottle, manifold dispenser or auto-washer.
10. Microtiter plate reader capable of reading an OD at a wavelength of 450 nm.

3 Methods

3.1 Ethics Statement

All experimental procedures with mice were approved by the Texas Biomed Institutional Animal Care and Use Committee (IACUC) and Institutional Biosafety Committee (IBC), and performed under Biosafety Level 3 (BSL3) and Animal BSL3 (ABSL3) laboratories.

3.2 Intranasal Infection

1. Transfer the specific-pathogen-free, 4–5-weeks-old, female and male K18 hACE2 transgenic mice to the ABSL3 facility. Place a maximum of 5 mice per cage and identify each animal using an ear punch code or any other approved method (e.g., tail tattoo). Label each mouse cage according to the virus and dosage to be used.
2. Prepare the SARS-CoV-2 inoculum under aseptic conditions. Calculate the amount of SARS-CoV-2 needed to prepare the viral dilution applying the following formula: (Titer of the virus [PFU] needed /stock viral titer) × final number of animals. Complete a total volume of 50 μl dose/mouse using sterile PBS. Keep the diluted viral inoculum on ice until use. Aliquot the total amount needed of PBS to be used for the inoculation of the mock-infected group (Fig. 1a).
3. Using a scale, record the bodyweight of each mouse at 0 dpi corresponding to the bodyweight/survival group (Fig. 1a).

4. Anaesthetize each mouse by placing the animal inside the anesthesia chamber of a Veterinary Anesthesia System and provide oxygen mixed with 3% isoflurane (Fig. 1a). Take the mouse out of the anesthesia chamber to proceed with the viral infection when the animal takes 2–3 s between breaths. Verify that the mouse is fully anaesthetized by toe pinch to see a lack of pedal withdrawal reflex. If the mouse is not fully anaesthetized, it will cough up the virus, resulting in inconsistent infection (Fig. 1a).
5. Place the mouse on hands in a dorsal recumbent position. Lift its head in an upright position to ensure the inoculum goes down through the airways. Infect intranasally (i.n.) the mouse using 50 μl of the viral inoculum. Make sure to eject the solution slowly but constantly and check the mouse is inhaling the inoculum by verifying the inoculum drop disappearing.
6. Hold the mouse in an upright position until its breathing rhythm gradually returns to normal. Make sure it is breathing as the viral inoculum can block the respiratory tract. Return the mouse to its home cage in a dorsal recumbent position. If any sign of respiratory distress is observed, help the mouse to breathe by holding the animal vertically and induce impact-driven respiration. Keep monitoring until the mouse regains consciousness (2–5 min after being anaesthetized). Place some heating pads underneath the cage to keep the animals warmed while the anesthesia effect disappears.

3.3 Characterization of Viral Pathogenesis

3.3.1 Evaluation of Morbidity and Mortality (Fig. 1b)

1. After viral infection, record the weight of mice once a day during the next 14 consecutive days at approximately the same time that the viral infection was performed, minimizing weight fluctuations such as food ingestion/water uptake, and keep monitoring for any clinical signs. Humanely euthanize mice that lose over 20% of their initial body weight or show severe clinical signs such as lethargic behavior with rough fur, hunched appearance, immobile, social distancing, and/or if it does not eat or drink (Fig. 1b).
2. After 14 days (end-point), humanely euthanize the mice that survive the viral infection according to the standard procedures at your institution/your local IACUC (Fig. 2).

3.3.2 Necropsy (Fig. 1c)

1. At the established end-points (2 and 4 dpi, or at the end of your study), euthanize mice according to IACUC approved protocol, such as a lethal dose of Fatal-Plus (200 μl) via intraperitoneal route (i.p.) injection. (*see* **Note 4**).
2. Lay the euthanized mouse in a dorsal recumbent position on the clean necropsy board and disinfect the fur over the thorax and abdomen spraying 70% ethanol on it evenly. Using 18G

needles, restrain the animal on the necropsy board by the extremities. Take the scissors and cut the center skin of the body from the base of the abdomen to the chin, detaching the skin from the membrane that supports internal organs (do not tear apart the membrane). Make 1 cm incision from the abdomen and open the rib cage by cutting the ribs laterally from the costal cartilage through the diaphragm to the top of the sternum. Keep cutting until reaching the middle line of the animal toward the mouth without damaging the trachea. Remove the ribs and sternum and examine the thoracic cavity including all the lobes of the lung (*see* **Note 5**).

3. Lung isolation: Remove the whole lower respiratory tract from the start of the trachea to the lungs and place it on a dissecting board. Trim the collected lower respiratory tract to remove heart, fat, bronchus-associated lymphoid tissue (BALT) and unnecessary connective tissue.
4. Spleen isolation: Using forceps, gently pull the spleen free of the peritoneum, tearing or cutting the connective tissue attached to it. Avoid cutting into the intestine as this organ is not sterile.
5. Brain isolation: Using scissors, remove the head from the mouse body by cutting between cranium and atlas. Cut to remove also the skin of the head. Hold the cranium with the dissecting forceps and cut the cranium along the sagittal suture with scissors. Collect the brain from the inside part of the two halves of the cranium and place it on a dissecting board.
6. Nasal turbinate: Cut the junctions between the maxilla and the mandible, discard the mandible. With the scissors, make two cuts at the ends of the zygomatic arcs and remove them. Remove the eyes with the help of the dissecting forceps. The nasal turbinates are surrounded by the anterior bones of the skull including the nasal, maxilla, palatine, zygomatic, and ethmoid bones. Using the scissors, collect the whole anterior bones and nasal turbinate and place them on a dissecting board.

3.3.3 Gross Pathology

Lung Images: Place the lungs on a clean dissecting board for photographing and obtain the lung images from dorsal and ventral view to evaluate gross lung pathology (Fig. 3a, b).

3.3.4 Distribution of Tissues for Fixation and Homogenization

1. Transfer a half of each organ (left lung lobe, nasal turbinate and brain) to a fixative solution to inactivate following Subheading 3.5 (see below Histopathology section).
2. Place the other half of the lung (right lobes), nasal turbinate and brain, and the whole spleen in the corresponding sterile Precellys homogenizer tube filled with 1 ml of PBS.

3. Homogenize tissues in the Precellys tissue homogenizer using the following parameters: 5000 × *g*, 20 s, once. This allows a well-homogenized tissue to be obtained. If a different tissue homogenizer is used, the appropriate parameters to homogenize tissues must be standardized according to the case.
4. Transfer the tissue homogenates to sterile 1.5 ml microcentrifuge polypropylene tubes and centrifuge at 6000 × *g*, 5 min, 4 °C.
5. Collect the supernatant and transfer it to a new 1.5 ml polypropylene tube. Store at −80 °C until use for viral titers determination or cytokines/chemokines evaluation.

3.4 Viral Titration

1. The day before the in vitro infection for the viral titration, seed 96-well plates with Vero E6 cells (1 × 10^4 cells/well) with cell growth media. Maintain the cells at 37 °C, 5% CO_2. On the same day, thaw the homogenized tissue supernatant to be tested for the viral titration.
2. On the day of the viral titration, and before starting the in vitro viral infection, check the confluence of Vero E6 cells under the microscope to confirm a monolayer is formed.
3. In a different 96-well plate, prepare a ten-fold dilution of homogenized tissue supernatant in an empty, sterile 96-well plate using infection media. Briefly, add 90 μl of infection media to the 96-well plate rows A to H, and add 10 μl of the supernatant of each homogenized sample to wells of the row A. Using a multichannel micro dispenser, transfer 10 μl from row A to B, and mix ~10 times. Repeat this process from row B to row C, and subsequently until row H, changing the tips between dilutions to prevent the transfer of residual virus in the tips. Discard 10 μl from the dilution already mixed in row H so that each of the 96-well plate wells has 90 μl of diluted sample. Each sample should be tested at least in duplicate.
4. On the 96-well plate containing Vero E6 cells seeded, remove the tissue culture medium using a vacuum aspirator-multichannel adapter.
5. Proceed with the in vitro infection transferring 50 μl/well of the serially diluted tissue homogenized supernatants to the Vero E6 cells seeded in the 96-well plate starting from the lower dilutions (row H) to the higher dilutions (row A), so that it is not necessary to change the tips between different rows.
6. Incubate the 96-well plates with infected cells for 1 h at 37 °C in 5% CO_2 to allow viral adsorption. After this period of time, remove the inoculum using a vacuum aspirator-multichannel adapter and overlay with postinfection media containing 1%

Avicel. Then, incubate infected Vero E6 cells for 24 h at 37 °C, 5% CO_2.

7. At 24 h postinfection (hpi), remove infectious media and completely submerge the plate in 10% NBF for 24 h at 4 °C for fixation/inactivation (shown to inactivate SARS-CoV-2).
8. Upon fixation/inactivation, transfer the plate from the BSL3 to a BSL2 laboratory following approved standard procedures, remove residual 10% NBF by gently washing with dH_2O before staining.
9. Proceed with the staining protocol [26]. Wash cells gently 3× with 100 μl/well of PBS and permeabilize the cells with 100 μl/well of 0.5% Triton X-100 dissolved in PBS. Incubate at room temperature for 15 min in the biosafety cabinet.
10. Wash the cells three times with 100 μl/well of PBS and incubate fixed cells with blocking solution (100 μl/well of 2.5% BSA in PBS) at 37 °C for 1 h.
11. Remove the blocking solution. Add 50 μl/well of 1 μg/ml primary antibody solution (anti-NP monoclonal antibody 1C7C7) diluted in 1% BSA in PBS. Incubate the cells at 37 °C for 1 h.
12. Remove the primary antibody solution, wash with 100 μl/well of PBS, three times. Prepare the biotinylated anti-mouse antibody [VECTASTAIN® ABC-HRP Kit, Peroxidase (Mouse IgG); Vector Laboratory] following the manufacturer's instructions. Briefly, add 75 μl of normal blocking serum stock and 25 μl of biotinylated secondary antibody stock to 5 ml of PBS per 96-well plate. Add 50 μl/well of biotinylated antibody solution to each well and incubate for 30 min at 37 °C.
13. Remove the secondary biotinylated antibody solution and wash each well three times with 100 μl/well of PBS. Prepare VECTASTAIN ABC Reagent as described in manufacturer's instructions (VECTASTAIN ® ABC-HRP Kit, Peroxidase (Mouse IgG; Vector Laboratory). Add 50 μl each of Reagent A (Avidin, ABC) and Reagent B (Biotinylated HRP, ABC) to 5 ml of PBS for one 96-well plate. Mix the solution thoroughly and add 50 μl/well of VECTASTAIN ABC Reagent and incubate for 30 min at 37 °C.
14. Wash cells three times with 100 μl/well of PBS. Dry the plate by tapping on paper towels. Prepare developing solution by following manufacturer's instructions [DAB Substrate Kit, Peroxidase (HRP), with Nickel; Vector Laboratory]. Add 50 μl of developing a solution to each well and wait for ~3–5 min to visualize viral plaques.

15. Remove the developing solution to stop the reaction and wash with PBS. Make sure not to develop for too long once the plaques are evident to prevent the cells from turning black.
16. Count the stained positive cells using a CTL ImmunoSpot plate reader and counting software (or your alternative method) to calculate the viral titer. Calculate the viral titer expressed in PFU/ml from the dilution with 30 to 50 positive stained cells using the formula: [(Mean No of plaques] × 20 × 1/dilution (Fig. 3c).

3.5 Histopathology

This protocol describes methods corresponding to the preparation of tissues after necropsies of animals and their histopathological analysis.

1. As mentioned above (in the distribution of tissues for fixation and homogenization section), separate half of each organ (left lung lobe, nasal turbinate and brain) intended to be inactivated. Determine the dimension of each tissue using a precision ruler to calculate the volume of 10% NBF to be used. Determine the length, width, and thickness of the tissue. The sample cannot exceed the following dimensions: 2.5 cm × 2.5 cm × 0.6 cm (thickness) due to the decontamination protocol approved using 10% NBF which allows that the 10% NBF completely penetrates the tissue.
2. Determine the volume of 10% NBF solution using the following formula: Length × Width × Thickness = Tissue Volume × 15 (Volume of NBF per volume of tissue) (*see* **Note 6**).
3. Fully submerge the infected tissue in a conical tube using the appropriate volume of 10% NBF solution according to the formula described above. If the volume needed is less than the maximum capacity of the conical tube, such tube must be completely full of NBF. Coat tube threads and gasket with 10% NBF prior to closing the tube's cap.
4. Store samples at the BSL3 laboratory submerged in 10% NBF for 7 days at room temperature prior to removal to a lower BSL laboratory (*see* **Note 7**).
5. Follow your approved institutional biosafety standard procedures for removal of tissues in 10% NBF from the BSL3 and transfer them to the corresponding Pathology Laboratory for subsequent histopathological analysis.
6. Proceed with the paraffin processing and embedding of tissues. Fixed tissues are processed using a Tissue Tek VIP tissue processor by dehydration through a series of graded alcohols, clearing through one change of 50:50 absolute alcohol–xylene mixture and two changes of xylene, concluding with the

infiltration of tissues with paraffin wax using Parapro XLT infiltration and embedding media.

7. The paraffin-embedded blocks are then cut using a Microm HM325 rotary microtome at a thickness of 4 microns and mounted onto microscope slides by use of a flotation water bath set between 46 °C and 48 °C. Unstained slides are dried in a 60 °C oven for 1 h before proceeding with Hematoxylin and Eosin (H&E) staining protocol.
8. Proceed with the H&E staining using a Varistain Gemini Automated Slide Stainer. Deparaffinization of tissue sections is performed through three changes of xylene (incubation time for each xylene step is 3 min, 2 min, and 2 min, respectively), two changes of absolute alcohol (incubation time for each absolute alcohol step is 1 min), two changes of 95% alcohol (incubation time for each 95% alcohol step is 1 min) and rinse in running water for 30 s.
9. Perform the hematoxylin staining for 2 ½ min, followed by rinsing in running water for 1 min. Remove the excess of stain using High-Def solution for 45 s, rinsing with running water for 1 min. Continue bluing the hematoxylin stain using Reserve Bluing Reagent for 1 min. Rinse in running water for 1 min and add 95% alcohol for 15 s. Tissue sections are then stained with eosin for 2 ½ min.
10. Dehydration is performed through three changes of absolute alcohol (incubation time for each absolute alcohol step is 1 min), a 50:50 absolute alcohol–xylene (v/v) mixture for 30 s and cleared in three changes of xylene (incubation time for each xylene step is 1 min) before coverslipping.
11. Upon completion of the H&E staining, tissue sections are evaluated using light microscopy in a blinded manner by a board-certified veterinary pathologist. Images are acquired using a Bx46 microscope and the CellSense vr. 1.18 Life science Imaging software.

3.6 Immunohistochemistry

To perform antibody staining, paraffin wax must be removed from the sample and the sample must be rehydrated (*see* **Note 8**).

1. Place the slides in a 56–60 °C oven for 1 h (*see* **Note 9**).
2. Transfer slides to the xylene bath and perform three changes of xylene for 5 min each.
3. Drain off excess liquid and rehydrate slides in three changes of fresh absolute ethanol for 2 min each.
4. Drain off excess liquid and place slides in fresh 95% ethanol for 2 min.
5. Drain off excess liquid and place slides in fresh 80% ethanol for 2 min.

6. Place the slides in deionized H_2O.
7. Microwave double antigen retrieval procedure: For the first antigen retrieval, place slides rack in a microwave-resistant 250 ml glass beaker containing TRIS-based Antigen unmasking solution, pH 9.0 (Vector Labs H-3301), that also contains 0.01% Tween-20. Make sure slides are fully covered with the solution.
8. Operate the microwave oven on 20% power (700 watts) for 11–12 min for one beaker and for about 20 min for 2 beakers (*see* **Note 10**). Leave slides in the beaker before proceeding to the next step.
9. Second Antigen Retrieval: Take two 250 ml size glass beakers, one filled with deionized H_2O and another filled with 0.01 M Citrate-based Antigen unmasking solution, pH 6.0 (Vector Laboratories, H-3300), each up to 200 ml volume. Place deionized H_2O and a 0.01 M citrate-based solution containing beakers in the microwave oven.
10. Operate the microwave oven on high power about 3–4 min (*see* **Note 10**).
11. Take out slides from the antigen unmasking Tris-based solution beaker, drain off excess solution, and transfer it to the microwave boiled deionized H_2O beaker to wash the slides only (e.g., 30 s).
12. Next, transfer slides from the deionized H_2O beaker to the microwave boiled Antigen unmasking Citrate-based solution beaker and let cool slowly at room temperature for at least 30 min.
13. After this period, transfer slides into a beaker containing Tris Buffered Saline (TBS).
14. For chromogenic staining, wash sections in TBS for 5 min.
15. Use a hydrophobic pen to draw a large circle around the sample, taking care not to touch the sample. This creates a hydrophobic boundary so that a smaller volume of antibody solution can be used and, if desired, allows multiple sections on one slide to be stained with different antibodies.
16. To prevent nonspecific binding of the antibody to tissues, block each section with 100–400 μl Background punisher (blocking solution) for 40 min at room temperature in a humidified chamber.
17. To stain hACE2, use anti-hACE2 as a primary antibody: Remove blocking solution and add 100–400 μl primary antibody diluted in recommended antibody diluent (by the manufacturer) to each section for 1 h at room temperature in a humidified chamber.

18. Equilibrate MACH 3 detection reagent kept at 4 °C to room temperature. (Use detection kit of the same species as the source of the primary antibody).
19. Remove primary antibody solution and wash sections in wash buffer three times for 10 min each.
20. Cover section with 1 to 3 drops of MACH 3 probe as needed. Incubate in a humidified chamber for 20 min at room temperature.
21. Wash sections 3× with wash buffer for 10 min each.
22. Cover section with 1 to 3 drops of MACH 3 alkaline phosphatase polymer as needed. Incubate in a humidified chamber for 20 min at room temperature.
23. Wash sections 3× with wash buffer for 10 min each.
24. Development: Add 1 drop of Warp Red Chromogen to 2.5 ml of Warp Red Buffer mix well before use. Apply the Warp Red mixture to the tissue section. Visualize staining of tissue under a microscope using a bright-field illumination. Incubate 5–20 min, stop reaction when you observe a red coloration in the cells indicating a positive reaction.
25. Wash sections in deionized water 2× for 5 min each.
26. Stained tissue mounted with mounting media with DAPI, fluorescent stain that binds strongly to adenine–thymine-rich regions in DNA, used to visualize nucleus and distinct cells).
27. Warp Red is highly fluorescent and will not fade. Use with a Texas Red filter.
28. Repeat antigen retrieval procedures from **steps** 7 to **13**, for fluorescence IHC.
29. For fluorescence staining, dry slides as much as possible without touching the cells or tissues.
30. Circle tissue with a wax pen, put slides inside a slide Coplin jar, and add PBS-FSG.
31. Place slides in humidifying chamber (such as a black slide box, cover the bottom with a shallow layer of tap water) for the following steps.
32. Cover the tissues with 10% normal goat serum for 60 min (approx. 200 μl/slide).
33. Prepare primary antibody in optimal concentration. Use 10% normal goat serum to dilute the antibody; for example, if you have 5 slides and you need 200 μl of liquid to cover each one, you will need 1 ml total. Adjust numbers to reflect the quantity of slides you are staining and be certain about all calculations.
34. Centrifuge the primary antibody diluted in 10% normal goat serum at 13,000 × g for 1 min (use microfuge).

35. Remove normal goat serum from slides (e.g., dump it onto a paper towel). Do not rinse tissue!
36. Add the appropriate concentration of primary antibody, diluted in 10% normal goat serum.
37. Incubate at room temperature for 60 min, in black slide box.
38. Fill a Coplin jar with PBS-FSG-TX100.
39. Using a transfer pipette, rinse tissue thoroughly and place slides in the Coplin jar.
40. Place on the rocker for 10 min.
41. Repeat 3×.
42. Fill another Coplin jar with PBS-FSG. Using a transfer pipette, rinse the tissue thoroughly and place slides in the Coplin jar.
43. Place on the rocker for 10 min.
44. Centrifuge the secondary antibody diluted in 10% Normal Goat Serum at 13,000 × *g* for 1 min.
45. Apply enough of the secondary antibody solution (approximately 200 μl) to all slides.
46. Incubate at room temperature, for 60 min, in a black slide box.
47. Fill a Coplin jar with PBS-FSG-Tx100. Using a transfer pipette, rinse tissue thoroughly and place slides in the Coplin jar.
48. Place on the rocker for 10 min.
49. Repeat 3×.
50. Fill another Coplin jar with PBS-FSG. Using a transfer pipette, rinse the tissue thoroughly and place slides in the Coplin jar until you can check them under the microscope.
51. Record your results and suggestions, as you examine each slide.
52. If you desire more antigen colors for these slides, you may return to **step 5** and repeat the protocol. Be sure that the antibodies are compatible to be distinguished by secondary antibody detection.
53. Once your slides have all desired antigens labelled, mount slides in antiquenching media.

3.7 Determination of Cytokines/Chemokines

The determination of cytokines and chemokines is performed using a custom multiplex panel mouse magnetic bead Luminex assay, which is summarized below according to the instructions provided by the manufacturer.

1. Prepare reagents and standards provided in the kit following the instructions of the manufacturer (*see* **Note 11**). Use the supernatant from homogenized tissues (nasal turbinate, lung,

brain and spleen) to prepare a dilution 1:2 using the Calibrator Diluent RD6–52 provided in the kit (*see* **Note 12**).

2. Add 50 μl per well of standard or diluted tissue supernatant to the 96-well microplate provided in the kit (*see* **Note 13**).
3. Resuspend the diluted Microparticle Cocktail (provided in the kit and diluted according to the manufacturer's instructions) by inversion or vortex (*see* **Note 14**). Add 50 μl to each well of the Microparticle Cocktail (*see* **Note 13**). Cover with a foil plate sealer and incubate for 2 h at room temperature on a horizontal orbital microplate shaker set at 800 ± 50 rpm (*see* **Note 15**).
4. Wash the 96-well microplate by applying a magnetic device to the bottom of such plate and allow 1 min before removing the liquid. Fill each well with 100 μl of Wash Buffer (provided in the kit and diluted according to the manufacturer's instructions) and allow 1 min before removing the liquid again. Perform this step 3× (*see* **Note 16**).
5. Add 50 μl per well of diluted Biotin-Antibody Cocktail (provided in the kit and diluted according to manufacturer's instructions) (*see* **Note 13**). Cover with a foil plate sealer and incubate for 1 h at room temperature on a shaker set at 800 ± 50 rpm (*see* **Note 15**).
6. Repeat once the wash step (*see* **Note 16**).
7. Add 50 μl to each well of diluted Streptavidin-PE (Streptavidin–phycoerythrin provided in the kit and diluted according to the manufacturer's instructions) (*see* **Note 14**). Cover with a foil plate sealer and incubate for 30 min at room temperature on a shaker set at 800 ± 50 rpm (*see* **Note 15**).
8. Repeat the wash step (*see* **Note 16**).
9. Proceed with the decontamination process of the Luminex plate for removal of the BSL3 laboratory. Bring into the cabinet a brand-new plastic black 96-well plate and label it specifying the assay name and date (*see* **Note 17**).
10. Remove the Luminex plate from the magnetic device (after the final wash step) and resuspend the beads by adding 150 μl per well of 1% Formaldehyde in PBS. Mix content immediately by pipetting up and down 4 to 6 times to prevent bead agglutination.
11. Transfer the resuspended beads to the brand-new labelled 96-well plate keeping the same well position. Use different tips for each row/column and keep the same order as the initial plate to avoid confusions.

12. Cover the plate with a plate sealer or film and aluminum foil. Incubate overnight inside the biosafety cabinet in the dark at room temperature.
13. After overnight decontamination, wipe the external surface of the plate with 70% ethanol and place it inside a sealable plastic bag.
14. Follow approved institutional biosafety standard procedures for removal of the plate from the BSL3 laboratory and bring it to the BSL2 laboratory.
15. Proceed with washing process of the plate and data acquisition. Wipe the external surface of the 96-well plate again. Bring the Washing Buffer (provided in the Luminex kit) to room temperature.
16. Wash the plate adding 200 μl per well of Washing Buffer and using the magnetic device as previously described. Remove and transfer the liquid to a container with 1% bleach solution.
17. Repeat the wash step once.
18. Resuspend the microparticles by adding 125 μl per well of Luminex 1× sheath fluid or appropriate buffer and volume described in the commercial kit protocol. Mix the beads by pipetting up and down 5 to 6 times, using different tips for each column.
19. Read using a Luminex 100/200 System (*see* **Notes 18** and **19**).

3.8 Type I IFN-α Assay

The determination of IFN-α (Type I) is performed using a Mouse IFN-α (Type I) ELISA assay, which is summarized below according to the manufacturer's instructions (*see* **Note 2**).

1. Prepare the reagents and standards provided in the kit following the manufacturer's instructions (*see* **Notes 20–22**). Use the supernatant from homogenized tissues (nasal turbinate, lung, brain and spleen) to prepare a dilution 1:2 using the Sample Buffer provided in the kit (*see* **Notes 23** and **24**).
2. Determine the number of microplate strips required to test the desired number of samples, besides the number of wells needed to run blanks and standards (*see* **Note 25**). Remove extra microtiter strips from the frame, seal in the foil bag provided and store at 2–8 °C. Unused strips can be used in later assays.
3. Add 100 μl per well of standard, diluted samples or blank to the precoated microtiter plate (provided in the kit).
4. Add 50 μl per well of diluted Antibody Solution (provided in the kit and diluted according to the manufacturer's instructions) completing a total volume of 150 μl per well, changing pipette tips between each addition.

5. Cover with a plate sealer (provided in the kit) and incubate for 1 h at room temperature (22–25 °C) on a shaker at 450 rpm (*see* **Note 26**).
6. Transfer the plate to 4 °C and incubate 20–24 h without shaking.
7. After 20–24 h, empty the contents of the plate by inverting and shaking over a sink and blotting the plate on lint-free absorbent paper; tap the plate. Wash the wells four times by adding a minimum of 300 μl of diluted Wash Solution (provided in the kit and diluted according to the manufacturer's instructions) for each wash step.
8. Add 100 μl per well of diluted horseradish peroxidase (HRP) Solution (provided in the kit and diluted according to the manufacturer's instructions). Cover with a plate sealer (provided in the kit) and incubate for 2 h at room temperature (22–25 °C) on a shaker at 450 rpm (*see* **Notes 26** and **27**).
9. Repeat the wash step once.
10. Add 100 μl per well of the Tetramethyl-benzidine (TMB) Substrate Solution (provided in the kit). Incubate 15 min in the dark at room temperature (22–25 °C). Do not use a plate sealer during the incubation. Do not shake.
11. Without emptying the wells and without washing them, add 100 μl per well of Stop Solution (provided in the kit).
12. Read the plate absorbance at 450 nm within 5 min using a microplate reader.
13. The interferon titer in the samples can be determined by plotting obtained optical densities (ODs) using a 4-parameter fit for the standard curve. Blank ODs should be subtracted from standards and sample ODs to eliminate the background.

3.9 Type III IFN-λ Assay

The determination of IFN-λ (Type III) is performed using a Mouse IFN-λ (Type III) ELISA assay, which is summarized below according to the manufacturer's instructions (*see* **Notes 2** and **28**).

1. Coat a 96-well microplate with 100 μl per well of the Capture Antibody (provided in the kit and diluted according to the manufacturer). Seal the plate and incubate overnight at room temperature (22–25 °C) (*see* **Notes 20, 21**, **24**, **25**, **29** and **30**).
2. Aspirate each well and wash with Wash Buffer by adding 400 μl per well. Repeat this process 2× to complete three total washes. After the last wash, remove any remaining Wash Buffer by aspirating or by inverting the plate and shaking over a sink and blotting the plate on lint-free absorbent paper; tap the plate (*see* **Note 31**).

3. Block the plate by adding 300 μl per well of Reagent Diluent. Incubate at room temperature (22–25 °C) for a minimum of 1 h (*see* **Note 30**).
4. Repeat the wash step once (*see* **Note 31**).
5. Prepare the standards provided in the kit following the manufacturer's instructions (*see* **Note 22**). Use the supernatant from homogenized tissues (nasal turbinate, lung, brain and spleen) to prepare a dilution 1:2 using the Reagent Diluent recommended (*see* **Note 32**).
6. Add 100 μl per well of sample or standards diluted in Reagent Diluent. Cover with a plate sealer and incubate 2 h at room temperature (22–25 °C).
7. Repeat the wash step once (*see* **Note 31**).
8. Add 100 μl per well of the Detection Antibody (provided in the kit and diluted according to the instructions of the manufacturer). Cover with a new plate sealer and incubate 2 h at room temperature (22–25 °C).
9. Repeat the wash step once (*see* **Note 31**).
10. Add 100 μl per well of the streptavidin–horseradish peroxidase (HRP, provided in the kit and diluted according to the manufacturer's instructions). Cover the plate with a new plate sealer and incubate for 20 min at room temperature (22–25 °C). Avoid placing the plate in direct light.
11. Repeat the wash step once (*see* **Note 31**).
12. Add 100 μl per well of Substrate Solution to each well. Incubate for 20 min at room temperature (22–25 °C). Avoid placing the plate in direct light.
13. Add 50 μl of Stop Solution to each well. Gently tap the plate to ensure thorough mixing.
14. Determine ODs of each well immediately, using a microplate reader set to 450 nm. If wavelength correction is available, set to 540 nm or 570 nm. If wavelength correction is unavailable, subtract readings at 540 nm or 570 nm from the readings at 450 nm (*see* **Note 33**).
15. Average the duplicate readings for each standard, control, and sample and subtract the average zero standard OD. Create a standard curve by reducing the data using computer software capable of generating a four-parameter logistic (4-PL) curve-fit. If samples have been diluted, the concentration read from the standard curve must be multiplied by the dilution factor.

4 Notes

1. The volume of the 10% NBF solution must be 15–20 times the volume of tissue. For inactivation, tissues must be incubated with NBF solution for 7 days before removing from the BSL3 facility [27].
2. The detection of IFN-α (Type I) and IFN-λ (Type III) is performed using separated commercial kits because such analytes are not part of the Luminex custom multiplex panel.
3. -The use of high quality BSA for the Reagent Diluent is critical for the optimum performance of the kit. Impurities such as proteases, binding proteins, soluble receptors or other interfering substances can be found to varying degrees in virtually all BSA preparations and can inhibit or interfere with the detection of certain analytes. If the standard curve appears suppressed, consider evaluating a different preparation of BSA.
4. Euthanasia using carbon dioxide (CO_2) overdose is not recommended because it induces artifact lung pathological changes such as pulmonary congestions and hemorrhage, and perivascular and peribronchiolar edema [28, 29].
5. To avoid contamination of samples, clean and disinfect the dissecting tools between tissues and between animals by submerging the tools in the 250 ml beaker with 70% ethanol. When removed from the beaker, let them air-dry each time before use.
6. Examples of calculation of NBF volume per sample dimensions/volume for rodent tissues: Example 1: 1 cm (length) × 1 cm (width) × 0.6 cm (thickness) = 0.6 cm^3 = 0.6 ml (tissue volume) × 15 (volume of NBF per volume of tissue) = 9 ml of 10% NBF solution. Recommended to use a 15 ml conical tube and fill it up to 15 ml of 10% NBF to assure the tissue is fully submerged. Example 2 (maximum tissue volume allowed): 2.5 cm (length) × 2.5 cm (width) × 0.6 cm (thickness) = 3.75 cm^3 = 3.75 ml (tissue volume) × 15 (volume of NBF per volume of tissue) = 57 ml of 10% NBF solution.
7. According to the literature, 4% formaldehyde solution penetrates at a rate of 2.4 mm/24 h. Given a sample thickness of 6 mm, complete penetration by the fixating agent will be obtained 6 mm/2.4 mm per 24 h = 2.5 days [30].
8. Do not allow slides to dry at any time during this procedure as this can lead to inconsistent staining.
9. The oven temperature must not exceed 60 °C.

10. Stop when the solution starts boiling and use care with hot solutions.
11. Bring all reagents to room temperature before use. When mixing or reconstituting protein solutions, always avoid foaming.
12. The kit provides a plate layout to record standards and samples assayed.
13. To avoid cross-contamination, change pipette tips between additions of each standard level, between sample additions and between reagent additions. Also, use separate reservoirs for each reagent.
14. Protect microparticles and Streptavidin-PE from light at all times to prevent photobleaching.
15. To ensure accurate results, proper adhesion of plate sealers during incubation steps is necessary.
16. Complete removal of liquid during washing steps is essential for good performance. Do NOT blot; this may cause a loss of microparticles).
17. Before proceeding with this step, decontaminate the whole surface of the biosafety cabinet with 2% Wex-Cide solution and 70% ethanol, or with the approved disinfectants available in your BSL3 facility.
18. Adjust the probe height setting on the analyzer to avoid puncturing the plate as well as other parameters such as sample volume, bead type, microparticle region, and count/region, according to the manufacturer's instructions. Calibrate the analyzer using the proper reagents for superparamagnetic microparticles (refer to instrument manual).
19. If samples have been diluted, the concentration read from the standard curve must be multiplied by the dilution factor. If samples generate values higher than the highest standard, further dilute the samples with calibrator diluent and repeat the assay.
20. For retention of full activity, all reagents should be kept at 2–8 °C in the dark.
21. Deionized or distilled H_2O should be used for the preparation of all reagents.
22. All dilutions should be made with polypropylene tubes and pipette tips. Pipette tips should be changed between each dilution tube.
23. It is important to run the standard curve in the same matrix as your samples unless you have demonstrated that the matrix does not affect the signal.
24. The kit provides a plate layout to record standards and samples assayed.

25. It is recommended to run standards, blanks, and samples in duplicate. At least two control wells (wells with Sample Buffer only) should be used for each assay; these control values should be subtracted from all readings prior to any calculations or plots of the data.
26. All incubations should be performed in a closed chamber at room temperature (22–25 °C) keeping the plate away from drafts and other temperature fluctuations.
27. Prepare the HRP Solution within 15 min prior to use.
28. This kit detects both Lambda 2 and 3 with similar reactivity.
29. Bring all reagents to room temperature (22–25 °C), before use. Avoid microbial contamination of reagents and buffers because it may interfere with the sensitivity of the assay. Buffers containing a large quantity of protein should be made under sterile conditions and stored at 2–8 °C or be prepared fresh daily. The type of enzyme, substrate, and the concentrations of capture/detection antibodies used can be varied to create an immunoassay with different sensitivity and dynamic range.
30. Use a fresh reagent reservoir and new pipette tips for each step.
31. Complete removal of liquid at each step is essential for good performance.
32. It is important that the diluents selected for reconstitution and for dilution of the standard reflect the environment of the samples being measured. The diluent suggested in this protocol should be suitable for most cell culture supernatant samples. Validate diluents for specific sample types prior to use.
33. The reading subtraction will correct for optical imperfections in the plate. Readings made directly at 450 nm without correction may be higher and less accurate.

Acknowledgments

We thank Texas Biomedical Research Institute Pathology Laboratory members, specially Dolores Renee Escalona and Drs. Olga Gonzalez, Anna Allue Guardia, and Shalini Gautam, for their support providing detailed information about the discussed protocols. Jun-Gyu Park and Paula A. Pino contributed equally to this study.

References

1. Rokni M, Ghasemi V, Tavakoli Z (2020) Immune responses and pathogenesis of SARS-CoV-2 during an outbreak in Iran: comparison with SARS and MERS. Rev Med Virol 30(3): e2107. https://doi.org/10.1002/rmv.2107
2. Yang X, Yu Y, Xu J, Shu H, Xia J, Liu H, Wu Y, Zhang L, Yu Z, Fang M, Yu T, Wang Y, Pan S, Zou X, Yuan S, Shang Y (2020) Clinical course and outcomes of critically ill patients with SARS-CoV-2 pneumonia in Wuhan, China: a

single-centered, retrospective, observational study. Lancet Respir Med 8(5):475–481. https://doi.org/10.1016/S2213-2600(20)30079-5

3. Glass WG, Subbarao K, Murphy B, Murphy PM (2004) Mechanisms of host defense following severe acute respiratory syndrome-coronavirus (SARS-CoV) pulmonary infection of mice. J Immunol 173(6):4030–4039. https://doi.org/10.4049/jimmunol.173.6.4030
4. Subbarao K, McAuliffe J, Vogel L, Fahle G, Fischer S, Tatti K, Packard M, Shieh WJ, Zaki S, Murphy B (2004) Prior infection and passive transfer of neutralizing antibody prevent replication of severe acute respiratory syndrome coronavirus in the respiratory tract of mice. J Virol 78(7):3572–3577. https://doi.org/10.1128/jvi.78.7.3572-3577.2004
5. Roberts A, Paddock C, Vogel L, Butler E, Zaki S, Subbarao K (2005) Aged BALB/c mice as a model for increased severity of severe acute respiratory syndrome in elderly humans. J Virol 79(9):5833–5838. https://doi.org/10.1128/JVI.79.9.5833-5838.2005
6. Chen J, Lau YF, Lamirande EW, Paddock CD, Bartlett JH, Zaki SR, Subbarao K (2010) Cellular immune responses to severe acute respiratory syndrome coronavirus (SARS-CoV) infection in senescent BALB/c mice: CD4+ T cells are important in control of SARS-CoV infection. J Virol 84(3):1289–1301. https://doi.org/10.1128/JVI.01281-09
7. Roberts A, Vogel L, Guarner J, Hayes N, Murphy B, Zaki S, Subbarao K (2005) Severe acute respiratory syndrome coronavirus infection of golden Syrian hamsters. J Virol 79(1):503–511. https://doi.org/10.1128/JVI.79.1.503-511.2005
8. McCray PB Jr, Pewe L, Wohlford-Lenane C, Hickey M, Manzel L, Shi L, Netland J, Jia HP, Halabi C, Sigmund CD, Meyerholz DK, Kirby P, Look DC, Perlman S (2007) Lethal infection of K18-hACE2 mice infected with severe acute respiratory syndrome coronavirus. J Virol 81(2):813–821. https://doi.org/10.1128/JVI.02012-06
9. Tseng CT, Huang C, Newman P, Wang N, Narayanan K, Watts DM, Makino S, Packard MM, Zaki SR, Chan TS, Peters CJ (2007) Severe acute respiratory syndrome coronavirus infection of mice transgenic for the human angiotensin-converting enzyme 2 virus receptor. J Virol 81(3):1162–1173. https://doi.org/10.1128/JVI.01702-06
10. van den Brand JM, Haagmans BL, Leijten L, van Riel D, Martina BE, Osterhaus AD, Kuiken T (2008) Pathology of experimental SARS coronavirus infection in cats and ferrets. Vet Pathol 45(4):551–562. https://doi.org/10.1354/vp.45-4-551
11. Chu YK, Ali GD, Jia F, Li Q, Kelvin D, Couch RC, Harrod KS, Hutt JA, Cameron C, Weiss SR, Jonsson CB (2008) The SARS-CoV ferret model in an infection-challenge study. Virology 374(1):151–163. https://doi.org/10.1016/j.virol.2007.12.032
12. Roberts A, Deming D, Paddock CD, Cheng A, Yount B, Vogel L, Herman BD, Sheahan T, Heise M, Genrich GL, Zaki SR, Baric R, Subbarao K (2007) A mouse-adapted SARS-coronavirus causes disease and mortality in BALB/c mice. PLoS Pathog 3(1):e5. https://doi.org/10.1371/journal.ppat.0030005
13. Gu H, Chen Q, Yang G, He L, Fan H, Deng YQ, Wang Y, Teng Y, Zhao Z, Cui Y, Li Y, Li XF, Li J, Zhang NN, Yang X, Chen S, Guo Y, Zhao G, Wang X, Luo DY, Wang H, Yang X, Li Y, Han G, He Y, Zhou X, Geng S, Sheng X, Jiang S, Sun S, Qin CF, Zhou Y (2020) Adaptation of SARS-CoV-2 in BALB/c mice for testing vaccine efficacy. Science 369(6511):1603–1607. https://doi.org/10.1126/science.abc4730
14. Oladunni FS, Park JG, Pino PA, Gonzalez O, Akhter A, Allue-Guardia A, Olmo-Fontanez A, Gautam S, Garcia-Vilanova A, Ye C, Chiem K, Headley C, Dwivedi V, Parodi LM, Alfson KJ, Staples HM, Schami A, Garcia JI, Whigham A, Platt RN 2nd, Gazi M, Martinez J, Chuba C, Earley S, Rodriguez OH, Mdaki SD, Kavelish KN, Escalona R, Hallam CRA, Christie C, Patterson JL, Anderson TJC, Carrion R Jr, Dick EJ Jr, Hall-Ursone S, Schlesinger LS, Alvarez X, Kaushal D, Giavedoni LD, Turner J, Martinez-Sobrido L, Torrelles JB (2020) Lethality of SARS-CoV-2 infection in K18 human angiotensin-converting enzyme 2 transgenic mice. Nat Commun 11(1):6122. https://doi.org/10.1038/s41467-020-19891-7
15. Moreau GB, Burgess SL, Sturek JM, Donlan AN, Petri WA, Mann BJ (2020) Evaluation of K18-hACE2 mice as a model of SARS-CoV-2 infection. Am J Trop Med Hyg 103(3):1215–1219. https://doi.org/10.4269/ajtmh.20-0762
16. Jiang RD, Liu MQ, Chen Y, Shan C, Zhou YW, Shen XR, Li Q, Zhang L, Zhu Y, Si HR, Wang Q, Min J, Wang X, Zhang W, Li B, Zhang HJ, Baric RS, Zhou P, Yang XL, Shi ZL (2020) Pathogenesis of SARS-CoV-2 in transgenic mice expressing human angiotensin-converting enzyme 2. Cell 182(1):50–58.e8. https://doi.org/10.1016/j.cell.2020.05.027

17. Sun SH, Chen Q, Gu HJ, Yang G, Wang YX, Huang XY, Liu SS, Zhang NN, Li XF, Xiong R, Guo Y, Deng YQ, Huang WJ, Liu Q, Liu QM, Shen YL, Zhou Y, Yang X, Zhao TY, Fan CF, Zhou YS, Qin CF, Wang YC (2020) A mouse model of SARS-CoV-2 infection and pathogenesis. Cell Host Microbe 28(1):124–133. e4. https://doi.org/10.1016/j.chom.2020.05.020

18. Chandrashekar A, Liu J, Martinot AJ, McMahan K, Mercado NB, Peter L, Tostanoski LH, Yu J, Maliga Z, Nekorchuk M, Busman-Sahay K, Terry M, Wrijil LM, Ducat S, Martinez DR, Atyeo C, Fischinger S, Burke JS, Slein MD, Pessaint L, Van Ry A, Greenhouse J, Taylor T, Blade K, Cook A, Finneyfrock B, Brown R, Teow E, Velasco J, Zahn R, Wegmann F, Abbink P, Bondzie EA, Dagotto G, Gebre MS, He X, Jacob-Dolan C, Kordana N, Li Z, Lifton MA, Mahrokhian SH, Maxfield LF, Nityanandam R, Nkolola JP, Schmidt AG, Miller AD, Baric RS, Alter G, Sorger PK, Estes JD, Andersen H, Lewis MG, Barouch DH (2020) SARS-CoV-2 infection protects against rechallenge in rhesus macaques. Science 369(6505):812–817. https://doi.org/10.1126/science.abc4776

19. Singh DK, Singh B, Ganatra SR, Gazi M, Cole J, Thippeshappa R, Alfson KJ, Clemmons E, Gonzalez O, Escobedo R, Lee TH, Chatterjee A, Goez-Gazi Y, Sharan R, Gough M, Alvarez C, Blakley A, Ferdin J, Bartley C, Staples H, Parodi L, Callery J, Mannino A, Klaffke B, Escareno P, Platt RN 2nd, Hodara V, Scordo J, Gautam S, Vilanova AG, Olmo-Fontanez A, Schami A, Oyejide A, Ajithdoss DK, Copin R, Baum A, Kyratsous C, Alvarez X, Ahmed M, Rosa B, Goodroe A, Dutton J, Hall-Ursone S, Frost PA, Voges AK, Ross CN, Sayers K, Chen C, Hallam C, Khader SA, Mitreva M, Anderson TJC, Martinez-Sobrido L, Patterson JL, Turner J, Torrelles JB, Dick EJ Jr, Brasky K, Schlesinger LS, Giavedoni LD, Carrion R Jr, Kaushal D (2021) Author correction: responses to acute infection with SARS-CoV-2 in the lungs of rhesus macaques, baboons and marmosets. Nat Microbiol 6(3):413. https://doi.org/10.1038/s41564-021-00867-2

20. Yu J, Tostanoski LH, Peter L, Mercado NB, McMahan K, Mahrokhian SH, Nkolola JP, Liu J, Li Z, Chandrashekar A, Martinez DR, Loos C, Atyeo C, Fischinger S, Burke JS, Slein MD, Chen Y, Zuiani A, Lelis FJN, Travers M, Habibi S, Pessaint L, Van Ry A, Blade K, Brown R, Cook A, Finneyfrock B, Dodson A, Teow E, Velasco J, Zahn R, Wegmann F, Bondzie EA, Dagotto G, Gebre MS, He X, Jacob-Dolan C, Kirilova M, Kordana N, Lin Z, Maxfield LF, Nampanya F, Nityanandam R, Ventura JD, Wan H, Cai Y, Chen B, Schmidt AG, Wesemann DR, Baric RS, Alter G, Andersen H, Lewis MG, Barouch DH (2020) DNA vaccine protection against SARS-CoV-2 in rhesus macaques. Science 369(6505): 806–811. https://doi.org/10.1126/science.abc6284

21. Rathnasinghe R, Strohmeier S, Amanat F, Gillespie VL, Krammer F, Garcia-Sastre A, Coughlan L, Schotsaert M, Uccellini M (2020) Comparison of transgenic and adenovirus hACE2 mouse models for SARS-CoV-2 infection. bioRxiv. https://doi.org/10.1101/2020.07.06.190066

22. Bao L, Deng W, Huang B, Gao H, Liu J, Ren L, Wei Q, Yu P, Xu Y, Qi F, Qu Y, Li F, Lv Q, Wang W, Xue J, Gong S, Liu M, Wang G, Wang S, Song Z, Zhao L, Liu P, Zhao L, Ye F, Wang H, Zhou W, Zhu N, Zhen W, Yu H, Zhang X, Guo L, Chen L, Wang C, Wang Y, Wang X, Xiao Y, Sun Q, Liu H, Zhu F, Ma C, Yan L, Yang M, Han J, Xu W, Tan W, Peng X, Jin Q, Wu G, Qin C (2020) The pathogenicity of SARS-CoV-2 in hACE2 transgenic mice. Nature 583(7818):830–833. https://doi.org/10.1038/s41586-020-2312-y

23. Zheng J, Wong LR, Li K, Verma AK, Ortiz ME, Wohlford-Lenane C, Leidinger MR, Knudson CM, Meyerholz DK, McCray PB Jr, Perlman S (2021) COVID-19 treatments and pathogenesis including anosmia in K18-hACE2 mice. Nature 589(7843): 603–607. https://doi.org/10.1038/s41586-020-2943-z

24. Golden JW, Cline CR, Zeng X, Garrison AR, Carey BD, Mucker EM, White LE, Shamblin JD, Brocato RL, Liu J, Babka AM, Rauch HB, Smith JM, Hollidge BS, Fitzpatrick C, Badger CV, Hooper JW (2020) Human angiotensin-converting enzyme 2 transgenic mice infected with SARS-CoV-2 develop severe and fatal respiratory disease. JCI Insight 5(19): e142032. https://doi.org/10.1172/jci.insight.142032

25. Winkler ES, Bailey AL, Kafai NM, Nair S, McCune BT, Yu J, Fox JM, Chen RE, Earnest JT, Keeler SP, Ritter JH, Kang LI, Dort S, Robichaud A, Head R, Holtzman MJ, Diamond MS (2020) Publisher correction: SARS-CoV-2 infection of human ACE2-transgenic mice causes severe lung inflammation and impaired function. Nat Immunol 21(11):1470. https://doi.org/10.1038/s41590-020-0794-2

26. Park JG, Oladunni FS, Chiem K, Ye C, Pipenbrink M, Moran T, Walter MR, Kobie J, Martinez-Sobrido L (2020) Rapid in vitro

assays for screening neutralizing antibodies and antivirals against SARS-CoV-2. J Virol Methods 287:113995. https://doi.org/10.1016/j.jviromet.2020.113995
27. Alfson KJ, Griffiths A (2018) Development and testing of a method for validating chemical inactivation of Ebola virus. Viruses 10(3):126. https://doi.org/10.3390/v10030126
28. Boivin GP, Bottomley MA, Schiml PA, Goss L, Grobe N (2017) Physiologic, Behavioral, and histologic responses to various euthanasia methods in C57BL/6NTac male mice. J Am Assoc Lab Anim Sci 56(1):69–78
29. Fawell JK, Thomson C, Cooke L (1972) Respiratory artefact produced by carbon dioxide and pentobarbitone sodium euthanasia in rats. Lab Anim 6(3):321–326. https://doi.org/10.1258/002367772781006185
30. Start RD, Layton CM, Cross SS, Smith JH (1992) Reassessment of the rate of fixative diffusion. J Clin Pathol 45(12):1120–1121. https://doi.org/10.1136/jcp.45.12.1120

Chapter 17

Immunohistochemical Detection of SARS-CoV-2 Antigens by Single and Multiple Immunohistochemistry

Silvia Lonardi, Mattia Bugatti, Arianna Valzelli, and Fabio Facchetti

Abstract

Severe acute respiratory syndrome coronavirus 2 (SARS-CoV-2) infection can be demonstrated in tissue sections by immunohistochemistry (IHC), which has the power to localize in bright field specific antigens in cells and tissues. The use of double or triple immunostains is capable of highlighting which cells are infected and/or the relationship of infected cell with other cells and tissue structures. In addition, immunoenzymatic multi-staining permits the simultaneous identification, localization, and enumeration of different cellular epitopes. Moreover, this method improves analytical precision, decreasing the time required for morphometric quantification, maximizing the information obtained from a single slide of paraffin-embedded tissue.

Key words SARS-CoV-2, Immunohistochemistry Double/Triple immunohistochemistry, Antibodies

1 Introduction

1.1 Detection of SARS-CoV-2 by Immunohistochemistry (IHC)

Immunohistochemistry (IHC) can be applied to confirm the diagnosis of viral or bacterial infections difficult to diagnose with certainty using H&E or histochemical staining alone. A plethora of antibodies specific for viruses or bacteria were described for skin pathology [1–3]. The most widely used antibodies against pathogens in surgical pathology are Epstein–Barr virus (LMP-1, EBNA-1), Herpes viruses (HSV-1/2 and HHV-8), Cytomegalovirus (CMV), Hepatitis B Virus (HBsAg and HBcAg), Polyomaviruses (SV40 and MCPyV), *Treponema pallidum*, *Toxoplasma gondii*, and *Bartonella henselae*.

During the COVID-19 pandemic, several papers have been published where SARS-CoV-2 was visualized on formalin-fixed, paraffin-embedded (FFPE) tissues using IHC and/or in situ

Silvia Lonardi and Mattia Bugatti contributed equally to this work.

Justin Jang Hann Chu et al. (eds.), *SARS-CoV-2: Methods and Protocols*,
Methods in Molecular Biology, vol. 2452, https://doi.org/10.1007/978-1-0716-2111-0_17,

hybridization (ISH) (reviewed in Schwartz et al. [4], and first described by [5–12]). Sensitivity of IHC and ISH are similar [13], but IHC has the advantage of lower costs, enabling it to be usable for screening of large cohort of cases. Moreover, IHC is more suitable to identify in addition to SARS-CoV-2 antigens, and other cell markers, thus adding useful information on cell distribution and interaction in inflamed tissues. For example, Schaefer et al. used double IHC on lung tissues providing evidence of the occurrence of Nucleocapsid (N) antigen in TTF-1-positive pneumocytes [14]. In the study by Facchetti et al., we analyzed the relationship between SARS-CoV-2 Spike (S) or N protein infected placental tissue and their relationship with several types of inflammatory cells, showing that N antigen, in addition to syncytiotrophoblast, also occurs in maternal inflammatory cells, as well as in villous Hofbauer cells [15].

1.2 Immunohistochemical Detection of SARS-CoV-2 Antigens in FFPE

IHC often requires antigen retrieval by pretreating the tissue sections to unmask antigens hidden by formalin fixation. This step can be done by different procedures, such as microwaving the tissue in appropriate buffers (heat-induced epitope retrieval, HIER) or by the use of proteolytic enzymes (e.g., pepsin). SARS-CoV-2 antigen retrieval requires HIER using a microwave oven or water bath (as detailed in Subheading 3.3). Different antibodies have been applied to detect SARS-CoV-2 antigens (listed in Table 1). We used the antibodies raised in rabbit, the monoclonal antibody anti-S (clone 007, cat. Num. 40150-R007) and polyclonal antibody anti-N (cat. Num. 40588-T62) from Sino Biological. We have also obtained very bright stains of the Spike protein using a in-house mouse monoclonal antibody (clone 14b-5) provided by Dr. Marina Cella, Washington University, St. Louis, MO, USA. As a secondary step procedure, the EnVision System horseradish peroxidase (HRP) labeled polymer Anti-Rabbit (Agilent Technologies) was used to reveal the Sino Biological antibodies, or the Novolink polymer detection system (Leica Microsystems) for the anti-Spike clone 14b-5. The reaction was finally developed using 3,3′ diaminobenzidine (DAB) as chromogen.

1.3 Double and Triple IHC

The immunohistological demonstration of multiple antigens in a single tissue section has many applications and is commonly used to elucidate the distribution and eventual coexpression of different antigens in the same cell, the topographic relationship of antigenically defined cell populations, and to correlate phenotypic information with functional or prognostic markers [16].

The basic principles for multiple antigen staining are identical to those for single staining, with the additional precaution to avoid cross-reaction (crosstalk) between the different staining sequences. The crosstalk is not expected when the primary antibodies originate from different species and the secondary antibodies are species-

Table 1
Antibodies shown to identify SARS-CoV/SARS-CoV-2 antigens or SARS-CoV-2 antigens in immunohistochemistry

SARS-CoV/SARS-CoV-2 or SARS-CoV-2 * Spike (S) and Nucleocapsid (N)	Species/ clone (&)	Source	Catalog number	References
S*	R/007	Sino Biological (Beijing, China)	40150-R007	[4, 15]
S	R/P	Sino Biological	40150-T62-COV2	[11]
S	M/1A9	GeneTex (Irvine, CA, USA)	GTX632604	[4, 6, 9, 10]
N	R/P	Thermo Scientific (Waltham, MA, USA)	PA1-41098	[4, 6, 8, 10]
N*	R/P	Sino Biological	40588-T62	[4, 15]
N	R/P	Novus Biologicals (Littleton, CO, USA)	NB100-56576	[5, 7, 13, 14]
N*	R/P	Abclonal (Woburn, MA, USA)	A20021	[4]
N	M/B46F	Thermo Scientific	MA1-7404	[12]
N*	M/1C7	Bioss Antibodies (Woburn, MA, USA)	BSM-41411M	[12]
N	M/8G8A	Bioss Antibodies	BSM-49131M	[12]
N	M/4B21	Creative Diagnostics (NY, USA)	MA1-7404 SARS	[10]
N	M/6F10	BioVision (Milpitas, CA, USA)	A2060	[10]
N	M/05	Sino Biological	40143-MM05	[4, 11]

&: M = mouse, R = rabbit, P = polyclonal

specific and developed using distinct chromogens [17]. However, nowadays most of the commercially available primary antibodies are IgG1 raised in mouse or rabbit, the double immune stain must be performed "consecutively" [17, 18] and it is mandatory to avoid "the cross-talk" between the two immune reactions. For this purpose, different procedures have been proposed, such as to apply preformed complexes of the primary and secondary enzyme-conjugated antibodies [19, 20], to use a secondary polyclonal monovalent F(ab) antibody in the first step to block the epitopes on the primary antibody, thus preventing access of successive antibodies [21], or to employ commercially available blockers between the two immune reactions [22].

In our experience, most successful results are obtained with (a) the "chromogen sheltering effect," consisting in the fact the

DAB deposition at the end of the first reaction acts as a shelter of the previously applied antibodies, rendering them unavailable for the subsequent immune reagents [17, 23, 24] and (b) the "immunoglobulins denaturation by antigen retrieval procedure," a practical, and affordable approach to denature residual immunoglobulins stitched to the tissue (antigen-antibody complex) from the first IHC run, obtained by subsequent heating the sections in a microwave oven [22, 25, 26]. This process first causes the denaturation of the F(ab) region and then of the Fc region of the previously applied antibodies, and it is even capable to strip the antibodies if the microwave heating is prolonged [27–29].

A relevant factor for a successful double or triple IHC consists in the careful selection of the chromogenic sequence, which should be made according to the cellular distribution of the antigens. Optimal multistaining results are achieved when the two or three antigens localize in different cellular compartments of the same cell (e.g., nuclear, cytoplasmic, and membrane) or in different cells. Nevertheless, DAB should preferentially be used first, especially if the antigen to be detected is located in the nucleus or if it is the weaker and/or the less diffusely expressed of the two/three antigens to be tested. When DAB labels a cytoplasmic antigen, it may interfere with the other antigen/s if located in close proximity to them, either in the cytoplasm or on the cell membrane [22]. In addition, the antigens which are more diffusely expressed in the sample and/or the antibodies having higher sensitivity, should be revealed as second or third in double and triple IHC.

According to our experience, the preferable double chromogen stain procedure includes DAB as the first step (brown stain) and Ferangi Blue (Biocare Medical) or Vector Blue (Vector Laboratories) as the second step (blue stain). In the case of triple IHC ImmPACT Vector Red (Vector Laboratories), chromogen is used as the third step (magenta). Light nuclear counterstain is performed using diluted Mayer's hematoxylin or Nuclear Fast Red.

If two or even three antigens have been located in the same cell compartment, in order to avoid steric hindrance due to the previous chromogen deposit, it is possible to apply "sequential immunostainings" using a nonpermanent chromogen (such as 3-Ammino-9-etylcarbazole, AEC), which after digitalization of the first immunostain can be removed using 95% ethanol, followed by the second immune reaction and so on [30–32]. At the end of the procedure, the serially digitalized images can be overlapped using image processing software [31, 33]. It has also been proposed to elute, after the chromogen removal, high affinity antibodies ($>10^{-9}$ KD) using specific solutions (e.g., containing 2-ME/SDS) [31, 32, 34].

2 Materials

2.1 Common Solutions

- Distilled water.
- Xylene.
- Ethanol 100% and 95%.
- Methanol + H_2O_2 (0.03%) solution (for endogenous peroxidase blocking); may also use a peroxidase blocking solution available from various manufacturers (the percentage of hydrogen peroxide should be below 0.1%).
- Antigen retrieval buffers (EDTA-based at different pH):
  - pH 9.0: Trizma Base 1.2 g, Ethylenediaminetetraacetic acid (EDTA) 0.366 g, 1 L distilled water.
  - pH 8.0: 0.372 g EDTA, 1 L distilled water, NaOH 1 N to pH 8.0;
- Wash buffer: Tris–HCl (6.06 g Trizma base, 7.92 g NaCl, 0.5 mL Tween 20, 1 L distilled water, 3.4 mL HCl) pH 7.4–7.6. Phosphate-buffered saline (PBS) should be avoided because phosphates act as a competitive inhibitor to AP enzymes.
- Antibody diluent solution (100 mL Tris–HCl, 5 mL NaN_3, 1 g bovine serum albumin-BSA [Fraction V, Capricorn Scientific, Ebsdorfergrund, Germany]).
- Mayer's hematoxylin (Bio-Optica, Milan, Italy), Nuclear Fast Red (Bio-Optica).
- Xylene-based mounting medium (e.g., PER-TEK, Kaltek, Padua, Italy).
- Glycerol gelatin (Sigma-Aldrich, St. Louis, MO, USA).

2.2 Common Material

- Charged slides to promote tissue retention (e.g., SuperFrost Plus slides from Bio-Optica).
- Hydrophobic pen to enclose the tissue, necessary to detain the reagents on the site of reaction (e.g., Dako pen from Agilent Technologies, Santa Clara, CA, USA).
- Humidified chambers for slides incubation.
- Coverslips to protect the stained sections.

2.3 Common Laboratory Equipment

- Oven, fridge and freezer, chemical hood, analytical balance, pH meter, microwave oven, water bath, vortex, microcentrifuge, micropipettes and caps, tubes, microscope.

2.4 Specific Reagents for SARS-CoV-2 Immunohistochemistry

- Primary antibodies: SARS-CoV-2 Spike antibody (rabbit Mab, clone 007, Sino Biological, Beijing, China) and SARS-CoV-2 Nucleocapsid antibody (polyclonal rabbit, Sino Biological), mouse anti-Spike antibody, clone 14b-5.
- Secondary antibodies: polymers against mouse and/or rabbit immunoglobulins. They are preferred to biotinylated antibodies for their increased sensitivity [35]: EnVision System-HRP Labeled polymer (Agilent Technologies) and Novolink Polymer Detection System (Leica Microsystems, Wetzlar, Germany) were HRP-linked secondary; MACH 4 Universal AP-Polymer Kit (Biocare Medical, Concord, CA, USA) is an AP-conjugated polymer.
- Chromogens: DAB (Leica Microsystems); Ferangi Blue (Biocare Medical).

3 Methods

3.1 Slide Preparation

- Cool FFPE blocks on cooling plate (from 0 °C to −10 °C).
- Cut 4 μm thick sections and mount them on charged slides (*see* **Notes 1**).
- Dry sections overnight at room temperature (RT) or in a 37 °C oven. Alternatively, dry slides at 60 °C in oven for 10–15 min before IHC stain.

3.2 Deparaffinization and Rehydration

- Immerse slides in xylene 20 min (at least 3 changes in different jars).
- Dip slides in 100% ethanol for 10 min (2 changes).
- Immerse slides in methanol + H_2O_2 (0.03%) solution 20 min.
- Dip slides in 95% ethanol for 5 min.
- Wash slides with tap water for a couple of minutes, and then in distilled water for 5 min.

After hydration, the section should never dry up to the end of the staining procedure.

3.3 Antigen Retrieval

Optimal antigen retrieval conditions for each antibody must be determined previously.

For Sino Biological antibodies, SARS-CoV-2 antigens were retrieved using the microwave oven Max setup (*see* Subheading 3.3.1) and pH 9.0 EDTA buffer (*see* Subheading 2.1). For the Spike antibody clone 14b-5, HIER was performed using thermostatic bath (*see* Subheading 3.3.2) and pH 8.0 buffer (*see* Subheading 2.1).

3.3.1 Microwave Oven

- Place slides in plastic jar with proper retrieval buffer pH 9.0 or pH 8.0. The jar must be placed in recipient filled with distilled water to avoid buffer evaporation.
- We apply different setups for microwave HIER. Max setup: carry out 2 cycles at 1000 W followed by 3 cycles at 750 W. Common setup: heat up the buffer for 5 min, dip the slides in and carry out 3 cycles at 750 W. Verify that the buffer covers the slides along the full procedure and refill if necessary.

  For the SARS-CoV-2 Sino Biological antigens we adopted the Max setup.
- Let the slides cool in the buffer (to quicken this step put the jar on ice) and then wash with wash buffer for 5 min.

3.3.2 Thermostatic Bath

- Set the thermostatic bath at 98 °C.
- Heat up the buffer for a few minutes.
- Put the slides in for 40 min.
- Let the slides cool in their buffer (to quicken this step put the jar on ice) and then wash with wash buffer for 5 min.

3.4 Primary Antibody Application

- Optimal antibody titration, incubation time and temperature for each antibody must be determined previously.
- Dilute the primary antibody with antibody diluent solution. SARS-CoV-2 S and N were diluted 1:100 (both Spike abs) and 1:3500 respectively.
- Incubate the diluted antibody for 60 min RT in humidified chamber. The amount of antibody to be applied may be reduced by surrounding the section using a hydrophobic pen.
- Rinse slides with wash buffer for 5 min (*see* **Note 2**).

3.5 Secondary Antibody Application

- SARS-CoV-2 Sino Biological antibodies were revealed applying the Envision System-HRP for 30 min (Agilent Technologies), while 14b-5 anti-Spike was revealed using the Novolink polymer (Leica Microsystems).
- Rinse slides with wash buffer for at least 5 min.

3.6 Chromogen Application

- Prepare DAB chromogen according to the datasheet specification. DAB is very stable, could be also used a few hours after preparation.
- Incubate slides with chromogen for 5 min.
- Wash in tap water followed by distilled water.
- Counterstain slides in Mayer's hematoxylin for 10 s.
- Wash slides with tap water to blueish hematoxylin.

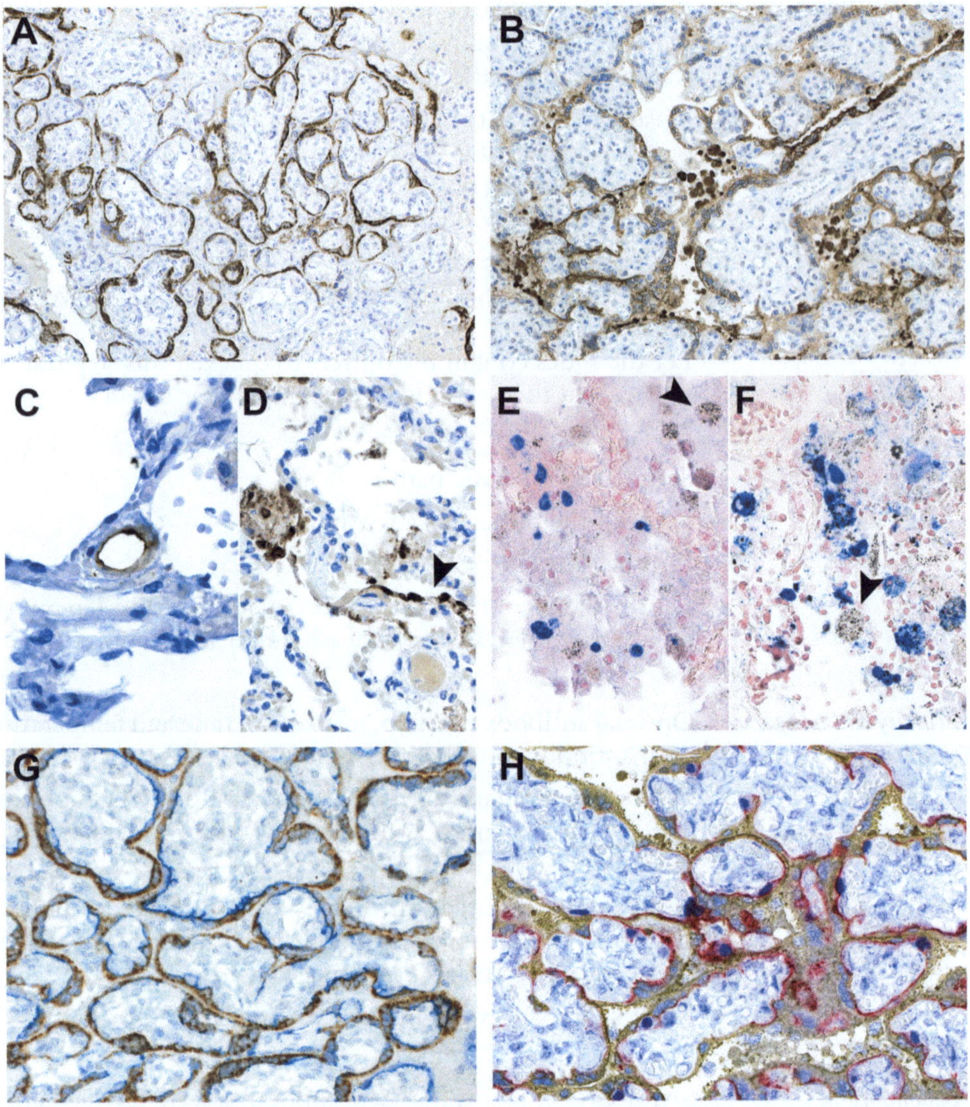

Fig. 1 FFPE placental (**a**, **b**) and autoptic lung tissue (**c**–**f**) from COVID-19 patients, single immunostained for SARS-CoV-2 Spike (**a**, **c** and **e**) and Nucleocapsid antigens (**b**, **d** and **f**) developed with DAB (brown, **a**–**d**) and Ferangi Blue (blue, **e**, **f**). The strong expression of the viral antigens is recognized along the syncytiotrophoblasts (**a**, **b**) and Nucleocapsid is also positive in intervillous macrophages (**b**). In the lung, SARS-CoV-2 antigens mark endothelial cells (Spike, **c**), pneumocytes (arrow, **d**) and alveolar macrophages (**d**–**f**). Note in E and F, several pigmented alveolar macrophages that are negative for SARS-CoV-2 antigens (arrows). FFPE placental tissue -from a COVID-19 woman- double (**g**) and triple (**h**) immunostained for SARS-CoV-2 Spike (developed using DAB, brown) coupled with E-cadherin (Ferangi Blue, blue)(**g**) and Nucleocapsid antigens (DAB, brown) with GATA-3 (Ferangi Blue, blue) and E-cadherin (ImmPACT Vector Red, magenta)(**h**). In **g**, note the sharp distinction between the two antigens, revealing the membranous expression of E-cadherin in the syncytiotrophoblasts and the cytoplasmatic positivity of the Spike antigen in the same cells. Also in **h** the three antigens are easily distinguishable. GATA-3 stains occasional syncytiotrophoblast cells; due to very strong nuclear GATA-3 expression, the light nuclear counterstaining with hematoxylin does not impair immune reactivity evaluation. Antibodies used: mouse anti-Spike antibody, clone 14b-5 and Sino Biological

- Rinse in distilled water.
- Dip slides in ethanol 95% for 5 min.
- Immerse slides in 3 subsequent washes of 100% ethanol, 5 min each.
- Immerse slides in xylene (twice, 5 min each).
- Mount slides with coverslip using xylene-based mounting medium.

3.7 Double Immunohistochemistry

Once the first step chromogen is completed let the slide in distilled water. The second antibody staining run can be performed in the same or the following day (*see* **Notes 3–8**).

3.7.1 Second Run: Antigen Retrieval

A new HIER is performed for the second antibody, using the pH and setting condition standardized for that antibody (*see* Subheading 3.3).

3.7.2 Second Run: Primary Antibody

- *See* Subheading 3.4. Antibodies used are detailed in the legend of Fig. 1.

3.7.3 Second Run: Secondary Antibody

- Apply species specific secondary antibody according to datasheet specifications. All the antibodies we used for the second run were raised in mouse or rabbit and were revealed using MACH 4 Universal AP-Polymer Kit (Biocare Medical).

3.7.4 Second Run: Chromogen

- Prepare Ferangi Blue chromogen (according to datasheet) immediately before use.
- Incubate slides with chromogen for 10 min.
- After 10 min live check the chromogenic reaction under the microscope. In case of weak staining lengthen the chromogen incubation, for a maximum of additional 10 min.
- Put slides in distilled water to stop the reaction.

3.7.5 Counterstaining and Mounting

- Counterstain the nuclei using either Mayer's hematoxylin for 5–10 s or Nuclear Fast Red for 4 min.
- Wash slides with tap water, followed by distilled water.

Fig. 1 (continued) Nucleocapsid antibody (detailed in Materials and Methods sections), GATA-3 (HIER using microwave oven, max setup, in pH 8.0 buffer, prediluted, mouse L50-823, from Cell Marque, Rocklin, CA, USA) and E-cadherin (HIER using thermostatic bath in pH 8.0 buffer, dilution 1:25, mouse antibody clone 4A2C7 from Invitrogen, Carlsbad, CA, USA). Mayer's hematoxylin was used as nuclear counterstain in **a–h** (diluted in **g** and **h**), Nuclear Fast Red in **e**, **f**. Original magnification 200× (**a**, **b**), 400× (**d–h**) 600× (**c**)

- Apply coverslip using hot glycerol gelatin as mounting medium (Ferangi Blue must to be mounted in aqueous solution to prevent fading).
- Let the slide dry and then use nail varnish to seal off the coverslip.

3.8 Triple Immunohistochemistry

Once the second step chromogen is completed let the slide in distilled water. The third antibody staining run can be performed in the same or the following day (*see* **Notes 3–8**).

3.8.1 Third Run: Antigen Retrieval

- *See* Subheadings 3.3 and 3.7.1.

3.8.2 Third Run: Primary Antibody

- *See* Subheading 3.4.

3.8.3 Third Run: Secondary Antibody

- *See* Subheading 3.3 and 3.7.3.

3.8.4 Third Run: Chromogen

- Prepare ImmPACT Vector Red chromogen (according to datasheet) immediately before use.
- Incubate slides with chromogen for 20 min.
- Live check chromogenic deposition under the microscope after 20 min. In case of weak staining lengthen chromogen incubation, for a maximum of additional 20 min.
- Put the slides in distilled water to stop reaction.

3.8.5 Counterstaining and Mounting

- *See* Subheading 3.7.5 and Fig. 1.

4 Notes

1. FFPE tissues should be properly fixed (formalin fixation time within 24 h) to guarantee a homogeneous staining of the multiple markers. The use of freshly cut sections especially for double/triple IHC is highly recommended in order to prevent antigen loss. Thick sections may increase the background noise.
2. After hydration, the section should never dry up to the end of the staining procedure. After rinsing the slides, discharge the excess of washing solution without drying the section and load immediately the following reagent.
3. In the case sections containing pigments such as melanin, anthracosis or iron deposition, blue or red chromogens are required.

4. In multiple immunostains, the less represented antigen should be revealed using the most sensitive secondary antibody (e.g., two-step enzyme-conjugated polymers) and DAB as chromogen. DAB is the ideal chromogen, due to its stability, insolubility, and sensitivity. The other chromogens are less stable and can be faded by subsequent antigen unmasking, the mounting in xylene can be challenging [22].
5. As a general rule the minimum HIER should be performed. Antibodies that lose sensitivity after repeated HIER should be applied first in multiple immunostains [22].
6. In multiple immunostains chemical antigen retrieval (e.g., protease) should be avoided.
7. Endogenous AP can produce a background noise when using AP chromogen substrates in tissues like kidney, intestine, bone (osteoblasts), lymphoid tissue and placenta. Generally, endogenous AP blocking is recommended in frozen tissue sections [36], while for FFPE it is recommended to test first the occurrence of endogenous AP through incubation with chromogen alone, and only in the case of positivity AP blocking with levamisole should be applied.
8. In multiple immunostains, whether the second and/or the third antibody result fader compared to single stains, their titration might be increased, or two subsequent fresh chromogen incubations can be applied.

Acknowledgments

SL and AV are supported by Fondazione Beretta (Brescia). Silvia Lonardi and Mattia Bugatti contributed equally to this work.

References

1. Molina-Ruiz AM, Santonja C, Rütten A et al (2015) Immunohistochemistry in the diagnosis of cutaneous viral infections--part I. Cutaneous viral infections by herpesviruses and papillomaviruses. Am J Dermatopathol 37(1):1–14. https://doi.org/10.1097/DAD.0000000000000203
2. Molina-Ruiz AM, Santonja C, Rütten A et al (2015) Immunohistochemistry in the diagnosis of cutaneous viral infections- part II: cutaneous viral infections by parvoviruses, poxviruses, paramyxoviridae, picornaviridae, retroviruses and filoviruses. Am J Dermatopathol 37(2):93–106. https://doi.org/10.1097/DAD.0000000000000200
3. Molina-Ruiz AM, Cerroni L, Kutzner H et al (2015) Immunohistochemistry in the diagnosis of cutaneous bacterial infections. Am J Dermatopathol 37(3):179–193. https://doi.org/10.1097/DAD.0000000000000227
4. Schwartz DA, Baldewijns M, Benachi A et al (2020) Chronic histiocytic intervillositis with trophoblast necrosis are risk factors associated with placental infection from coronavirus disease 2019 (COVID-19) and intrauterine maternal-fetal severe acute respiratory syndrome coronavirus 2 (SARS-CoV-2) transmission in liveborn and stillborn infants. Arch Pathol Lab Med 145(5):517–528. https://doi.org/10.5858/arpa.2020-0771-SA

5. Hecht JL, Quade B, Deshpande V et al (2020) SARS-CoV-2 can infect the placenta and is not associated with specific placental histopathology: a series of 19 placentas from COVID-19-positive mothers. Mod Pathol 33(11): 2092–2103. https://doi.org/10.1038/s41379-020-0639-4
6. Sauter JL, Baine MK, Butnor KJ et al (2020) Insights into pathogenesis of fatal COVID-19 pneumonia from histopathology with immunohistochemical and viral RNA studies. Histopathology 77(6):915–925. https://doi.org/10.1111/his.14201
7. Martines RB, Ritter JM, Matkovic E et al (2020) Pathology and pathogenesis of SARS-CoV-2 associated with fatal coronavirus disease, United States. Emerg Infect Dis 26(9): 2005–2015. https://doi.org/10.3201/eid2609.202095
8. Remmelink M, De Mendonça R, D'Haene N et al (2020) Unspecific post-mortem findings despite multiorgan viral spread in COVID-19 patients. Crit Care 24(1):495. https://doi.org/10.1186/s13054-020-03218-5
9. Colmenero I, Santonja C, Alonso-Riaño M et al (2020) SARS-CoV-2 endothelial infection causes COVID-19 chilblains: histopathological, immunohistochemical and ultrastructural study of seven paediatric cases. Br J Dermatol 183(4):729–737. https://doi.org/10.1111/bjd.19327
10. Carossino M, Ip HS, Richt JA (2020) Detection of SARS-CoV-2 by RNAscope® in situ hybridization and immunohistochemistry techniques. Arch Virol 165(10):2373–2377. https://doi.org/10.1007/s00705-020-04737-w
11. Liu J, Babka AM, Kearney BJ et al (2020) Molecular detection of SARS-CoV-2 in formalin-fixed, paraffin-embedded specimens. JCI. Insight 5(12):e139042. https://doi.org/10.1172/jci.insight.139042
12. Best Rocha A, Stroberg E, Barton LM et al (2020) Detection of SARS-CoV-2 in formalin-fixed paraffin-embedded tissue sections using commercially available reagents. Lab Investig 100(11):1485–1489. https://doi.org/10.1038/s41374-020-0464-x
13. Massoth LR, Desai N, Szabolcs A et al (2021) Comparison of RNA in situ hybridization and immunohistochemistry techniques for the detection and localization of SARS-CoV-2 in human tissues. Am J Surg Pathol 45(1):14–24. https://doi.org/10.1097/PAS.0000000000001563
14. Schaefer IM, Padera RF, Solomon IH et al (2020) In situ detection of SARS-CoV-2 in lungs and airways of patients with COVID-19. Mod Pathol 33(11):2104–2114. https://doi.org/10.1038/s41379-020-0595-z
15. Facchetti F, Bugatti M, Drera E et al (2020) SARS-CoV2 vertical transmission with adverse effects on the newborn revealed through integrated immunohistochemical, electron microscopy and molecular analyses of placenta. EBioMedicine 59:102951. https://doi.org/10.1016/j.ebiom.2020.102951
16. Krenacs T, Krenacs L (1994) Immunogoldsilver staining (IGSS) for immunoelectron microscopy and in multiple detection affinity cytochemistry. In: Gu J, Hacker GW (eds) Modern methods in analytical morphology. Plenurn, New York, pp 225–251
17. Krenacs T, Krenacs L, Raffeld M (2010) Multiple antigen immunostaining procedures. Methods Mol Biol 588:281–300. https://doi.org/10.1007/978-1-59745-324-0_28
18. Vermi W, Lonardi S, Morassi M et al (2009) Cutaneous distribution of plasmacytoid dendritic cells in lupus erythematosus. Selective tropism at the site of epithelial apoptotic damage. Immunobiology 214(9–10):877–886. https://doi.org/10.1016/j.imbio.2009.06.013
19. Krenacs T, Uda H, Tanaka S (1991) One-step double immunolabeling of mouse interdigitating reticular cells: simultaneous application of preformed complexes of monoclonal rat antibody M J -8 with horseradish peroxidase-linked anti-rat immunoglobulin and of monoclonal mouse anti Ia antibody with alkaline phosphatase coupled anti-mouse immunoglobulins. J Histochem Cytochem 39:1719–1723
20. Morris TJ, Stanley EF (2003) A simple method for immunocytochemical staining with multiple rabbit polyclonal antibodies. J Neurosci Methods 127:149–155. https://doi.org/10.1016/s0165-0270(03)00119-5
21. Negoescu A, Labat-Moleur F, Lorimier P et al (1994) F(ab) secondary antibodies: a general method for double immunolabeling with primary antisera from the same species. Efficiency control by chemiluminescence. J Histochem Cytochem 42(3):433–437. https://doi.org/10.1177/42.3.7508473
22. Osman TA, Øijordsbakken G, Costea DE et al (2013) Successful triple immunoenzymatic method employing primary antibodies from same species and same immunoglobulin subclass. Eur J Histochem 57(3):e22. https://doi.org/10.4081/ejh.2013.e22
23. Sternberger LA, Joseph SA (1979) The unlabeled antibody method. Contrasting color staining of paired pituitary hormones without antibody removal. J Histochem Cytochem 27:

1424–1429. https://doi.org/10.1177/27.11.92498

24. Krenacs T, Laszik Z, Dobo E (1989) Application of immunogold-silver staining and immunoenzymatic methods in multiple labeling of human pancreatic Langerhans islet cells. Acta Histochem 85:79–85. https://doi.org/10.1016/S0065-1281(89)80102-3
25. Lan HY, Mu W, Nikolic-Paterson DJ et al (1995) A novel, simple, reliable, and sensitive method for multiple immunoenzyme staining: use of microwave oven heating to block antibody crossreactivity and retrieve antigens. J Histochem Cytochem 43:97–102. https://doi.org/10.1177/43.1.7822770
26. Chan A, Matias MA, Farah CS (2011) A novel and practical method using HRPpolymer conjugate and microwave treatment for visualization of 2 antigens raised from the same or different species in paraffin-embedded tissues. Appl Immunohistochem Mol Morphol 19: 376–383. https://doi.org/10.1097/PAI.0b013e31820251c0
27. Vermeer AW, Norde W (2000) The thermal stability of immunoglobulin: unfolding and aggregation of a multi-domain protein. Biophys J 78(1):394–404. https://doi.org/10.1016/S0006-3495(00)76602-1
28. Dominguez E, Perez MD, Calvo M (1997) Effect of heat treatment on the antigen-binding activity of anti-peroxidase immunoglobulins in bovine colostrum. J Dairy Sci 80: 3182–3187. https://doi.org/10.3168/jds.S0022-0302(97)76290-8
29. Tornehave D, Hougaard DM, Larsson L (2000) Microwaving for double indirect immunofluorescence with primary antibodies from the same species and for staining of mouse tissues with mouse monoclonal antibodies. Histochem Cell Biol 113:19–23. https://doi.org/10.1007/s004180050002
30. Nakata T, Suzuki N (2012) Chromogen-based immunohistochemical method for elucidation of the coexpression of two antigens using antibodies from the same species. J Histochem Cytochem 60(8):611–619. https://doi.org/10.1369/0022155412449348
31. Giurisato E, Lonardi S, Telfer B et al (2020) Extracellular-regulated protein kinase 5-mediated control of p21 expression promotes macrophage proliferation associated with tumor growth and metastasis. Cancer Res 80(16):3319–3330. https://doi.org/10.1158/0008-5472.CAN-19-2416
32. Gatta LB, Melocchi L, Bugatti M et al (2019) Hyper-activation of STAT3 sustains progression of non-papillary basal-type bladder cancer via FOSL1 regulome. Cancers (Basel) 11(9): 1219. https://doi.org/10.3390/cancers11091219
33. Glass G, Papin JA, Mandell JW (2009) SIMPLE: a sequential immunoperoxidase labeling and erasing method. J Histochem Cytochem 57(10):899–905. https://doi.org/10.1369/jhc.2009.953612
34. Gendusa R, Scalia CR, Buscone S et al (2014) Elution of high-affinity (>10-9 KD) antibodies from tissue sections: clues to the molecular mechanism and use in sequential immunostaining. J Histochem Cytochem 62(7):519–531. https://doi.org/10.1369/0022155414536732
35. Taylor CR (2014) Immunohistochemistry in surgical pathology: principles and practice. Methods Mol Biol 1180:81–109. https://doi.org/10.1007/978-1-4939-1050-2_5
36. Bulman AS, Heyderman E (1981) Alkaline phosphatase for immunocytochemical labelling: problems with endogenous enzyme activity. J Clin Pathol 34(12):1349–1351. https://doi.org/10.1136/jcp.34.12.1349

Chapter 18

Measuring Neutralizing Antibodies to SARS-CoV-2 Using Lentiviral Spike-Pseudoviruses

Sabari Nath Neerukonda, Russell Vassell, Carol D. Weiss, and Wei Wang

Abstract

Assays measuring neutralizing antibodies (nAbs) against SARS-CoV-2 are used to evaluate serological responses after SARS-CoV-2 infection and the potency of therapeutic antibodies and preventive vaccines. It is therefore imperative that neutralization assays be sensitive, specific, quantitative, and scalable for high throughput. Pseudoviruses are excellent surrogates for highly pathogenic viruses such as SARS-CoV-2 because they can be safely used to measure nAbs in a biosafety level-2 laboratory. In addition, pseudoviruses allow for easy introduction of mutations to study the effect of amino acid changes in the spike protein. In this chapter, we describe a recently optimized assay for measuring neutralizing antibodies to SARS-CoV-2 that uses a HIV-based lentiviral vector pseudotyped with the spike glycoprotein of SARS-CoV-2 to infect 293T cells stably expressing ACE2 and TMPRSS2.

Key words Neutralizing antibodies, SARS-CoV-2, Pseudovirus, Luciferase reporter, Assay optimization

1 Introduction

In late 2019, a novel coronavirus, later renamed to severe acute respiratory syndrome coronavirus 2 (SARS-CoV-2), emerged in China and subsequently spread across the globe to cause a pandemic. SARS-CoV-2 uses angiotensin-converting enzyme 2 (ACE2) as its receptor for host cell binding and entry. The viral surface trimeric glycoprotein known as spike (S) engages ACE2 via its receptor binding domain (RBD). During biosynthesis, the S protein is proteolytically cleaved into the S1 subunit, which contains the RBD, and the S2 subunit, which contains domains needed for fusion [1–4]. For successful viral entry into the target cells, the S protein requires an additional priming step at the S2′ site by cellular proteases, such as transmembrane serine protease 2 (TMPRSS2) or cathepsins B and L [2]. This priming step allows the S protein to acquire a conformation that can mediate fusion between virus and

Justin Jang Hann Chu et al. (eds.), *SARS-CoV-2: Methods and Protocols*,
Methods in Molecular Biology, vol. 2452, https://doi.org/10.1007/978-1-0716-2111-0_18,

host cell membrane. While the region around the RBD is a major target of neutralizing antibodies (nAbs) induced by infection or vaccination, antibodies to other regions of S can also neutralize virus by interfering with the S protein fusion function [5]. This can occur by preventing S2′ priming by cellular proteases or interfering with S protein fusogenic conformational changes.

The neutralization assay described here uses pseudoviruses as surrogates for replicating SARS-CoV-2 [6]. Pseudoviruses are made using defective viral genomes and therefore only undergo a single round of infection [7]. However, pseudoviruses can faithfully recapitulate the early steps of the viral life cycle, including attachment, fusion, and delivery of the pseudoviral genomic contents into the target cell. Production of HIV-based pseudoviruses involves encapsulating HIV reverse-transcriptase (RT) and integrase (INT) along with the firefly luciferase (*fluc*) reporter gene that is flanked by long terminal repeats (LTRs). As nascent capsids buds from the plasma membrane, spike proteins are incorporated on the pseudovirus surface. When S-pseudoviruses infect 293T cells that stably express the ACE2 receptor and TMPRSS2 protease (293T-ACE2/TMPRSS2), RNA encoding the *fluc* reporter gene is released into the cell and reverse-transcribed into DNA that gets integrated into the host cell genome by the HIV INT [8]. Expression of the *fluc* gene is measured as relative luminescence units (RLU) in a simple luciferase activity assay. Luciferase activity is directly proportional to the pseudovirus titer in the inoculum over a wide range of values.

Stocks of S-pseudoviruses are produced in 293T cells by cotransfection with three plasmids: (1) an expression plasmid encoding a codon-optimized spike gene, (2) a plasmid encoding a partial HIV genome containing core packaging *gag* and *pol* genes but lacking the *env*, and (3) a transfer vector encoding the *fluc* gene, Ψ-RNA packaging signal, and 5′- and 3′-flanking HIV LTRs. Following transfection, pseudovirus particles bearing S protein are released into the supernatant of 293T cells. Harvested S-pseudoviruses are then used for neutralization assays. S-pseudoviruses are preincubated with serial dilutions of nAbs of a test sample prior to inoculation on target 293T-ACE2/TMPRSS2 cells. After incubation for 48 h, neutralization is scored by the percent inhibition of luciferase activity in S-pseudoviruses pretreated with nAbs compared to control S-pseudovirus without nAb treatment. The assays are performed in 96-well plates for high throughput capacity.

2 Materials

2.1 Cells, Plasmids and Reagents

1. Dulbecco's Modified Eagle's Medium (DMEM) with 4.5 g/L glucose and sodium pyruvate without L-glutamine (Corning, Corning, NY) supplemented with 10% heat inactivated fetal

bovine serum (FBS, R&D Systems, Minneapolis, MD), 25 mM HEPES (Gibco/BRL, Bethesda, MD), 2 mM L-glutamine (Corning, Corning, NY), 1× MEM nonessential amino acids (Corning, Corning, NY), and 1× penicillin and streptomycin (Corning, Corning, NY). Complete medium should be stored at 4 °C.

2. Gibco Opti-MEM™ Reduced Serum Medium (ThermoFisher Scientific, Waltham, MA).
3. 0.05% Trypsin, 0.53 mM EDTA 1×, (Corning, Corning, NY). Store at 4 °C.
4. Dulbecco's phosphate-buffered saline (D-PBS), no calcium chloride, no magnesium chloride. (Corning, Corning, NY).
5. Luciferase Assay System (Promega, Madison, WI). Reconstitute one vial of lyophilized Luciferase Assay Substrate with 105 mL of Luciferase Assay Buffer Solution. After the substrate has dissolved completely (about 1 min), mix gently and distribute to 25 mL conical polypropylene tubes and store at −70 °C or use immediately. Thaw in a room temperature water bath in the dark immediately before each use. Excess reagent may be stored at −70 °C and used once more.
6. Luciferase Cell Culture Lysis 5× Reagent (Promega, Madison, WI).
7. FuGENE-6 ®Transfection Reagent (Promega, Madison, WI).
8. 293T/17 cells (American Type Culture Collection, Manassas, VA).
9. 293T-ACE2/TMPRSS2 cells (BEI Resources catalog #NR-55293) [6] (*see* **Note 1**).
10. pCMVΔR8.2, second-generation packaging vector expressing HIV-Gag-Pol (Addgene Plasmid #12263).
11. pCSFLW, lentiviral plasmid expressing firefly luciferase [7, 9].
12. pcDNA3.1-S, codon-optimized spike gene of SARS-COV2 Wuhan-Hu-1 isolate (GenBank accession: YP_009724390.1) (*see* **Notes 2** and **3**).
13. Trypan Blue Dye (Bio-Rad, Hercules, CA).
14. Reference sera.
15. Reference monoclonal antibody.

2.2 Consumables

1. Polypropylene sterile conical tubes:
   (a) Falcon™ 5 mL Round-Bottom Tubes.
   (b) Falcon™ 15 mL Conical Centrifuge Tubes.
   (c) Falcon™ 50 mL Conical Centrifuge Tubes.
2. 1.7-mL microcentrifuge tubes

3. Dual Chamber Cell Counting Slides for TC10™/TC20™ Cell Counter (Bio-Rad, Hercules, CA).
4. Sterile reagent reservoirs, (Thomas Scientific. Swedesboro, NJ).
5. Nunc™ 96-Well Polypropylene 1.3 mL DeepWell™ Storage Plates (ThermoFisher Scientific, Waltham, MA).
6. Thermo Scientific™ Nunc™ Sealing Tapes (ThermoFisher Scientific, Waltham, MA).
7. 100 mm Falcon Tissue Culture Dish
8. Greiner Bio-One™ CellStar™ 96-Well, Cell Culture-Treated, Flat-Bottom Microplate, White (Greiner Bio-One, Monroe, NC).
9. Acrodisc® Syringe Filters with Supor® Membrane, Sterile - 0.45 μm, 25 mm Diameter (Pall Corporation, Port Washington, NY).

2.3 Equipment

1. Class II biological safety cabinet.
2. Light microscope.
3. TC20™ Automated Cell Counter (Bio-Rad, Hercules, CA).
4. Scientific rocker.
5. Spectramax Plate Reader (Molecular Devices, LLC. San Jose, CA).
6. Thermo Scientific™ Sorvall™ Legend™ XTR Centrifuge TX-1000 (Fisher Scientific).

3 Methods

The neutralization assay is designed to evaluate five test samples in an eight-point dilution series that runs in duplicate in a single 96-well plate. Samples that are positive for neutralization typically exhibit a sigmoidal inhibition curve with a linear area of the curve between 20% and 80% inhibition in RLU. The nAb titers are typically reported as 50% inhibitory dilution (ID_{50}) for serum or plasma test samples and 50% inhibitory concentration (IC_{50}) for monoclonal antibody test samples, which lie at the midpoint of the linear portion of a neutralization curve. A positive control sample with a known neutralization titer to assure assay integrity (e.g., WHO international NIBSC standard; code 20/136) and a negative control sample to exclude nonspecific background activity must be included [6].

3.1 Splitting Cells

1. Aspirate the culture medium from a confluent monolayer of cells in a 100 mm culture dish. Remove residual serum by rinsing the monolayer with 3 mL of sterile PBS.

2. Slowly add 3 mL of 0.05% trypsin–EDTA solution to cover the cell monolayer. Incubate at 37 °C in a 5% CO_2–95% air environment for 5–10 min.
3. Add 3 mL of complete DMEM medium and suspend the cells by gentle pipette action. Count cells.
4. Seed new 100 mm culture dishes with approximately 2.5×10^6 cells in 10 mL of complete DMEM. Cells should be split approximately 1:6 about every 3 days.

3.2 Production of Spike Pseudoviruses by Transfection in 293T Cells

1. Seed 5×10^6 293T cells in a 100 mm culture dish containing 10 mL complete DMEM 1 day prior to transfection. Incubate overnight at 37 °C in a 5% CO_2–95% air environment. Monolayers should be 50–80% confluent at the time of transfection.
2. Combine 4 μg of codon-optimized pcDNA3.1-S plasmid DNA, 5 μg pCSFLW plasmid DNA, and 5 μg pCMVΔR8.2 in a 1.5-mL sterile microcentrifuge tube. Adjust the final volume to 50 μL with Opti-MEM.
3. In a second 5 mL sterile polypropylene tube, add 547 μL Opti-MEM. Add 56 μL of FuGENE-6 reagent directly into the Opti-MEM without contacting the sides of the tube. Mix gently. Incubate for 5 min.
4. Transfer the entire plasmid DNA contents from the first tube to the second tube containing the FuGENE-6 solution. Mix by pipette action or brief vortexing. Incubate at room temperature for 30 min to allow DNA–liposome complexes to form.
5. Transfer the DNA–FuGENE-6 reaction mixture dropwise to a monolayer of 293T cells that is 50–80% confluent in a 100 mm dish containing 10 mL of complete DMEM. Swirl the contents gently to allow uniform distribution across the cell monolayer. Incubate at 37 °C in a 5% CO_2–95% air environment for 2 days to permit plasmid DNA entry and pseudovirus production.
6. Collect the pseudovirus-containing culture supernatant.
7. Filter the supernatant using a 0.45-micron syringe low protein-binding filter. Store the filtered pseudovirus preparations in 1 mL aliquots at −70 °C.

3.3 Determination of Pseudovirus Infectivity Titer

1. Seed 3×10^4 293 T-ACE2/TMPRSS2 cells per well in all wells of a 96-well flat-bottom culture plate 1 day prior to pseudovirus infection. Incubate overnight at 37 °C in a 5% CO_2–95% air environment.
2. On the day of infection, place 240 μL complete DMEM per well in all wells from rows B-H (one column per pseudovirus to be titered) of a 96-well deep-well storage plate (*see* **Note 4**).
3. Place 400 μL complete DMEM in row A of a 96-well deep-well storage plate as shown in Fig. 1. Add 100 μL pseudovirus and mix to achieve a total volume of 500 μL at 1:5 final dilution.

	1	2	3	4	5	6	7	8	9	10	11	12
A	Dil 1	Dil 1	Dil 1	Dil 1	Dil 1	Dil 1	Dil 1	Dil 1	Dil 1	Dil 1	Dil 1	Dil 1
B	Dil 2	Dil 2	Dil 2	Dil 2	Dil 2	Dil 2	Dil 2	Dil 2	Dil 2	Dil 2	Dil 2	Dil 2
C	Dil 3	Dil 3	Dil 3	Dil 3	Dil 2	Dil 3	Dil 3	Dil 3	Dil 3	Dil 2	Dil 3	Dil 3
D	Dil 4	Dil 4	Dil 4	Dil 4	Dil 4	Dil 4	Dil 4	Dil 4	Dil 4	Dil 4	Dil 4	Dil 4
E	Dil 5	Dil 5	Dil 5	Dil 5	Dil 5	Dil 5	Dil 5	Dil 5	Dil 5	Dil 5	Dil 5	Dil 5
F	Dil 6	Dil 6	Dil 6	Dil 6	Dil 6	Dil 6	Dil 6	Dil 6	Dil 6	Dil 6	Dil 6	Dil 6
G	Dil 7	Dil 7	Dil 7	Dil 7	Dil 7	Dil 7	Dil 7	Dil 7	Dil 7	Dil 7	Dil 7	Dil 7
H	CC	CC	CC	CC	CC	CC	CC	CC	CC	CC	CC	CC

Fig. 1 Standard assay template for determining infectivity. Deep-well storage template for serial dilution of 12 pseudoviruses (A1-A12) per 96-well plate. *CC* cell control wells (cells only), *Dil* dilution

4. Perform serial twofold dilutions (i.e., transfer 240 μL, mixing each time) for a total of seven dilutions. Discard 240 μL from the seventh dilution. Wells in row H serve as controls for background luminescence.
5. Aspirate the complete DMEM from the 96-well flat-bottom culture plate containing HEK293T-ACE2/TMPRSS2 cells using a sterile multichannel aspirator.
6. Carefully transfer 100 μL of pseudovirus in duplicate, from deep-well storage to 96-well flat-bottom culture plate. Change pipette tips in between each step to minimize carryover.
7. Incubate at 37 °C in a 5% CO_2–95% air environment for 48 h.
8. Remove inoculum from all wells using a multichannel aspirator. Dispense 25 μL 1× cell culture lysis reagent to each well. Incubate at room temperature for 30 min on a rocker set on low speed to allow complete cell lysis.
9. Read the firefly luciferase activity in a luminometer at 570 nm with injection of 50–100 μL of luciferase assay substrate, 2 s delay and 1 s read.
10. Calculate the RLU/mL as a measure of pseudovirus titer. Wells ≥100 times background are considered positive for the calculation. Luminescence resulting from 293T cellular background typically ranges between 40 and 70 RLU.

3.4 Neutralization Assay

1. Seed 3×10^4 293 T-ACE2/TMPRSS2 cells per well in all wells of a 96-well flat-bottom culture plate 1 day prior to neutralization assay. Incubate overnight at 37 °C in a 5% CO_2–95% air environment.

	1	2	3	4	5	6	7	8	9	10	11	12
A	Dil 1	Dil 1	Dil 1	Dil 1	Dil 1	VC	CC					
B	Dil 2	Dil 2	Dil 2	Dil 2	Dil 2	VC	CC					
C	Dil 3	Dil 3	Dil 3	Dil 3	Dil 2	VC	CC					
D	Dil 4	Dil 4	Dil 4	Dil 4	Dil 4	VC	CC					
E	Dil 5	Dil 5	Dil 5	Dil 5	Dil 5	VC	CC					
F	Dil 6	Dil 6	Dil 6	Dil 6	Dil 6	VC	CC					
G	Dil 7	Dil 7	Dil 7	Dil 7	Dil 7	VC	CC					
H	Dil 8	Dil 8	Dil 8	Dil 8	Dil 8	VC	CC					
test sample	1	2	3	4	5							

	1	2	3	4	5	6	7	8	9	10	11	12
A	Dil 1	Dil 1	Dil 1	Dil 1	Dil 1	Dil 1	Dil 1	Dil 1	Dil 1	Dil 1	VC	CC
B	Dil 2	Dil 2	Dil 2	Dil 2	Dil 2	Dil 2	Dil 2	Dil 2	Dil 2	Dil 2	VC	CC
C	Dil 3	Dil 3	Dil 3	Dil 3	Dil 2	Dil 3	Dil 3	Dil 3	Dil 3	Dil 2	VC	CC
D	Dil 4	Dil 4	Dil 4	Dil 4	Dil 4	Dil 4	Dil 4	Dil 4	Dil 4	Dil 4	VC	CC
E	Dil 5	Dil 5	Dil 5	Dil 5	Dil 5	Dil 5	Dil 5	Dil 5	Dil 5	Dil 5	VC	CC
F	Dil 6	Dil 6	Dil 6	Dil 6	Dil 6	Dil 6	Dil 6	Dil 6	Dil 6	Dil 6	VC	CC
G	Dil 7	Dil 7	Dil 7	Dil 7	Dil 7	Dil 7	Dil 7	Dil 7	Dil 7	Dil 7	VC	CC
H	Dil 8	Dil 8	Dil 8	Dil 8	Dil 8	Dil 8	Dil 8	Dil 8	Dil 8	Dil 8	VC	CC
test sample	1	1	2	2	3	3	4	4	5	5		

Fig. 2 Standard assay templates. Top: Deep-well storage template for serial dilution of five test samples per 96-well plate. Bottom: Template for measuring the neutralization titer of five test samples per 96-well plate. *CC* cell control wells (cells only), *VC* virus control wells (virus and cells but no test sample), *Dil* dilution

2. Using the deep-well storage plate layout shown in Fig. 2, place 120 μL of complete DMEM in all wells of column 6 (virus control) and column 7 (cell control). Place 120 μL in all wells of columns 1–5, rows B–H. Place 240 μL of test samples diluted in complete DMEM to achieve 2× initial concentration in row A (*see* **Note 5**); first test sample in column 1, second test sample in column 2, third test sample in column 3, fourth test sample in column 4, fifth test sample in column 5 (*see* **Notes 5** and **6**).

3. Mix and repeat the transfer of 120 μL of test samples from row A through row H (these are serial twofold dilutions). After final

transfer and mixing is complete, discard 120 μL from the wells in columns 1–5, row H into a waste container of disinfectant (*see* **Notes 5** and **6**).

4. Thaw assay stocks of S-pseudoviruses in an ambient temperature water bath. When completely thawed, dilute the virus in complete DMEM to achieve a 2× concentration with 1,000,000-2,000,000 RLU/mL. Transfer the pseudovirus suspension into a sterile reagent reservoir.
5. Dispense 120 μL of spike pseudoviruses to all wells in columns 1–6, rows A through H to achieve a final 1× concentration of pseudovirus + test sample mixture (columns 1–6) and virus control (pseudovirus + complete DMEM). Gently mix up and down 5 times with a pipette. Seal the plates with optical sealing tape and incubate at 37 °C in a 5% CO_2–95% air environment for 1 h.
6. Aspirate complete DMEM from 96-well plate flat-bottom culture plate seeded with 293T-ACE2/TMPRSS2 cells on the previous day. Dispense 100 μL of pseudovirus + test sample mixture in duplicate wells from columns 1 through 10. Dispense 100 μL pseudovirus and cell control mixtures to column 11 and 12 respectively. Cover plates and incubate at 37 °C in a 5% CO_2–95% air environment for 48 h.
7. Remove 100 μL of inoculum/media from all wells using a multichannel aspirator. Dispense 25 μL 1× cell culture lysis reagent to all wells. Incubate at room temperature for 30 min on a rocker set on low speed to allow complete cell lysis.
8. Read the firefly luciferase activity in a luminometer at 570 nm with injection of 50–100 μL of Luciferase assay substrate, 2 s delay and 1 s read.
9. Calculate the percent neutralization at each test sample dilution by first taking the difference in average RLU from the duplicates between the test sample wells (cells + test sample + pseudovirus) and cell control wells (cells only, column 12), then dividing this result by the difference in average RLU between virus control (cell + virus, column 11) and cell control wells (column 12), followed by subtracting from 1 and multiplying by 100. The nAb titers are expressed as the test sample dilution (serum and plasma) or concentration (monoclonal antibody) required to reduce RLU by 50%. Nonlinear regression curve fit analysis over the test sample dilutions can be performed in GraphPad Prism 7 or equivalent software to calculate ID_{50} or IC_{50} of test samples.
10. Assay performance is considered acceptable when: (1) the mean RLU for the virus control is ≥200× the mean RLU of the cell control, (2) the neutralization titer of positive control test sample is within threefold of the expected titer, and (3) there is lack of nonspecific inhibition (*see* **Note 7**).

4 Notes

1. 293T-ACE2/TMPRSS2 cells described in this protocol are stable for approximately 20 passages. Subsequent passages exhibit gradual loss in ACE2 and/or TMPRSS2 expression resulting in reduction of infectivity of S-pseudoviruses. Therefore, it is advisable to generate low passage cell banks. Freeze cells in 90% fetal calf serum and 10% DMSO. On thawing, pellet cells in a large volume of medium, plate in fresh medium, and incubate overnight. The next day, remove medium and dead cells, and replace with fresh medium.
2. Other mammalian vectors instead of pcDNA3.1 can be used. Use codon-optimized transgene to increase spike protein expression in mammalian cells. Amino acid variations in spike can be tested by using plasmids with the desired mutations in the S gene. All plasmids need to be transformed into DH5α cells or similar cells with an appropriate selection antibiotic. Plasmids can be purified using maxiprep purification protocols.
3. A plasmid encoding vesicular stomatitis virus G protein (pcDNA3.1-VSV-G) or an empty pcDNA3.1 vector could be used instead of pcDNA3.1-S to generate VSV-G pseudovirus or bald pseudovirus without Env (ΔEnv) for use as controls.
4. While a polycation, such polybrene, is typically used at 8 μg/mL final concentration to minimize charge-repulsion between the virus and cells to enhance virus attachment, we did not observe an effect of polybrene on S-pseudovirus infectivity.
5. The neutralization assay procedure described here in the Fig. 2 plate layout requires test sample dilutions to be performed at 2× initial concentration. For instance, to measure nAb titers that are in the range of 1:40 to 1:5120, initial 2× concentrations range from 1:20 to 1:2560 respectively. After mixing and incubating with an equal volume of 2× pseudovirus, a final 1× concentration is achieved. A different range of titers may be measured by altering test sample dilutions and plate layout. For instance, to detect low levels of nAbs, a starting dilution of 1:10 or lower may be used. Alternatively, if the test sample contains higher levels of nAbs, a higher starting dilution or an increased serial dilution interval (e.g., fourfold serial dilution) may be used. On the other hand, plate layout can be changed to measure a wider range of 12-point titers by performing serial dilutions from 1 to 12 rather than A–H.
6. We recommend all serum and plasma samples to be heat-inactivated at 56 °C for 30 min prior to use in the assay in order to destroy complement.

7. Factors in serum and plasma that exhibit nonspecific antiviral activity or cell toxicity can result in false positives results in the neutralization assay. For instance, serum and plasma of HIV-infected individuals on antiretroviral therapy (ART) may contain reverse transcriptase or integrase inhibitors in adequate concentrations to inhibit respective enzyme functions needed for *fluc* reporter expression in a neutralization assay [6, 10]. We recommend screening all sera and plasma for nonspecific neutralization using pseudoviruses bearing an envelope protein from amphotropic murine leukemia virus or VSV as an additional control to identify nonspecific inhibition.

References

1. Coutard B, Valle C, de Lamballerie X, Canard B, Seidah NG, Decroly E (2020) The spike glycoprotein of the new coronavirus 2019-nCoV contains a furin-like cleavage site absent in CoV of the same clade. Antivir Res 176:104742. https://doi.org/10.1016/j.antiviral.2020.104742
2. Hoffmann M, Kleine-Weber H, Schroeder S, Kruger N, Herrler T, Erichsen S, Schiergens TS, Herrler G, Wu NH, Nitsche A, Muller MA, Drosten C, Pohlmann S (2020) SARS-CoV-2 cell entry depends on ACE2 and TMPRSS2 and is blocked by a clinically proven protease inhibitor. Cell 181(2):271–280.e8. https://doi.org/10.1016/j.cell.2020.02.052
3. Wrapp D, Wang N, Corbett KS, Goldsmith JA, Hsieh C-L, Abiona O, Graham BS, McLellan JS (2020) Cryo-EM structure of the 2019-nCoV spike in the prefusion conformation. bioRxiv:2020.2002.2011.944462. https://doi.org/10.1101/2020.02.11.944462
4. Neerukonda SN, Katneni U (2020) A review on SARS-CoV-2 virology, pathophysiology, animal models, and anti-viral interventions. Pathogens 9(6):426. https://doi.org/10.3390/pathogens9060426
5. Amanat F, Stadlbauer D, Strohmeier S, Nguyen THO, Chromikova V, McMahon M, Jiang K, Arunkumar GA, Jurczyszak D, Polanco J, Bermudez-Gonzalez M, Kleiner G, Aydillo T, Miorin L, Fierer DS, Lugo LA, Kojic EM, Stoever J, Liu STH, Cunningham-Rundles C, Felgner PL, Moran T, Garcia-Sastre A, Caplivski D, Cheng AC, Kedzierska K, Vapalahti O, Hepojoki JM, Simon V, Krammer F (2020) A serological assay to detect SARS-CoV-2 seroconversion in humans. Nat Med 26(7):1033–1036. https://doi.org/10.1038/s41591-020-0913-5
6. Neerukonda SN, Vassell R, Herrup R, Liu S, Wang T, Takeda K, Yang Y, Lin T-L, Wang W, Weiss CD (2020) Establishment of a well-characterized SARS-CoV-2 lentiviral pseudovirus neutralization assay using 293T cells with stable expression of ACE2 and TMPRSS2. bioRxiv:2020.2012.2026.424442. https://doi.org/10.1101/2020.12.26.424442
7. Zufferey R, Dull T, Mandel RJ, Bukovsky A, Quiroz D, Naldini L, Trono D (1998) Self-inactivating lentivirus vector for safe and efficient in vivo gene delivery. J Virol 72(12): 9873–9880. https://doi.org/10.1128/JVI.72.12.9873-9880.1998
8. Millet JK, Tang T, Nathan L, Jaimes JA, Hsu H-L, Daniel S, Whittaker GR (2019) Production of pseudotyped particles to study highly pathogenic coronaviruses in a biosafety level 2 setting. J Vis Exp (145):e59010. https://doi.org/10.3791/59010
9. Almasaud A, Alharbi NK, Hashem AM (2020) Generation of MERS-CoV pseudotyped viral particles for the evaluation of neutralizing antibodies in mammalian sera. Methods Mol Biol 2099:117–126. https://doi.org/10.1007/978-1-0716-0211-9_10
10. Montefiori DC (2009) Measuring HIV neutralization in a luciferase reporter gene assay. In: Prasad VR, Kalpana GV (eds) HIV protocols. Humana Press, Totowa, NJ, pp 395–405. https://doi.org/10.1007/978-1-59745-170-3_26

Part IV

Antivirals and Vaccines

Chapter 19

Antiviral Strategies Against SARS-CoV-2: A Systems Biology Approach

Erica T. Prates, Michael R. Garvin, Piet Jones, J. Izaak Miller, Kyle A. Sullivan, Ashley Cliff, Joao Gabriel Felipe Machado Gazolla, Manesh B. Shah, Angelica M. Walker, Matthew Lane, Christopher T. Rentsch, Amy Justice, Mirko Pavicic, Jonathon Romero, and Daniel Jacobson

Abstract

The unprecedented scientific achievements in combating the COVID-19 pandemic reflect a global response informed by unprecedented access to data. We now have the ability to rapidly generate a diversity of information on an emerging pathogen and, by using high-performance computing and a systems biology approach, we can mine this wealth of information to understand the complexities of viral pathogenesis and contagion like never before. These efforts will aid in the development of vaccines, antiviral medications, and inform policymakers and clinicians. Here we detail computational protocols developed as SARS-CoV-2 began to spread across the globe. They include pathogen detection, comparative structural proteomics, evolutionary adaptation analysis via network and artificial intelligence methodologies, and multiomic integration. These protocols constitute a core framework on which to build a systems-level infrastructure that can be quickly brought to bear on future pathogens before they evolve into pandemic proportions.

Key words Antiviral, SARS-CoV-2, COVID-19, Systems Biology, Multiomics, Pandemic

1 Introduction

As of this writing, COVID-19 has surpassed 130 million cases, leading to the death of nearly 3 million people worldwide. While it took a century to gain a clear mechanistic understanding of the H1N1 virus and the 1918 pandemic, the global scientific community has produced a stunning picture of the SARS-CoV-2 and COVID-19 in just under a year. As a result, vaccines are being deployed worldwide, new therapeutics are being developed, and

Erica T. Prates and Michael R. Garvin contributed equally to this work.

Justin Jang Hann Chu et al. (eds.), *SARS-CoV-2: Methods and Protocols*,
Methods in Molecular Biology, vol. 2452, https://doi.org/10.1007/978-1-0716-2111-0_19,

FDA-approved pharmaceuticals are being repurposed. Arguably the most critical factor in the successes we have achieved to date has been the open flow of data and scientific insights, which is nearly equal to the scale and distribution of the virus itself. The challenge is no longer generating data, but rather extracting and integrating virus-centric, host-centric, and virus–host interaction insights to understand all aspects of viral pathogenesis, that is, from a systems level. It is of paramount important that we learn from this experience so that we can prevent the next contagious disease from becoming an epidemic, let alone one of pandemic proportions.

Computational systems biology is a rapidly advancing field that leverages high-performance computing (HPC), machine learning (ML), and artificial intelligence (AI) algorithms to mine complex and diverse data sets to extract critical networks of information. A crucial first step in this system-level approach is to assemble a multidisciplinary team and provide infrastructure and common language for rapid and efficient communication among members. Here we describe a systems-level antiviral strategy built by a team with expertise in pathogen detection, molecular evolution, structural biology, AI algorithm development, multiomics integration, and clinical medicine. Herein we detail five methods (Fig. 1) to be employed *synergistically* to collect, analyze, and report a diversity of information underlying SARS-CoV-2 pathogenesis toward establishing promising strategies to restrain the viral spread. Ultimately, we devised this material to be applicable against the next potential outbreak. Future efforts will likely increase the use of AI and ML approaches to provide more accurate predictive models in shorter timescales as pathogens emerge in the human population.

1.1 Pathogen Detection

Identification of the pathogen from early clinical samples is crucial for limiting the expansion of the virus. Furthermore, the detection of opportunistic bacterial and fungal pathogens and their impact on the host-microbiome are important factors that can influence treatments and patient outcomes. Here, taxonomic identification was performed on nonhost RNASeq with a parallelized version of Kraken2, ParaKraken [1], using HPC facilities to cover an extensive range of potential taxa and analyze a large set of samples. Identification of a broad range of taxa at large scales may allow for the detection of emergent, potentially novel pathogens and assessment of putative zoonotic events. For individuals with limited access to HPC resources, Kraken2 can be used with a more focused range of taxa on fewer samples. After taxonomic classification is performed, subsequent analyses are needed to mitigate potential false positives, identify potential microbial dysbiosis, and determine the potential enrichment of putative pathogens. Furthermore, RNA transcripts assigned to a particular taxon can then be extracted and assembled for additional analyses if there is sufficient read depth. These assembly-based analyses may be informative for understanding

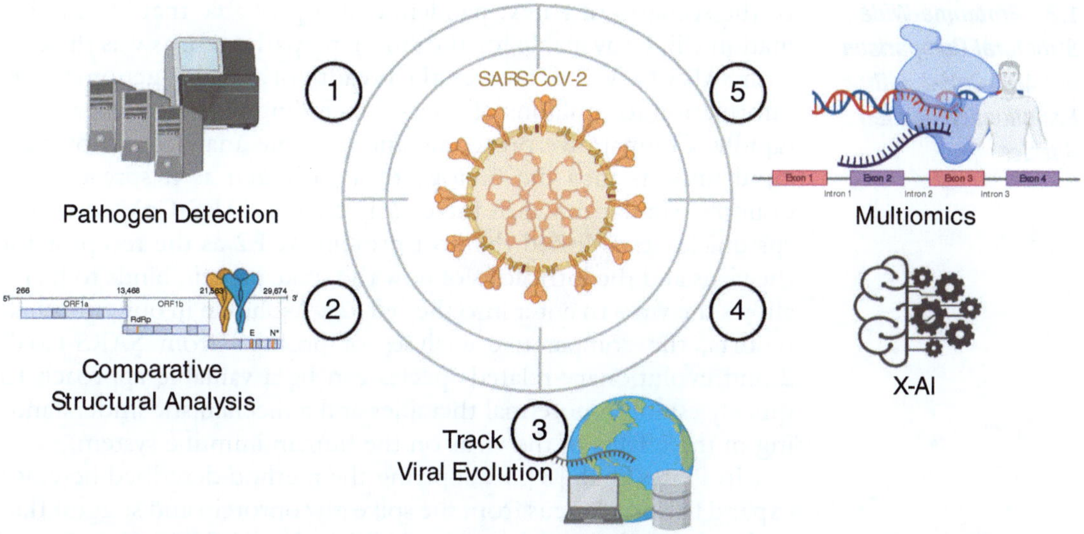

Fig. 1 Systems Biology strategies against SARS-CoV-2. (**1**) The use of HPC for pathogen detection from nonhost RNA-Seq and identification of co-occurring microbial communities; (**2**) Proteome-wide structural comparison of SARS-CoV-2 with evolutionary-related species to identify potential molecular features determining pathogenicity and ideal targets for broad-spectrum antivirals; (**3**) Tracking SARS-CoV-2 evolution; (**4**) Explainable artificial intelligence predictive models (X-AI) such as iterative Random Forest Leave One Out Prediction (iRF-LOOP) and Random Intersection Trees (RIT) to determine feature importance and interaction for various applications such as identifying which mutations may be coevolving; (**5**) Understanding COVID-19 pathogenesis using multiomics integration and using disease and biological function gene ontologies (GOs)

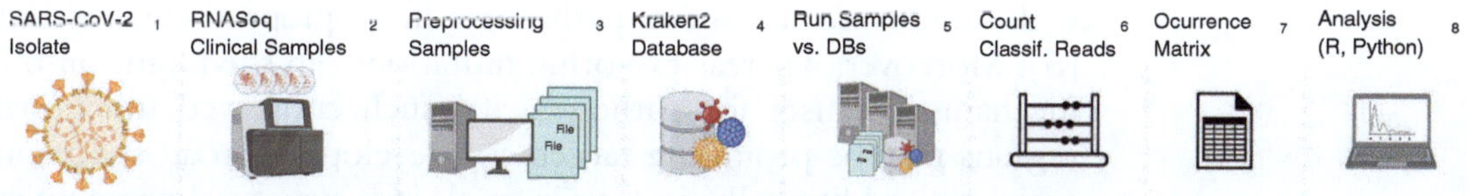

Fig. 2 Workflow for Pathogen Detection: Strategy Overview—(**1**) Obtain genome sequence of pathogen isolate; (**2**) Obtain bulk RNASeq clinical samples; (**3**) Preprocessing samples; (**4**) Create or download database with whole-genome sequences of pathogens of interest; (**5**) Compare RNASeq clinical samples against the database; (**6**) Count the classified reads; (**7**) Occurrence matrix; (**8**) Run data analysis

viral evolution (in the case of a viral-identified taxon) or to improve classification. In Fig. 2, we give an overview of the process, which is detailed in Subheading 3.1.

A community analysis of the microbial taxa is important to determine the potential for dysbiosis. Pathogenicity of viruses and viral-associated disease outcomes are strongly influenced by microbial dysbiosis. Therefore, understanding the community structure allows for the potential to influence treatment, and improve patient outcomes.

1.2 Proteome-Wide Structural Comparison of SARS-CoV-2 with Evolutionary-Related Species

In the scenario of a next pandemic, it is probable that little information will be available for the emerging pathogen, as was the case with SARS-CoV-2. However, the breadth of available genome-level information for millions of species, including viruses, is expanding rapidly. Comparative proteome and genome analyses can provide rapid insights into the biology of a pathogen as it spreads. For example, similarities with the closely related SARS-CoV coronavirus quickly established the host protein ACE2 as the receptor for the virus and the intricacies of how the spike protein binds to it and allows the virus to enter into the cell [2]. As shown in our published reports, the comparative analyses of proteins from SARS-CoV-2 and evolutionary-related species can be a valuable approach to quickly establish potential therapies and a mechanistic understanding of the effects of the virus on the human immune system.

In Prates et al. [3], by applying the method described here, we expand the usual focus from the spike glycoprotein and suggest that molecular differences between SARS-CoV and SARS-CoV-2 proteomes in other regions, such as in the nsp1 and nsp3 proteins, likely have a significant contribution due to their distinct pathogenic profiles. On the other hand, based on the comparison with another coronavirus, the porcine epidemic diarrhea virus (PEDV), we suggest that the highly conserved binding site of the SARS-CoV-2 main protease may be able to bind and cleave the NF-kB essential modulator [4], possibly resulting in an additional mechanism of circumventing the activation of the host immune response by NF-kB signaling—a hypothesis that has been recently associated with microvascular brain pathology in a preprint manuscript [5]. Moreover, whereas exploring mutations can shed light on the mechanistic causes for pathogenicity, such conserved functional regions may be promising targets for developing broad-spectrum antivirals. Additionally, in Garvin et al. [6], structural proteomics was applied in synergy with median-joining network (MJN) analysis (Subheadings 1.3 and 3.3) to unravel the likely molecular basis of adaptive mutations, and to identify understudied mutation sites that may have major pathogenic consequences.

The sensitive mutations that we aim to find with such comparative structural analysis are not always clearly detectable. For example, Zhang et al., showed that a single peripheral mutation involving residues of similar properties (Q33E) in human Pin1 caused a significant reduction of protein thermostability [7]. Therefore, we note that although the present method does not lead to conclusive results per se, it is a valuable approach to identify likely key mutations for phenotypic variation and, with that, establish priorities for further investigation through more extensive computational and experimental techniques. Additionally, the integration of structural proteomics with other 'omics layers of information is crucial for enhanced robustness of the proposed hypotheses regarding the functional impact of specific mutations.

1.3 Tracking SARS-CoV-2 Evolution

The mutations occurring as the SARS-CoV-2 virus spreads across the globe into millions of human (and, in many cases, nonhuman) hosts represent potentially adaptive responses. These changes have implications for drug and vaccine development throughout the pandemic. They can also provide an important means of real-time tracking of the spread of strains that cause varying disease severity and may affect the use of antiviral treatments or vaccines. A case in point is a "variant of concern" known as the B.1.1.7 lineage, listed as VOC-202012/01 by the CDC. It was first detected in the United Kingdom in November 2020 and is suspected to be the strain that is overwhelming medical facilities and increasing mortalities across the globe. Much attention has been focused on a single mutation in the spike protein (N501Y) as the cause of its dominance, but the geospatial distribution and temporal appearance of cosegregating mutations so far have not been considered systematically. The availability of hundreds of thousands of sequences of the virus and corresponding metadata allows one to track its spatial and temporal molecular evolution. Unlike phylogenetic trees, MJN of haploid (typically) nonrecombining genomes such as the SARS-CoV-2 virus facilitates the visualization of many valuable layers of information simultaneously, which can provide valuable insights into variants such as the B.1.1.7 lineage as it spreads. We have developed a computational systems biology pipeline to ingest, annotate, curate, interpret and display these diverse data types in the context of viral molecular evolution. Most of the methods we have employed have never before been used on this scale and therefore, we provide detailed notes on how individuals with limited access to HPC systems can execute this pipeline on typical user workstations, desktops or laptops.

1.4 Explainable Artificial Intelligence Models

The main advantage of using Explainable Artificial Intelligence (X-AI) algorithms over traditional linear algorithms or Black Box AI is that X-AI is able to combine the accuracy and efficiency of modelling complex systems (like Black Box AI) while maintaining the ability to produce results that are human-interpretable (like traditional methods). X-AI methods such as iterative Random Forest (iRF) [8], iterative Random Forest Leave One Out Prediction (iRF-LOOP) [9], and Random Intersection Trees (RIT) [10] are used to determine feature importance and interaction. By applying iRF-LOOP with RIT on the SARS-CoV-2 virus mutations across samples, we gain a better understanding of which mutations are associated and may be coevolving. This allows for the generation of hypotheses, such as on compensatory mutations or if specific mutations are causative for higher mutation rates in other parts of the sequence. Although our application here is to address the molecular evolution of the SARS-CoV-2 virus, these methods can be used on highly diverse data types.

1.5 Understanding COVID-19 Pathogenesis Through Gene Ontology and Multiomics Network Analysis

Understanding the mechanism underlying pathology is critical to developing new treatment strategies against SARS-CoV-2 infection and the resulting COVID-19 disease. Analysis from publicly available transcriptomics datasets from SARS-CoV-2 patients can be integrated with existing databases of human gene expression (e.g., HumanNet [11] and Genome-Tissue Expression Project, GTEx [12]) to obtain mechanistic insights on pathogenesis. Differentially expressed genes caused by a viral infection can be categorized by using protein function and phenotype ontologies to identify common biological pathways involved in the course of the disease. Incorporating viral–host protein interaction networks into downstream graph traversal analyses can also identify genes of interest that may be differentially expressed in the host due to the direct binding of viral to host proteins. Furthermore, integrating drug-to-target networks (e.g., DrugBank, ChEMBL) with differentially expressed genes involved in pathogenesis can suggest putative treatments based on biologically informed results.

2 Materials: Computational Resources, Software, and Packages

The SARS-CoV-2 dataset was analyzed using the supercomputing capacity of Summit and Rhea on the Oak Ridge Leadership Computing Facility (OLCF) supercomputer platform at the Oak Ridge National Laboratory's (ORNL). Summit is composed of 4608 compute nodes, each equipped with 512 GB of DDR4 memory for use by the two 22-core IBM POWER9 processors as well as six NVIDIA Volta V100 graphics processing units (GPUs). Routine calculations were performed on a standard laptop or desktop. Table 1 provides a list of the main publicly available software packages/libraries used in the described methods.

3 Methods

3.1 Pathogen Detection and Analysis

3.1.1 Pathogen Detection

1. Obtain the genomic sequences of a viral isolate of interest. The sequence of a closely related virus could also be used.
2. Obtain bulk RNASeq clinical samples from targets of interest with appropriate controls if needed.
3. Download the human genome or transcriptome from the NCBI web portal (assembly database, GRCh38).
4. Preprocessing-samples:
   (a) Assess the quality of the RNASeq data using FastQC (https://www.bioinformatics.babraham.ac.uk/projects/fastqc/). Perform adapter and quality score trimming (*see* **Note 1**). Align the RNASeq reads to the host

Table 1
Main publicly available software packages/libraries used

Resource	Reference/Source
CLC Genomics (viewer is freeware)	[digitalinsights.qiagen.com/downloads/product-downloads/]
Cytoscape	[cytoscape.org]
data.table	https://cran.r-project.org/web/packages/data.table/index.html
DESeq2	https://bioconductor.org/packages/release/bioc/html/DESeq2.html
EdgeR	https://bioconductor.org/packages/release/bioc/html/edgeR.html
igraph	[https://igraph.org/r/; https://igraph.org/python/]
iterative Random Forest (iRF)	[https://github.com/Jromero1208/RangerBasediRF, https://cran.r-project.org/web/packages/iRF/index.html]
Kraken2	[https://ccb.jhu.edu/software/kraken2/]
Pathview	[https://bioconductor.org/packages/release/bioc/html/pathview.html, https://pathview.uncc.edu/]
PHI-base	[http://www.phi-base.org/]
PopArt	[popart.otago.ac.nz/index.shtml]
plyr	[https://cran.r-project.org/web/packages/plyr/index.html]
Random Intersection Trees (RIT)	[https://rdrr.io/cran/FSInteract/man/RIT.html]
Samtools	[http://www.htslib.org/]
Scikit-Bio	[http://scikit-bio.org/docs/0.5.0/index.html]
SRA Tool Kit	[https://github.com/ncbi/sra-tools]
STAR	[https://github.com/alexdobin/STAR]
TrRosetta	[https://yanglab.nankai.edu.cn/trRosetta/, https://github.com/gjoni/trRosetta]
Vegan	[https://cran.r-project.org/web/packages/vegan/index.html]
VEuPathDB	[https://veupathdb.org]
Visual Molecular Dynamics (VMD)	[https://www.ks.uiuc.edu/Research/vmd/]

transcriptome using BWA [13], or against the whole genome using STAR [14] (*see* **Note 2**). Extract the unmapped reads—these will be sequence samples used in taxonomic identification (*see* **Note 3**). Alternatively, CLC Genomics Workbench (v. 20.0.3, Qiagen, Hilden, Germany) can perform trimming using default parameters and alignment.

(b) Determine the quality of the alignment (*see* **Note 4**).

(c) Splitting Sequence Samples: If the samples are too big, it may be necessary to split the sequence samples into various files to process them in parallel, reducing overall execution time. Be sure to keep their original IDs while renaming the series.

5. Use a prebuilt database from Kraken2 (Table 1) if it contains the viral isolates of interest. Otherwise, build a custom database by utilizing publicly available whole-genome sequences. Remember to add the sequences of the viral isolates in addition to targets/pathogens of interest (*see* **Note 5**).
6. Run all the unmapped sequenced clinical samples against the Kraken2 database (*see* **Note 6**). If the sequence samples were split in the preprocessing phase, it is necessary to merge the resulting files to obtain a valid result with all counts for the taxonomic identification of a given single sample.
7. Count the number of classified reads for a given taxon for each respective sample. Kraken2 uses the letters C/U at the beginning of each line in the output to identify if a read was classified or unclassified (*see* **Note 7**).
8. Generate an occurrence matrix after filtering for fungal, bacteria, and/or viral taxa. Here, in the matrix, each row represents a taxon and each column a sample, and the values indicate the number of reads classified as the given taxa in the particular sample (*see* **Note 8**).

3.1.2 Data Analysis

This analysis can be performed using a scripting language of choice, such as R or Python.

1. Identify potential sequence contaminants from the occurrence matrix. These may include for example PhiX174microvirus. These potential contaminants can be discarded.
2. From the occurrence matrix, generate a bar plot of the number of taxa identified per sample. The bar plot should be generated for the lowest level of specificity (*see* **Note 7**). This will indicate samples that may be outliers. A sample can be considered an outlier if it has an abnormal number (either too large or small) of taxa identified for that sample. Discard these outlier samples.
3. Normalize data to account for library size biases and generate relative abundance values (*see* **Note 9**).
4. Determine if there is sufficient confidence in using the data quantitatively or qualitatively. This can be determined by the original quality assessment of the sequencing run together with the library size count (*see* **Note 10**).
5. Aggregate data to a specific taxonomic level, for example, phylum. Calculate a sample-based distance matrix, thereby resulting in a sample by the sample matrix. Where each value

in the matrix represents the dissimilarity between the respective samples. Perform an ordination analysis on the samples, such as Principal Coordinate Analysis (PCoA), and plot the result (*see* **Note 11**).

6. Based on the metadata and dispersion in the PCoA plot, factors that may drive the dispersion of the plot can be investigated for statistical significance. This is done by a PERMANOVA analysis (*see* **Note 12**).
7. Perform an alpha-diversity analysis over the samples. Any replicates of samples can be averaged at this step if desired. Standard alpha-diversity indices are Shannon, Simpson, and Chao1 (*see* **Note 13**).
8. Repeat **steps 13–15** for different taxonomic levels, to better understand the data.
9. Obtain a list of known or potential pathogens from publicly available databases, such as PHI-base and VEuPathDB (Table 1).
10. Determine the influence of potential pathogens on the microbial communities. Perform **steps 13** and **14** on a taxa basis (previously the analysis was performed on a sample basis). Classify taxa as either potentially pathogenic or not. Use this in a similar way as the sample metadata was used in **steps 13** and **14** (*see* **Note 14**).

3.2 Proteome-Wide Structural Comparison of SARS-CoV-2 with Evolutionary-Related Species

1. Establish the reference genomes and perform proteome-wide primary sequence comparison (*see* **Note 15**): The genome MK062179 can be used for SARS-CoV, for example, and NC_045512 for SARS-CoV-2 (both available on NCBI website) [15]. Download from NCBI the amino acid sequences in FASTA format of all mature SARS-CoV and SARS-CoV-2 proteins. Use Clustal Omega [16] or ClustalW2 (available for download from http://www.clustal.org/clustal2/) for pairwise sequence alignment. Export in FASTA format. List the position and identity of all mutations (substitutions, deletions, and additions) (Fig. 3a).
2. Gather information on protein topology and structure–function relationships from the literature related to query proteins and homologs (*see* **Note 16**). UniProt Knowledgebase (UniProtKB) is a valuable resource for locating integrated protein sequence and related functional information [18]. Generate a *feature profile* for each protein by mapping the applicable features enumerated below to the respective residue position(s) in the amino acid sequence. Complement this data curation using recommended prediction methods or servers, listed in parenthesis.

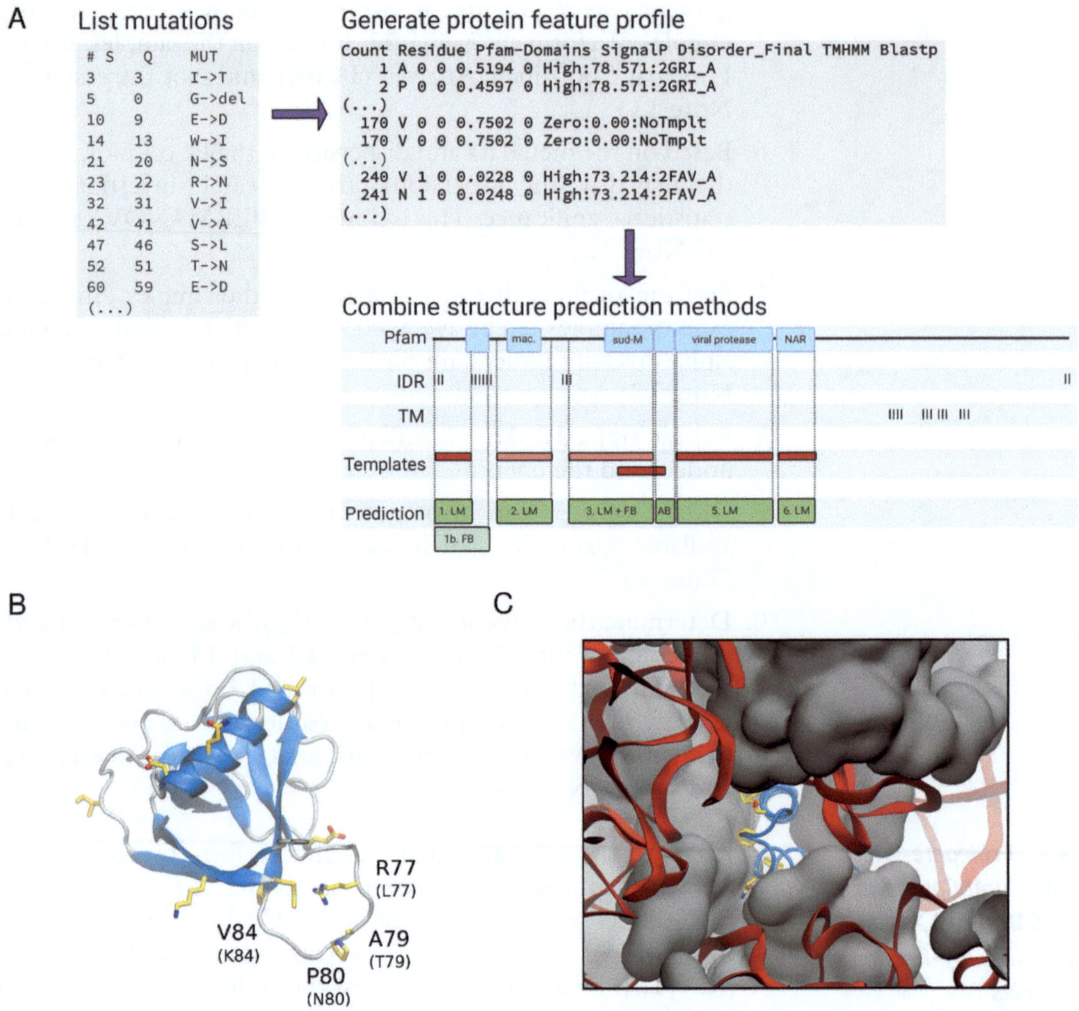

Fig. 3 Main steps for comparative protein structure analysis. (**a**) Sequence parsing and structure prediction of SARS-CoV-2 nonstructural protein 3 (nsp3). List site mutations using a text editor of choice (*see* **step 1**). The "S" column corresponds to the mutation sites in the subject sequence (here, SARS-CoV nsp3) and "Q" column corresponds to the mutation sites in the query sequence (here, SARS-CoV-2 nsp3). Next, collect and assemble a table (the protein feature profile) with functional and structural information for the query sequence from available public repositories (*see* **step 2**). Then, partition the sequence into logical sub-regions that can be modeled independently (*see* **step 6**). Choose the appropriate combination of structure prediction methods, as exemplified (LM: local modeling, FB: fragment-based modeling, AB: ab initio modeling). Find the detailed decision process for this example case in [3]. We note that this example represents a moment prior to the release of experimentally solved structures of SARS-CoV-2 nsp3. (**b**) SARS-CoV-2 nonstructural protein 1 (nsp1, a.a. 13–127, PDB ID 7K3N) with the β3–4 loop (a.a. 76–81) built in silico. Nonconservative substitutions relative to SARS-CoV are depicted in licorice representation (*see* **step 9**). In Prates et al. [3], it is suggested that the substitutions in the β3–4 loop, namely, Leu^{77}Arg, Thr^{79}Ala, Asn^{80}Pro, and Lys^{84}Val, may directly impact pathogenicity. (**c**) C-terminal fragment of SARS-CoV-2 nsp1 bound to rabbit 40S ribosome complex (protein domains are represented in grey and the rRNA, in red, PDB ID 7JQB) [17]. Nonconservative substitutions are found in the region and may affect the interaction of nsp1 with 40S

Table 2
Classification of potential conservative substitutions applied in our studies. Amino acids in brackets have to be analyzed in the structural context

Type	Amino acid
Glycine	G
Aromatic	Y, W, [H], [F]
Nonpolar	A, I, V, L, M, [F], [P]
Polar, hydrogen bond interacting	S, T, [C]
Amidic, hydrogen bond interacting	N, Q
Acidic	D, E
Basic	R, H, K

(a) Signal peptide (signalP) [19].

(b) Intrinsically disordered regions (SPOT-disorder) [20, 21].

(c) Protein domains (PFAM) [22]

(d) Transmembrane regions (TMHMM) [23].

(e) Experimentally solved regions, either of the query sequence itself or of homologs: Use the Protein BLAST (blastp) [24] server to search for the available solved structures. Configure the blastp query to use the Protein Data Bank (PDB) as the search set and use the expected threshold of 0.001. Register the PDB identification code (s) (PDB ID) of the sequence(s) with the highest alignment score and its corresponding identity value relative to the query sequence.

(f) Known posttranslational modification (PTM) sites, such as phosphorylation, ubiquitination, O/N-glycosylation, palmitoylation, disulfide bridges, proteolytic cleavage, and sumoylation [25].

(g) Catalytic and auxiliary catalytic residues.

(h) Other key functional sites identified with mutagenesis experiments.

3. Query each feature server or use a local software package, as appropriate, with the protein sequence as input and obtain and parse the results into a per residue array of values. Depending on the feature, the entry for each residue could be presence/absence, a category type, or a quantitative measure (Fig. 3a). Create a protein sequence profile file for marking per residue feature information, with a column for each feature. Thus, the first two columns would have the residue numbers and amino acid codes, respectively, followed by the feature columns. This

file can then be used for visual inspection or as an input to a script for the application of a set of rules (like those described in **step 6**) for partitioning the sequence into logical, independent regions to which the appropriate structure prediction methods can be assigned.

4. To date, most of SARS-CoV and SARS-CoV-2 proteins are at least partially experimentally solved, including some of the protein complexes they form with host proteins. The summary of available structures is frequently updated on SARS-CoV-2 NCBI resources [26]. Download solved structures from the PDB website (https://www.rcsb.org/).
5. Create a code (for example, in python) to easily shift residue sequence numbers (columns 23–26) of PDB files—it will be helpful in different steps of this protocol.
6. If a predicted structured region is not yet experimentally solved, the feature profile generated in **step 2** for each protein can be used to define the optimum combination of the state-of-the-art methods of protein structure prediction, in a case-by-case manner (Fig. 3a). The following main decision steps are suggested:
   (a) Identify the regions that are amenable for modelling along the protein sequence. An overlap with templates and/or predicted domains, in contrast to predicted intrinsically disorder, indicates stable structured regions.
   (b) Search for models generated with well-established methods of protein structure prediction. We particularly recommend the models predicted with the AlphaFold2 system (*see* **Note 17**), which was used to solve understudied SARS-CoV-2 proteins, such as, M, nsp2, nsp4, nsp6, and the C-terminal domain of nsp3 [27].
   (c) For regions of high identity (>70%) between target and template: In this case, map the substitutions on the template structure to evaluate their likely structural impact, as described in **step 9**. If all the substitutions are structurally conservative (*see* **Note 18**), they can be locally modelled (LM) directly on the template structure, as well as any short missing loop (*see* **Note 19**) (Fig. 3b). However, follow the next item (d) for targets involving structurally nonconservative substitutions, including deletions and additions on structured regions or long-missing loops (>10 amino acid residues). Check for missing loops (not terminals) on the header of the template PDB. Build the short missing loops using RosettaRemodel [28], following these main steps.

- Install Rosetta as described on available documentation online (http://www.rosettacommons.org/). We have been using Rosetta 3.10.
- Generate a blueprint file from the starting PDB:
  - rosetta/tools/remodel/getBlueprintFromCoords.pl -pdbfile [starting pdb] -chain [chain id] > [blueprint file],
- Edit the generated blueprint file to build the loop (*see* Fig. 2 in Huang et al. [28] and documentation in [29]) (*see* **Note 20**).
- Create the flag file defining the input/output files and the parameters to run RosettaRemodel. Find an example of a flag file under "[rosetta_path]/demos/tutorials/loop_modeling" in the Rosetta folder. In a multicore machine, run RosettaRemodel with:
  - [rosetta_path]/main/source/bin/remodel.mpi.linuxgccrelease @flag_missing_loops.

After building the missing loops, renumber residue IDs according to the target protein sequence. Then, model substitutions using Rosetta fixbb application [30]. The first step for that is to change the name of the residues to be mutated on the PDB file (columns 18–20). Then, remove the lines corresponding to its side chain atoms. This will be the input to run fixbb as described in *demos* of the Rosetta documentation online.

(d) Medium identity (30–70%) between target and template, water-soluble region: Use the I-TASSER suite [29] for fragment-based structure prediction (FB) (*see* **Note 21**). Restrict the length of disordered terminals that do not overlap with the templates to not more than five amino acid residues. By doing so, the I-TASSER quality metrics (c-score) will better reflect the prediction accuracy of conserved structured domains.

(e) Low identity (<30% or no template found) between target and template, water-soluble regions: Submit the protein sequence region to deep learning-guided ab initio modeling (AB) using the trRosetta workflow (*see* **Note 22**) [31]. The sequence can be submitted to the trRosetta webserver or the software can be downloaded for local use (Table 1).

(f) Transmembrane regions of medium to low identity between target and template or template free: Submit the corresponding sequence region to AB modelling using the C-I-TASSER server.

(g) As shown in Fig. 3a, one may combine the approaches above to model regions overlapping templates of different sizes and identities.

7. Collect information about the biological assembly of each protein. This can often be found on the Structure Summary of the corresponding PDB webpage. The oligomeric state of a protein or the complex it forms with other biomolecule(s) may be built by structural alignment with available structures (Fig. 3c).

8. Structural alignment of homologous proteins can be done using the *multiseq* plugin of Visual Molecular Dynamics (VMD) [32, 33]. The alignment of specific regions of the proteins that do not necessarily have the same size can be done using LovoAlign [34]. In this case, the selection of the regions to be aligned can be done by identifying it in the column of occupancy (columns 55–60) of the PDB files, for example.

9. Locate the mutation sites on the solved/predicted protein structures: For each protein, load the structures on the VMD program. Set display mode to *orthographic*, use *NewCartoon* representation for the full protein, and display mutation sites using the *licorice* representation (Fig. 3b, *see* **Note 23**). Visual analysis can help to predict if the mutation will affect the local/global protein structure and dynamics. Consider checking the following aspects for each mutation site:

   (a) It is a buried or a surface-exposed site. For more quantitative analysis, count the number of contacts of specific residues using the VMD *Timeline* plugin (*calc. Inter-sel. contacts*). Define selected atoms as pairs "{resid [residue number] and name CB}" and "{protein}". Set a distance cutoff of 4 Å. One can assume that buried residues have a higher number of contacts than exposed residues and define a cutoff between the two types.

   (b) It likely affects the interaction with other residues, such as hydrogen bonds, salt bridges, π-stacking, and hydrophobic contacts. As a preliminary analysis, one may leave out conservative substitutions (*see* **Note 18**, Table 2). On the VMD Display window, press "1" and click on a residue to see its identity and number. Distances and angles within and between residues can also be measured (activate measurement pressing "2" and "3," respectively) to check the possibility of interaction with neighboring residues (*see* **Note 24**).

   (c) It likely forms or subtracts a disulfide bridge (*see* **Note 24**).

   (d) It adds or subtracts a PTM site.

(e) It likely affects the local secondary structure and, thus, protein flexibility. Check residue-specific secondary structure propensity [35].

(f) It is located on the interface with a neighboring protomer in the biological assembly (*see* **step** 7).

Taken together, the assembled feature profile and this structural analysis can be used to identify likely functionally and/or structurally nonconservative substitutions (*see* **Note 25**).

3.3 Tracking the Evolution of SARS-CoV-2

1. Obtain weekly download sequences in FASTA format along with corresponding metadata. Process them using the following steps and append them to previous sequences (*see* **Note 26**).
2. Align full sequences with MAFFT [36] to an established reference genome (NC_045512 for SARS-CoV-2) (see the manual for MAFFT).
3. Upload aligned sequences into the software in which alignments can be trimmed (*see* **Notes 27** and **28**).
4. Trim sequences to the start and stop codons (nsp1 start site and ORF10 stop codon).
5. Export aligned sequences and generated a count of mutations per site (*see* **Notes 29** and **30**).
6. Generate a count of "nonreference" mutations (i.e., the number of individuals that differ from the reference per site). Subtract this number and the number of "N" at each site from the total number of sites. Remove sites with fewer than ten variable (i.e., nonreference) sites (*see* **Note 31**).
7. Remove likely noise. First, divide the "N" counts at each site by the total number of variable sites and remove the sites in which the "N" count is more than 20% of the total variable sites. Next, calculate the total number of non-"N" for the remaining sites and remove those with fewer than ten non-"N" counts. From this final set of "true variable sites," remove all individuals (rows) that have an "N" at any site. The remaining data set should maximize the number of important variable sites in the context of molecular evolution and maximize the sample size.
8. Create the *Variable Site File.* Format the file so that column 1 is the GISAID accession number, column 2 is a "null" character such as a period (".") and the remaining columns are the variable sites. Have a header for each column. Save the file as a tab-delimited file, open it in BBBedit or any text editor of choice (*see* **Note 29**) and remove tab spaces ("find and replace" "\t" by nothing, ""). Replace the period character in column 2 with a tab space. This produces a 2-column file where the first

column is a unique identifier and the second is a concatenated sequence of all variable sites—*the haplotype*—for downstream analyses.

9. Open the file in R using “header = TRUE” and produce an object that consists of the haplotype column. Remove all duplicate sequences using the “unique” function. This is the list of unique haplotypes for the entire population of sequences. This can also be done in Excel™ using the “remove duplicates” tool but with large numbers of variable sites, it may overload the system.
10. Create the *Unique Haplotype File*. Label each of the haplotypes (Hap001, Hap002, etc.). This will produce a 2-column file in which column 1 is the haplotype name and column 2 is the haplotype sequence.
11. Create the *Haplotype Network File*. Using R, open the *Variable Site File* and the *Unique Haplotype File*. Make sure that the columns with haplotype sequences in each of the two files have the same header name. Use the “join_all” function in the package *plyr* to “left join” the *Variable Site File* to the *Unique Haplotype File* using the header of the haplotype sequence column. This will produce a new file in which column 1 is the GISAID accession, column 2 is the null character, column 3 is the haplotype sequence, and column 4 is the haplotype name. If using Excel™, the VLOOKUP formula may work but with large numbers of variable sites, it may take a very long time on a standard personal computer.
12. Create the *Median-Joining File*. Produce a new object in R that consists of columns 3 and 4 from the *Haplotype Network File*. Remove duplicate haplotypes using the haplotype name and change to NEXUS format (*see* **Note 32**). One should now have a file of unique haplotype sequences and the n haplotypes labels.
13. *Cytoscape Network File*: Open the *Median-Joining File* in the software PopArt [37], produce a median-joining network setting epsilon to 0. Export the network as a table—column 1 is a haplotype, column 2 is an edge, and column 3 is the haplotype that is connected to the one listed in column 1.
14. Create the *MetaTable File* for upload to Cytoscape [38]. Open the metadata file downloaded from GISAID, and the *Variable Site File* in R. Use the *plyr* package and “left join” the GISAID metadata csv file to the *Variable Site File* using the GISAID accession number.
15. Calculate *Success Metric*: Generate counts of individuals per haplotype, number of geographic regions that a haplotype is found in, and the number of days it has persisted (*see* **Note 33**). Divide the number of individuals by the number of days and

then by the number of geographic regions (*see* **Note 34**). This is the absolute success number. Unitize the metric by dividing each of these numbers by the lowest success number (1 will then be the lowest number).

16. *Cytoscape MetaTable File*: Generate a file from the *MetaTable File* and the *Success metric* data with the haplotype name, the number of individuals with that haplotype, and the success metric of each haplotype.
17. *Cytoscape Network*: Load the *Cytoscape Network File* in Cytoscape and define source, edge, and target nodes (columns 1, 2, and 3 respectively). Import the *Cytoscape MetaTable File*. Cytoscape is very flexible and one can incorporate any information of interest, as long as it is imported with the same haplotype naming structure as the *Cytoscape Network File*.
18. Reduce feedback loops due to homoplasy by removing haplotypes with fewer than 10 individuals. See Cytoscape manual for further information on network formatting.

3.4 X-AI Driven Predictive Models

3.4.1 Run iRF-LOOP

1. Format data into sample-row and feature-column format. Remove sample IDs.
2. For each feature, create and run an iRF (*see* **Notes 35** and **36**) model using an iRF codebase, either C++ [9] for big data or R [8] for small data (*see* **Note 37**).
3. Normalize each feature importance score for each iRF model by dividing each feature importance value by the sum of that model's feature importance.
4. Add a "Dependent Variable" Column to the normalized feature importance files.
5. Combine all the normalized feature importance files together (*see* **Note 38**).

3.4.2 Run RIT

1. Run RIT [10] on the path files from each of the resulting iRF models, producing the sets of interacting features in each model (*see* **Notes 39–41**).
2. Analyze the sets for anomalies, such as features that appear in many sets or always appear with specific other features.

3.4.3 iRF-LOOP and RIT Applied to Molecular Evolution

By running iRF-LOOP on the mutations in the sequences of SARS-CoV-2 virus one can obtain information regarding site coevolution.

1. Format the virus sequence data such that each row is a unique sequence and each column is a mutation from a given reference, with the matrix cell values being binary—zero for "does not contain this mutation," and one for "does contain this mutation."

2. Run iRF-LOOP (*see* **steps 1–5**) on the mutation data file.
3. Create a graphical network of the results. Mutations that act as a destabilizer, triggering several new mutations, will be connected toward them and mutation hotspots will be pointed at by the number of preceding mutations.
4. Run RIT (*see* **steps 6** and 7).

3.5 Understanding Effects of SARS-CoV-2 Infection on Hosts

Transcriptomics from Infected Patient and Control Samples

1. Obtain samples and reference genome as described in Subheading 3.1, **step 2**. PRJNA605983 includes nine bronchoalveolar lavage fluid samples from patients infected with SARS-CoV-2. NCBI's SRA run selector provides a table of metadata for each sample. This study does not include control samples; therefore, the metadata are used to identify PRJNA434133, which includes forty control samples sequenced using similar methods. Download the FASTQ data files using NCBI's SRA Toolkit software [39]:
   (a) Find the study on the SRA Run Selector site (acc = PRJNA605983):
   - https://www.ncbi.nlm.nih.gov/Traces/study/.

   (b) Select Accession List to download the list of runs as a text file (SRR_Acc_list.txt).
   (c) Download the SRA archive files. This command will download an SRA compressed file for each run in the study:
   - prefetch --option-file SRR_Acc_list.txt,

   (d) For each run, extract the fastq files from the SRA compressed file with the following.
   - fasterq-dump SRR11092056.sra.
2. Align the sample reads to the reference. The SARS-CoV-2 genome can be used as a reference (MN908947) in addition to the human reference genome (GRCh38).
3. Once the sample data and reference genomes have been downloaded, align reads to the reference as described in Subheading 3.1, **step 4**. Alternatively, create a combined reference with alignment software such as STAR [14] to simplify assigning reads.
4. After alignment, reads mapped predominantly to repetitive regions can be discarded. This includes, for example, reads that map to Alu elements, pseudogenes, and repetitive sequences should be removed.
5. Following alignment and read counting, counts should be normalized using the Trimmed Mean of M-values (TMM)

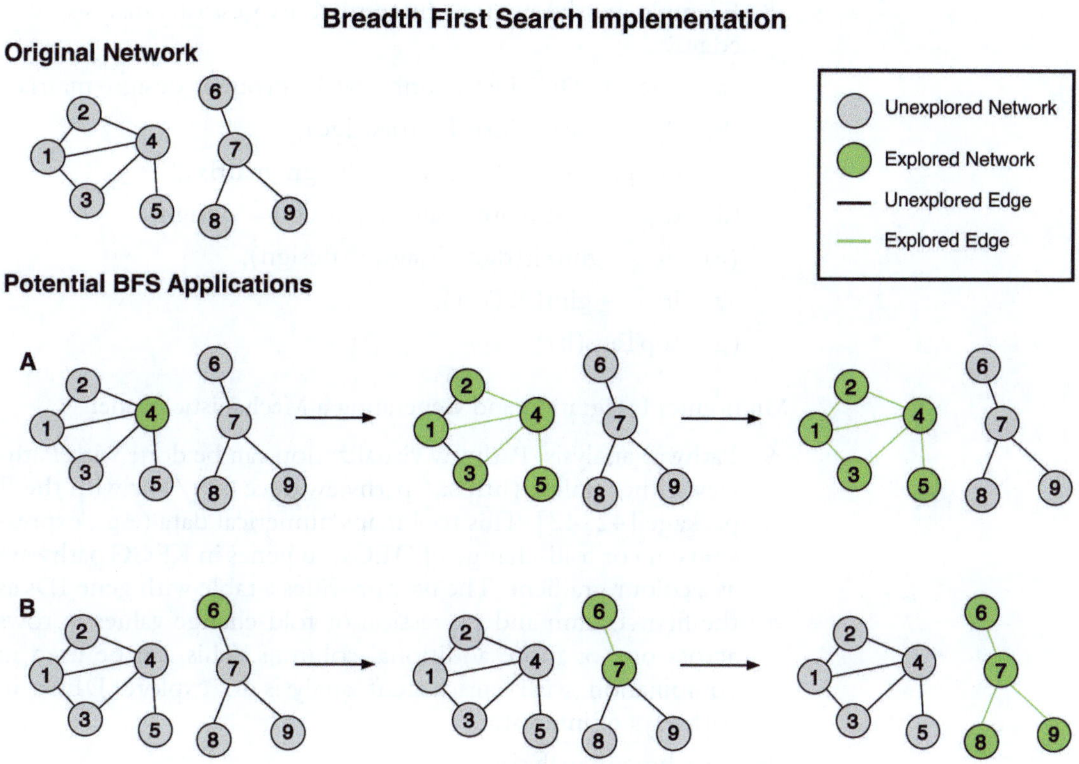

Fig. 4 BFS on gene-to-gene network data. Two separate BFS implementations are shown. By implementing BFS from a list of seed genes of interest, a community of genes predictive of those seed genes can be extracted. Assuming nodes 4 and 6 in the above network are the genes of interest, implementing BFS with each as a seed (examples *A* and *B*, respectively) yields two separate communities with predictive power respective to these genes

method for cross-study comparisons (i.e., if infected and healthy samples are from different sources). This method is available in the edgeR package via the "calcNormFactors" function [40].

6. Ordination plots (e.g., PCA or UMAP) can be used for exploring the initial results; these methods project high-dimensional data (n-dimensions = n-genes) into fewer dimensions (typically 2–3) for visual inspection. Particular attention should be paid to any outlier samples that may need to be removed before differential expression analysis.
7. Differential gene expression (DGE) analysis can be done with edgeR or DESeq2 [41]. Inspect transcript-level expression values for differentially expressed genes (DEGs) of interest. Remove transcripts mapping to artifacts that were not removed automatically, as well as those that encode a truncated form of the protein. Once these are removed, repeat the analysis to verify reproducibility.

8. Example workflow for differential expression analysis with edgeR:
   (a) dge < – DGEList(count_matrix, group = design_matrix),
   (b) dge < – calcNormFactors(dge),
   (c) design <– model.matrix(~design_matrix),
   (d) dge < – estimateDisp(dge, design = design),
   (e) fit <– glmFit(dge, design = design),
   (f) lrt < – glmLRT(fit),
   (g) topTags(lrt).

Multiomics Integration and Generating a Mechanistic Model

9. Pathway analysis. Pathway visualization can be done with Pathview either online (https://pathview.uncc.edu/) or with the R package [42, 43]. This tool maps numerical data (e.g., expression data or fold-change of DEGs) to genes in KEGG pathways as a colour gradient. The user provides a table with gene IDs as the first column and expression or fold-change values in rows across one or more additional columns. This can be used in combination with enrichment analysis to explore DEGs in pathways of interest.
10. Enrichment analysis.
    (a) In order to derive biological relevance, lists of differentially expressed genes can be tested for functional enrichment through various ontologies (i.e., biological processes, phenotypes, coexpression, diseases, and transcription factor binding sites) in the ToppGene Suite (toppgene.cchmc.org) [44], which integrates Gene Ontology with various databases, including PANTHER [45], KEGG [46], Reactome [47], MSigDB [48], and DrugBank [49] (*see* **Note 42**).
    (b) On the ToppGene web portal, access ToppFun for functional enrichment lists. Copy and paste lists of DEGs by HGNC symbol, Entrez ID, Ensembl ID, or RefSeq ID (*see* **Note 43**).
    (c) Under Calculations, select among the several options (e.g., GO, Human Phenotype, Domain Pathway etc.) for ontology features. Ontology enrichment thresholds (p-values and p-value correction) can also be adjusted as desired. Default values for this section can be used if exploring a gene list for the first time (all features, FDR-corrected p-value cutoff <0.05).
    (d) Compare reported clinical symptoms from the disease to be studied to ontologies from ToppFun. In particular, the GO: Biological Process, Human Phenotype, and

Disease ontologies can be used to determine which genes contribute to certain pathological processes. Select "Genes from Input" to identify which genes from the input list contributed to the ontology and "Genes in Annotation" to see the complete list of genes in the ontology (*see* **Note 44**).

11. Integrating omics datasets through network analysis.

(a) Another strategy for integrating results from differential gene expression analysis is to use gene lists as starting genes (also known as "seed genes") for graph traversal using graphs composed of multiple gene-to-gene networks. To build a multilayer network, first download existing published networks (e.g., HumanNet [11] or STRING [50]) containing gene-to-gene edges from various lines of experimental evidence, including: high-throughput omics datasets, protein–protein interaction data, and genes cocited in the same publication.

(b) Predictive-expression networks (PEN) can be useful components for integrated analyses and graph traversals.

- Genetic data are obtained through the RNASeq V8 data sets hosted by the Genotype-Tissue Expression (GTEx) project [12].

  Extract individual parent tissue (SMTS) and accurate tissue site (SMTSD) RNAseq sample ID data from the V8 Annotations Sample Attributes data file. Genetic data per tissue type can then be matched with corresponding sample IDs in the Gene TPM data set.

- For a tissue, apply iRF-LOOP, following the steps in Subheading 3.4 (**steps 1–5**), with each individual corresponding to a sample-row and the genes to the feature-columns (*see* **Note 45**).

- See **Note 38** for the generation of a graphical network from the total concatenated importance files. The generated network contains directed edges between any two genes X and Y. Edges are weighted such that the greater its weight, the higher the probability of gene X predicting gene Y (*see* **Note 46**).

(c) Once all desired network layers are downloaded, read in the files using R with header = FALSE. Be sure that all gene IDs are consistent across all networks (do not mix Ensembl IDs with Entrez IDs, RefSeq IDs, etc.) Create each network layer as an independent graph using the *graph_from_data_frame* function from the igraph package (*see* **Note 47**).

(d) Using the *union* function from the igraph package in R to generate a gene-to-gene network composed of multiple network layers.

(e) Read in lists of differentially expressed genes identified from the transcriptomic analysis. Ensure that these "seed gene" IDs are in the same format as all network layers. Using the parameters "unreachable = FALSE, order = TRUE, rank = TRUE", use the *bfs* function from the igraph package to perform a breadth-first search (BFS) of the combined graph layers from each seed gene in the list. *See* Fig. 4 as an example of how BFS uses seed genes to explore communities with predictive power.

(f) Use highly ranked genes from BFS (e.g., top 50–100 ranked genes) and identify if these genes are differentially expressed in COVID-19 samples. Repeat the "Pathway Analysis" and "Enrichment Analysis" sections to glean biological or pathological insights.

12. Integrating host–viral protein–protein interaction networks. Download existing viral–host protein–protein interaction networks. These networks contain edges from high-throughput experiments used to determine which SARS-CoV-2 proteins interact with human proteins, with gene IDs used as edges [51]. Human host genes of interest identified via differential expression or pathway, enrichment, or network analysis can be cross-referenced with these networks to determine if SARS-CoV-2 proteins can directly affect corresponding human proteins.

13. Repurposing approved drugs based on output.

(a) Using genes of interest identified from **steps 1–12**, putative treatments can be proposed using approved drugs that target the proteins encoded by these genes.

(b) Download publicly available drug-to-drug target protein networks: examples include DrugBank [49] and ChEMBL [52]. Import these networks into R, Python, or Excel™.

(c) Search for edges containing genes of interest to identify possible drugs of interest that interfere with viral pathogenesis (*see* **Note 48**).

4 Notes

1. Adapter and poly(A) tail sequences need to be removed from the reads. The Phred [14, 53] quality score of reads varies based on several factors. A suggested minimum threshold to

apply is 20. This threshold corresponds to an estimated base call accuracy of 99%.

2. Using a splice-aware aligner (such as STAR) for RNASeq data is preferred. However, alternative strategies can include a pseudoaligner, such as Kallisto or Salmon, against the transcriptome. If these options are not available, a genome aligner (for example BWA) can be used against the transcriptome.
3. Some aligners allow for the option to output unmapped reads as separate files. Use this option if available. Alternatively, Samtools (Table 1) can extract the unmapped reads using the command:

   samtools view -f 4 sample.bam > unmapped.sam.
4. Numerous potential biases can impact alignment. It is therefore impossible to do an exhaustive quality assessment. However, a few factors can be considered, such as: Are there any dramatic differences in the percentage of mapped reads between samples? How many reads are in the sample (i.e., library size)? Do the reads from a particular sample align preferentially to repetitive or low complexity regions? Is there a high percentage of mismatches among the aligned reads for a sample? Considering these can help determine if the mapping of a sample is reliable.
5. If computing capacity is limited, we suggest that a small set of genomes be used when constructing the custom database. The building of a Kraken2 database can be computationally intensive, and the resultant database may be too large for the subsequent classification. Furthermore, if a new database is being built, the FASTA sequences need to be processed to include the taxaID in the sequence identification line. Please refer to the Kraken2 manual for the required format (https://github.com/DerrickWood/kraken2/wiki/Manual).
6. If the flag --use-names is used, the subsequent taxa identification may be easier. Without this flag, Kraken2 will use the numeric taxaID, which would then have to be manually parsed.
7. For the particular scientific question, it is important to determine the lowest level of taxonomic identification that is needed. For example, at the strain level, it may be unlikely that the particular strain in the database is also present within the clinical sample. However, taxa of the corresponding genera or species may be present. It is therefore up to the user to determine the level of specificity required for their analysis. We suggest an initial analysis at the species level together with additional analysis at the genera level.
8. We suggest that the taxonomic names include the full classification from kingdom to phylum all the way down to the level of

specificity. This makes it easier to aggregate the occurrence matrix to a different taxonomic resolution.

9. There are several approaches to adjust for library size biases. We use a basic per sample normalization approach by dividing each taxon read count by the total number of reads in that sample. The normalized values are then scaled to a reasonable range by multiplying by a constant factor, such as 10e4. The scaling factor is important as some downstream analyses may require integer values. To obtain these values, we take the floor of the respective values, for example, 42.7 becomes 42.
10. The threshold for choosing between qualitative or quantitative is hypothesis and data-dependent. We suggest choosing a minimum read count threshold (or relative abundance threshold) and doing an initial qualitative analysis. This can be followed by a quantitative analysis if needed.
11. There are several software options for this. We suggest using either the Scikit-Bio package in Python or the vegan package in R. The process is quite straightforward and is generally detailed by the package itself. For the distance matrix either UniFrac distance can be used (if the full phylogeny is available) or Bray-Curtis dissimilarity. Colour the respective samples using metadata to investigate the resultant pattern in the plot.
12. The choice of analysis is dependent on the nature of the data and how appropriate the model assumption may be. In general, PERMANOVA is used when an Analysis of Similarity (ANOSIM) is performed to determine the significance of group variables (metadata). There are appropriately named functions (anosim) to do this in both Scikit-Bio and vegan.
13. Both the Scikit-Bio and vegan packages have documentation on how to perform these diversity analyses using the respective package.
14. There are several potential caveats at this step. The most likely scenario is that the number of putative pathogens may be small compared to the non-pathogens, thus leading to a class imbalance. In this case, several runs of downsampling the non-pathogens can be used in a bootstrapping approach to investigate statistical significance from the ANOSIM analysis. Alternatively, for matrix values within a reasonable range, a Fisher Test can be used. Finally, a descriptive approach may also be appropriate without a statistical analysis. An example of the latter would be clustering the dissimilarity results from a taxa-based Unifrac/Bray–Curtis calculation and then seeing where pathogen/nonpathogen cluster.
15. Here, as an example, we consider the comparative structural proteomics of SARS-CoV and SARS-CoV-2, as in Prates et al. [3]. Circulating in humans since, at least, December 2019,

researchers have now registered more than 360,000 SARS-CoV-2 variants (GISAID database) [54]. Here, the choice of using as the reference genome the first released sequence of SARS-CoV-2 (NC_045512) is based on the assumption that it has all the molecular features that drive its extremely differing spread rate relative to SARS-CoV [55].

16. Besides conducting a literature review using online engines for broad literature research, such as Google Scholar, PubMed Central, and Web of Knowledge, we particularly recommend looking for information about protein structure–function relationships at the literature cited on the Protein Data Bank (PDB) webpage of the protein structures related to the study.

17. The structures predicted with AlphaFold2, a newly developed deep learning-based engine, are particularly recommended as it showed a remarkable superior performance relative to other methods in the last Critical Assessment of Structure Prediction competition (CASP14). For instance, the predicted structure of ORF3a was solved by AlphaFold2 despite being a challenging target due to the unavailability of related sequences. In AlphaFold2, a folded protein is treated as a "spatial graph," where residues are the nodes and edges connect the residues in close proximity. Via an attention-based neural network system and multiple sequence alignment (MSA) with evolutionary-related sequences, the system develops strong predictions of the protein's underlying physical structure and can efficiently determine highly accurate structures. Additionally, AlphaFold2 can predict which parts of predicted protein structure are reliable using an internal confidence measure.

18. Here, we classify the mutations that meet at least one of the following aspects as structurally nonconservative: it breaks/form disulfide bridges, it involves residues with different physicochemical properties (potential conservative substitutions are shown in Table 2), it likely adds/subtracts key interactions, or it breaks the local secondary structure. On the other hand, surface exposed conservative mutations that change the PTM pattern or the intermolecular interactions in a protein complex are also sites that may worth further investigation regarding their functional effects, even though they do not necessarily affect protein structure.

19. LM can be easily performed using the CHARMM-GUI interface [56], through the Solution Builder tool, for example. However, here we describe the use of Rosetta applications as they can be run locally and integrate a pipeline for proteome-wide structure comparison.

20. Note that position numbering in the blueprint file starts with 1 and it is not discontinuous on the missing loop regions.

Thus, the blueprint numbering will not match the residue number of the PDB file. We suggest creating a copy of the template PDB file with renumbered residue positions starting with 1 to help identify the position in the blueprint file in which the loop will be inserted.

21. The newer C-I-TASSER workflow [57], which applies deep-learning for contact prediction generates more accurate models than I-TASSER, but a standalone package of C-I-TASSER is not yet available. Here, the choice of using the I-TASSER suite is due to the possibility of performing embarrassingly parallel runs of multiple targets in local machines. I-TASSER was ranked the best function prediction in CASP9 and the workflow has been continuously among the top prediction methods in the subsequent contests. In general, the method provides the correct global topology for the cases described in **step 6** (Subheading 3.2), which is sufficient for the proposed goals. The estimated quality of the predicted topology is assessed with the TM-score (TM-score > 0.5 indicates a correct topology). However, further refinement is highly advised [58] to use the I-TASSER predicted models on molecular dynamics simulations, as even minor local inaccuracies can be propagated once velocities are assigned to the atoms, leading to major conformational deviations.

22. The trRosetta (transform-restrained Rosetta) workflow generates protein structure models through two main steps: (1) the prediction of inter-residue orientations and distances via a deep residual-convolutional neural network which takes an MSA as input, and (2) energy minimization via a fast Rosetta model building protocol using distance and orientation restraints derived from (1).

23. Visual inspection is a relatively simple and important step in studying protein structure–function relationships. Therefore, getting familiar with the many user interface components and molecular drawing methods of VMD (or another preferred software) is an important part of the procedure described here. For visual clarity, we recommend using a graphical representation that depicts the secondary structure backbone and the side-chain atoms of the mutation sites. For example, one can set the following.

    (a) Protein backbone: Drawing method—*NewCartoon*; Coloring Method—*ColorID 13* (mauve).

    (b) Mutation sites: Drawing method—*Licorice*; Coloring Method—*Name* (using orange for C atoms—the color of atom types can be defined in Graphics > Colours); select only heavy side-chain atoms in Selected Atoms with resid [mutation site numbers] and not name C O N and noh.

24. Consider the estimated quality/resolution of the protein structure being used to make any inference about possible interactions between amino acid residues. For example, a typical hydrogen bond may be identified adopting as geometric criteria a cutoff of 3.0 Å for donor-acceptor distance and 20° for acceptor-donor-H angle. Similarly, a S-S distance cutoff of 3.0 Å is applied for disulfide bonds in the PDB database. These conditions should be relaxed depending on the resolution of the model. Evaluate if such criteria would be achieved with side-chain dihedral rotations or translational motions of residues in flexible regions like loops and coils.
25. We emphasize that the present method does not aim to provide, per se, conclusive information about the functional effects of site mutations. Ideally, it should instead be applied in conjunction with other omics layers to gather multiple lines of evidence for the hypotheses raised and guide further experimental and in silico studies. For the latter, we point out the DynaMut server [59] for quantitative estimation of the impact of specific mutations of interest on protein flexibility and stability, for example. Moreover, a valuable part of the current efforts against the SARS-CoV-2 pandemic is the search for potential antiviral compounds targeting viral proteins with virtual screening. Approaches relying on ensemble molecular docking and machine-learning-based scoring functions have been described [60]. To expand the study of proteins of higher interest using such an approach, which involves molecular docking and molecular dynamics simulations, achieving the highest possible accuracy of predicted models cannot be overlooked, despite the longer times it will demand. Additionally, it is possible that AlphaFold2, once publicly available, may reduce the need for sequence partitioning, described in **step 6** in Subheading 3.2. However, CASP14 overall results indicate that AlphaFold2 did require human input in some edge cases and nearly one-third of its predictions did not achieve comparable quality to experimental structures. Also, other methods did outperform AlphaFold2 in a few cases. Therefore, preprocessing sequences may still be valuable to rapidly identify and dissect edge cases.
26. This molecular evolution protocol is based on our published analysis [6] and the pipeline we provide above includes steps leading to an improved signal to noise ratio. In addition, because bioinformatic expertise may differ among readers, we provide steps to perform this in widely used software such as Microsoft Excel™.
27. If the number of aligned sequences is greater than 3000, work with smaller batches if using a standard personal computer.

28. Many sequence alignment viewers including CLC Genomics do not export the insertion character, so change "-" to "d" before uploading. Although CLC Genomics is a commercial product, the Viewer is a free resource with many options for manipulating alignments and sequences. This graphical user interface (GUI) presents sequences in a manner that makes multiple sequence alignments easy to edit.

29. Export the file as .csv for Excel or .txt for the R environment (https://cran.r-project.org/). To easily manipulate large files, we recommend the text editors BBBedit, 010Editor, or Textwrangler. Count all IAPUC codes that are not "A," "T," "G," or "C" as "N" unless the "-" character was changed to "d." If using Excel™, save the .csv file exported from CLC Genomics as a tab-delimited text file. Open it in a text editor and replace each nucleotide label by itself followed by a tab character (e.g., find "A" and replace it with "A\t"). This generates a file from which site counts can be easily calculated (*see* **Note 30**).

30. Generating site counts can be done in R or Excel™. If using R, large files are easily uploaded with the "fread" function in the *data.table* package. In Excel™, there is a limit of 16,000 columns. Therefore, split the alignment in half prior to exporting it from CLC Genomics. Then, perform site counts, reduce to variable sites, and merge the halves once the number of sites has been reduced. A useful package for manipulating files is *plyr* in R using the "join_all" function that allows for defining a common feature to join files. In Excel™, the VLOOKUP function (see Microsoft manual) can be used to merge files, but with large files this may overload a less powerful desktop or laptop.

31. If using Excel™, the most straightforward means is to have the NC_045512 reference sequence in the first row and use the COUNTIF function for each column. Then subtract the total number of rows (using COUNTA). This is the total number of nonreference sites that includes "N" sites.

32. Formatting files for different software packages is a tedious task in Data Science. If using CLC Genomics Workbench, import a FASTA alignment of the haplotypes with their corresponding haplotype names and then export it as a NEXUS formatted file that PopArt will accept. Do *not* use the dash ("-") character in the haplotype name; it generates an unreadable file in PopArt. Check the example files of PopArt for comparison (*see* PopArt manual). The median-joining algorithm can rapidly increase execution time on a computer system if more than 1000 haplotypes are used. The processing can be significantly accelerated if rare haplotypes are removed (e.g., less than five occurrences). This also reduces homoplasy loops that can occur from either

back mutations to an ancestral state or recombination among strains. There are two other packages for generating median-joining networks. NETWORK is an older, windows-based software that is efficient, but will not consider more than 500 variable sites and its graphical interface is not as versatile as PopArt. The R package PEGAS has a median-joining algorithm implementation but due to the quadratic nature of its complexity time, the execution time can grow quickly, not converging if there are any sizable number of sites or sequences. The same occurs with PopArt, but it is currently the best option.

33. In Excel™, create a file from the MetaTable in which column 1 is the GISAID accession, column 2 is the haplotype name, and column 3 is the sample date of the GISAID accession. Sort by ascending date and then use the tool "remove duplicates," using the haplotype name as the character that identifies duplicates. One date for each haplotype will be left, which will be the earliest sample registered. Now repeat this procedure, but this time sort by *descending* date. This will return the latest date for each haplotype. Use either *plyr* in R or VLOOKUP in Excel™ to create a file with column 1 as the haplotype name, column 2 as the earliest date sampled, and column 3 as the latest. Subtract column 2 from column 1. This is the half-life of the haplotype.
34. Create a file from the MetaTable in which column 1 is the GISAID accession, column 2 is the haplotype name, and column 3 is the geographic region sampled. Remove all duplicates using both haplotype name and geographic region sampled. Use the COUNTIF function in Excel™ to count the numbers of each haplotype remaining. This is the count of countries that the haplotype is found in.
35. Iterative Random Forest (iRF) is an X-AI method that takes advantage of the benefits of the classic machine learning method Random Forest and expands upon it with a few extra steps, including Random Intersection Trees (RIT). These steps increase the explainability and add the ability to determine feature interaction.
36. iRF-LOOP uses iRF to build relationships within a feature set by determining the importance of each feature when predicting each other feature, while including the background information of the other features in the dataset.
37. The C++ version is designed for use on HPC systems. This version should be used on a large, distributed system when a desktop or laptop struggles to load or analyze the data set.
38. Optional: Generate a graphical network in a tool such as Cytoscape, treating the "Feature" column and the

"Dependent Variable" column as node labels and the normalized feature importance values as edge weights.

39. RIT is a method to quickly and efficiently find sets of features that co-occur in a dataset by using beneficial properties innate to decision trees.
40. The FSInteract R package, which contains RIT, is loaded when the iRF R module is loaded.
41. To use the RIT R package with the C++ version of iRF requires the additional step of transforming the iRF-produced pathfile into the appropriate format by keeping only the feature names on each row from the pathfile and removing all other information.
42. While this analysis focused on transcriptomic results, the same logic can be applied to results using protein IDs (Uniprot, Entrez, or RefSeq IDs) in order to glean biological understanding from proteomics data.
43. To identify significant ontologies if many genes are differentially expressed, try different combinations of gene lists to identify possibly significant phenotype/biological function/diseases associated with differentially expressed genes. Examples of different thresholds to try: top 50 or 100 upregulated genes (ranked by lowest p-value with log fold change >0.00), top 50 or 100 downregulated genes (ranked by lowest p-value with log fold change <0.00), top 50 or 100 differentially expressed genes (combined upregulated and downregulated genes). Thresholds for significant differential expression can also be adjusted ($p < 0.05$, $p < 0.01$, etc.).
44. Integration and interpretation of results are crucial for making conclusions about the molecular basis of pathology. While including more networks and ontologies can help facilitate this process, human input is necessary to establish new connections. In this respect, having multiple individuals participate in the integration process with varied biological backgrounds can result in a more holistic understanding of the viral pathology.
45. Only tissue-specific data sets with greater than 100 individual samples should be used for the application of iRF-LOOP to ensure enough variance exists within the data set.
46. The question is raised as to whether using a PEN as opposed to using a traditional correlation network is necessary. PENs generated from iRF-LOOP have previously illustrated a greater likelihood of finding more biologically relevant Gene Ontology annotation edges than correlation networks [9].
47. While this protocol uses the R programming environment, Python also implements igraph, which can be used to build multiplex networks.

48. If necessary, to convert gene or protein IDs for DrugBank or other networks, using tools such as gProfiler (https://biit.cs.ut.ee/gprofiler/convert) and Biomart (http://www.biomart.org/) can be used.

5 Conclusions and Future Directions

Here, we provide a series of protocols that we recommend being used in an integrative approach to understanding the current, expanding SARS-CoV-2 pandemic. This series of methods should provide a framework for rapid deployment when the next pandemic begins to emerge.

We are continually improving this series and recommend several additions. For example, the GISAID database has been a critical and necessary tool to combat the current pandemic. Continued support from the global scientific community can ensure that this becomes a central repository for all virus data that can expand to include sex, age, days since symptom onset, list of specific symptoms, zip code-equivalent, a rank of severity on a scale of 1–10, and full nucleotide sequence data, preferably from long-read sequencing technology. We also recommend improved graph traversal strategies for our data integration approaches. While algorithms such as breadth-first search can help identify closely related genes that contribute to pathology, more advanced techniques may also be useful for understanding the mechanism of COVID-19 disease. One such strategy that may prove useful is the guilt-by-association algorithm Random Walk with Restart (RWR), a graph traversal method compatible with multiplex networks. We have recently developed a new version of this approach to integrate it into this workflow. Prior versions of this technique have been successfully used to discover biologically relevant genes contributing to human pathologies [61, 62], and an open-source R package is available [61]. Moreover, in addition to using network-based analyses on homogeneous networks, generating heterogeneous networks (e.g., combining gene-to-gene networks with gene-to-drug target and host gene-to-viral gene networks) may facilitate biological understanding by reducing the required human input.

A critical component that we are currently implementing is the extraction of relevant information from longitudinal electronic health record (EHR) data at a national level in the Veterans Affairs Healthcare System. This rich data comprises comprehensive and highly diverse data on the patient's health such as drug prescription fills and refills, diagnostic and procedure codes, laboratory test results, patient demographic data, open field progress notes, among others. The depth and breadth of these are well suited for the use of advanced X-AI tools such as iRF and RIT to identify important signals related to COVID-19 prognosis. For example, iRF and RIT can be used to identify FDA-approved drugs for immediate repurposing in phase 2 and 3 clinical trials and those that may worsen disease-related symptoms or increase the probability of mortality. Further, iRF and RIT are a powerful combination of tools for identifying comorbidities, vital signs, and critical laboratory abnormalities that increase the risk of death in COVID-19 positive patients, providing means for clinicians to triage cases. Similarly, these methods will be critical in both characterizing the Long Covid Phenotype in survivors of the disease, also known as Long Hauler Syndrome, and determining who is at greatest risk of long-term complications of COVID-19.

Finally, environmental factors also contribute to the spread of SARS-CoV-2 and COVID-19 severity. As with large scale EHR data, global-scale environmental data are available for integration into this workflow. For example, iRF predictive models can be created at county-level granularity using environmental, demographic, and other features to predict disease outcomes. Application of RIT to these results can determine which sets of features interact, even across feature types (i.e., an interaction between an environmental and demographic feature). Together, these described methods allow for a comprehensive systems biology view of an ongoing pandemic and can help guide clinical practices and provide the scientific community with novel hypotheses to be tested, shortening the time required to find solutions.

Acknowledgments

This research funded by the Laboratory Directed Research and Development and Seed funding programs of Oak Ridge National Laboratory (LOIS:10074,10124) and the DOE Office of Science through the National Virtual Biotechnology Laboratory (NVBL), a consortium of DOE national laboratories focused on response to COVID-19, with funding provided by the Coronavirus CARES Act. This research used resources of the Oak Ridge Leadership Computing Facility (OLCF) and the Compute and Data Environment for Science (CADES) at the Oak Ridge National Laboratory, which is supported by the Office of Science of the U.S. Department of Energy under Contract No. DE-AC05-00OR22725. Figures were generated with Biorender and VMD. Erica T. Prates and Michael R. Garvin contributed equally to this work.

Author Contributions: *Erica Prates* (Conceptualization, Investigation, Methodology, Visualization, Writing—original draft, review and editing), *Michael Garvin* (Conceptualization, Investigation, Methodology, Visualization, Writing——original draft, review and editing), *Piet Jones* (Investigation, Methodology, Writing—original draft, review and editing, visualization), *J. Izaak Miller* (Writing—original draft, Writing—review and editing), *Kyle*

A. Sullivan (Investigation, Writing—original draft, Writing—review and editing), *Ashley Cliff* (Investigation, Methodology, Writing—original draft, Writing—review and editing), *Joao Gabriel Felipe Machado Gazolla* (Investigation, Methodology, Visualization, Writing—original draft, Writing—review and editing), *Manesh Shah* (Investigation, Methodology, Writing—original draft, review and editing), *Angelica M. Walker* (Investigation, Methodology, Writing—original draft, review and editing), *Matthew Lane* (Investigation, Methodology, Writing—original draft, review and editing), *Christopher Rentsch* (Investigation, Methodology, Writing—original draft, review and editing), *Amy Justice* (Investigation, Methodology, Writing—original draft, review and editing), *Mirko Pavicic* (Writing—original draft, review and editing), *Jonathon Romero* (Investigation, Methodology, Writing—original draft, review and editing), *Daniel Jacobson* (Conceptualization, Formal Analysis, Funding acquisition, Methodology, Supervision, Writing—original draft, Writing—review and editing).
This work was also funded by the United States Government.

References

1. Garcia BJ, Labbé JL, Jones P et al (2018) Phytobiome and transcriptional adaptation of Populus deltoides to acute progressive drought and cyclic drought. Phytobiom J 2:249–260
2. Wang Q, Zhang Y, Wu L et al (2020) Structural and functional basis of SARS-CoV-2 entry by using human ACE2. Cell 181:894–904.e9
3. Prates ET, Garvin MR, Pavicic M et al (2020) Potential pathogenicity determinants identified from structural proteomics of SARS-CoV and SARS-CoV-2. Mol Biol Evol 38(2):702–715
4. Wang D, Fang L, Shi Y et al (2016) Porcine epidemic diarrhea virus 3C-like protease regulates its interferon antagonism by cleaving NEMO. J Virol 90:2090–2101
5. Lampe J, Wenzel J, Müller-Fielitz H, et al (2020) The SARS-CoV-2 main protease Mpro causes microvascular brain pathology by cleaving NEMO in brain endothelial cells. https://www.researchsquare.com/article/rs-86988/v1. Accessed 19 Jan 2021
6. Garvin MR, Prates ET, Pavicic M et al (2020) Potentially adaptive SARS-CoV-2 mutations discovered with novel spatiotemporal and explainable AI models. Genome Biol 21:304
7. Zhang M, Case DA, Peng JW (2018) Propagated perturbations from a peripheral mutation show interactions supporting WW domain thermostability. Structure 26:1474–1485.e5
8. Basu S, Kumbier K, Brown JB et al (2018) Iterative random forests to discover predictive and stable high-order interactions. Proc Natl Acad Sci U S A 115:1943–1948
9. Cliff A, Romero J, Kainer D et al (2019) A high-performance computing implementation of iterative random forest for the creation of predictive expression networks. Genes 10:996
10. Shah R, Meinhausen N (2014) Random intersection trees. J Mach Learn Res 15:629–654
11. Hwang S, Kim CY, Yang S et al (2019) HumanNet v2: human gene networks for disease research. Nucleic Acids Res 47:D573–D580
12. The GTEx Consortium (2015) The Genotype-Tissue Expression (GTEx) pilot analysis: Multitissue gene regulation in humans. Science 348:648–660
13. Li H, Durbin R (2009) Fast and accurate short read alignment with burrows-wheeler transform. Bioinformatics 25:1754–1760
14. Dobin A, Davis CA, Schlesinger F et al (2013) STAR: ultrafast universal RNA-seq aligner. Bioinformatics 29:15–21
15. NCBI SARS-CoV-2 Resources (2021) Bethesda (MD): National Library of Medicine (US), National Center for Biotechnology Information. https://www.ncbi.nlm.nih.gov/sars-cov-2/. Accessed 15 Jan 2021
16. Madeira F, Park YM, Lee J et al (2019) The EMBL-EBI search and sequence analysis tools APIs in 2019. Nucleic Acids Res 47:W636–W641

17. Yuan S, Peng L, Park JJ et al (2020) Nonstructural protein 1 of SARS-CoV-2 is a potent pathogenicity factor redirecting host protein synthesis machinery toward viral RNA. Mol Cell 80:1055–1066.e6
18. UniProt Consortium (2019) UniProt: a worldwide hub of protein knowledge. Nucleic Acids Res 47:D506–D515
19. Almagro Armenteros JJ, Tsirigos KD, Sønderby CK et al (2019) SignalP 5.0 improves signal peptide predictions using deep neural networks. Nat Biotechnol 37:420–423
20. Hanson J, Paliwal KK, Litfin T et al (2019) SPOT-Disorder2: improved protein intrinsic disorder prediction by ensembled deep learning. Genomics Proteomics Bioinformatics 17:645–656
21. Nielsen JT, Mulder FAA (2019) Quality and bias of protein disorder predictors. Sci Rep 9: 5137
22. El-Gebali S, Mistry J, Bateman A et al (2019) The Pfam protein families database in 2019. Nucleic Acids Res 47:D427–D432
23. Krogh A, Larsson B, Heijne G von, et al (2001) Predicting transmembrane protein topology with a hidden Markov model: application to complete genomes. J Mol Biol 305:567–580
24. Altschul SF, Gish W, Miller W et al (1990) Basic local alignment search tool. J Mol Biol 215:403–410
25. Fung TS, Liu DX (2018) Post-translational modifications of coronavirus proteins: roles and function. Future Virol 13:405–430
26. SARS-CoV-2 conserved domains and 3D structures (2021) Bethesda (MD): National Library of Medicine (US), National Center for Biotechnology Information. https://www.ncbi.nlm.nih.gov/Structure/SARS-CoV-2.html. Accessed 2 Jan 2021
27. Jumper J, Tunyasuvunakool K, Kohli P, Hassabis D, The AlphaFold Team (2020) Computational predictions of protein structures associated with COVID-19, Version 3, DeepMind website4. https://deepmind.com/research/open-source/computational-predictions-of-protein-structures-associated-with-COVID-19. Accessed 2 Jan 2020
28. Huang P-S, Ban Y-EA, Richter F et al (2011) RosettaRemodel: a generalized framework for flexible backbone protein design. PLoS One 6: e24109
29. Yang J, Yan R, Roy A et al (2015) The I-TASSER suite: protein structure and function prediction. Nat Methods 12:7–8
30. Kuhlman B, Baker D (2000) Native protein sequences are close to optimal for their structures. Proc Natl Acad Sci U S A 97: 10383–10388
31. Yang J, Anishchenko I, Park H et al (2020) Improved protein structure prediction using predicted interresidue orientations. Proc Natl Acad Sci U S A 117:1496–1503
32. Roberts E, Eargle J, Wright D et al (2006) MultiSeq: unifying sequence and structure data for evolutionary analysis. BMC Bioinformatics 7:382
33. Humphrey W, Dalke A, Schulten K (1996) VMD: visual molecular dynamics. J Mol Graph 14(33–8):27–28
34. Martínez L, Andreani R, Martínez JM (2007) Convergent algorithms for protein structural alignment. BMC Bioinformatics 8:306
35. Koehl P, Levitt M (1999) Structure-based conformational preferences of amino acids. Proc Natl Acad Sci U S A 96:12524–12529
36. Katoh K, Misawa K, Kuma K-I et al (2002) MAFFT: a novel method for rapid multiple sequence alignment based on fast Fourier transform. Nucleic Acids Res 30:3059–3066
37. Leigh JW, Bryant D (2015) Popart: full-feature software for haplotype network construction. Methods Ecol Evol 6:1110–1116
38. Shannon P, Markiel A, Ozier O et al (2003) Cytoscape: a software environment for integrated models of biomolecular interaction networks. Genome Res 13:2498–2504
39. Leinonen R, Sugawara H, Shumway M et al (2011) The sequence read archive. Nucleic Acids Res 39:D19–D21
40. Robinson MD, McCarthy DJ, Smyth GK (2010) edgeR: a Bioconductor package for differential expression analysis of digital gene expression data. Bioinformatics 26:139–140
41. Love MI, Huber W, Anders S (2014) Moderated estimation of fold change and dispersion for RNA-seq data with DESeq2. Genome Biol 15:550
42. Luo W, Brouwer C (2013) Pathview: an R/Bioconductor package for pathway-based data integration and visualization. Bioinformatics 29:1830–1831
43. Luo W, Pant G, Bhavnasi YK et al (2017) Pathview Web: user friendly pathway visualization and data integration. Nucleic Acids Res 45:W501–W508
44. Chen J, Bardes EE, Aronow BJ et al (2009) ToppGene suite for gene list enrichment analysis and candidate gene prioritization. Nucleic Acids Res 37:W305–W311
45. Mi H, Ebert D, Muruganujan A et al (2021) PANTHER version 16: a revised family classification, tree-based classification tool, enhancer

regions and extensive API. Nucleic Acids Res 49:D394–D403

46. Kanehisa M, Furumichi M, Tanabe M et al (2017) KEGG: new perspectives on genomes, pathways, diseases and drugs. Nucleic Acids Res 45:D353–D361
47. Fabregat A, Korninger F, Viteri G et al (2018) Reactome graph database: efficient access to complex pathway data. PLoS Comput Biol 14:1–13
48. Liberzon A, Birger C, Thorvaldsdóttir H et al (2015) The molecular signatures database (MSigDB) hallmark gene set collection. Cell Syst 1:417–425
49. Wishart DS, Feunang YD, Guo AC et al (2018) DrugBank 5.0: a major update to the DrugBank database for 2018. Nucleic Acids Res 46: D1074–D1082
50. Szklarczyk D, Gable AL, Lyon D et al (2018) STRING v11: protein–protein association networks with increased coverage, supporting functional discovery in genome-wide experimental datasets. Nucleic Acids Res 47:D607–D613
51. Gordon DE, Jang GM, Bouhaddou M et al (2020) A SARS-CoV-2 protein interaction map reveals targets for drug repurposing. Nature 583:459–468
52. Nowotka MM, Gaulton A, Mendez D et al (2017) Using ChEMBL web services for building applications and data processing workflows relevant to drug discovery. Expert Opin Drug Discov 12:757–767
53. Ewing B, Green P (1998) Base-calling of automated sequencer traces using phred. II. Error probabilities. Genome Res 8:186–194
54. Elbe S, Buckland-Merrett G (2017) Data, disease and diplomacy: GISAID's innovative contribution to global health. Glob Chall 1:33–46
55. Wu F, Zhao S, Yu B et al (2020) A new coronavirus associated with human respiratory disease in China. Nature 579:265–269
56. Qi Y, Cheng X, Han W et al (2014) CHARMM-GUI PACE CG builder for solution, micelle, and bilayer coarse-grained simulations. J Chem Inf Model 54:1003–1009
57. Zheng W, Zhang C, Li Y et al (2021) Folding non-homology proteins by coupling deep-learning contact maps with I-TASSER assembly simulations. Cell Rep Methods 1(3): 100014
58. Heo L, Arbour CF, Feig M (2019) Driven to near-experimental accuracy by refinement via molecular dynamics simulations. Proteins 87: 1263–1275
59. Rodrigues CHM, Pires DEV, Ascher DB (2018) DynaMut: predicting the impact of mutations on protein conformation, flexibility and stability. Nucleic Acids Res 46:W350–W355
60. Batra R, Chan H, Kamath G et al (2020) Screening of therapeutic agents for COVID-19 using machine learning and ensemble docking studies. J Phys Chem Lett 11:7058–7065
61. Valdeolivas A, Tichit L, Navarro C et al (2019) Random walk with restart on multiplex and heterogeneous biological networks. Bioinformatics 35:497–505
62. Li L, Wang YS, An L et al (2017) A network-based method using a random walk with restart algorithm and screening tests to identify novel genes associated with Menière's disease. PLoS One 12:1–19

Chapter 20

Neutralization Assay for SARS-CoV-2 Infection: Plaque Reduction Neutralization Test

Sandra Westhaus and Holger F. Rabenau

Abstract

Neutralization assays are often used as part of research and diagnostics to detect neutralizing antibodies and to determine a possible protective antibody titer after infection or vaccination. Here we describe a conventional plaque reduction neutralization test (PRNT) to check the presence of antibodies against SARS-CoV-2 in patient samples (serum or plasma).

Key words Plaque reduction neutralization test, SARS-CoV-2, Neutralizing antibody, Neutralizing assay, Caco-2 cell line

1 Introduction

As a result of a pathogen invasion, the human body develops a number of defense mechanisms to fight against infections. Besides innate immunity as first line of defense, the production of pathogen-specific antibodies and establishment of memory B cells belongs to the adaptive immunity. Therefore, antibodies are evidence of a previous infection [1].While many assays allows the detection and quantification of antibodies, they do not provide any information about the neutralizing capacity of these antibodies and their protective effect against (re)infection with a pathogen. To investigate neutralizing characteristics, various methods for detection and quantification of antibodies reactive against viral proteins like hemagglutination inhibition assay, indirect antibody staining, or plaque reduction neutralization test (PRNT) are available. PRNT was first described in the late 1950 for detection of arbovirus plaques using agar overlay and neutral red staining and was used to measure serum antibodies neutralizing titers [2]. In general, this assay allows for detection and quantification of neutralizing antibodies of plasma or serum samples by determining the percentage of reduction of virus infectivity. The assay is not only accepted as a

Justin Jang Hann Chu et al. (eds.), *SARS-CoV-2: Methods and Protocols*,
Methods in Molecular Biology, vol. 2452, https://doi.org/10.1007/978-1-0716-2111-0_20,

conventional assay in serological diagnosis, but is also used in vaccine development to determine the protective capacity of the immune response induced by the vaccine candidate [3].

The severe acute respiratory syndrome (SARS) coronavirus (CoV)-2 (SARS-CoV-2) was first identified as a novel coronavirus (nCoV) in December 2019 in Wuhan, China and is the causative agent for Coronavirus Disease 2019 (Covid-19) [4]. Within weeks, the virus spread all over the world and to date nearly 100 million cases with more than 2 million deaths have been confirmed worldwide [5]. SARS-CoV-2 belongs to the family of beta-coronavirus and is an enveloped virus with a single-stranded, positive-orientated RNA genome [6]. Individuals infected with SARS-CoV-2 often remains asymptomatic or show only mild symptoms but severe clinical courses with fatal outcome occur frequently promoted by limited treatment options [7]. Infection with SARS-CoV-2 led to the induction of the adaptive immune system and the formation of virus-specific antibodies targeting the nucleocapsid (N) or spike (S) protein of SARS-CoV-2. In line with this, the S protein has shown to be highly immunogenic and is the main target for neutralizing antibodies [8]. Rapid development and approval of various prophylactic vaccines have taken a first step toward combating the pandemic but limited knowledge about the persistence of immunity and effectiveness against emerging virus variants requires further investigation [9]. A neutralization assay is a cheap, simple, and quick method to investigate the capacity of antibodies newly formed upon SARS-CoV-2 infection or prophylactic vaccination against the virus.

2 Materials

Prepare reagents for cell culture on a clean safety cabinet and store them at 4 °C (unless indicate otherwise). To prepare medium for cell maintenance and infection use heat-inactivate fetal bovine serum (FBS). Heat-inactivation is performed at 56 °C for 30 min in water bath before long-term storage at −20 °C.

2.1 Cell Culture

1. *Cell line*: Caco-2 (human colon carcinoma); adherent epithelia cells (biosafety level 1); established from the primary colon tumor (adenocarcinoma) of a 72-year-old Caucasian man in 1974. Alternatively Vero E6 cells can be used (*see* **Note 1**).
2. *Virus*: SARS-CoV-2 isolate FFM1 (isolated from a patient at University Hospital Frankfurt who was tested positive for SARS-CoV-2 by PCR) [10].
3. *Culture medium*: Minimum Essential Media (MEM) supplemented with 10% heat-inactivated FBS, 1% penicillin–streptomycin, and 2% L-glutamine.

4. *Infection medium*: Minimum Essential Media (MEM) supplemented with 1% heat-inactivated FBS, 1% penicillin–streptomycin, and 2% L-glutamine.
5. 1× Dulbecco's phosphate buffer saline (DPBS).
6. 0.05% (v/v) trypsin–EDTA solution.
7. 96-well cell culture plates with flat bottom, clear well
8. Incubator with 5% constant carbon dioxide (CO_2) supply and high humidity levels of at least 95%.
9. Transmitted light microscope.
10. Lab with Biosafety Level (BSL) 3 and safety cabinet with UV light.

2.2 Patient Samples

Patient samples (serum or plasma) are stored for up to 1 week at 4 °C. Long-term storage is performed at −20 °C.

1. Serum or plasma from patients potentially infected with SARS-CoV-2 from at least 14 days postinfection.
2. Serum or plasma from healthy donors (as negative control).
3. Serum or plasma from SARS-CoV-2 infected patient (as positive control).
4. Serum used in the assay is not heat-inactivated (*see* **Note 2**).

3 Methods

All experiments must be performed under sterile condition and conducted in a clean safety cabinet. Equipment and reagents used in the experiments must be sterile. Cells were maintained in 175 cm^2 cell culture flasks in a humidified incubator at 37 °C and 5% CO_2.

3.1 Cell Culture: Preparation of Cell Monolayer in 96-Well Plates—Biosafety Level 1

1. Caco-2 cells are grown to a confluent monolayer before prepared for neutralization test.
2. Remove expended culture medium and wash cell twice with 20 mL ambient 1× DPBS.
3. Add 2 mL of 0.05% trypsin–EDTA solution and incubate for 20–30 min to detach cells from the surface.
4. Resuspend cells with 70 mL culture medium by repeated pipetting.
5. Transfer 100 μL/well of cell suspension onto a flat bottom 96-well plate.
6. Gently rock the plate in a figure eight descriptive motion to achieve an even distribution of the cells.

7. Incubate cells for 96 h in a humidified incubator at 37 °C supplemented with 5% CO_2.
8. Allow the cells to reach 100% confluency before using for infection assay (*see* **Note 1**).

3.2 Preparation of Virus–Antibody Complex: Biosafety Level 1 and 3

Preparation of test sample dilution can be performed under biosafety level 1 in a clean safety cabinet if it can be excluded that the sample contains infectious material.

1. Let frozen patient samples (serum or plasma) or recombinant antibody slowly thaw at 4 °C.
2. Mix gently by vortexing for a few seconds.
3. Prepare a serial dilution for the test samples as followed (*BSL 1*):
   (a) Predilution of test sample and controls.
   - Add 200 μL/well of infection medium on a clean 96-well plate (flat bottom). Prepare duplicates (*or quadruplicates*) for each test sample.
   - Add 50 μL/well of test sample or control to prepare a 1:5 initial dilution. (Fig. 1a).
   - Mix carefully by repeated pipetting.

   (b) Serial dilution of test sample.
   - Add 50 μL/well of infection medium on a clean 96-well plate (flat bottom) prepare duplicated (*or quadruplicates*) for each test sample.
   - Add 50 μL/well of prediluted (1:5) test sample or control to prepare 1:10 dilution and carefully by repeated pipetting.
   - Serially dilute the test sample or control twofold preparing a final dilution of 1: 1280 (Fig. 1b). Change tips after every dilution step.
4. Prepare virus–antibody complex as followed (*BSL 3*) (*see* **Note 3**):
   (a) Slowly, thaw virus stock at 4 °C (*see* **Note 4**).
   (b) Mix virus stock with infection medium to prepare a virus suspension with a virus titer of 4.000 $TCID_{50}$/mL (= 100 $TCID_{50}$/cavity).
   (c) Mix 50 μL of virus suspension with 50 μL of diluted test sample or control and incubate virus–antibody mixture for 1 h at 37 °C (*see* **Note 5**).

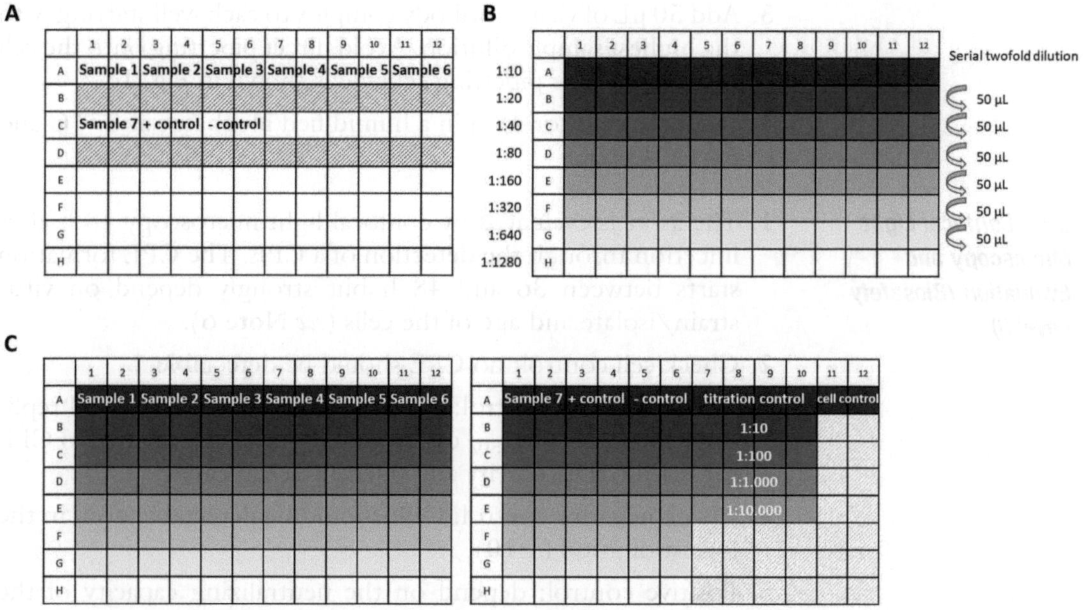

Fig. 1 Layout of 96-well plate for plaque reduction neutralization test (PRNT). (**a**) Predilution plate for initial dilution of test samples and controls. (**b**) Dilution plate for serial dilution of test samples and controls. (**c**) Layout for infection assay of two experimental replicates

3.3 Infection Assay: Biosafety Level 3

To verify the neutralization test, several controls have to be carried out.

- Cell control: infection medium only (no virus, no sample).
- Negative control: virus suspension, serum/plasma sample from health control or IgG control.
- Positive control: virus suspension, serum/plasma from SARS-CoV-2 infected patient or specific SARS-CoV-2 neutralizing antibody that was proved to neutralize virus infection.
- Titration control: serial dilution of virus suspension (no sample).
  - 100 $TCID_{50}$/cavity (100% CPE)
  - 10 $TCID_{50}$/cavity (100% CPE)
  - 1 $TCID_{50}$/cavity (0–100% CPE)
  - 0.1 $TCID_{50}$/cavity (0% CPE)
  - 0.01 $TCID_{50}$/cavity (0% CPE).

This "reverse titration" control should ensure that only 100 $TCID_{50}$/cavity were used in the assay.

1. Aspirate expended culture medium and wash cells once with ambient 1× DPBS.
2. Add 50 µL/well of fresh infection medium.

3. Add 50 μL of virus–antibody complex to each well starting with the highest sample dilution. Avoid direct pipetting onto the cell monolayer. The pipetting scheme is shown in Fig. 1c.
4. Incubate cells for 48 h in a humidified incubator at 37 °C and 5% CO_2.

3.4 Confocal Light Microscopy and Evaluation (Biosafety Level 3)

1. The assay is evaluated by confocal light microscopy 48 h after infection through the detection of a CPE. The CPE formation starts between 36 and 48 h but strongly depend on virus strain/isolate and age of the cells (*see* **Note 6**).
2. Check cell control: no CPE should be detectable.
3. Check titration control: CPE should be detectable in all replicates for 100 $TCID_{50}$/cavity and 10 $TCID_{50}$/cavity. No CPE for 0.1 $TCID_{50}$/cavity and 0.01 $TCID_{50}$/cavity.
4. Check negative control: CPE should be detectable even in the lowest dilution (1:10).
5. Positive control: depend on the neutralizing capacity of the serum antibody, no CPE should be detectable in three lowest dilutions (1:10, 1:20 and 1:40).
6. If all controls are valid, the samples are evaluated (Fig. 2) (*see* **Note 7**).
   (a) CPE in both wells of 1:10 dilution = *negative.*
   (b) CPE first in both wells of 1:20 dilution = *borderline.*
   (c) CPE first in both wells of 1:40 dilution or higher = *positive.*

4 Notes

1. Caco-2 cells are suitable for SARS-CoV-2 assay based on the capability to form CPE upon SARS-CoV-2 infection. An alternative cell line widely used in viral assay is Vero E6 kidney-derived cell line from the African Green Monkey (*Cercopithecus aethiops*).

   Caco-2 cells have low doubling time of app. 80 h. Therefore, split cells only once a week and keep at a low passage number. For optimal results, a 100% confluent cell monolayer in the 96-well plate is necessary. Allow cells to reach confluence by culture them 96 h, if necessary, longer.
2. Previous comparison of noninactivated and heat-inactivated patient sample did not reveal a difference in the efficacy of neutralization. Nevertheless, heat-inactivation of serum or plasma can be performed to inactivate complement factors. If heat-inactivation is performed let the sample cool down to room temperature.

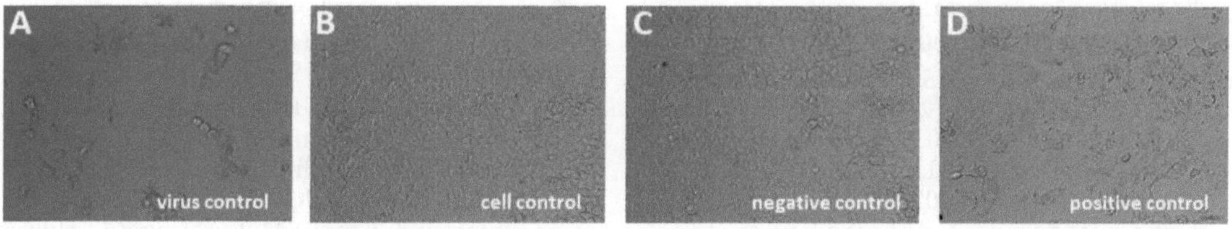

Fig. 2 Images made by confocal light microscopy from the cells incubated with (**a**) virus suspension (100 $TCID_{50}$/cavity), (**b**) infection medium only, (**c**) virus–antibody complex with serum from healthy control patient (as negative control) or (**d**) virus–antibody complex with serum from an SARS-CoV-2 infected patient (as positive control)

3. Use personal protection equipment (PPE) when working on infectious SARS-CoV-2 which is classified as a biosafety level 3 airborne pathogen. PPE should consist of one-time lab coat/ overall, two pairs of protection gloves, protection goggles, overshoes, hair bonnet, and an FFP 3 face mask.
4. Avoid multiple freeze–thaw cycles of virus stock to prevent loss of viral titer. Virus stock left over can be stored for up to 2 weeks at 4 °C.
5. Avoid extension of virus–antibody complex incubation period that can result in undesirable effects like partial virus inactivation [3]. To prevent any inadvertent effect in this context store virus–antibody complex at 4 °C after incubation at 37 °C for 1 h. For optimal results perform assay without any delay during incubation periods.
6. Check plates for CPE 48 h postinfection with SARS-CoV-2. If necessary, incubate plates for further 12–24 h to reach clear CPE in virus control that can be observed by light microscopy.
7. To determine inhibitory concentration that blocks SARS-CoV-2 infection by 50% or 90% (IC_{50} or IC_{90}) using for example recombinant antibodies plaques can be counted when using agarose or methyl cellulose overlay and crystal violet staining.

Acknowledgments

We would like to thank the kindly support by the technicians in diagnostic and research labs at the Institute of Medical Virology.

References

1. Kenneth Murphy CW (2016) PART I - an introduction to immunobiology and innate immunity. In: Janeway's immunobiology, 9th edn. Garland Science, New York, NY
2. Henderson JR, Taylor RM (1959) Arthropod-borne virus plaques in agar overlaid tube cultures. Proc Soc Exp Biol Med 101(2): 257–259. https://doi.org/10.3181/00379727-101-24902
3. Roehrig JTHJ, Barrett AD (2008) Guidelines for plaque-reduction neutralization testing of human antibodies to dengue viruses. Viral

Immunol 21(2):123–132. https://doi.org/10.1089/vim.2008.0007

4. Rabi FA, Al Zoubi MS, Kasasbeh GA, Salameh DM, Al-Nasser AD (2020) SARS-CoV-2 and coronavirus disease 2019: what we know so far. Pathogens 9(3):231. https://doi.org/10.3390/pathogens9030231
5. WHO (2020) https://covid19.who.int/
6. Zhou P, Yang X-L, Wang X-G, Hu B, Zhang L, Zhang W, Si H-R, Zhu Y, Li B, Huang C-L, Chen H-D, Chen J, Luo Y, Guo H, Jiang R-D, Liu M-Q, Chen Y, Shen X-R, Wang X, Zheng X-S, Zhao K, Chen Q-J, Deng F, Liu L-L, Yan B, Zhan F-X, Wang Y-Y, Xiao G-F, Shi Z-L (2020) A pneumonia outbreak associated with a new coronavirus of probable bat origin. Nature 579(7798):270–273. https://doi.org/10.1038/s41586-020-2012-7
7. WHO (2020) Coronavirus. https://www.who.int/health-topics/coronavirus#tab=tab_3
8. Ou X, Liu Y, Lei X, Li P, Mi D, Ren L, Guo L, Guo R, Chen T, Hu J, Xiang Z, Mu Z, Chen X, Chen J, Hu K, Jin Q, Wang J, Qian Z (2020) Characterization of spike glycoprotein of SARS-CoV-2 on virus entry and its immune cross-reactivity with SARS-CoV. Nat Commun 11(1):1620. https://doi.org/10.1038/s41467-020-15562-9
9. Weisblum Y, Schmidt F, Zhang F, DaSilva J, Poston D, Lorenzi JC, Muecksch F, Rutkowska M, Hoffmann HH, Michailidis E, Gaebler C, Agudelo M, Cho A, Wang Z, Gazumyan A, Cipolla M, Luchsinger L, Hillyer CD, Caskey M, Robbiani DF, Rice CM, Nussenzweig MC, Hatziioannou T, Bieniasz PD (2020) Escape from neutralizing antibodies by SARS-CoV-2 spike protein variants. eLife 9:e61312. https://doi.org/10.7554/eLife.61312
10. Hoehl S, Rabenau H, Berger A, Kortenbusch M, Cinatl J, Bojkova D, Behrens P, Boddinghaus B, Gotsch U, Naujoks F, Neumann P, Schork J, Tiarks-Jungk P, Walczok A, Eickmann M, Vehreschild M, Kann G, Wolf T, Gottschalk R, Ciesek S (2020) Evidence of SARS-CoV-2 infection in returning travelers from Wuhan. China N Engl J Med 382(13):1278–1280. https://doi.org/10.1056/NEJMc2001899

Chapter 21

Pseudovirus-Based Assays for the Measurement of Antibody-Mediated Neutralization of SARS-CoV-2

Corey Balinsky, Vihasi Jani, Peifang Sun, Maya Williams, Gabriel Defang, and Kevin R. Porter

Abstract

SARS-CoV-2 has emerged as a significant cause of morbidity and mortality worldwide. Virus neutralization assays are critical for the development and evaluation of vaccines and immunotherapeutics, as well as for conducting basic research into the immune response, spread, and pathogenesis of this disease. However, neutralization assays traditionally require the use of infectious virus which must be carefully handled in a BSL-3 setting, thus complicating the assay and restricting its use to labs with access to BSL-3 facilities. Pseudovirus-based assays are an alternative to the use of infectious virus. SARS-CoV-2 pseudovirus contains only the spike structural protein, and infection results in a single round of replication, thus allowing for the assay to be run safely under BSL-2 conditions. In this chapter, we describe protocols and considerations for the production and titration of lentivirus-based SARS-CoV-2 pseudovirus, as well as for running and analysis of FACS-based pseudovirus neutralization assays.

Key words Pseudovirus (PV), SARS-CoV-2, COVID-19, Neutralization, Antibody, Lentivirus, Spike, Protocols

1 Introduction

Severe acute respiratory syndrome coronavirus 2 (SARS-CoV-2) is the etiologic agent of COVID-19 and has emerged as a significant cause of morbidity and mortality worldwide. SARS-CoV-2 is a member of the Betacoronavirus genera of the *Coronaviridae* which also include SARS and MERS. The SARS-CoV-2 virion consists of a positive sense genome of about 29.9 kbp in length which is packaged by the Nucleocapsid (N) protein as well as by the

Justin Jang Hann Chu et al. (eds.), *SARS-CoV-2: Methods and Protocols*,
Methods in Molecular Biology, vol. 2452, https://doi.org/10.1007/978-1-0716-2111-0_21,

major structural proteins: membrane (M), envelope (E), and spike (S) [1, 2]. The SARS-CoV-2 S protein is responsible for virus binding to the Angiotensin-Converting Enzyme 2 (ACE2) receptor as well as for fusion with the plasma membrane. The S protein also serves as the major target for antibody-mediated neutralization of SARS-CoV-2, as well as for other SARS-related coronaviruses [3–6].

The Spike protein is a class I fusion protein that forms trimers on the surface of the virion and is composed of two subunits S1 and S2. S1 consists of an N-terminal domain, the receptor-binding domain (RBD), and subdomains 1 and 2; while S2 contains the fusion peptide (FP), heptad repeats, transmembrane domain, and the cytoplasmic tail [4, 7–9]. The binding of the spike protein RBD to the ACE2 receptor causes a conformational change, exposing a cleavage site which may be targeted by TMPRSS2 or other proteases, resulting in membrane fusion [10, 11]. SARS-CoV-2 also contains a furin cleavage site, which is thought to result in pre-activation of the S protein and to mediate cell–cell fusion [12, 13].

Due to its high pathogenicity and respiratory transmission, SARS-CoV-2 is designated as a BSL-3 agent, requiring specialized facilities for culture and handling. The pathogenicity and BSL-3 designation of SARS-CoV-2 complicates the performance of assays that require the use of live infectious virus, such as traditional antibody-mediated neutralization assays [14, 15], which are critical for the evaluation of vaccines and immunotherapeutics as well as for the study of virus-induced immunity. In order to circumvent this issue and to allow for analysis of serum neutralization under safer conditions, assays have been developed using replication defective pseudovirus (PV), in which the SARS-CoV-2 S protein is used to package a vector expressing a reporter gene, frequently a fluorescent protein or luciferase. These systems have the advantages of being safe and can be used under BSL-2 conditions. They also have the advantage of being malleable and can be easily modified to study different variants of the S protein [16, 17].

Pseudovirus-based systems have been established to study a number of pathogenic viruses including Ebola [18], Marburg [19], Lassa virus [20], Chikungunya [21], Nipah [22], as well as coronaviruses SARS and MERS [23, 24]. A number of pseudovirus-based systems have been recently developed for the study of SARS-CoV-2. These systems can be based on murine leukemia retrovirus (MLV) [25, 26], lentivirus [25, 27–29], or vesicular stomatitis virus (VSV) [30–33] and have been shown to correlate well with assays utilizing infectious SARS-CoV-2 [29, 30, 32, 34].

Here we describe protocols and key considerations for the production and titration of SARS-CoV-2 pseudotyped lentivirus particles, as well as for the running and analysis of FACS-based neutralization assays.

2 Materials

2.1 Plasmids

A number of lentivirus transfer and packaging plasmids have been described for pseudovirus production [25, 27, 29]. Here, we use a third-generation-based, HIV-1-derived, lentivirus system described by the Bloom lab [28] and based on the backbone developed in the laboratory of Dr. Richard C. Mulligan at Harvard Medical School. The plasmids have been generously made available through BEI Resources (NIAID, NIH).

1. pHAGE-CMV-ZsGreen (BEI Resources; NR-52520). Replication incompetent HIV-1-based vector containing the ZsGreen reporter gene.
2. Packaging plasmids separately expressing Gag-Pol, Rev., and Tat.
   (a) HDM-Hgpm2 (BEI Resources; NR-52517).
   (b) pRC-CMV-Rev1b (BEI Resources; NR-52519).
   (c) HDM-tat1b (BEI Resources; NR52518).

While pseudovirus particles can be produced using full-length S protein, C-terminal deletions of 18 or 19 amino acids result in higher titers [25, 30]. Higher titers can also be obtained using S derived from the D614G strain as opposed to the ancestral (D614) strain [35–37]. In addition, incorporation of a polymorphism at the furin cleavage site, observed after serial passage in cell culture, can result in increased PV titers [25, 38]. While the D614G or furin (R682Q) mutations do not appear to influence PV neutralization, some polymorphisms may alter antibody-mediated neutralization as has been shown with several spike variants [34, 39–41]. The S variant used should be chosen based on the experimental requirements. Here, we use codon-optimized spike protein with the C-terminal 19 amino acids deleted and containing the D614G mutation pSd19D614G; however, we have also obtained good results using a variant containing the R682Q mutation or a combination of both. A plasmid expressing the G protein from VSV can be used as a control to monitor PV production.

1. pCMV-Sd19D614G (NMRC). Codon optimized S(D614G) protein with a 19 aa C-terminal deletion, synthesized by GenScript USA Inc. (Piscataway, NJ).
2. pCMV-VSV-G (Addgene; 8454). VSV-G can be used as a positive control for PV production.

Plasmids were propagated using standard molecular biology techniques. All plasmids used here contained the ampicillin resistance gene and were grown in LB media. Plasmids were purified using QIAGEN Plasmid Plus Maxi Kit (Qiagen).

2.2 Transfection Reagents

DMEM, low glucose, pyruvate, HEPES (ThermoFisher Scientific), containing 10% FBS.

Opti-MEM I Reduced Serum Medium (ThermoFisher Scientific).

Lipofectamine 3000 transfection reagent (ThermoFisher Scientific).

PLUS reagent (ThermoFisher Scientific).

2.3 Cell Culture

Cell lines used for neutralization are another important consideration when performing lentivirus-derived SARS-CoV-2 PV assays, as expression of ACE2 is critical for infectivity [10, 42]. Here we use HEK293 cells stably transfected with ACE2 and subcloned based on maximal expression (NMRC; unpublished). However, a number of ACE2-expressing cell lines have been developed utilizing HEK293T [28] or BHK-21 [30] cells. ACE2-expressing cell lines are also commercially available or can be obtained through BEI Resources. Co-expression of TMPRSS2 may further enhance infectivity in certain cell lines [32]. Whichever cell line is chosen, care should be taken to ensure adequate ACE2 expression. We recommend testing cells periodically to ensure that expression has not decreased over time (*see* Subheading 3.3).

HEK-293T/17 cells (ATCC CRL-11268), used for PV production, are maintained in Dulbecco's Modified Eagle Medium (DMEM) and supplemented with 10% fetal bovine serum (FBS) and penicillin/streptomycin (PS).

HEK293-ACE2 (NMRC), used for PV assays, are maintained in DMEM supplemented with 10% FBS and PS, as well as with G418 (at a final concentration 0.4 mg/mL). As ACE2 expression is a critical factor for infectivity, cells should be periodically checked for ACE2 expression and positive controls monitored for any potential loss of infectivity. We have observed minimal loss of expression through passage 20. However, this will be dependent on the cell line used and should be monitored closely.

All cells are maintained at 37 °C in an incubator set to 5% CO_2.

Phosphate-buffered saline PBS (pH 7.4).

Fetal Bovine Serum (heat inactivated at 56 °C for 30 min).

DMEM (with 4.5 g/L D-Glucose and L-Glutamine).

Trypsin–EDTA (Quality Biological).

Penicillin–Streptomycin mixture (10,000 U/mL).

G418 (100 mg/mL) Sock Solution.

Human ACE-2 Alexa Fluor 647-conjugated Antibody (R&D Systems; FAB9332R).

2.4 Lab Equipment and Supplies

Flow cytometer capable of reading fluorescence. We use a BD FACSCanto II equipped with a BD High Throughput Sampler (HTS), and running FACSDiva software (Version 8.0.1).

Inverted microscope.

Cellometer Auto 2000 (Nexcelom Bioscience).

Cellometer PD100 counting chambers (Cat #: CHT4-PD100-003, Nexcelom Bioscience).

AO/PI (Cat #: CS-0106-5 mL, Nexcelom Bioscience) Nexelom ViaStain AOPI staining solution.

Pipettors (10, 20, 200, and 1000).

Multichannel Pipettors (5–50 and 50–300 μL volume ranges).

Sterile Pipette tips (10 μL, 20 μL, 200 μL, and 1 mL).

Sterile multichannel pipettor basins.

Tissue culture flasks (T-75).

Flat bottom plates (6-well and 96-well).

Sterile 50-mL conical tubes.

Sterile 15-mL conical tubes.

0.45-μM filters and syringe units.

3 Methods

3.1 Production of SARS-CoV-2 Pseudovirus

Optimal conditions and kinctics for production and collection of PV will vary depending on culturing conditions as well as on the reporter, packaging and spike vectors used. Transfection conditions and collection times should be determined empirically before beginning large-scale production. Production can be scaled up either by pooling multiple wells of a 6-well plate or by performing the transfection in T-75 flasks.

The day prior to transfection, plate HEK-293T/17 cells (ATCC CRL-11268) at 1×10^6 cells per well in DMEM containing 10% FBS and PS. The number of cells should be adjusted to achieve 70–90% confluency at the time of transfection.

The following day, remove media and gently replace with 1 mL of fresh low-glucose DMEM containing 10% FBS.

Prepare mixture for co-transfection of cells with reporter, packaging and spike plasmids.

1. To 250 μL of Opti-MEM I Reduced Serum Media (ThermoFisher; Cat# 31985062), add 2 μg of the lentivirus backbone expressing ZsGreen (pHAGE2-CMV-ZsGreen-W; NR-52520) and 0.44 μg of each of the helper plasmids expressing HIV Gag-Pol (HDM-Hgpm2; NR-52517), Tat (HDM-tat1b; NR-52518), and Rev (pRC-CMV-Rev1b;

NR-52520). Add 0.68 μg of pSd19D614G (or desired spike variant). Briefly vortex and spin down (*see* **Note 1**).

2. Add 5 μL of P3000 reagent, vortex and spin down using a benchtop centrifuge.
3. Add 8 μL of Lipofectamine 3000, vortex and spin down to collect contents. Incubate the reaction at room temperature for 20 min. This corresponds to a 1:2 ratio of DNA (4 μg) to lipofectamine 3000 (8 μL) (*see* **Note 2**).
4. Add the Lipid–DNA complexes dropwise to each well. Add solution gently so as to not disrupt the monolayer. Gently rock the plate back and forth to disperse the complexes. Return the plate(s) to the 37 °C incubator, maintained at 5% CO_2 and incubate overnight (approximately 16 h). If desired, media can be replaced 4 h after the addition of Lipid–DNA complexes.
5. The following day, remove the media and replace with 2 mL of low-glucose DMEM containing 10% FBS. Add media dropwise so as to not disturb the monolayer. Return plates to the 37 °C incubator.
6. Approximately 48 h after replacing the media (64 h post transfection), remove media from the 6-well plate and filter through a 0.45-μm low protein binding syringe filter (*see* **Note 3**). Optimal collection times may vary depending on conditions. We recommend collecting PV at several time points to determine the time point at which maximum titers can be obtained.
7. PV can then be aliquoted out and stored at −80 °C until use (*see* **Note 4**).

3.2 Determining the Titer of Pseudovirus Preparations

PV titers should be determined before beginning neutralization experiments. Titering should be performed using the cell type used for the neutralization assay as infectivity can vary widely based on cell characteristics and ACE2 expression (***see*** Subheading 3.3 **on ACE2 expression**).

1. Thaw PV on ice and gently mix by inversion. Spin briefly to collect contents and keep PV on ice until ready for use (*see* **Notes 5** and **6**).
2. In a 96-well plate, perform duplicate twofold serial dilutions of the PV in DMEM containing 10% FBS. We use a final volume of 200 μL/well (100 μL of PV dilution and 100 μL of cells) and include cell-only controls to aid in gating during analysis. To avoid possible edge effects, use only the internal wells of the 96-well plate for titrations (Fig. 1a). The outer wells can be filled with PBS or media (*see* **Note** 7).
   (a) Add 200 μL PBS or DMEM to outer wells of 96-well plate(s).
   (b) Add 100 μL DMEM containing 10% FBS to rows 3–11.

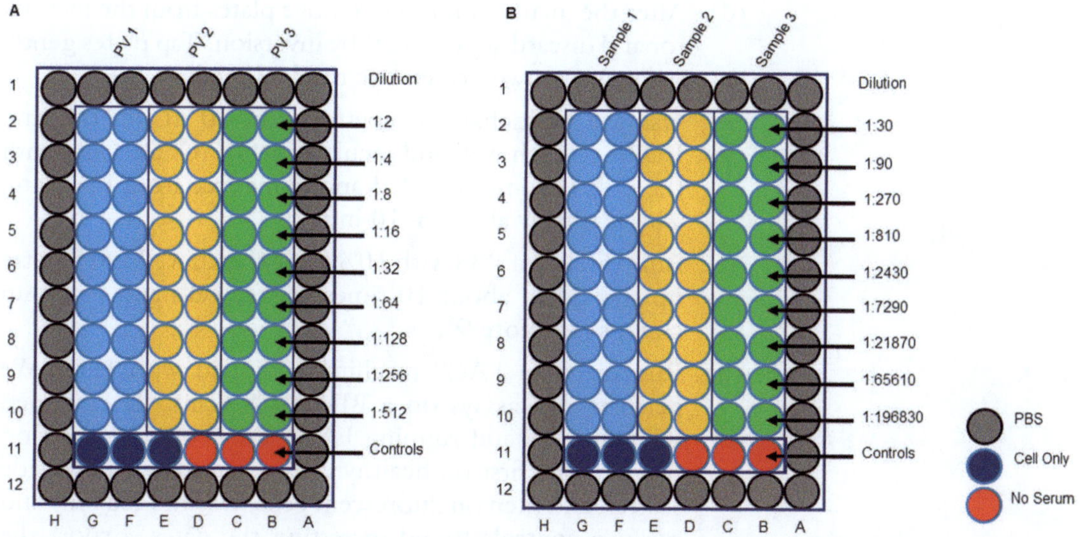

Fig. 1 Example plate layouts for (**a**) pseudovirus titration, where PV was titered in duplicate in 96-well plates at 1:2 dilutions, and (**b**) neutralization assays, where neutralizing antibody was diluted 1:3 prior to the addition of PV. In both the plates, outer wells contain PBS, and positive and negative controls are depicted in row 11

(c) Add 200 μL PV (in duplicate) to the second row.

(d) Make 8 twofold dilutions of PVs, by removing 100 μL from the second row to the third row and pipetting up and down 10 times to mix, changing pipette tips between each dilution.

(e) Continue twofold dilutions to the tenth row (remove 100 μL from this row and discard).

(f) Add an additional 100 μL of DMEM containing 10% FBS to the 11th row as a cell only control.

3. Prepare HEK-293-ACE2 or cells of choice. Cells should be healthy and confluent at the time of use, typically, we use our cells between passage 3 and 20.

(a) Remove media from T-75 and trypsinize for 5 min at 37 °C until cells come free. Break up cells by pipetting and transfer to a centrifuge tube with an equal volume of complete media. Centrifuge at 1000 RPM for 5 min. Remove media and resuspend in complete media.

(b) Check viability and count cells. Cells should be >90% viable. Dilute cells in complete media so as to add 2×10^4 cells per well. We dilute to 2×10^5 cells per mL and add 100 μL to each well (*see* **Note 8**).

(c) Add cells to each well of the 96-well plate and incubate at 37 °C 5% CO_2 for 48 h. While we usually develop the plates after 48 h, plates can be run between 36 and 72 h.

(d) After the incubation time, remove plates from the incubator and discard supernatant by inversion. Tap plates gently on a paper towel to remove residual media.

(e) Using a multichannel pipettor, add 50 μL of Trypsin–EDTA to each well and incubate at 37 °C/5% CO_2 until the cells become rounded and come free from the plate. This is usually about 5–10 min.

(f) Add 150 μL of PBS with 10% FBS to each well and pipette up and down about 10 times to free cells and break up clumps (*see* **Note 9**).

(g) Run plate on FACS machine equipped with HTS. We perform our assays on a BD FACS Canto II equipped with an HTS and running FACS Diva software (Version 8.0.1). Gate first on healthy cells using forward and side scatter, and then on fluorescent cells. Use the cell only and positive controls to aid in setting the gates. Export the percent fluorescent cells for analysis.

4. Calculate PV Titer.

Calculate PV titer using more than one dilution and averaging the results. Some data points, especially at higher dilutions, may be nonlinear. Therefore, we select three or four data points that follow a linear progression and are at the lower end of the dilution range for our calculations and average the results [28, 43]. The equation below can be used to calculate IU/mL.

$$\text{IU/mL} = (\%\text{positive cells}/100) \times (\text{number of cells per well})/(\text{volume of PV containing media in mL})$$

5. Determining Appropriate Working Dilution and Compliance with the "Percentage Law"

The optimal working concentration is at the user's discretion. However, the most reproducible data will be derived from PV concentrations that adhere to the percentage law [44, 45], which states that since antibody-mediated neutralization typically obeys the law of mass action, as long as the antibody concentration is in excess to the amount of PV, the neutralization (EC50) value that is obtained in response to the given antibody concentration will remain constant across a range of virus or pseudovirus titers. To ensure compliance with the percentage law, perform neutralization assays (*see* Subheading 3.4) using a range of PV concentrations. EC50s should remain consistent throughout a range of titers. Monitor results for a shift in the EC50, indicating that the percentage law has been violated (Fig. 2). For a detailed discussion of mass action and the percentage law, and how it pertains to neutralization assays, *see* [43, 46–49].

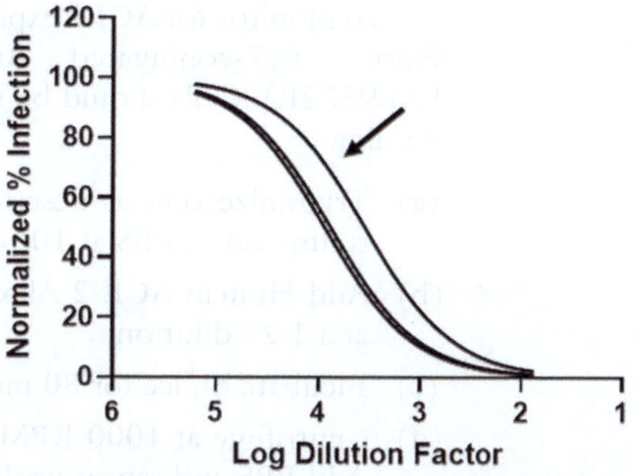

Fig. 2 Example demonstrating "percentage law." The same serum sample was run using four different PV concentrations. Sigmoidal dose–response curves of normalized percent infection as a function of log dilution factor are shown. The arrow indicates the dose–response curve, where the "percentage law" has been violated, evident by a shift of the dose–response curve

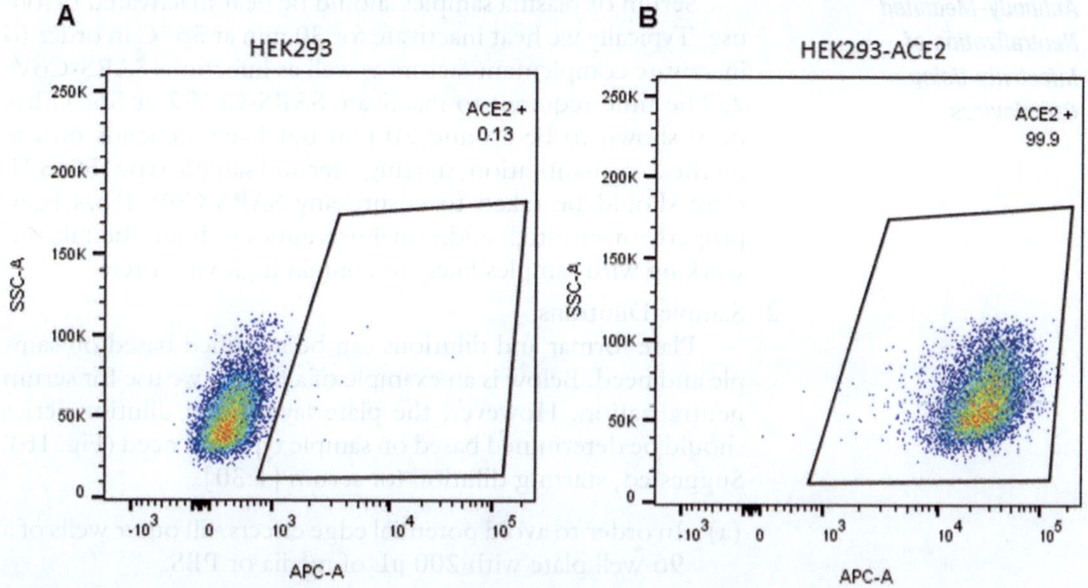

Fig. 3 Dot plot showing ACE2 expression in HEK293 (**a**) and HEK293-ACE2 (**b**) cell lines, where the *X*-axis depicts APC-labeled ACE2 antibody, and the *Y*-axis depicts side scatter. The shift in fluorescence intensity between **a** and **b** is representative of expression of ACE2. The percentage of ACE2-expressing cells is shown in the upper right corner of each panel

3.3 Monitoring of ACE2 Expression in Cell Lines

As discussed previously, ACE2 expression is a critical factor for infectivity of lentivirus-based pseudovirus. Therefore, it is advisable to check cell lines for ACE2 expression to ensure assay consistency and reliability (Fig. 3). Loss of ACE2 expression should also be considered if encountering an unexpected drop in infectivity.

To monitor for ACE2 expression, we use Human ACE-2 Alexa Fluor® 647-conjugated Antibody (R&D Systems, Cat# FAB9332R). Cells should be healthy and confluent at the time of staining.

(a) Trypsinize cells as described in Subheading 3.2, **step 3** and resuspend in PBS at 10^6 cells per mL.

(b) Add Human ACE-2 Alexa Fluor® 647-conjugated Antibody, at a 1:25 dilution.

(c) Incubate on ice for 30 min.

(d) Centrifuge at 1000 RPM for 5 min and remove supernatant. Add PBS and repeat wash two additional times.

(e) Resuspend in PBS and run samples in FACS, gating first on healthy cells, and then on the fluorescent conjugate.

The majority of cells should be ACE2 positive.

3.4 Measurement of Antibody-Mediated Neutralization of Infectivity Using Pseudovirus

1. Sample Preparation.

   Serum or plasma samples should be heat inactivated before use. Typically we heat inactivate for 30 min at 56 °C in order to inactivate complement factors as well as infectious SARS-CoV-2. The time required to inactivate SARS-CoV-2 at 56 °C has been shown to be around 30 min but likely depends on the method of quantitation, starting titer and sample type [50–52]. Care should be taken to ensure any SARS-CoV-2 has been properly inactivated; additional precautions should be taken if working with samples likely to contain high viral titers.

2. Sample Dilutions.

   Plate format and dilutions can be modified based on sample and need. Below is an example of a format we use for serum neutralization. However, the plate layout and dilution series should be determined based on sample type and need (Fig. 1b). Suggested, starting dilution for serum [1:30].

   (a) In order to avoid potential edge effects, fill outer wells of a 96-well plate with 200 μL of media or PBS.

   (b) Add 50 μL of complete media to the internal 60 wells of the plate.

   (c) Prepare starting dilution of sample and add to first well of the plate. Adjust starting volume so that the final volume (after removing diluted sample) will be 50 μL. For example, if making 1:3 dilutions, the first well should contain a starting volume of 75 μL.

   (d) Prepare serial dilutions by removing desired volume from first well (i.e., 25 μL) and transferring to the second well. Pipette up and down to mix, change pipette tips, and repeat to the final well of the series. After adding sample to the final well, remove 25 μL (or volume diluted) and discard in order to maintain consistent volumes. In each plate include cell only and PV only controls.

3. Neutralization Assay.
   (a) Prepare desired dilution of PV (discussed in Subheading 3.2) with complete media, mix gently and spin down. Using a multichannel pipette, add 50 μL to each well. Mix plate gently on a flat surface and transfer to a 37 °C 5% CO_2 incubator. Incubate for 30 min. During this incubation period, prepare cells for use (*see* **Note 10**).

       Remove HEK293-ACE2 cells from incubator. Remove media from flask and add 5 mL of trypsin–EDTA. Incubate at 37 °C/5% CO_2 for 5 min.

       Add 5 mL of DMEM containing 10% FBS and pipette up and down to break up clumps. Transfer to a 15-mL tube and centrifuge at 1000 RPM for 5 min. Remove supernatant and resuspend in 10 mL DMEM containing 10% FBS.
   (b) Resuspend cells in complete media and measure viability and cell density using a cell counter. Dilute cells in complete media to final concentration of 2×10^5 cells per mL (*see* **Note 11**).
   (c) After the 30 min incubation, add 100 μL of the diluted cells (2×10^4 cells) to each well using a multichannel pipette and return plates to the 37 °C incubator for 48 h (*see* **Note 12**).
   (d) After incubation time, prepare plates as previously described (Subheading 3.2, **step 3**), by dumping media into a waste container and firmly inverting on a paper towel to remove residual media. Add 50 μL of Trypsin/EDTA and incubate at 37 °C for about 5–10 min or until cells come free. Resuspend in 150 μL of PBS supplemented with 10% FBS and pipette up and down to fully suspend and separate cells. If desired, cells can be fixed by adding 32% paraformaldehyde (PFA) to a final concentration of 1.2% and visualized at a later time.

       Plates can then be analyzed using a flow cytometer as described in Subheading 3.2, **step 2**, by gating first on healthy cells using forward and side scatter, and then on fluorescent cells. Use the cell only and no serum controls to aid in setting the gates. Then export data as percent fluorescent-positive cells and transfer to GraphPad Prism (Version 8.3.1) for further analysis.

4. Graphing and Analysis of Neutralization Data Using GraphPad Prism.

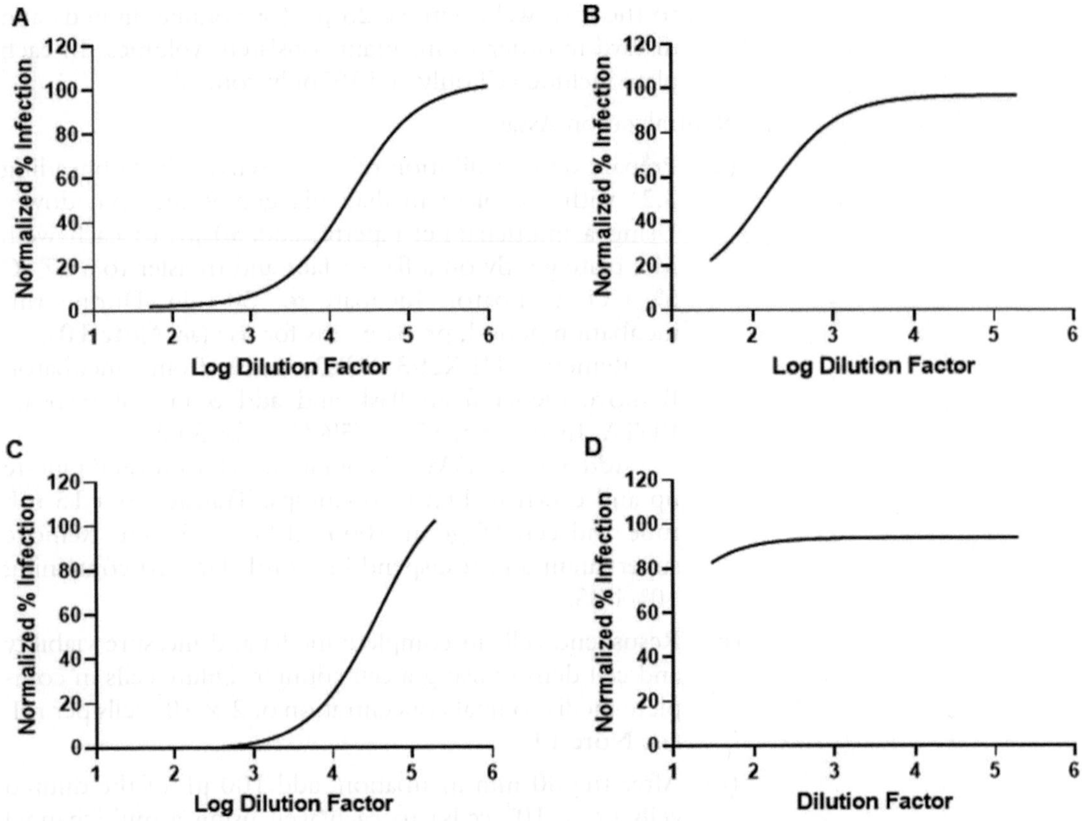

Fig. 4 Examples of frequently observed dose–response curves. Here, normalized percent infection is plotted as a function of log dilution factor. (**a**) Sigmoidal dose–response with top and bottom plateau. (**b**) Dose–response with no bottom plateau, signifying that antibody at the lowest dilution is not sufficient to completely neutralize the PV. (**c**) Dose–response with no top plateau, signifying that neutralization is occurring at the highest dilution. Here, better data quality would be achieved by repeating with higher dilutions. (**d**) No neutralization observed

If performing replicate experiments select the option to enter replicates side-by-side in subcolumns. Plot the percent positive cells against the log-transformed dilution factor (or antibody concentration).

Results can be normalized to aid in analysis, using the Normalize function in the Analyze Data tab. Here, the percentage of positive cells in the absence of neutralizing antibody would become 100%, and the cell only control (frequently 0 already) would become 0%.

Perform a nonlinear regression analysis and examine the results. When examining the data it is important to examine both the dose–response neutralization curve as well as the statistical output provided by Prism.

The dose–response curve will frequently display a sigmoidal shape which plateaus at the top (signifying dilutions at which neutralization is no longer evident). The percent positive cells at this region of the curve should be equivalent to that observed in the no serum control. If no plateau evident, it may be necessary to repeat the experiment with additional dilutions to obtain meaningful data (Fig. 4).

Optimally, a bottom plateau should also be evident, signifying that all the PV has been neutralized. However, this will not always be possible due to sample limitations and the ability of the serum/antibody to neutralize. In this case, the bottom of the curve can be constrained to 0. However, care should be taken when applying constraints as it may not be appropriate in every situation such as if the bottom of the curve appears to plateau before reaching 0. For a detailed discussion on this, *see* [43, 53], as well as guidance provided on the Prism software package.

In addition to the characteristics of the dose–response curve, it is also important to examine the statistical output provided by Prism. The 95% confidence interval (CI) and the coefficient of determination (R^2) are both useful metrics from which to determine the overall quality of the data. In some cases an analysis may yield a 50% neutralizing titer (NT50), represented in Prism as an EC50, but a closer examination of the dose–response curve or CI and R^2 values will reveal poor data quality. Care should be taken to carefully examine all aspects of the data before making a determination of its significance.

4 Notes

1. As a positive control for PV production, a plasmid expressing the VSV G protein can be substituted for the SARS-CoV-2 Spike protein. Typically, this will result in higher titers that can be easily visualized via FACS or using a fluorescent microscope.
2. If scaling up to T-75 flasks, we increase the transfection volumes by a factor of 6.
3. If scaling up to T-75 flasks, the supernatant containing the PV is then filtered through a Steriflip vacuum-driven filtration system (EMD Millipore).
4. While we tend to aliquot our PV for single use, we have observed that PV can be freeze thawed at least four times with minimal loss of activity.
5. While we have observed minimal loss of infectivity of PV on ice or at room temperature, prolonged maintenance at room temperature will result in loss of infectivity over time, and care

should be taken to minimize the amount of time the PV is maintained at elevated temperatures.

6. We have observed that incubation at or above 56 °C will result in rapid inactivation of PV.
7. We have observed a small but measurable difference in infectivity around the edges of the plate. To avoid edge effect, the outside edges can be filled with media or PBS.
8. We check viability using 1:1 AO/PI, but the method is at the user's discretion.
9. Check cells under microscope to ensure cells are separated (especially around edges of the well). Incomplete freeing/separation of cells can result in poor data quality.
10. Incubation of up to an hour will not affect the assay; however, longer incubations should be avoided due to loss of PV infectivity.
11. Cells should be >90% viable at the time of use. As discussed in Subheading 3.3, care should be taken to ensure cells are expressing adequate levels of ACE2.
12. We typically visualize plates after 48 h. However, plates can be visualized between 36 and 72 h after addition of PV.

Disclaimers

The views expressed in this chapter reflect the results of research conducted by the author and do not necessarily reflect the official policy or position of the Department of the Navy, Department of Defense, the United States Government, nor the Henry M. Jackson Foundation for the Advancement of Military Medicine, Inc.

This work was supported/funded by work unit number: A1436, DHP RDT&E supplemental COVID funding.

References

1. Kim D, Lee JY, Yang JS, Kim JW, Kim VN, Chang H (2020) The architecture of SARS-CoV-2 transcriptome. Cell 181(4):914–921.e10. https://doi.org/10.1016/j.cell.2020.04.011
2. Hu B, Guo H, Zhou P, Shi ZL (2020) Characteristics of SARS-CoV-2 and COVID-19. Nat Rev Microbiol 19(3):141–154. https://doi.org/10.1038/s41579-020-00459-7
3. Jiang S, Hillyer C, Du L (2020) Neutralizing antibodies against SARS-CoV-2 and other human coronaviruses. Trends Immunol 41(5):355–359. https://doi.org/10.1016/j.it.2020.03.007
4. Ou X, Liu Y, Lei X, Li P, Mi D, Ren L, Guo L, Guo R, Chen T, Hu J, Xiang Z, Mu Z, Chen X, Chen J, Hu K, Jin Q, Wang J, Qian Z (2020) Characterization of spike glycoprotein of SARS-CoV-2 on virus entry and its immune cross-reactivity with SARS-CoV. Nat Commun 11(1):1620. https://doi.org/10.1038/s41467-020-15562-9
5. Mossel EC, Huang C, Narayanan K, Makino S, Tesh RB, Peters CJ (2005) Exogenous ACE2 expression allows refractory cell lines to support severe acute respiratory syndrome coronavirus replication. J Virol 79(6):3846–3850.

https://doi.org/10.1128/JVI.79.6.3846-3850.2005

6. Jia HP, Look DC, Shi L, Hickey M, Pewe L, Netland J, Farzan M, Wohlford-Lenane C, Perlman S, McCray PB Jr (2005) ACE2 receptor expression and severe acute respiratory syndrome coronavirus infection depend on differentiation of human airway epithelia. J Virol 79(23):14614–14621. https://doi.org/10.1128/JVI.79.23.14614-14621.2005
7. Ashhurst AS, Tang AH, Fajtova P, Yoon M, Aggarwal A, Stoye A, Larance M, Beretta L, Drelich A, Skinner D, Li L, Meek TD, McKerrow JH, Hook V, Tseng CK, Turville S, Gerwick WH, O'Donoghue AJ, Payne RJ (2020) Potent in vitro anti-SARS-CoV-2 activity by gallinamide A and analogues via inhibition of cathepsin L. bioRxiv. https://doi.org/10.1101/2020.12.23.424111
8. Padmanabhan P, Desikan R, Dixit NM (2020) Targeting TMPRSS2 and cathepsin B/L together may be synergistic against SARS-CoV-2 infection. PLoS Comput Biol 16(12): e1008461. https://doi.org/10.1371/journal.pcbi.1008461
9. Duan L, Zheng Q, Zhang H, Niu Y, Lou Y, Wang H (2020) The SARS-CoV-2 spike glycoprotein biosynthesis, structure, function, and antigenicity: implications for the design of spike-based vaccine immunogens. Front Immunol 11:576622. https://doi.org/10.3389/fimmu.2020.576622
10. Hoffmann M, Kleine-Weber H, Schroeder S, Kruger N, Herrler T, Erichsen S, Schiergens TS, Herrler G, Wu NH, Nitsche A, Muller MA, Drosten C, Pohlmann S (2020) SARS-CoV-2 cell entry depends on ACE2 and TMPRSS2 and is blocked by a clinically proven protease inhibitor. Cell 181(2):271–280.e8. https://doi.org/10.1016/j.cell.2020.02.052
11. Fuentes-Prior P (2020) Priming of SARS-CoV-2 S protein by several membrane-bound serine proteinases could explain enhanced viral infectivity and systemic COVID-19 infection. J Biol Chem 296:100135. https://doi.org/10.1074/jbc.REV120.015980
12. Hoffmann M, Kleine-Weber H, Pohlmann S (2020) A multibasic cleavage site in the spike protein of SARS-CoV-2 is essential for infection of human lung cells. Mol Cell 78(4): 779–784.e5. https://doi.org/10.1016/j.molcel.2020.04.022
13. Jaimes JA, Millet JK, Whittaker GR (2020) Proteolytic cleavage of the SARS-CoV-2 spike protein and the role of the novel S1/S2 site. iScience 23(6):101212. https://doi.org/10.1016/j.isci.2020.101212
14. Muruato AE, Fontes-Garfias CR, Ren P, Garcia-Blanco MA, Menachery VD, Xie X, Shi PY (2020) A high-throughput neutralizing antibody assay for COVID-19 diagnosis and vaccine evaluation. bioRxiv. https://doi.org/10.1101/2020.05.21.109546
15. Amanat F, White KM, Miorin L, Strohmeier S, McMahon M, Meade P, Liu WC, Albrecht RA, Simon V, Martinez-Sobrido L, Moran T, Garcia-Sastre A, Krammer F (2020) An in vitro microneutralization assay for SARS-CoV-2 serology and drug screening. Curr Protoc Microbiol 58(1):e108. https://doi.org/10.1002/cpmc.108
16. Li Q, Wu J, Nie J, Zhang L, Hao H, Liu S, Zhao C, Zhang Q, Liu H, Nie L, Qin H, Wang M, Lu Q, Li X, Sun Q, Liu J, Zhang L, Li X, Huang W, Wang Y (2020) The impact of mutations in SARS-CoV-2 spike on viral infectivity and antigenicity. Cell 182(5): 1284–1294.e9. https://doi.org/10.1016/j.cell.2020.07.012
17. Greaney AJ, Loes AN, Crawford KHD, Starr TN, Malone KD, Chu HY, Bloom JD (2021) Comprehensive mapping of mutations in the SARS-CoV-2 receptor-binding domain that affect recognition by polyclonal human plasma antibodies. Cell Host Microbe 29(3): 463–476.e6
18. Chen Q, Tang K, Zhang X, Chen P, Guo Y (2018) Establishment of pseudovirus infection mouse models for in vivo pharmacodynamics evaluation of filovirus entry inhibitors. Acta Pharm Sin B 8(2):200–208. https://doi.org/10.1016/j.apsb.2017.08.003
19. Zhang L, Li Q, Liu Q, Huang W, Nie J, Wang Y (2017) A bioluminescent imaging mouse model for Marburg virus based on a pseudovirus system. Hum Vaccin Immunother 13(8): 1811–1817. https://doi.org/10.1080/21645515.2017.1325050
20. Li Q, Liu Q, Huang W, Wu J, Nie J, Wang M, Zhao C, Zhang L, Wang Y (2017) An LASV GPC pseudotyped virus based reporter system enables evaluation of vaccines in mice under non-BSL-4 conditions. Vaccine 35(38): 5172–5178. https://doi.org/10.1016/j.vaccine.2017.07.101
21. Wu J, Zhao C, Liu Q, Huang W, Wang Y (2017) Development and application of a bioluminescent imaging mouse model for Chikungunya virus based on pseudovirus system. Vaccine 35(47):6387–6394. https://doi.org/10.1016/j.vaccine.2017.10.007
22. Nie J, Liu L, Wang Q, Chen R, Ning T, Liu Q, Huang W, Wang Y (2019) Nipah pseudovirus system enables evaluation of vaccines in vitro

and in vivo using non-BSL-4 facilities. Emerg Microbes Infect 8(1):272–281. https://doi.org/10.1080/22221751.2019.1571871

23. Han DP, Kim HG, Kim YB, Poon LL, Cho MW (2004) Development of a safe neutralization assay for SARS-CoV and characterization of S-glycoprotein. Virology 326(1):140–149. https://doi.org/10.1016/j.virol.2004.05.017
24. Zhao G, Du L, Ma C, Li Y, Li L, Poon VK, Wang L, Yu F, Zheng BJ, Jiang S, Zhou Y (2013) A safe and convenient pseudovirus-based inhibition assay to detect neutralizing antibodies and screen for viral entry inhibitors against the novel human coronavirus MERS-CoV. Virol J 10:266. https://doi.org/10.1186/1743-422X-10-266
25. Johnson MC, Lyddon TD, Suarez R, Salcedo B, LePique M, Graham M, Ricana C, Robinson C, Ritter DG (2020) Optimized pseudotyping conditions for the SARS-COV-2 spike glycoprotein. J Virol 94(21): e01062-20. https://doi.org/10.1128/JVI.01062-20
26. Pinto D, Park YJ, Beltramello M, Walls AC, Tortorici MA, Bianchi S, Jaconi S, Culap K, Zatta F, De Marco A, Peter A, Guarino B, Spreafico R, Cameroni E, Case JB, Chen RE, Havenar-Daughton C, Snell G, Telenti A, Virgin HW, Lanzavecchia A, Diamond MS, Fink K, Veesler D, Corti D (2020) Cross-neutralization of SARS-CoV-2 by a human monoclonal SARS-CoV antibody. Nature 583(7815):290–295. https://doi.org/10.1038/s41586-020-2349-y
27. Huang SW, Tai CH, Hsu YM, Cheng D, Hung SJ, Chai KM, Wang YF, Wang JR (2020) Assessing the application of a pseudovirus system for emerging SARS-CoV-2 and re-emerging avian influenza virus H5 subtypes in vaccine development. Biomed J 43(4):375–387. https://doi.org/10.1016/j.bj.2020.06.003
28. Crawford KHD, Eguia R, Dingens AS, Loes AN, Malone KD, Wolf CR, Chu HY, Tortorici MA, Veesler D, Murphy M, Pettie D, King NP, Balazs AB, Bloom JD (2020) Protocol and reagents for pseudotyping lentiviral particles with SARS-CoV-2 spike protein for neutralization assays. Viruses 12(5):513. https://doi.org/10.3390/v12050513
29. Yang R, Huang B, Ruhan A, Li W, Wang W, Deng Y, Tan W (2020) Development and effectiveness of pseudotyped SARS-CoV-2 system as determined by neutralizing efficiency and entry inhibition test in vitro. Biosaf Health 2(4):226–231. https://doi.org/10.1016/j.bsheal.2020.08.004
30. Xiong HL, Wu YT, Cao JL, Yang R, Liu YX, Ma J, Qiao XY, Yao XY, Zhang BH, Zhang YL, Hou WH, Shi Y, Xu JJ, Zhang L, Wang SJ, Fu BR, Yang T, Ge SX, Zhang J, Yuan Q, Huang BY, Li ZY, Zhang TY, Xia NS (2020) Robust neutralization assay based on SARS-CoV-2 S-protein-bearing vesicular stomatitis virus (VSV) pseudovirus and ACE2-overexpressing BHK21 cells. Emerg Microbes Infect 9(1): 2105–2113. https://doi.org/10.1080/22221751.2020.1815589
31. Nie J, Li Q, Wu J, Zhao C, Hao H, Liu H, Zhang L, Nie L, Qin H, Wang M, Lu Q, Li X, Sun Q, Liu J, Fan C, Huang W, Xu M, Wang Y (2020) Establishment and validation of a pseudovirus neutralization assay for SARS-CoV-2. Emerg Microbes Infect 9(1):680–686. https://doi.org/10.1080/22221751.2020.1743767
32. Oguntuyo KY, Stevens CS, Hung CT, Ikegame S, Acklin JA, Kowdle SS, Carmichael JC, Chiu HP, Azarm KD, Haas GD, Amanat F, Klingler J, Baine I, Arinsburg S, Bandres JC, Siddiquey MNA, Schilke RM, Woolard MD, Zhang H, COVIDAR Argentina Consortium, Duty AJ, Kraus TA, Moran TM, Tortorella D, Lim JK, Gamarnik AV, Hioe CE, Zolla-Pazner-S, Ivanov SS, Kamil JP, Krammer F, Lee B. Quantifying absolute neutralization titers against SARS-CoV-2 by a standardized virus neutralization assay allows for cross-cohort comparisons of COVID-19 sera. medRxiv [Preprint]. 2020. https://doi.org/10.1101/2020.08.13.20157222. Update in: mBio. 2021;12(1): PMID: 32817961; PMCID: PMC7430605
33. Almahboub SA, Algaissi A, Alfaleh MA, ElAssouli MZ, Hashem AM (2020) Evaluation of neutralizing antibodies against highly pathogenic coronaviruses: a detailed protocol for a rapid evaluation of neutralizing antibodies using vesicular stomatitis virus pseudovirus-based assay. Front Microbiol 11:2020. https://doi.org/10.3389/fmicb.2020.02020
34. Schmidt F, Weisblum Y, Muecksch F, Hoffmann HH, Michailidis E, Lorenzi JCC, Mendoza P, Rutkowska M, Bednarski E, Gaebler C, Agudelo M, Cho A, Wang Z, Gazumyan A, Cipolla M, Caskey M, Robbiani DF, Nussenzweig MC, Rice CM, Hatziioannou T, Bieniasz PD (2020) Measuring SARS-CoV-2 neutralizing antibody activity using pseudotyped and chimeric viruses. bioRxiv. https://doi.org/10.1101/2020.06.08.140871
35. Zhang L, Jackson CB, Mou H, Ojha A, Rangarajan ES, Izard T, Farzan M, Choe H (2020) The D614G mutation in the SARS-CoV-

2 spike protein reduces S1 shedding and increases infectivity. bioRxiv. https://doi.org/10.1101/2020.06.12.148726

36. Plante JA, Liu Y, Liu J, Xia H, Johnson BA, Lokugamage KG, Zhang X, Muruato AE, Zou J, Fontes-Garfias CR, Mirchandani D, Scharton D, Bilello JP, Ku Z, An Z, Kalveram B, Freiberg AN, Menachery VD, Xie X, Plante KS, Weaver SC, Shi PY (2020) Spike mutation D614G alters SARS-CoV-2 fitness. Nature 592(7852):116–121. https://doi.org/10.1038/s41586-020-2895-3
37. Hu J, He C-L, Gao Q-Z, Zhang G-J, Cao X-X, Long Q-X, Deng H-J, Huang L-Y, Chen J, Wang K, Tang N, Huang A-L (2020) The D614G mutation of SARS-CoV-2 spike protein enhances viral infectivity and decreases neutralization sensitivity to individual convalescent sera. bioRxiv. https://doi.org/10.1101/2020.06.20.161323
38. Ogando NS, Dalebout TJ, Zevenhoven-Dobbe JC, Limpens R, van der Meer Y, Caly L, Druce J, de Vries JJC, Kikkert M, Barcena M, Sidorov I, Snijder EJ (2020) SARS-coronavirus-2 replication in Vero E6 cells: replication kinetics, rapid adaptation and cytopathology. J Gen Virol 101(9):925–940. https://doi.org/10.1099/jgv.0.001453
39. Kemp SA, Collier DA, Datir R, Ferreira I, Gayed S, Jahun A, Hosmillo M, Rees-Spear C, Mlcochova P, Lumb IU, Roberts DJ, Chandra A, Temperton N, Sharrocks K, Blane E, Briggs J, van Gils MJ, Smith K, Bradley JR, Smith C, Doffinger R, Ceron-Gutierrez L, Barcenas-Morales G, Pollock DD, Goldstein RA, Smielewska A, Skittrall JP, Gouliouris T, Goodfellow IG, Gkrania-Klotsas E, Illingworth C, McCoy LE, Gupta RK (2020) Neutralising antibodies in spike mediated SARS-CoV-2 adaptation. medRxiv. https://doi.org/10.1101/2020.12.05.20241927
40. Weisblum Y, Schmidt F, Zhang F, DaSilva J, Poston D, Lorenzi JC, Muecksch F, Rutkowska M, Hoffmann HH, Michailidis E, Gaebler C, Agudelo M, Cho A, Wang Z, Gazumyan A, Cipolla M, Luchsinger L, Hillyer CD, Caskey M, Robbiani DF, Rice CM, Nussenzweig MC, Hatziioannou T, Bieniasz PD (2020) Escape from neutralizing antibodies by SARS-CoV-2 spike protein variants. elife 9: e61312. https://doi.org/10.7554/eLife.61312
41. Thomson EC, Rosen LE, Shepherd JG, Spreafico R, da Silva Filipe A, Wojcechowskyj JA, Davis C, Piccoli L, Pascall DJ, Dillen J, Lytras S, Czudnochowski N, Shah R, Meury M, Jesudason N, De Marco A, Li K, Bassi J, O'Toole A, Pinto D, Colquhoun RM, Culap K, Jackson B, Zatta F, Rambaut A, Jaconi S, Sreenu VB, Nix J, Jarrett RF, Beltramello M, Nomikou K, Pizzuto M, Tong L, Cameroni E, Johnson N, Wickenhagen A, Ceschi A, Mair D, Ferrari P, Smollett K, Sallusto F, Carmichael S, Garzoni C, Nichols J, Galli M, Hughes J, Riva A, Ho A, Semple MG, Openshaw PJM, Baillie JK, Rihn SJ, Lycett SJ, Virgin HW, Telenti A, Corti D, Robertson DL, Snell G (2020) The circulating SARS-CoV-2 spike variant N439K maintains fitness while evading antibody-mediated immunity. bioRxiv. https://doi.org/10.1101/2020.11.04.355842
42. Yan R, Zhang Y, Li Y, Xia L, Guo Y, Zhou Q (2020) Structural basis for the recognition of SARS-CoV-2 by full-length human ACE2. Science 367(6485):1444–1448. https://doi.org/10.1126/science.abb2762
43. Mukherjee S, Pierson TC, Dowd KA (2014) Pseudo-infectious reporter virus particles for measuring antibody-mediated neutralization and enhancement of dengue virus infection. Methods Mol Biol 1138:75–97. https://doi.org/10.1007/978-1-4939-0348-1_6
44. Andrewes CH, Elford WJ (1933) Observations on anti-phage sera. I: "the percentage law". Br J Exp Pathol 14(6):367–376
45. Klasse PJ, Sattentau QJ (2002) Occupancy and mechanism in antibody-mediated neutralization of animal viruses. J Gen Virol 83(Pt 9): 2091–2108. https://doi.org/10.1099/0022-1317-83-9-2091
46. Klasse PJ (2014) Neutralization of virus infectivity by antibodies: old problems in new perspectives. Adv Biol 2014:157895. https://doi.org/10.1155/2014/157895
47. Pierson TC, Sanchez MD, Puffer BA, Ahmed AA, Geiss BJ, Valentine LE, Altamura LA, Diamond MS, Doms RW (2006) A rapid and quantitative assay for measuring antibody-mediated neutralization of West Nile virus infection. Virology 346(1):53–65. https://doi.org/10.1016/j.virol.2005.10.030
48. Cai W, Tang ZM, Wen GP, Wang SL, Ji WF, Yang M, Ying D, Zheng ZZ, Xia NS (2016) A high-throughput neutralizing assay for antibodies and sera against hepatitis E virus. Sci Rep 6:25141. https://doi.org/10.1038/srep25141
49. Chen M, Chang JS, Nason M, Rangel D, Gall JG, Graham BS, Ledgerwood JE (2010) A flow cytometry-based assay to assess RSV-specific neutralizing antibody is reproducible, efficient and accurate. J Immunol Methods 362(1–2): 180–184. https://doi.org/10.1016/j.jim.2010.08.005

50. Pastorino B, Touret F, Gilles M, de Lamballerie X, Charrel RN (2020) Heat inactivation of different types of SARS-CoV-2 samples: what protocols for biosafety, molecular detection and serological diagnostics? Viruses 12(7):735. https://doi.org/10.3390/v12070735
51. Wang TT, Lien CZ, Liu S, Selvaraj P (2020) Effective heat inactivation of SARS-CoV-2. medRxiv. https://doi.org/10.1101/2020.04.29.20085498
52. Batéjat C, Grassin Q, Manuguerra J-C, Leclercq I (2020) Heat inactivation of the severe acute respiratory syndrome coronavirus 2. bioRxiv. https://doi.org/10.1101/2020.05.01.067769
53. McCoy LE, Falkowska E, Doores KJ, Le K, Sok D, van Gils MJ, Euler Z, Burger JA, Seaman MS, Sanders RW, Schuitemaker H, Poignard P, Wrin T, Burton DR (2015) Incomplete neutralization and deviation from sigmoidal neutralization curves for HIV broadly neutralizing monoclonal antibodies. PLoS Pathog 11(8):e1005110. https://doi.org/10.1371/journal.ppat.1005110

Chapter 22

Cytopathic Effect (CPE)-Based Drug Screening Assay for SARS-CoV-2

Yan Ling Ng, Chee Keng Mok, and Justin Jang Hann Chu

Abstract

Identification of an effective antiviral for the treatment of COVID-19 is considered one of the holy grails in the bid to end the pandemic. However, the novelty of SARS-CoV-2, along with the little knowledge available about its infection characteristics at the beginning of this pandemic, challenges the scientific world on how one may be able to promptly identify promising drug candidates from a myriad of compound libraries. Here, we describe a cytopathic effect (CPE)-based drug screening assay for SARS-CoV-2 which allows for rapid assessment of drug compound libraries through pre- or posttreatment drug screening procedures and evaluation using a light microscope. By comparing the virus-induced CPE of the drug-treated cells against the vehicle and drug controls, potent drug candidates can be quickly identified for further downstream studies.

Key words Drug screening, Antiviral, Cytopathic Effect

1 Introduction

Over the last two decades, the emergence of SARS-CoV, MERS-CoV, and SARS-CoV-2 outbreaks has highlighted the human vulnerability to emerging infectious diseases [1]. To date, the COVID-19 pandemic caused by SARS-CoV-2 has resulted in over 180 million reported cases and nearly 4 million deaths globally, posing a major health threat owing to its unprecedented spread efficiency [2]. As the number of cases and variants with increasing virulence and transmissibility continues to rise [3], the identification of potent antiviral drugs and expansion of therapeutic arsenal against the novel coronavirus is considered one of the holy grails in ending the pandemic.

With the urgency of the pandemic making de novo drug discovery almost impossible, drug repositioning becomes an attractive proposition for rapid identification of promising antiviral drug candidates, especially if the safety profiles of the said drugs have

Justin Jang Hann Chu et al. (eds.), *SARS-CoV-2: Methods and Protocols*,
Methods in Molecular Biology, vol. 2452, https://doi.org/10.1007/978-1-0716-2111-0_22,

already been established [4, 5]. With the repurposing success story of remdesivir and its subsequent FDA approval for treatment of COVID-19 [6], it is hoped that drug repurposing would become a major strategy in dealing with future disease outbreaks involving novel etiological agents.

The first step in drug repositioning typically involves the screening of compound libraries. Here, the drug screening success is limited by the quality and chemical space covered by the screening libraries [7]. Antiviral drug screening may also be further divided into pretreatment and posttreatment screen, depending on whether the compounds inhibit the earlier steps (e.g., virus entry), or the later steps of the virus life cycle (e.g., replication) [8, 9]. In a pretreatment screen, the cells are treated with a number of drug candidates prior to virus infection while in a posttreatment screen, the cells are infected with a virus first before being treated with various drug candidates [9]. As such, it is important to determine whether a compound library is to be put through a pretreatment or a posttreatment screen to ensure result accuracy. For instance, ACE2 inhibitors compound library is typically screened using pretreatment assays, since ACE2 is the cognate receptor of the SARS-CoV-2 spike protein required for virus entry. Putting an ACE2 inhibitor compound library through a posttreatment screen may produce false negatives since ACE2 primarily works in the earlier steps of the life cycle.

The cytopathic effect (CPE)-based drug screening is an assay routinely used in the identification of valuable hit compounds for DENV, ZIKV, and SARS-CoV-2 [10–12]. In this assay, cells which have undergone either pre- or posttreatment of drugs are compared against cells treated with a vehicle control (e.g., DMSO) or a known effective antiviral drug based on the extent of virus-induced CPE appearing on the cells in the well. Cells which are treated with a vehicle control would typically produce extensive virus-induced CPE, such as rounding of cells and syncytium formation. On the contrary, cells which are treated with a drug known to inhibit virus replication would maintain its regular cell morphology, akin to the mock-infected cells. From here, hits are identified and selected based on the reduction of CPE and >50% inhibition in duplicate wells in comparison to vehicle control (Fig. 1), making this assay simple to perform and interpret even for a researcher with no prior drug discovery knowledge.

In this chapter, we describe the methods for performing a CPE-based drug screening assay to identify promising hits against SARS-CoV-2 using four compound libraries: an FDA-approved compound library, natural product library, flavonoids compound library and ACE2-inhibitor compound library. We were able to identify a total of 121 compounds out of 2000 compounds from four libraries using this method, demonstrating the power and robustness of this assay.

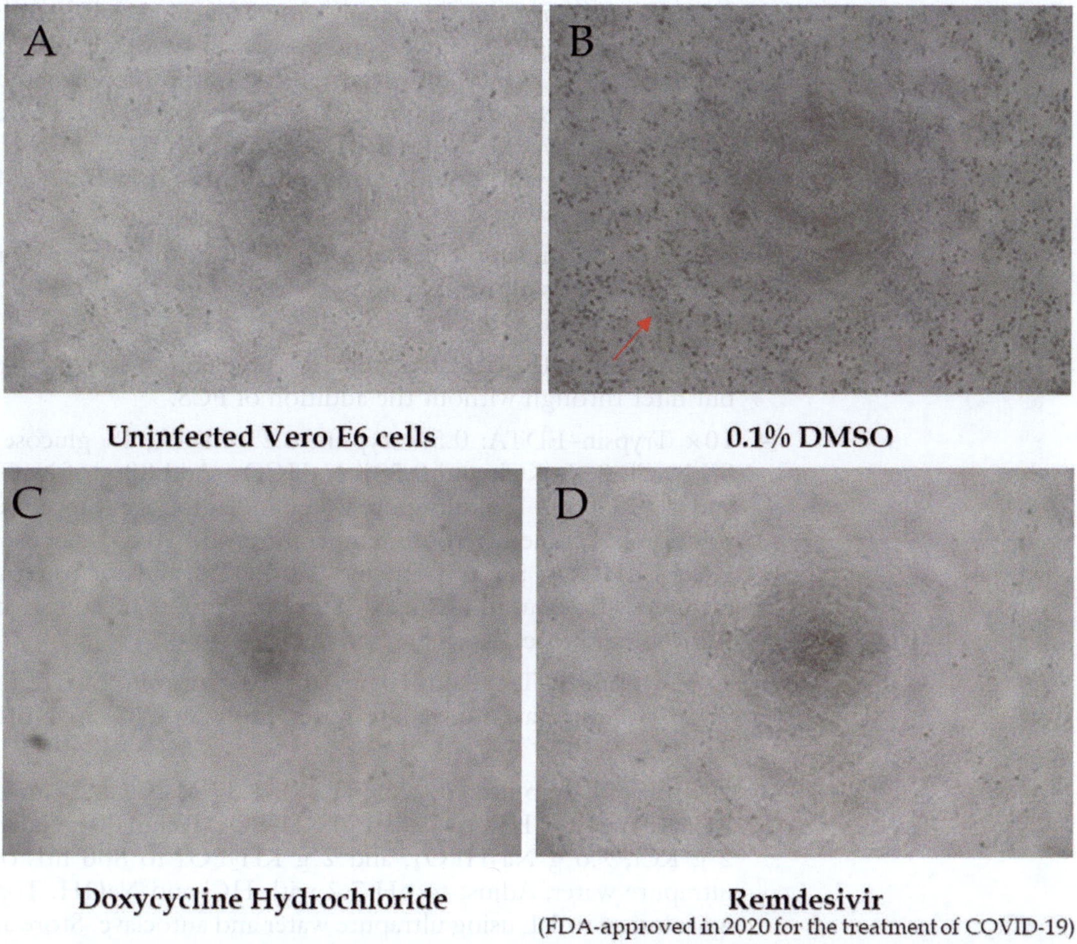

Fig. 1 Cytopathic-effect (CPE) based drug screening in Vero E6 cells. Vero E6 cells were either pre- or posttreated with drug compounds and infected with SARS-CoV-2 at the multiplicity of infection (MOI) of 1. (**a**) represents uninfected, untreated Vero E6 cells. (**b**) represent cells treated with 0.1% DMSO and infected with SARS-CoV-2 as a vehicle control. Virus-induced CPE is observed by the rounding of cells (red arrow) distributed over the well. (**c**) represents cells posttreated with 10 μM doxycycline hydrochloride and infected with SARS-CoV-2. Note the reduction of virus-induced CPE in comparison to vehicle control. (**d**) represents cells treated with 100 μM of remdesivir which was included as a positive control. All images were taken at 4 days postinfection using a light microscope

2 Materials

Prepare all solutions using ultrapure water (prepared by purifying deionized water, to attain a sensitivity of 18 MΩ-cm at 25 °C). Prepare all materials in a biosafety cabinet (BSC) to ensure sterility and store all reagents as at indicated temperatures. Diligently follow all waste disposal regulations when disposing of SARS-CoV-2 contaminated materials.

2.1 Culture Media

1. DMEM medium supplemented with 10% fetal calf serum (FCS). Weigh 2 g of $NaHCO_3$ and dissolve in 1 L of autoclaved ultrapure water. Once dissolved, add 1 bottle of powdered DMEM into the same bottle and mix well. To prepare 500 mL of DMEM with 10% FCS, add 50 mL of FCS to 450 mL of DMEM solution and filter through a 0.22 μM Corning filter. Store at 4 °C.
2. DMEM medium supplemented with 2% FCS. Instructions for preparation is as above, but add 10 mL of FCS to 490 mL of DMEM solution and filter through.
3. Serum-free DMEM. Instructions for preparation is as above but filter through without the addition of FCS.
4. 10× Trypsin–EDTA: 0.5% trypsin, 0.2% EDTA, 1% glucose, 8% NaCl, 0.4% KCl, and 0.58% $NaHCO_3$. Add 80 g of NaCl and 4 g KCl to 1 L of ultrapure water and autoclave. Once the solution has cooled to room temperature, add 10 g d-glucose, 5.8 g $NaHCO_3$, 5 g trypsin, and 2 g EDTA, and stir using a magnetic stirrer until dissolved. Store in aliquots of 10 mL at −20 °C (*see* **Note 1**).
5. 1× Trypsin–EDTA. Add 10 mL of 10× trypsin–EDTA to 90 mL of autoclaved ultrapure water. Store at 4 °C (*see* **Note 2**).
6. 10× phosphate buffered saline (PBS): 1.37 M NaCl, 27 mM KCl, 43 mM Na_2HPO_4, 14.7 mM KH_2PO_4. Weigh 80 g NaCl, 2 g KCl, 7.6 g Na_2HPO_4, and 2 g KH_2PO_4 to 800 mL of ultrapure water. Adjust to pH 7.2 with HCl and NaOH. Top up solution to 1 L using ultrapure water and autoclave. Store at room temperature.
7. 1× PBS. Add 50 mL of 10× PBS to 450 mL of ultrapure water and autoclave. Store at 4 °C.
8. 10% formalin: Add 100 mL formalin (37–40% stock solution) to 900 mL of ultrapure water. Store at room temperature.

2.2 Cells and Virus

1. Vero E6 cells.
2. SARS-CoV-2-containing cell supernatants.

2.3 Compound Libraries

1. ACE2-targeted compound library (TargetMol).
2. Natural product compound library.
3. Flavonoids library (TimTec).
4. FDA-approved drug library (SelleckChem).
5. Dimethyl sulfoxide (DMSO).

2.4 Consumables

1. 10 mL serological pipettes
2. 96-well plates

3. Micropipettes (P20, P200, P1000, and multichannel pipettes).
4. Pipette tips.
5. Solution reservoirs.
6. 4% Melsept disinfectant solution: Add 200 mL of Melsept to 4.8 L of tap water.

2.5 Equipment

1. BSC.
2. Light microscope.
3. Lock & Lock boxes.
4. Heat block.
5. Hemocytometer.
6. Autoclaved glass bottles.
7. Counter.
8. CO_2 incubator.

3 Methods

Carry out all procedures in the BSC and diligently follow all safety procedures as prescribed by the laboratory. To minimize prolonged exposure to SARS-CoV-2, cell seeding and drug preparation procedures are typically performed in a biosafety level 2 laboratory before securely transferred to a biosafety level 3 laboratory for SARS-CoV-2 infection and postinfection analysis. Here, the screening is performed using 96-well plates.

3.1 Generation of SARS-CoV-2 Virus Stock

1. Generate a working pool of SARS-CoV-2 virus stock (*see* this chapter).
2. Aliquot the virus stock in 300 μL and 600 μL volumes in a cryovial.
3. Surface-decontaminate each cryovial by wiping down with 4% Melsept and store each tube in a cryobox. Surface-decontaminate the cryobox and place them in a basket.
4. Transfer the cryobox into the −80 °C freezer till further use.
5. Titer virus stock by performing plaque assay (*see* this chapter).

3.2 Seeding Vero E6 Cells on 96-Well Plates

1. Warm all cell culture media, PBS and trypsin–EDTA in a water bath at 37 °C prior to cell culture.
2. Check the Vero E6 cells under a light microscope for confluency (~90%).
3. Remove spent cell culture media from a confluent flask of Vero E6 cells (*see* **Note 3**).

4. Wash the cell monolayer once with 5 mL of 1× PBS (*see* **Note 4**).
5. Add 2 mL of 1× trypsin–EDTA and incubate at room temperature for 1–3 min (*see* **Note 5**).
6. Gently knock the side of the flask to dislodge cells.
7. Add 8 mL of DMEM with 10% FCS to neutralize trypsin–EDTA. Mix well by pipetting the cell suspension up and down several times to ensure complete neutralization of trypsin–EDTA and even dispersion of cells (*see* **Note 6**).
8. Pipet 20 μL of Vero E6 cell suspension onto a hemocytometer and count cells under a light microscope using a counter. Calculate the appropriate dilution factor for a seeding density that would result in a confluency of >80% after overnight incubation (*see* **Note 7**).
9. Dilute cells in DMEM with 10% FCS in a centrifuge tube or an autoclaved glass bottle, depending on the number of plates to be seeded. Pipet to ensure even dispersion of cells (*see* **Note 8**). Dispense some cell suspension onto the solution reservoir.
10. Using a multichannel pipette, dispense 100 μL of diluted cell suspension into each of the 96-well plates (*see* **Note 9**).
11. Leave the seeded plates in the BSC or bench top for at least 15 min to allow cells to settle (*see* **Note 10**).
12. Incubate plates overnight at 37 °C with 5% CO_2 to allow attachment in the BSL-2 laboratory.

3.3 Preparation of Compound Libraries

1. Determine the molecular weight and pack size of each compound in the library of interest by referring to the datasheet provided by the library manufacturer (*see* **Note 11**).
2. Calculate the volume of 100% DMSO required to prepare a 10 mM master stock solution using the molarity equation (*see* **Note 12**).
3. Further dilute the 10 mM stock solution to 100 μM using serum-free DMEM (*see* **Note 13**). This is known as the working stock.
4. Store compounds at −20 °C till further use.

3.4 Pretreatment Screen

1. On the day of the pretreatment drug screen, remove and thaw the working stock of compounds from the −20 °C freezer.
2. Dilute the working stock further to 20 μM in a 96-well plate (*see* **Note 14**).
3. Include 0.2% DMSO and 200 μM remdesivir on the 96-well plate as vehicle and positive controls, respectively.
4. Remove the incubated cells from the CO_2 incubator and transfer into the BSC.

5. Pipet 100 μL of diluted compounds from the 96-well plates in **step 2** to the cells to achieve the final concentration of 0.1% DMSO, 100 μM remdesivir and 10 μM of drug compounds.
6. Incubate the cells for 2 h at 37 °C in the CO_2 incubator.
7. After 2 h, remove the media and wash the cells twice with 1× PBS and replace with DMEM with 2% FCS.
8. Place the plates containing the treated cells inside a sturdy and leak proof secondary container such as a Lock & Lock box and transport them to the BSL-3 laboratory.
9. Once inside the BSL-3, place the Lock & Lock box containing the treated cells into a basket and place the plates into the CO_2 incubator and prepare the BSC.
10. Place the virus-containing cryobox from the −80 °C freezer into a basket and transfer the cryobox into the BSC.
11. Retrieve the desired vials of virus required and place the vials on a heat block to thaw at 37 °C. Surface-decontaminate the cryobox by wiping down with 4% Melsept and return the cryobox to the −80 °C freezer.
12. Dilute the virus stock using DMEM with 2% FCS to achieve the multiplicity of infection (MOI) of 1 (*see* **Note 15**). Gently resuspend using a serological pipette and pipet the virus-containing media to a solution reservoir.
13. Dispose of all used filtered pipette tips, serological pipettes, and cryovials by rinsing and aspirating with 4% Melsept in the stainless-steel liquid waste container. Discard the used pipette tip and cryovial directly into the liquid waste.
14. Open the CO_2 incubator and lock all sides of the Lock & Lock box containing the pretreated cells and place them in a basket. Transfer the box into the BSC.
15. Remove the media in each well using a multichannel pipette and dispose of into the liquid waste container. Dispose of all used filtered pipette tips as above.
16. Add 50 μL of diluted SARS-CoV-2 virus supernatant to each well of the 96-well plate and 50 μL of DMEM with 2% FCS to mock-infected wells. Dispose of all used filtered pipette tips as above.
17. Wipe the surface of the plate with 4% Melsept and place the plates into the Lock & Lock box. Surface-decontaminate the Lock & Lock box with 4% Melsept and lock all sides.
18. Place the Lock & Lock box into a basket and return the plates to the rear end of the CO_2 incubator. Once placed inside, unlock two sides of the Lock & Lock box and incubate the plates for 1 h.

19. After 1 h, open the CO_2 incubator and lock all sides of the Lock & Lock box and place them in a basket. Transfer the box into the BSC.
20. Using a multichannel pipette, remove the virus supernatant and wash the cells with 1× PBS thrice. Dispose of all used filtered pipette tips as above.
21. Add 100 μL of DMEM with 2% FCS to each well.
22. Surface-decontaminate the plates and Lock & Lock box as above and return the plates back to the CO_2 incubator.
23. Incubate the cells for 4 days at 37 °C, 5% CO_2.

3.5 Posttreatment Screen

1. On the day of the posttreatment drug screen, remove and thaw the working stock of compounds from the −20 °C freezer.
2. Dilute the working stock of FDA-approved and flavonoid compound libraries further to 10 μM and 50 μM in a 96-well plate, respectively.
3. Include 0.1% DMSO and 100 μM remdesivir on the 96-well plate as vehicle and positive controls, respectively.
4. Place the plates containing the cells and drugs inside a sturdy and leak-proof secondary container such as a Lock & Lock box and transport them to the BSL-3 laboratory.
5. Once inside the BSL-3, place the Lock & Lock box containing the cells into a basket and place the plates into the CO_2 incubator and prepare the BSC. Leave the plate of drugs at room temperature till later use.
6. Place the virus-containing cryobox from the −80 °C freezer into a basket and transfer the cryobox into the BSC.
7. Retrieve the desired vials of virus required and place the vials on a heat block to thaw at 37 °C. Surface-decontaminate the cryobox by wiping down with 4% Melsept and return the cryobox to the −80 °C freezer.
8. Dilute the virus stock using DMEM with 2% FCS to achieve the multiplicity of infection (MOI) of 1. Gently resuspend using a serological pipette and pipet the virus-containing media to a solution reservoir.
9. Dispose of all used filtered pipette tips and serological pipettes by rinsing and aspirating with 4% Melsept in the stainless-steel liquid waste container. Discard the used pipette tips, serological pipette, and cryovial directly into the liquid waste.
10. Open the CO_2 incubator and lock all sides of the Lock & Lock box containing the pretreated cells and place them in a basket. Transfer the box into the BSC.

11. Remove the media in each well using a multichannel pipette and dispose of into the liquid waste container. Dispose of all used filtered pipette tips as above.
12. Add 50 μL of diluted SARS-CoV-2 virus supernatant to each well of the 96-well plate and 50 μL of DMEM with 2% FCS to mock-infected wells. Dispose of all used filtered pipette tips as above.
13. Wipe the surface of the plate with 4% Melsept and place the plates into the Lock & Lock box. Surface-decontaminate the Lock & Lock box with 4% Melsept and lock all sides.
14. Place the Lock & Lock box into a basket and return the plates to the rear end of the CO_2 incubator. Once placed inside, unlock two sides of the Lock & Lock box and incubate the plates for 1 h.
15. After 1 h, open the CO_2 incubator and lock all sides of the Lock & Lock box and place them in a basket. Transfer the box into the BSC.
16. Using a multichannel pipette, remove the virus supernatant and wash the cells with 1× PBS thrice. Dispose of all used filtered pipette tips as above.
17. Pipet 100 μL of diluted compounds from the FDA-approved and flavonoid compound libraries in **step 2** to the cells.
18. Surface-decontaminate the plates and Lock & Lock box as above and return the plates back to the CO_2 incubator.
19. Incubate the cells for 4 days at 37 °C, 5% CO_2.

3.6 Hit selection Via CPE Analysis

1. At 4 days postinfection, open the CO_2 incubator and lock all sides of the Lock & Lock box and place them in a basket. Transfer the box into the BSC.
2. Dispense an appropriate volume of 10% formalin onto a solution reservoir.
3. Remove virus supernatant and pipet 100 μL of 10% formalin to each well. Virus supernatant may also be transferred to a cryotube for plaque assays, if required.
4. Surface-decontaminate the plates using 4% Melsept as above and place the plates into a clear Ziplock bag.
5. Place the Ziplock bags containing the plates into a Lock & Lock box and surface-decontaminate as above.
6. Turn on the light microscope and examine the infected cells using light microscopy.
7. Observe the differences in cell viability caused by virus-induced CPE. To begin comparing the different wells, first observe the 0.1% DMSO (vehicle control) and remdesivir (positive control)-treated wells. DMSO-containing wells, being

untreated, would show massive cell death represented by the rounding of cells following virus infection. In contrast, remdesivir-treated wells would show little to no rounding of cells similar to the mock-infected cells.

8. Keeping in mind the extent of CPE effects of the control wells, next compare the CPE of each drug candidate against the DMSO controls.
9. Hits can be identified and selected based on the reduction of CPE and >50% inhibition in duplicate wells in comparison to vehicle control. Toxic compounds and drug compounds would not score as positive hits as they would increase cell death.

4 Notes

1. Trypsin undergoes autolysis in solution and slowly loses its enzymatic activity over time, especially upon repeated freeze–thaw cycles and storage at temperatures above −20 °C. As such, the shelf life of trypsin–EDTA upon storage at −20 °C is 1 year. Aliquoting 10× trypsin–EDTA in smaller volumes eliminate the need for repeated freeze-thawing and help prolongs the life span of each batch of concentrated trypsin–EDTA made.
2. To ensure that the enzymatic activity remains at sufficient levels needed to dislodge cells, prepare small volumes (100 mL) of 1× trypsin–EDTA and store at 4 °C for use in cell culture. Larger volumes may be prepared if trypsin–EDTA is to be rapidly used up (within 4 weeks).
3. For drug screening, any cell line that is permissive to SARS-CoV-2 infection and result in observable virus-induced CPE can be utilized, such as CRFK and COS cells. If alternative cell lines are to be used, cell seeding densities and incubation periods stated in this protocol would have to be modified and specifically optimized to suit the growth rates of these substitute cell lines.
4. Washing the cells with 1× PBS helps remove all traces of FCS, which contains inhibition of trypsin. When adding PBS into the flask, do dispense to the bottom corner of the flask and not directly onto the cell monolayer. Directly dispensing PBS onto cells may cause cells to dislodge from the monolayer. Gently rock the flask back and forth several times to wash the cells before discarding the PBS.
5. Repeated warming of trypsin–EDTA at 37 °C may result in the loss of enzymatic activity. As such, the incubation time required to dislodge the cells become progressively longer, sometimes longer than 5 min in some cases. During incubation with

trypsin–EDTA, the side of the flask should be gently knocked against the palm of the hand to check whether cells are sufficiently trypsinized. If trypsinization is sufficient, cells should detach in clumps upon gentle knocking and the cell monolayer will be visibly disrupted. In addition, cells will also appear rounded when observed under a light microscope. When this occurs, the flask should be gently knocked until the entire monolayer is dislodged. If trypsinization is insufficient (i.e., gently knocking does not disturb the monolayer), the incubation time should be lengthened before knocking the flask further. Vigorous banging of the flask should be avoided as this may induce cell death due to mechanical stress. Flasks can also be placed in a 37 °C incubator to speed up trypsinization.

6. Vero E6 cells tend to clump if the cell suspension is not mixed properly. In addition, trypsinizing the cells for too long or using freshly prepared trypsin may result in visible clumps upon neutralization. In such cases, pipetting numerous times until the clumps become smaller or nonexistent should suffice to obtain a cell suspension that can be used for seeding.
7. For drug screening using 96-well plates, Vero E6 cells seeded at a seeding density of 1×10^4 cells per 100 μL would likely produce a confluency of at least 80% after overnight incubation.
8. Thorough mixing is necessary to ensure even seeding. If cells are seeded in clumps, cell growth will be distributed unevenly across the well, with some regions either becoming over- or underconfluent. This may lead to uneven virus infection and difficulties in analyzing virus-induced CPE postinfection later in the drug screening process.
9. The pipetting method has to be optimized to ensure even seeding of cells. Some researchers resuspend the cell suspension several times first using a multichannel pipette and expulse the cell suspension by slowing pushing the plunger all the way down till it hits the first stop consistently for all cell dispensing steps. Pipetting methods differ among researchers and individual researchers should explore varying methods to find one that is easy for them. It is important that cells are seeded evenly, without clumping in the middle or one side of the well to ensure reproducibility and accuracy of results.
10. Moving the plates immediately upon seeding may also result in clumping of cells. Leaving them in the BSC to allow the cells to settle is ideal. However, the internal airflow of some BSCs produces significant vibrations on the BSC floor, which may result in unnecessary agitation of the cell suspension and cell clumping. In such cases, plates should be taken out and placed on a stable bench top for cells to settle. After cells are settled to the bottom of each well, plates can be viewed under a light

microscope to observe seeding density and whether cells are uniformly distributed prior to incubation in the CO_2 incubator.

11. The delivery of compound libraries by suppliers are usually accompanied with an Excel file or datasheet describing the compounds supplied, arrangement of the compound library, and the pack size (e.g., 1 mg/5 mg) of each compound. Knowing the molecular weight and the pack size of each compound is necessary to calculate the volume of solvent required to dissolve the compound into a stock solution for storage.
12. The molarity calculator equation is Mass (g) = Concentration (mol/L) × Volume (L) × Molecular Weight (g/mol). Do use an appropriate solvent specified by your compound library supplier to dissolve your compounds. In the case of the compound libraries used in this chapter, 100% DMSO was recommended by the supplier for dissolving the compounds.
13. Repeated freezing and thawing cycles may cause the compounds to degrade over time. To maintain the stability and chemical integrity of the drug compounds, drugs are usually dissolved as soon as it arrives into master stock concentrations of 10 mM in an appropriate solvent. Once dissolved, the drugs will be aliquoted into smaller volumes for long-term storage until further use to minimize the number of freeze–thaw cycles. Before performing a drug screen, the required number of plates containing the compounds will then be thawed to generate working concentrations of the drug by diluting the stock concentration volumes to serum-free media. If required, these working concentrations of drug may be further diluted, when necessary.
14. In the pretreatment assay, one may prepare the drug compounds on the 96-well plate at a concentration that is double of the desired concentration required for the drug treatment to skip the step of removing the cell culture media. For this step, we have prepared 0.2% DMSO, 200 μM, and 20 μM concentrations of the vehicle control, remdesivir control and drug compounds respectively on the 96-well plate. During pretreatment, 100 μL of each solution is added to the plate already containing 100 μL of cell culture media to achieve the final concentration of 0.1% DMSO, 100 μM remdesivir and 10 μM of drug compounds.
15. Special consideration for disposing of waste containing SARS--CoV-2 are given in this experiment. At the end of the experiment, all wastes produced from this experiment are immediately wrapped up for autoclaving and disposal according to waste management section in Biosafety Manual of the BSL-3 laboratory.

Acknowledgments

We are grateful to the Yong Loo Lin School of Medicine BSL-3 Core Facility for their support with this work. This work was supported by the following grants: (1) NUHSRO/2020/066/NUSMedCovid/01/BSL3 Covid Research Work NUHSRO/2020/050/RO5 + 5/NUHS-COVID/4; (2) Ministry of Education, Singapore MOE2017-T2-2-014; (3) Ministry of Education, Singapore MOE2017-T2-2-014; (4) Singapore NMRC Centre Grant Program—Diabetes, Tuberculosis and Neuroscience CGAug16M009; (5) Ministry of Health MOH-COVID19RF2-000 and (6) Sino-Singapore Cooperation for Evaluating the Effectiveness and Application of Guangxi Zhuang/Yao Medicines Against COVID-19 (No. GUIKE AB20036001).

References

1. Zhu Z, Lian X, Su X, Wu W, Marraro GA, Zeng Y (2020) From SARS and MERS to COVID-19: a brief summary and comparison of severe acute respiratory infections caused by three highly pathogenic human coronaviruses. Respir Res 21:224
2. Batalha PN, Forezi LSM, Lima CGS, Pauli FP, Boechat FCS, de Souza MCBV et al (2021) Drug repurposing for the treatment of COVID-19: pharmacological aspects and synthetic approaches. Bioorg Chem 106:104488
3. Abdool Karim SS, de Oliveira T (2021) New SARS-CoV-2 variants – clinical, public health, and vaccine implications. N Engl J Med 384(19):1866–1868
4. Simonis A, Theobald SJ, Fätkenheuer G, Rybniker J, Malin JJ (2021) A comparative analysis of remdesivir and other repurposed antivirals against SARS-CoV-2. EMBO Mol Med 13:e13105
5. WHO Solidarity Trial Consortium, Pan H, Peto R, Henao-Restrepo A, Preziosi M, Sathiyamoorthy V et al (2021) Repurposed antiviral drugs for Covid-19 – interim WHO solidarity trial results. N Engl J Med 384:497–511
6. Rubin D, Chan-Tack K, Farley J, Sherwat A (2020) FDA approval of Remdesivir – a step in the right direction. N Engl J Med 383:2598–2600
7. Gong Z, Hu G, Li Q, Liu Z, Wang F, Zhang X et al (2017) Compound libraries: recent advances and their applications in drug discovery. Curr Drug Discov Technol 14:216–228
8. Day CJ, Bailly B, Guillon P, Dirr L, Jen FEC, Spillings BL et al (2021) Multidisciplinary approaches identify compounds that bind to human ACE2 or SARS-CoV-2 spike protein as candidates to block SARS-CoV-2-ACE2 receptor interactions. mBio 12:e03681–e03620
9. Mok CK, Ng YL, Ahidjo BA, Lee RCH, Loe MWC, Liu J et al. Calcitriol, the active form of vitamin D, is a promising candidate for COVID-19 prophylaxis. bioRxiv. https://doi.org/10.1101/2020.06.21.162396
10. McCormick KD, Liu S, Jacobs JL, Marques ETA Jr, Sluis-Cremer N, Wang T (2012) Development of a robust cytopathic effect-based high-throughput screening assay to identify novel inhibitors of dengue virus. Antimicrob Agents Chemother 56:3399–3401
11. Dabaja MZ, de Oliveira LE, de Oliveira DN, Guerreiro TM, Melo CFOR, Morishita KN et al (2018) Metabolic alterations induced by attenuated Zika virus in glioblastoma cells. Cell Biosci 8:47
12. Chen CZ, Shinn P, Itkin Z, Eastman RT, Bostwick R, Rasmussen L et al (2020) Drug repurposing screen for compounds inhibiting the cytopathic effect of SARS-CoV-2. Front Pharmacol 11:592737

Part V

Biorisk and Mitigation Measurements

Chapter 23

Strengthening Biorisk Management in Research Laboratories with Security-Sensitive Biological Agents Like SARS-CoV-2

Sabai Phyu, Tessy Joseph, and Margarida Goulart

Abstract

In this chapter, we discuss potential incidents associated with SARS-CoV-2 experimental work in high containment research laboratories. The risk landscape in high containment laboratories is changing due to the strong innovation drive of the life sciences research. Thus, the WHO has recommended life sciences organizations to incorporate good research practices and ethical principles into a risk-based approach of the biorisk management (BRM). Currently, BRM systems in high containment laboratories are predominantly steered by operational personnel and laboratory professional. It is well known that without having a systematic approach and leadership support from the organization, the BRM system in the high containment laboratory will not be sustainable. Even though the roles of organizations and their leadership in establishing the BRM system are spelt out in many international standards, guidance documents and national legislations, operational aspects of these roles are rarely discussed.

It is therefore important for everyone to understand about their roles in organizational processes (communication, decision, and performance evaluation) involved in implementation of BRM related operational activities. In this chapter, discussion is based on operational activities of four main organizational behaviors that are considered to have strengthened BRM systems in high containment laboratories: (1) displaying a visible commitment and support to the BRM system from different levels of management, (2) developing a competent and responsible workforce with BRM technical skills and problem identification/solving skills, (3) integrating learning and improvement principles into the BRM system, and (4) enhancing the continuous motivation of laboratory personnel to avoid complacency. The categorization of these organizational behaviors is based on the International Atomic Energy Agency's principles and guidance for strengthening the safety and security culture in nuclear facilities. Furthermore, we encourage the laboratory management to identify gaps in processes and activities related to those organizational behaviors so that one could rapidly address biosafety and biosecurity vulnerabilities in high containment laboratories.

Key words Biorisk management, Biosafety, Biosecurity, Security-sensitive or high-risk biological agents, Organizational behaviors

Justin Jang Hann Chu et al. (eds.), *SARS-CoV-2: Methods and Protocols*,
Methods in Molecular Biology, vol. 2452, https://doi.org/10.1007/978-1-0716-2111-0_23,

1 Introduction

Containment laboratories are essential for carrying out research activities with security-sensitive or high-risk biological agents (BAs) [1]. Definition of security-sensitive or high-risk BAs and toxins and definitions of other relevant commonly used words and phrases in this chapter can be referred to Table 1. On 31 December 2019, the world learnt about the outbreak of respiratory disease caused by severe acute respiratory syndrome coronavirus 2 (SARS-CoV-2). The virus is one of the security-sensitive agents originating from animals and capable of human-to-human transmission [9]. Naturally, given the societal impact of the pandemic that followed, scientists around the world are eager to carry out research on SARS-CoV-2 but the activity is not exempt from risks. Since the beginning of the SARS-CoV-2 pandemic, diagnostic and research laboratories around the world kicked off intensive research into the novel virus and followed the WHO Interim Laboratory Biosafety Guidance [10] related to coronavirus disease 2019 (Covid-19); this guidance has been updated frequently. According to the guidance, nonpropagative diagnostic laboratory work (e.g., sequencing, nucleic acid amplification tests) is recommended to be conducted at a facility equivalent to BSL-2 laboratories. Virus propagative work (for example, virus culture, isolation or neutralization assays) is recommended to be conducted at a containment laboratory with inward directional airflow (equivalent to BSL-3 laboratory) [10]. Subsequently, several national or institutional interim guidelines were published for handling SARS-CoV-2 in diagnostic and research laboratories and for transporting virus-containing samples. For instance, the Biosafety Branch of the Singapore Ministry of Health issued interim biosafety guidelines for laboratories (both diagnostic and research) and personnel handling samples or materials potentially contaminated with SARS-CoV-2 [11]. Some institutions such as the US Centre for Disease Control and Prevention (CDC) and several other institutions and universities have developed biosafety guidance for researchers working with SARS-CoV-2 [12, 13]. The main purpose of establishing those guidelines is primarily to protect against personnel and environmental contamination.

It is well known that the prevalence of laboratory-acquired infections (LAI) among laboratory personnel is underreported [14]. The number of LAI among laboratory personnel is also difficult to establish during the peak of epidemics and pandemics since it is not easy to establish the source of infection. LAI are a public health concern, as infected personnel may become a source of infection for their colleagues, family members and others. After the SARS-CoV outbreaks in 2003, LAIs were reported in a few research laboratories from Beijing, Taiwan, and Singapore [15–

Table 1
Definitions for commonly used words and phrases in the chapter

Word/phrase	Definition	Reference
Biosafety	The containment principles, technologies and practices that are implemented to prevent unintentional exposure to pathogens and toxins, or their accidental release.	[1]
Laboratory biosecurity	The protection, control and accountability for valuable biological materials within laboratories, in order to prevent their unauthorized access, loss, theft, misuse, diversion or intentional release.	[2]
Biorisk	The probability that a particular adverse event (accidental infection or unauthorized access, loss, theft, misuse, diversion or intentional release), possibly leading to harm, will occur.	[2]
Bioethics	The study of ethical, social, and legal issues that arise in biomedicine and biomedical research. The study of the ethical and moral implications of biological discoveries, biomedical advances and their applications, as in the fields of genetic engineering and drug research.	[3, 4]
Dual-use research of concern (DURC)	Knowledge and technologies generated by legitimate life sciences research that may also be appropriated for illegitimate intentions and applications. Research that could possibly be misapplied to pose a threat to public health, animal health, or the environment is referred to as "DURC."	[4, 5]
Security-sensitive or High-risk biological agents and toxins (BAs/toxin)	BAs/toxins listed under the First and Second Schedule of Biological Agents and Toxins Act (BATA), Singapore Ministry of Health (MOH); Federal Select Agent Program control list under USA Human Health Services (HHS) and USA Department of Animals (USDA); and list of human and animal BAs and toxins for export control under the Australia Group. These BAs/toxins could be used as weapon and can pose a risk to national security.	[6–8]
Life sciences	All sciences that deal with organisms, including humans, animals, and plants including but not limited to biology, biotechnology, genomics, proteomics, bioinformatics, pharmaceutical, and biomedical research and techniques.	[4]

17]. Fortunately, those LAIs did not cause further outbreaks. In recent years, a couple of containment laboratory incidents in the USA [18, 19] have made the public concerned about unintentional release of BAs from high containment laboratories. Such public concerns can be valid if biorisks associated with dangerous viruses like SARS-CoV-2 are not mitigated properly. "Biorisk" definition can be referred to Table 1.

2 SARS-CoV-2 Research Laboratory Biorisk Concerns

2.1 Biosafety Risks

Potential biosafety related risks arising from SARS-CoV-2 experiments carried out in containment laboratories could be LAI among researchers as well as laboratory and environmental contamination. Further spread of the infection to the public and animals after LAI or direct contamination from the laboratory could also happen.

(a) LAI and other risks: The majority of LAI and laboratory contamination is due to human errors [14, 20]. Several factors are included in human error, such as a lack of experience in laboratory practices, a lack of understanding of laboratory practices, absent-mindedness and inadequacies in adhering to biosafety practices [14, 20]. LAI due to facility or equipment failure in containment laboratories is relatively rare [14].

Due to the Covid-19 pandemic, many countries have expanded the diagnostic laboratory capacities at a "speed of light" [21]. Large-scale expansion of experienced laboratory staff was not possible on such a short time. The main concern was that allowing less experienced laboratory personnel to handle SARS-CoV-2 could increase LAI since there is evidence that laboratory personnel with less technical training are prone to have more workplace accidents than properly trained personnel [20]. So far, no SARS-CoV-2 LAI case has been reported in the public domain. Organizations with Covid-19 research laboratories need to pay attention on establishing effective biosafety/biosecurity risk mitigation measures in the laboratory (Details on biosecurity management can be seen in the next chapter).

(b) Infectious animals: Many reports showed that this virus can be found in some domesticated animals (cats and dogs), farm animals (mink) and zoo animals (lions and tigers) [22]. There is no evidence that animals play a significant role in the spread of SARS-CoV-2 to humans, but limited evidence shows that the virus can transmit from human to farm animals like mink and from infected mink to human [23]. If the virus happens to transmit aggressively among domesticated animals or farm animals, the virus may persist in the environment. Since SARS-CoV-2 can infect various animal hosts, it may end up generating more potential pandemic strains.

(c) Unexpected risks (unforeseeable or unimaginable risks): Previous studies have shown that experimental results from life sciences research are not always predictable [24–26]. Some examples of possible unexpected outcomes from SARS-CoV-2 experiments are the emergence of highly virulent or easily transmissible SARS-CoV-2 virus, the emergence of resistant strains of the SARS-CoV-2 virus (to disinfectants, UV, or potential antivirals), or strains with altered host range or altered mode of transmission of the infection. If there are changes in the virulence of transmissibility of SARS-CoV-2, researchers need to study further to ensure that current knowledge about the virus is still relevant (e.g., incubation period, clinical features, susceptibility to currently applied disinfectants and medical treatment). Unexpected risks may be associated with potential negative consequences on human, society and the environment [27]. Thus, researchers who carry out SARS-CoV-2 research need to be responsible for managing the unexpected risks associated with their research. Organizations need to prepare a system to assist researchers who are working in such experimental work with unexpected or unknown risks.

2.2 Biosecurity Risks

Definition of "Biosecurity risks" in laboratory (Laboratory biosecurity risks) can be referred to Table 1, Potential SARS-CoV-2 related biosecurity risks could be:

(a) Unauthorized access to SARS-CoV-2 and to materials contaminated with the virus by theft or diversion and misusing: Unauthorized acquisition of the virus by criminals could happen in laboratory settings, or during the transportation of materials containing the virus, if the laboratory supervisors and organizations do not have a systematic management of biosecurity risks (details on biosecurity management can be seen in the next chapter). Since wild type SARS-CoV-2 has possible characteristics of a weaponizable BA, it can be attractive to terrorist or violent extremist organizations. Calls from some terrorist groups to spread the SARS-CoV-2 in the community have been reported recently [28]. Releasing this virus to the environment intentionally (contaminating public places with the virus) may or may not cause a serious public health threat since the virus is quite vulnerable to heat, humidity, and UV light [29, 30]. However, this virus could be used as a fear-inducing agent by terrorist organizations. In addition to the wild type of SARS-CoV-2 related risks, laboratory operators should pay attention to other potential risks, such as the theft or diversion of unexpected experimental materials as described above in Subheading 2.1, point c or deliberate releasing of SARS-CoV-2 infected experimental animals from laboratories.

(b) Unauthorized access to confidential information on SARS-CoV-2 research projects leading to theft, diversion, or misuse of that information: this can include SARS-CoV-2 related experimental information such as unexpected experimental results as described in Subheading 2.1, point a, information on the storage place of those viruses, list of researchers who have access to those viruses or who have the knowledge and technical know-how to design more dangerous virus, and transportation route of viruses between the laboratories could be stolen and misused by laboratory personnel (insider threat) or nonlaboratory personnel.

(c) Misusing new and emerging technologies or researcher's knowledge and know-how (intangible research assets): nowadays, researchers can use new and emerging technologies like synthetic biology, nanotechnology, and molecular biology for exploring vaccine or therapeutic development, diagnostic tool development, the transmissibility of virus, the ability of the virus to evade host immune system, and so on [31]. The same technologies can be used for designing hypervirulent or more pathogenic or easily transmissible SARS-CoV-2. A few rogue researchers may misuse intangible assets if their motivation is based on financial gain or political or ideological reasons. Thus, research institutions should keep in mind "dual-use research of concern" (Definition can be referred to Table 1) and insider threat [32] while reviewing SARS-CoV-2 experimental projects, especially where new technologies are involved.

3 Mitigating Biorisks in Containment Laboratories in International Level and Country Level

At the international level, biorisks posed by natural or accidental or deliberate events are governed by international regulations, resolutions and treaties. Their focus on laboratory biosafety and biosecurity related issues is explained in Table 2. Among four of the international frameworks in Table 2, the Biological Weapons Convention (BWC) covers laboratory biosafety/biosecurity; transportation of BAs and toxins between laboratories; awareness-raising of dual-use BAs, equipment and technologies; and awareness-raising of codes of conduct (COC) and ethical responsibility in life sciences. The International Health Regulations (IHR) cover the above first three elements. The United Nation Security Council resolution 1540 (UNSCR 1540) focuses on laboratory biosafety and biosecurity including transportation of BAs and toxins between laboratories. The International Air Transport Association (IATA) Dangerous Goods Regulations (DGR) is focusing only on transportation of BAs, toxins and materials containing them between laboratories.

Table 2
International regulations, resolution, and treaty guiding the evolvement of laboratory BRM systems of institutions or countries

No.	Legally binding international regulations, resolution, and treaty	Purpose	Sponsoring organization/ agency	Aim to strengthen the following issues: [a]Laboratory biosafety/ biosecurity risk management	BA/toxin (including dual use materials) transport associated risks	Awareness about dual use research	Awareness about ethical responsibility in life sciences research	Tool used for evaluating listed issues (column 5–8 of the table) in life sciences laboratory
1	International Health Regulations (IHA) 2005 [33]	**"To prevent, protect against, control and provide a public health response to the international spread of disease** in ways that are commensurate with and restricted to public health risks, and which avoid unnecessary interference with international traffic and trade." All States shall strengthen 8 core capacities to meet public health security. Under laboratory core, BRM is included.	World Health Organization (WHO)	Both aspects	Yes	Yes	No	Under "Prevention—Biosafety and biosecurity" (P6.1 and P6.2 session) of IHR Joint External Evaluation (JEE) form. (Refer to questionnaire stated in IHR (2005) Monitoring and Evaluation Framework. Joint External Evaluation Tool, IHR (2005)) [34]

(continued)

Table 2
(continued)

No.	Legally binding international regulations, resolution, and treaty	Purpose	Sponsoring organization/ agency	Aim to strengthen the following issues: [a]Laboratory biosafety/ biosecurity risk management	BA/toxin (including dual use materials) transport associated risks	Awareness about dual use research	Awareness about ethical responsibility in life sciences research	Tool used for evaluating listed issues (column 5–8 of the table) in life sciences laboratory
2	United Nation Security Council resolution 1540 (UNSCR 1540) [35]	All States shall **refrain from providing any form of support** to non-State actors that attempt to **develop, acquire, manufacture, possess, transport, transfer, or use nuclear, chemical, or biological weapons and their means of delivery,** in particular for terrorist purposes.	UNSC	Both aspects	Yes	No	No	1540 Assessment Matrix [36]
3	Biological Weapons Convention (BWC) [37, 38] It is designed	It is the first multilateral disarmament treaty focus on prohibition of BAs and toxin weapons,	United Nation (UN)	Yes	Yes	Yes	Yes	No tool is available for evaluating stated issues in the table. **Remark**: Confidence

	to supplement the 1925 Geneva Protocol.	as **States Parties to prohibit the development, production, stockpiling and use of bacteriological (biological) and toxin weapons**. It is to ensure that the **life sciences are used only for the benefit of humanity**. Focus to influence the **behavior of personnel rather than controlling technology or knowledge**.						building measure form needs to be submitted by countries annually. Submission is not legally binding, and a volunteer process. The primary aim is to build trust between states that no activities are taking place in breach of the convention. Thus, it is not a tool to measure the conducts of countries that have singed the BWC
4	International Air Transport Authority (IATA) IATA is a trade association of the world's airlines founded in 1945 [39]	IATA Dangerous Goods Regulations (DGR) was published in 1959 based on UN International Civil Aviation Organization (ICAO)'s Technical Instruction combined with variations and additional	IATA	Yes (Limited to packaging, transferring between laboratory personnel and a courier service provider)	Yes (Air transport only)	No	No	No evaluation tool was developed by IATA, but several institutions and countries have their own guidelines/ standards and checklist for shipping infectious materials via

(continued)

Table 2
(continued)

No.	Legally binding international regulations, resolution, and treaty	Purpose	Sponsoring organization/ agency	Aim to strengthen the following issues: [a]Laboratory biosafety/ biosecurity risk management	BA/toxin (including dual use materials) transport associated risks	Awareness about dual use research	Awareness about ethical responsibility in life sciences research	Tool used for evaluating listed issues (column 5–8 of the table) in life sciences laboratory
		requirements set by individual nations and airlines. ICAO recognizes IATA DGR as the **"field guide" for industry practical reference for the transport of DG including infectious substances via air**. It has been revised every year and is now 61st edition.						airlines based on IATA DGR

[a]Refer to the definitions in reference #1 and 2

Table 3
International guidelines and standards for designing, building, operating and managing high containment laboratories and setting up the BRM system

No.	Standards and guidelines	Published by	Focus
1	Laboratory Biosafety Manual, 3rd edition, 2004 [1]	WHO	Basic concept in laboratory biosafety to develop national code of practices for the safe handling of BAs and biosecurity concept. The 4th edition will be released soon
2	Bio-risk Management: Laboratory Biosecurity Guidance, 2006 [2]	WHO	Laboratory BRM, biosecurity and bioethics
3	Guidance on biosafety and biosecurity in veterinary laboratories and animal facilities, 2018 [40]	OIE	Biosafety and biosecurity of veterinary diagnostic laboratories and experimental animal laboratories
4	Responsible Life Sciences Research for Global Health Security: A Guidance Document, 2010 [4]	WHO	Carrying out high-quality life sciences research responsibly, safely and securely that could foster global health security and contribute to economic development, evidence-informed policy making, public trust and confidence in science
5	International Standard Organization (ISO) 35001: 2019 Biorisk management for laboratories and other related organizations [41] (Previously known as CWA 15793: Laboratory biorisk management standard)	ISO	Biorisk management system for laboratories (biosafety and biosecurity)
6	Guidance on regulations for the transport of infectious substances, 2019–2020 [42]	WHO	Transport of infectious substance by all modes of transport (road, rail, sea or air), both nationally and internationally. It is based on UN's recommendation on the Transport of Dangerous Goods: Model Regulations

International bodies such as World Health Organization (WHO), World Organization for Animal Health (OIE) and International Standardization Organization (ISO) have developed biosafety, biosecurity, biorisk management (BRM) guidelines and standards, for which the above stated international regulations, resolutions and treaty are used as guiding principles (Table 3). The same international regulations, conventions, treaties, standards and guidelines are also used in turn as references when countries develop biosafety- and biosecurity-related national legislation, regulations, standards and guidelines. Different countries have different governance and compliant structures for managing containment laboratories where SARS-CoV-2 research activities are carried out. For instance, containment laboratories and their related activities in Singapore should be certified annually by an external party, together with regulators, using the checklist based on the local legislation (Biological Agents and Toxins Act) [43]. In the USA, containment laboratories that registered with the Federal Select Agent Program are inspected by a regulator using the dedicated checklist [44]. Many EU countries also have specific legislation and regulations for building and operations of containment laboratories and protecting laboratory personnel under the umbrella of EU directive 2000/54/EC—Protection of workers from risks related to exposure to BAs at work [45] and EU Directive 2009/41/EC—Contained use of genetically modified microorganisms [46], and Council Regulation (EC) No. 428/2009—Setting up a Community regime for the control of exports, transfer, brokering, and transit of dual-use items and other directives [47]. Legislation and links can be accessed on the European Biosafety Association website (https://ebsaweb.eu/european-biosafety-biosecurity-legislation). Containment laboratories around the world will refer to different checklists for assessing their laboratory BRM strategies; for example, some institutions may use the WHO checklist [1], some may base theirs on Canada BSL-3 laboratory guidelines [48] or some may base on their own country checklist [49]. Regarding BAs/toxin transportation across the borders, some countries grouped together to harmonize the export control of BAs and dual use BA materials including equipment to ensure that exports do not contribute to the development of biological weapons. For instance, the Australia group composed of 42 countries and EU member states aim to harmonize export controls of chemical weapons precursors and biological weapons proliferation [8]. SARS-associated CoV (SARS-CoV) is included in the Australia group's export control list as well as [8] the USA Department of Health and Human Services and USDA select agents and toxins control list [7].

Some large initiatives, such as the Chemical Biological Radiological Nuclear Centres of Excellence of the European Union (EU CBRN CoE), which is active since 2010 and involves 62 countries in 8 regions in CBRN risk mitigation, include targeted actions on biorisk capacity and capability building. In the case of the EU CBRN CoE, the activities have encompassed establishing networks of biorisk management experts, providing training to personnel from participating countries in biosafety and biosecurity, and supporting the training of national laboratory experts in BRM [50]. Those activities reinforce a harmonization of the criteria for BRM and their implementation in research and diagnostic laboratories.

4 Current Biorisk Mitigation Strategies and Limitations

The WHO first introduced the overarching *Laboratory Biorisk Management* (BRM) *Strategy* in 2006. The currently available BRM approach is composed of biosafety, laboratory biosecurity and ethical responsibility to mitigate biorisks [2]. The BRM is the most used approach to minimize the likelihood of the occurrence of biorisks in laboratories [2]. According to the WHO, the responsibility of implementing a BRM strategies lies with the laboratories [2].

The current focus of the BRM strategy in the majority of laboratories is to manage the foreseeable and tangible risks (LAI, contaminating environment, loss/misuse/diversion/release of BAs/toxins, etc.) using a risk-based approach. With this approach, biosafety risk is determined by a combination of probability and consequences of the occurrence of a bioincident [2]. The probability arm of the biosafety risk is influenced by the availability of effective risk mitigation measures (i.e., effectiveness of BRM strategy). Biosecurity risk is much more difficult to be identified and mitigated since the probability arm of the risk analysis will be influenced not only by the vulnerability of a laboratory but also potential adversaries' intentions and capabilities [51]. For the majority of times, the "consequences" side of risk analysis is considered to be influenced by the characteristics of the hazard, people's experience in handling BAs/toxins and the host immune system. Thus, both biosafety and biosecurity standard risk analyses does not take into account human behavioral aspects even though organizations cannot exist without their staff.

5 Strengthening of the Current BRM Approach

5.1 Responsible Research Practices and Bioethics

As discussed earlier, current advances in life sciences technologies may cause dual use concerns involving unforeseen or unforeseeable risks (unexpected risks) with potential negative consequences on humans, society and the environment [27]. We can anticipate an emergence of unexpected risks in the near future, given the strong innovation drive of the life sciences industry and the likelihood that the implementation of adequate legislation, regulations, standards and guidelines may not be able to catch up with the speed of release and the scope of application of emerging technologies. Thus, the application of standard risk analysis methods alone is not sufficient for managing biorisks associated with security-sensitive BAs/toxins with potential dual uses. The WHO has recommended life sciences organizations to promote good research practices and ethical principles (maximizing public benefits, minimizing harm) through education and training and strengthening different security parameters of containment laboratory to fill up the gaps in existing BRM strategies [4]. Wagener and Bollaert have also proposed to add bioethics aspects into the standard BRM program for dealing with synthetic biology research with unknown risks [52]. Definition of "bioethics" can be referred to Table 1. It is therefore sensible to educate life sciences students and personnel on a range of bioethical issues associated with life sciences research that would facilitate in instilling safety and security culture of individuals and organizations.

5.2 Organizational Factors

Previous studies show that the majority of LAI or accidentally releases of BAs to the environment were caused by human errors such as slips and lapses in the handling of the BAs/toxins [14, 20, 53]. Slips and lapses could be made by even the most experienced, well trained and highly motivated people, and can occur if one's attention is diverted, even momentarily [54]. Thus, tightening rules and regulation or providing technical training to personnel who made such a mistake does not guarantee that this type of error will disappear. Reason [55] reported that high-reliability organizations (HROs) such as nuclear aircraft carriers, air traffic control towers and nuclear power plants, have developed a resilience management system to countermeasure human errors in the workplace. This system includes not only risk assessment but also an analysis of human behavior, instead of focusing on individual persons who made mistake. It is also believed that successful resilience against unexpected events in different industries (adaptive capacity in the face of adversity) are likely to be influenced by a combination of people armed with adaptive skills, plus a situation that is conducive to a successful response [56]. To foster resilience in the workplace, leaders in HROs emphasize the importance of working together in multidisciplinary teams and remove barriers to cross-functional collaboration [57]. They encourage continuous technical training in emergency responses against all possible failure scenarios including how to solve the problems and how to make decisions to manage the unexpected events [57]. In addition, the importance of reporting near miss and incidents, learning from mistakes (within or outside of organizations) and communicating the big picture to everyone in organizations is emphasized [57]. In short, HROs cultivate safety and security related values and individual behavior, and supportive organizational behavior.

Organizations are generally composed of "people," "technology," and "structure" and operate within an internal and external "environment" [58]. "People" employ "technology" in performing the tasks that they are responsible for, while the "structure" (i.e., formal relationship of people in organization) of the organization serves as a basis for coordinating all their different activities [58]. Numerous changes in the external environmental elements (e.g., government, the family, and other organizations) and several internal environmental structure and processes (e.g., how individual level, group level, and leadership interacts, communicates, and makes a decision) [58, 59] create demands on organizations and shape the effectiveness of an organization. All these elements are interdependent, and the organizational behavior and effectiveness of the organization are thus influenced greatly by the behavior of human [59].

The recently published international laboratory BRM standard, ISO 35001:2019 Biorisk management for laboratories and other related organizations [41], includes the organizational roles in supporting workplace safety and security behavior into the BRM approach. According to the ISO 35001, "*7.2.1 Behavioural factors and worker management: The organization shall address biorisks associated with human behaviour in the BRM plan, including how workers interact with the facility, its equipment, and coworkers.* ***The organization shall provide individual support and effective management of these behavioural factors.***"

6 Implementing an Effective BRM System by Incorporating Supportive Organizational Behavior into the Existing BRM Strategies

6.1 Roles of the Laboratory Management Team

In the context of containment laboratories, several stakeholders are involved in establishing an effective BRM system. To simplify the discussion, we only consider the organizational processes (communication, decision, and progress) of four major stakeholder groups who involved in the planning, implementing, monitoring of BRM related operational activities here: organization top management, laboratory management teams, biorisk management committee (BRMC), and supervisors (operations or scientific). The laboratory management team is usually composed of the operations team head, biosafety coordinator, and laboratory manager/director. The laboratory management teams should be familiar with international biosafety and biosecurity treaties/regulations, standards and guidelines, and national legislation and regulations. They should be well versed in organizational (bio)safety and (bio)security policies and guidance, and are responsible to implement and manage BRM system appropriately in research laboratories. In addition, their tasks include monitoring various operational activities, influencing/training the workforce (operations personnel and researchers), working collaboratively with stakeholders and acquiring input from them. Especially, they will need to discuss or share different issues such as laboratory safety/security policies, resource assistance, laboratory performance, and security with the top management and its representatives. Both the laboratory management team and supervisors have a responsibility to monitor and ensure that personnel behavior is according to expected behavior in laboratories. The responsibilities of BRMC can be found in ISO 35001

[41] and may vary from one institution to another and one country to another.

Generally speaking, the laboratory management team plays a critical role in establishing an effective BRM system in laboratories. Therefore, they also need to understand changing the risk landscape in containment laboratories where security-sensitive or high-risk BA research is carried out and how to weave supportive organizational behavior principle into their day-to-day operational activities to achieve an effective BRM system in the containment laboratory. So far, no guidance is available for the operationalization of this task.

6.2 Operationalization of Supportive Organizational Behavior Principles in the Advancement of an Effective BRM System

In this context, the laboratory management team should consider first which operational activities the workforce would recognize as indicating supportive organizational behavior. They should then address issues intertwined with these operational activities with the aim to establish an effective BRM system in the containment laboratory. The authors propose here four manifestations that could be considered as supportive organizational behavior. They are:

(a) Displaying a visible commitment and support to the BRM system from different levels of management.

(b) Developing a competent and responsible workforce with BRM technical skills and problem identification/solving skills (strengthening competency).

(c) Integrating learning and improvement principles into the BRM.

(d) Enhancing continuous motivation of laboratory personnel to avoid complacency.

These manifestations are based on the International Atomic Energy Agency (IAEA)'s principles and guidance for implementing safety and security culture in nuclear facilities [60].

Which operational activities to be implemented, how to implement them, why they need to be implemented and who could lead and facilitate to implement suitable operational activities are described in the text below and in Tables 4, 5, 6, and 7. This information is based on the authors' experience in BSL-3 operations and management for more than 10 years. Some operational activities described in the tables are based on the information described in the WHO biosafety manual [1], ISO 35001 [41], the self-scan tool kit and vulnerability scan toolkit developed by the Netherlands biosecurity office [61, 62], and WHO responsible conduct in life sciences [4]. In the first column of the tables, thematic questions are phrased according to the organizational processes (communication, decision and progress) involved in the implementation of operational activities [59]. In-depth discussions on laboratory operational activities that could be attributed for the

Table 4
Operational activities reflecting visible commitment and support to the BRM system from different levels of management

Thematic question to explore operational activities aiming to strengthen BRM system	BRM related operational activities. Some examples and notes are added for clarification purpose	Potential stakeholders who lead or facilitate to implement the policies or operational activities
1. How and what to communicate personnel clearly about (bio)safety/(bio)security values as the overriding priority in the laboratory?	• Communicate laboratory's values and beliefs in terms of (bio)safety and (bio)security to all stakeholders including researchers (e.g., "Risks are real and need to be taken seriously"). Consider including different topics—hazards and risks involved in experimental procedures, new risks, incidents, critical mitigation measures, regulations. **Note**: This message could be embedded in different modes of communications such as institutional goals, priority setting, laboratory policy statement posters, training sessions, continual education sessions, meetings, workshop, or appointment letters or motivating mass emailing to laboratory personnel. • Prepare and ensure that BRM policy is available for all relevant personnel. **Note**: The BRM policy could be drafted according to the guidance given in the ISO 35001.	Laboratory management team and supervisors
2. How should the organization demonstrate its commitments to health, safety and security including BRM issues?	• Appoint a dedicated top management personnel to oversee the biosafety and biosecurity related issues including high-containment laboratory.	Top management
	• Establish a policy for releasing a research funding from the organization (e.g., Do not support research activities without having committed (bio)safety and (bio)security mitigation measures for security-sensitive BA/toxin projects.)	Top management
	• Establish a policy to include researcher's BRM related performance as one of the decisions taking	Top management

(continued)

Table 4
(continued)

Thematic question to explore operational activities aiming to strengthen BRM system	BRM related operational activities. Some examples and notes are added for clarification purpose	Potential stakeholders who lead or facilitate to implement the policies or operational activities
	parameters for performance appraisal and promotion.	
	• Include potential critical biorisks related to containment laboratory in an organizational key risk profile (e.g., risk of contracting LAI and risk of spreading infectious disease to the community, loss of security-sensitive pathogens).	Top management
	• Initiate and promote discussions about critical laboratory biorisk issues in different types of institutional level meeting.	Top management
	• Establish research and publication policy for research work associated with "high-risk BAs/toxins" and "emerging technologies such as synthetic biology."	Top management
	• Appoint appropriate numbers of laboratory management team (laboratory director, biosafety officer, laboratory manager, operations head) to operate the laboratory.	Laboratory management BRMC and supervisors
	• Establish and implement health and safety program for personnel including externals (e.g., establish an occupational health program (OH) for personnel and externals if necessary. Prepare SOPs for previsiting, visiting, and post-visiting containment laboratory for externals). **Note**: SOPs are prepared for different groups of externals—collaborator researchers, waste contractor, delivering personnel for goods/other services, BAs courier service providers, facility and equipment maintenance service providers, and so on.	Laboratory management BRMC and supervisors

	• Support critical organizational processes needed for implementing effective BRM system. **Note**: Refer to points 4 and 5.	Laboratory management BRMC and supervisors
	• Ensure that laboratory leaders are competent in technical/administration and adherent to approved policies and SOPs. **Note**: Training on laboratory leaders can be referred to points 3–5 in Table 5.	Laboratory management BRMC and supervisors
3. How should the organization to ensure that the BRMC members take their roles and responsibility seriously?	• Inform BRMC members officially about their roles and responsibilities (e.g., Top management could issue an official appointment letter including roles and responsibility).	Top management and laboratory management
	• Acknowledge the contributions made by BRMC members toward a containment laboratory (e.g., The laboratory management informs about BRMC members' contribution to their supervisors or head of the department of BRMC members).	Laboratory management
	• Acknowledge the contributions made by BRMC members toward a containment laboratory (e.g., encouragement could be done by sending thank you note or letter).	Top management
	• Ensure that the BRMC members receive relevant training (e.g., BRM related international frameworks, national legislations and regulations for carrying out their tasks). **Note**: Explaining the BRMC members the bigger picture could assist them to see their involvement in international framework and national legislations/regulations.	Laboratory management
4. Which organizational support structure is necessary to maintain the effective BRM system?	• Hire essential personnel (at least Laboratory director, biosafety/biosecurity officer or biorisk management officer (BRMO), facility operations, facility engineer) with necessary experience for managing BRM related tasks in a laboratory. **Note**: Refer to point 1 in Table 5 for hiring	Top management, laboratory management

(continued)

Table 4
(continued)

Thematic question to explore operational activities aiming to strengthen BRM system	BRM related operational activities. Some examples and notes are added for clarification purpose	Potential stakeholders who lead or facilitate to implement the policies or operational activities
	processes of appropriate personnel. It is preferable to oversee both biosafety and biosecurity by a single person since it is important to understand where these two disciplines intersect and where diverge and how to apply an appropriate overlapping BRM program. Full time or part time appointment will depend on the size and workload of the laboratory.	
	• Remain available to hire additional experts or headcounts for a laboratory; for example, biorisk management advisor or experts in dual use research or experts in synthetic biology or genetic modification and so on may be needed from time to time to review the submitted project, while fulltime housekeeping personnel may need to be hired depending on the size and workload of the laboratory.	Top management, laboratory management
	• Remain available to reserve budget for containment laboratory emergency operations. **Note**: Not to miss out in estimating budget for dealing with emergency, for example engaging experts if biosafety/biosecurity committee does not have subject matter experts or decontamination due to a spill.	Top management, laboratory management
	• Recruit BRMO and researchers who are knowledgeable, skillful and experience in carrying out hands-on BSL-3 experience. **Note**: BRMO have the responsibility to manage both biosafety and biosecurity issues. Details can be referred to reference point 1 in Table 5.	Top management, laboratory management

	• Establish the BRMC for oversighting biosafety and biosecurity issues associated with containment laboratory.	Top management, laboratory management
	• Create a healthy workplace environment for workforce to perform well. **Note**: Refer to point 1 in Table 6 and point 3 in Table 7.	Top management, laboratory management
	• Motivate personnel using different methods. **Note**: Refer to Table 7. **Note**: All these activities cannot be successfully executed without proper funding and commitment from the top management.	Top management, laboratory management
5. How should the organization to set up to achieve effective BRM system in the laboratory?	• Lead/coordinate in designing and implementing effective BRM system (e.g., Laboratory director's signature is recorded in the BRM document). **Note**: Components of BRM system can be refereed to references in Table 3.	Laboratory management and BRMC
	• Lead/coordinate in carrying out risk analysis and management for all critical SOPs.	Laboratory management and BRMC
	• Align policies, SOPs, and work practices of containment laboratory with relevant international regulations/treaties, national legislations and regulations, and institutional health, (bio)safety, and (bio)security standards.	Laboratory management and BRMC
	• Lead/coordinate in designing and establishing biosafety/biosecurity training program for different professional groups working in the laboratories. **Note**: Training program should be composed of training, assessing and monitoring of competency (Refer to points 3, 4 and 5 in Table 5).	Laboratory management and BRMC
	• Establish mechanisms for monitoring and assessing BRM system including dual use research activities (e.g., self-assessing, inspections, auditing). **Note**: Refer to point 1 Table 6.	Laboratory management and BRMC
		Laboratory management, BRMC and supervisors

(continued)

Table 4
(continued)

Thematic question to explore operational activities aiming to strengthen BRM system	BRM related operational activities. Some examples and notes are added for clarification purpose	Potential stakeholders who lead or facilitate to implement the policies or operational activities
	• Lead or participate in monitoring process of laboratory BRM system (e.g., Laboratory inspection/auditing, certification/accreditation). • Lead in organizing awareness raising sessions for management, researchers, and BRMC members (New relevant international/national biosafety/biosecurity codes of conducts, regulations, issues, etc.).	Laboratory management, BRMC and supervisors

Table 5
Operational activities reflecting development of a competent and responsible workforce with BRM technical skills and problem identification/solving skills

Thematic question to explore operational activities aiming to strengthen BRM system	BRM related policies operational activities. Some examples and notes are added for clarification purpose	Potential stakeholders who lead or facilitate to implement the policies or operational activities
1. How to ensure that appropriate personnel (Laboratory manger/Operations head, BSO, researchers, BRMC members) are hired for a containment laboratory?	• Define the relevant (bio)safety and (bio)security core competency for laboratory personnel.	Laboratory management
	• Define roles, responsibility and accountability accordingly for respective positions (e.g., Prepare Job description document for each position). • Establish a policy and process for hiring personnel for a containment laboratory. For example, Interviewing in-person is critical. Verifying personal information (reference letters, educational background, previous jobs and job scopes), experiences (laboratory biosafety/biosecurity knowledge, problem solving skills), positive attitude (willingness to learn new technologies/methods, team spirit, etc.) are also important. **Note**: Communicate clearly to potential employees during an interview regarding back-ground check process, PPE routine and pros and cons, training process, and expectation behaviors at workplace. After the interview, candidates may change their mind to work in containment laboratories. Coordination between hiring manager(s) and HR personnel is essential.	Laboratory management and HR
2. How to reinforce personnel's responsibility and accountability in biosafety/biosecurity?	• Explain BRM associated roles and responsibility clearly to every new hired personnel; for example, it could be done verbally at an initial briefing session or training session. A written document should be handed to personnel properly as well. This activity should be documented properly. **Note**: Could also refer to point 5 below in this table (how to ensure laboratory personnel, laboratory management, BRMC members, and researchers take their roles and responsibility seriously).	Laboratory management
	• Emphasize the accountability (e.g., setting the biosafety/biosecurity related goals for each personnel and informing	Laboratory management

(continued)

Table 5 (continued)

Thematic question to explore operational activities aiming to strengthen BRM system	BRM related policies operational activities. Some examples and notes are added for clarification purpose	Potential stakeholders who lead or facilitate to implement the policies or operational activities
	them about their accountability for specific tasks or processes).	
	• Explain how personnel's tasks are relating to BRM system of the laboratory, safety/security policies of the organization, national legislations/regulations and international treaty/regulations (e.g., include this information during training). **Note**: Linking their activities with international framework and national legislations/regulations will encourage personnel to relate to their work in more meaningful way. It may motivate personnel to carry out their tasks more responsibly.	Laboratory management
3. How to ensure that the laboratory personnel receive effective BRM training properly?	• Design and ensure to provide comprehensive relevant biosafety/biosecurity training (training + assessment + continual training) to different workforce (laboratory management, researchers, housekeeping personnel, facility engineers and other contractors). **Note**: Different types of training (theory, practical, tabletop exercise, etc.) should be included based on the target populations and topics. Dedicated trainers must be appointed (Trainers should not be a convenient person available in the laboratory). Evidence of training document should be available.	Laboratory management
	• Evaluate biosafety/biosecurity competency of personnel using different assessment methods (e.g., MCQs or hands-on exercise/drills or interviewing methods) for different types of workforce). **Note**: Assessment records should be available. Personnel are not allowed to work in the laboratory without competency testing.	Laboratory management
	• Monitor and reinforce knowledge, skills and behaviors of laboratory personnel through constructive feedback (e.g., Use different assessment methods—MCQ or drills or group discussions or observing performance or inspection/auditing). **Note**: Personnel with good biosafety/biosecurity	Laboratory management

	knowledge, skills/behaviors should be acknowledged, and appropriate behavior should be reinforced through constructive feedback. Such feedback should be provided in one-to-one basis. Such practices will encourage personnel to relate to their work more deeply and wish to learn more.	
	• Prepare laboratory personnel to face unforeseen risks by introducing awareness raising in responsible life sciences research related topics (see point 5, this table), training personnel in cautionary strategies and attention to signals and warning (e.g., what could be signs of insider threat and how to respond). **Note**: Awareness raising about possible biosafety and biosecurity related threats (i.e., events have never happened in real life) and consequences associated with security-sensitive or high-risk BAs in the laboratory. Train personnel how to spot tell-tale sign of unexpected event and how to prevent not to escalate an such event.	Laboratory management
	• Compare own laboratory training program with training programs of other containment laboratories to figure out gaps. Then improve the training program if it is necessary.	Laboratory management and BRMC
4. How to facilitate laboratory management and senior operational personnel to be able carry out their task effectively in containment laboratories?	• Equip laboratory management and senior operational personnel with technical knowledge (e.g., providing training/refresher biosafety/biosecurity topics training, allowing them to participate in relevant conference/ workshop and meeting, etc.).	Top management and laboratory management
	• Cultivate values and beliefs of life sciences professional by educating them with relevant responsible research ethical conducts related to biosafety/biosecurity, synthetic biology, emerging technologies, dual-use research, and so on. **Note**: Refer to point 5 below.	Top management and laboratory management
	• Equip laboratory management and senior operational personnel with effective leadership skills; for example training could include "laboratory quality management, managing BRM system, effective problem solving, strategizing, service orientation, financing, personnel management, conflict resolution, personnel management. **Note**: Hiring external experts will need a support from the top management.	Top management and laboratory management

(continued)

Table 5
(continued)

Thematic question to explore operational activities aiming to strengthen BRM system	BRM related policies operational activities. Some examples and notes are added for clarification purpose	Potential stakeholders who lead or facilitate to implement the policies or operational activities
5. How to facilitate laboratory management, BRMC members and researchers in containment laboratories to become responsible persons in life sciences research?	• Awareness raising on integrity and professional responsibilities among researchers. **Note**: Refer to research integrity and researchers' responsibilities proposed by the World Conferences on Research Integrity Foundation (WCRIF)'s guidance. • Educate researchers in responsible life sciences research for weaponizable biological agents, dual-use research, synthetic biology and emerging technologies. **Note**: Laboratory management team should facilitate workforce to be trained in responsible life sciences research education. It may need assistance from the top management if an organization does not have a formal responsible life sciences research education.	Top management and laboratory management

Table 6
Operational activities reflecting integration of learning and improvement principles in BRM

Thematic question to explore operational activities aiming to strengthen BRM system	BRM related operational activities. Some examples and notes are added for clarification purpose	Potential stakeholders who lead or facilitate to implement the operational activities
1. How should the organization ensure to improve the BRM system continuously?	• Design and implement performance assessment and monitoring of laboratory BRM system properly (e.g., Using operational team's weekly facility inspection checklist, Biosafety officer's monthly facility safety and security checklist, checklists for different lab activities). **Note**: Performance assessment and monitoring process of BRM system will identify gaps and implement corrective action. Could refer to ISO 35001. • Periodically review BRM system including contingency plans and emergency drills, unforeseen/emerging biorisks of containment laboratory (e.g., Internal auditor, top management representative, and external auditor review annually BRM system and facility). **Note**: Biosecurity risks could change due to external environment and national risks. • Facilitate participatory approach in problem solving (e.g., Ensure to get input from diverse group of people for vetting laboratory SOPs/work practices including risk assessment.) **Note**: Assigning different reviewers, not assign only one person (BRMC members, biorisk experts, technical, operations, researcher) for vetting submitted SOPs/work practices. If discrepancy arises, the issues should be discussed carefully with BRMC members.	BRMC members, laboratory management and supervisors
	• Inform laboratory workforce clearly about expected (bio)safety/(bio)security workplace behavior (conformance vs. nonconformance, compliance vs. noncompliance). **Note**: This practice will assist personnel to understand what is right and what is wrong.	Laboratory management and supervisors

(continued)

Table 6
(continued)

Thematic question to explore operational activities aiming to strengthen BRM system	BRM related operational activities. Some examples and notes are added for clarification purpose	Potential stakeholders who lead or facilitate to implement the operational activities
	• Monitor workforce performance related to laboratory (bio)safety and (bio)security from time to time. Reinforcing positive behavior among them. **Note**: Without clearly informing personnel about expected behavior from the start, monitoring of performance cannot be done effectively. Consider providing appropriately awards and sanctions relating to (bio)safety/(bio)security work practices. Rewarding and sanction are executed based on institutional guidelines.	Top management, laboratory management, and supervisors
	• Create "just culture" environment to facilitate in establishing a user-friendly laboratory incident reporting system, reasonable investigation processes, sharing lessons learnt to reduce mistakes. **Note**: "Just culture" approach could avoid creating a blaming culture that could destroy the trust and transparency among laboratory workforces. • Create a healthy workplace environment such as encouraging a questioning attitude of laboratory personnel (e.g., Arrange a small group meeting so that personnel can raise biosafety/biosecurity related questions easily). **Note**: Laboratory director/ laboratory upper management should consistently appreciate personnel for asking questions and routinely discuss actual situations. This behavior of leadership will create a healthy workplace environment. • Encourage and support laboratory personnel in learning (Refer to point 3 in this table).	Top management, laboratory management, and supervisors
2. How to explore the vulnerability and weakness in containment laboratories?	• Establish near-miss reporting system in the laboratory and communicate clearly about the systems to	Laboratory management and supervisors

	workforce. Incorporate "just culture" environment in the system. **Note**: Explain definitions of near miss, what could be near miss scenarios, why near miss reporting is important, ensure them that reporting is not associated with reprimand or punishment. Refer to paragraph in point 1 of this table for incident reporting system.	
	• Establish a confidential reporting channel or anonymous reporting channel for communicating workforce (bio)safety and (bio)security concerns including unusual behavior of coworkers (e.g., If a laboratory technician notices that one of his colleagues does not follow SOP for doffing PPE upon exiting a BSL-3 laboratory or a BSL-3 laboratory technician has noticed one colleague from his group opened frequently inventory boxes containing freezer without a valid reason.). • Create transparent and trusty workplace environment that allows personnel to share about their experience in dealing with unexpected events in laboratories (see point 1, this table). **Note**: Such environment will encourage personnel to share their experience in workplaces without the fear of retribution and they will relate to their work more deeply.	Top management, laboratory management and supervisors
	• Encourage personnel to be vigilant in, responsible for and committed to laboratory (bio)safety and (bio)security. (e.g., Train personnel on attentive to details, guidance on spotting tell tell-tale signs and responding accordingly). **Note**: For further information, refer to point 3 of Table 5.	Laboratory management and supervisors
	• Having favorable traits of good leadership—Laboratory management and supervisors are technically competent; alert all the time for (bio)safety/(bio)security issues; adhere to all legislation, regulation, organizational polices and guidance, and laboratory SOPs (approachable, open,	Laboratory management and supervisors

(continued)

Table 6
(continued)

Thematic question to explore operational activities aiming to strengthen BRM system	BRM related operational activities. Some examples and notes are added for clarification purpose	Potential stakeholders who lead or facilitate to implement the operational activities
	creative, cooperative, visionary, etc.). **Note**: Leaders should "walk the walk." Gaining the trust of laboratory personnel is critical for laboratory management. For further information in building trust, refer to points 1 and 2 in this table.	
3. How to encourage laboratory personnel in learning?	• Create team cohesiveness environment in the laboratory by using different learning techniques (e.g., SOPs reviewing could be done by a diverse group of team members; emergency training could be done using peer–peer learning method or role play exercises). **Note**: Diverse workforce such as BRMC members, BR officer, operational personnel, and security personnel could assist more effective SOP development with extensive risk assessment and management. They become familiar to each other's working style, and they could share their experiences. Peer-peer learning, and role-paly exercise learning will facilitate in building trust among workforce as well.	Laboratory management and supervisors
	• Triger the curiosity of laboratory workers (e.g., Ask them to come up with the solutions for real-life cases/ problems of other laboratories or own laboratory). **Note**: This approach is powerful to motivate personnel, but solution comes out from the team needs to be guided by the laboratory management and supervisors to align with ultimate goal of the laboratory (i.e., to establish effective BRM system). In addition, if their proposed solutions are feasible and effective, these should be incorporated into the operational activities. Provide appreciation notice for personnel contributed to the issue.	Laboratory management and supervisors
	• Use different types of motivating drivers (*see* Table 7).	Laboratory management and supervisors

Table 7
Operational activities reflecting enhancement of continuous motivation of laboratory personnel to avoid complacency

Thematic question to explore operational activities aiming to strengthen BRM system	BRM related operational activities. Some examples and notes are added for clarification purpose	Potential stakeholders who lead or facilitate to implement the policies or operational activities
1. How to convince personnel that they have been involved in meaningful job with training options?	• Set the (bio)safety/(bio)security related goals for laboratory personnel and monitor and evaluate their contributions.	Laboratory management and supervisors
	• Help personnel to connect a big picture to their work (e.g., Explain relationship between their actions and the welfare of the whole organization, environment, public and the world by introducing them about the national legislation/regulations and international legislation/treaties). **Note**: For further information, refer to Tables 2 and 3.	Laboratory management
	• Assist laboratory personnel of their needs by laboratory management or operational staff or supervisors. (e.g., work together with researchers to assist them to speed up their research activities, regularly or ad hoc laboratory walkthrough is done and follow-up feedback appropriately).	Laboratory management and supervisors
	• Encourage personnel to become mastery in BRM related subjects (e.g., Train and coach laboratory biosafety officer to develop new skills and become competent in new assignment). **Note**: For further information about training related issues can be found in points 3–5 of Table 5.	Laboratory management and supervisors
	• Provide relevant training courses, meeting, conference, and so on to BRMC members and management personnel (administration, finance, BRM, biosecurity related to dual use research/new technology, etc.).	Laboratory management and supervisors

(continued)

Table 7
(continued)

Thematic question to explore operational activities aiming to strengthen BRM system	BRM related operational activities. Some examples and notes are added for clarification purpose	Potential stakeholders who lead or facilitate to implement the policies or operational activities
	Note: so that they could improve their skill and knowledge in their interested fields. See detailed discussion in training in points 3–5 of Table 5. • Seek assistance first from laboratory personnel instead getting from BRMC when biosafety/biosecurity related operational problems need to be solved. • Update laboratory personnel about laboratory performance (e.g., incidents, laboratory certification process plan and outcome of laboratory regulatory certification can be updated during regular update meeting). **Note**: This activity will be useful not only for monitoring the BRM system, but also to motivate personnel.	Laboratory management and supervisors
2. How should the organization recognize and value personnel's roles in enhancing BRM system?	• Establish career paths/opportunities for laboratory safety/security personnel in order to maintain the safety and security workforce and their competency. **Note**: Such plan could be initiated by the laboratory management and discussed with HR and subsequently receives the management's endorsement.	Laboratory management and HR (may or may not directly reporting to top management)
	• Provide rewards to or recognize personnel who are vigilant in, commit to, and responsible for (bio)safety and (bio)security improvements (e.g., Rewards could be monetary or nonmonetary such as changing job title, assigning new responsibility, sending a recognition letter	Laboratory management and HR

	[official appreciation letter or simple "thank-you note" or just tell him/her "thank you" verbally]).	
3. How should the organization to create a motivating work environment?	• Allow personnel to work in friendly, creative, and flexible work environment (e.g., they could present their work in conferences, seminars, and meeting, and could work with flexible work arrangement; colleagues respect each other and work together to reach common goals). • Create open, transparent, and trusty workplace with a good laboratory leader that will motivate personnel to communicate openly and inform incidents, potential risks, hazard/risk concerns, and so on. **Note**: Refer to points 1–3 of Table 6.	Top management, laboratory management, supervisors

manifestation of supportive organizational behaviors can be seen below in the text. In Tables 4, 5, 6, and 7, some examples and explanations are included for each operational activity to assist readers to understand why and how operational activities should be implemented.

(a) **Displaying a visible commitment and support to the BRM system from different levels of management**

Since safety and security is everyone responsibility, all stakeholders listed in this chapter, including laboratory personnel, have to share the same values and beliefs that (bio)-safety and (bio)security risks are real and important to be focused on. Without strong beliefs among personnel in the organization, an effective BRM system cannot be established. Previous organizational behavioral studies show that leadership exerts a direct effect on the safety climate of the organization (supportive organizational behavior) in high-risk workplaces such as oil rigs [63] and chemical laboratories [64]. Similarly, Randell noted that leaders of biomedical laboratories hold primary responsibility and accountability for the performance and success of the laboratory [65]. Franz has pointed out that instilling "trust" and providing a healthy and vibrant organizational culture by the leaders is critical for not only protecting personnel but also enhancing the productivity of the laboratory [66]. For more on "qualities of laboratory leaderships" and "effective leadership styles," refer to the book *Leadership basics for clinical laboratory professionals*, edited by Yenice and Randell [65, 67]. One of the main responsibilities of the top management and laboratory management is to cultivate an "open" and "trustful" environment for facilitating in implementation and continuous improvement of an effective BRM system. In the nuclear industry, health industry and HROs, creating a "trustful" and "open" workplace environment is essential for motivating personnel to report incidents [60, 68–70]. The detailed discussion on creating "trustful" and "open" workplace can be found in the text below, (c) Integrating *learning and improvement principles into the BRM.* The ISO 35001 recommends the top management to commit to the following elements: (1) providing adequate recourse; (2) prioritizing and communicating biosafety and biosecurity policy; (3) establishing performance expectations and integrating BRM throughout the organization; (4) determining causes of incidents and nonconformities and preventing recurrence; (5) identifying opportunities for improvement and prevention [41]. Table 4 shows some tips on how to operationalize those recommendations and how to reach out to top management to get assistance for establishing an effective BRM system.

(b) **Developing a competent and responsible workforce with BRM technical skills and problem identification/solving skills (strengthening competency)**

Laboratory management should systematically recruit the best people for the tasks and delegate responsibility and accountability to the appointed personnel. Points 1 and 2 in Table 5 show which activities can be implemented in the hiring process and how to reinforce personnel's responsibility and accountability in biosafety/biosecurity. With the support from the Human Resources (HR) office, the laboratory management should translate organizational hiring policies and guidance into more specific containment laboratory hiring procedures.

The vigilance and observational skills of the personnel are critical for security, including nuclear security [68]. Likewise, the laboratory management should encourage personnel working in containment laboratories with security-sensitive or high-risk BAs to be vigilant in, responsible for and committed to laboratory BRM practices. It is believed that personal behavior is influenced by his/her deep-seeded values both directly and indirectly through attitudes [71]. Human behavior does not happen in a vacuum. In the context of an organization, leaders of the laboratory should ensure that all personnel working in the laboratory have the same goals and trust each other. Evidence from the health care industry and defense industry shows that middle management personnel like laboratory management have the most influence on building the responsible and resilient workforce [72].

Strengthening technical knowledge and skills is a part of the quality management system in the laboratory [73]. Personnel working in containment laboratories need to be trained to acquire relevant knowledge and skills according to their responsibilities and tasks so that they can be able to identify human-, structure-, and facility/equipment-related vulnerabilities and weakness and mitigate the risks accordingly. Some outlines of a BRM training program for laboratory management, biorisk management committee, researchers can be found in points 3–5 in Table 5.

As discussed above, unforeseen/unexpected biorisks may arise in life sciences laboratories from using emerging technologies, despite risk based BRM approach and legislations. WHO has guidelines to educate researchers in good research practices and research integrity, and ethical considerations (collectively known as responsible conduct of research = RCR) to deal with potential risks posed by accidents or the deliberate misuse of life sciences research (Refer to point 4 WHO reference book in Table 3). Novossiolva et al. have suggested life sciences professional norms and code of conducts (COC) related to biosecurity to be included in RCR curricula for preventing and minimizing those type of biosecurity risks [74]. Professional norms of life sciences researchers could be based on research integrity and professional responsibilities proposed by the World Conferences on Research Integrity Foundation's guidance [75]. Currently available COC related to biorisk issues are designed to raise awareness of the dual-use potential of modern biotechnology among researchers and to facilitate the development of the culture of responsibility in the life sciences [74, 76] and align with objectives of the Biological and Toxin Weapons Convention (BTWC) (Refer to point 3 of Table 2). The BTWC eventually persuades the States to implement national management standards on biosafety and biosecurity and promote the development of training and education programs for personnel who have access to BAs and toxins and for those with the knowledge or capacity to modify such agents and toxins [37, 38]. Unfortunately, only a few institutions have the capability to deliver RCR courses related to ethically challenging dual use research, bioterrorism and bioweapons [77]. If such courses are not available, the laboratory management should play a key role in coordinating appropriate training in biosecurity education for personnel of containment laboratories. Proposed topics to be included in RCR training can be found in point 5 in Table 5. The RCR courses related to biorisks will allow scientists to be able to discuss, analyze, and resolve the potential dilemmas they may face in their research as well as to avoid possible misuse of the research. Such a practice is the best prevention approach against accidental or deliberate misuse of BAs and toxins and against potential unforeseen biorisks.

(c) **Integrating learning and improvement principles into the BRM**

The IAEA has published a guidance on the implementation of a nuclear security culture in an organization [68]. According to the guidance, managers of nuclear facilities are encouraged to instill a "learning and improvement" principle in the organization to guide decisions and behavior. According to ISO 35001, the laboratory management team should develop a BRM approach with the concept of continuous improvement through monitoring and assessing the performance of the BRM system [41]. This framework will assist in exploring the vulnerabilities and weaknesses in the BRM system and subsequently, fill up the gaps (risk management). Table 6 points 1 and 2 indicate how to incorporate learning and improvement principles in the BRM system. The different stakeholders should be involved in this process. For instance, the operations team and the laboratory management could carry out monitoring of day-to-day operational activities and assessing/evaluating laboratory BRM system performance at a fixed interval as described in the laboratory quality management program [73] and ISO 35001 [41]. In ISO 35001, the laboratory facility and the BRM system should be reviewed annually by representatives of the top management of the organization [41] suggesting that the accountability of this task lies with the top management. Feedback from those processes is useful in guiding relevant corrective action to eliminate nonconformities/noncompliance and their causes. In addition, laboratory BRM should include a contingency plan as an integral part of the BRM system for dealing with unforeseen events associated with dual-use research [78]. The contingency plan should include emergency response, crisis management and business continuity plan. Some incidents such as property damage, personal injury, different types of threats, or theft of BAs may require the involvement of external experts such as scientists, regulators, law enforcement personnel, health care personnel, and fire brigade personnel. Thus, the laboratory management should consider discussing with relevant sectors when developing the emergency and crisis response plan. In addition, the laboratory management should form a relevant multidisciplinary team to make decisions for identifying and mitigating some biosecurity risks. If it is necessary, it should approach the top management of the organization or national authorities to get their input in case the consequences of potential and unexpected risks are beyond the capacity of the laboratory management.

Additional critical operational activity of the laboratory management and supervisors is to inspire the workforce to report any incident, near misses or problems or potential risks observed in the laboratory. Without having a culture of reporting among the workforce, some hidden biorisks or vulnerable signs in the laboratory cannot be identified. According to Reason, management should create a "just culture" as an atmosphere of trust in which people are encouraged (even rewarded) for reporting their concerned safety-related information [70]. Reason also has pointed out that "trust" is the first step to be established in cultivating a reporting culture in HROs [55]. In the containment laboratory context, it is important to educate clearly the laboratory workforce on where the line is drawn between acceptable and unacceptable behavior in the laboratory [70] as well as the purpose and process of laboratory incident/near-miss investigations. When an investigation is done, the focus of the first question should be "What went wrong?" not "Who made a mistake?" as suggested by a recent publication [79] since the purpose of carrying out the investigation is to understand the root cause of the incident or near miss and address it appropriately. If the root cause is a violation of legislation or institutional standards/guidance/policies and critical laboratory SOPs or reckless behavior, sanctions should be imposed on those particular personnel appropriately [70]. If the root cause is unintentional human errors associated with system problems, both human and relevant systems should be addressed in a manner that is just and fair [65, 69]. This type of workplace venue will also promote lessons learnt from mistakes in the laboratory. Some operational considerations for the establishment of incident/near-miss reporting process and investigation are described in point 2 in Table 6. In addition to creating an open, trusted and transparent environment, laboratory management should promote a questioning attitude since it will support creation and innovation and foster the emergence of improvements to the healthy workplace, processes and procedures, and so on. [80].

ISO 35001 highlights the importance of providing refresher training and evaluating the performance of the BRM system [41]. In point 3 in Table 5 and point 3 in Table 7, some tips are given on how to use appropriate training methods and how to create a healthy workplace ("open" and "trustful" environment) to facilitate a continuous learning culture.

(d) **Enhance continuous motivation of laboratory personnel to avoid complacency**

Motivation is an inner psychological force that activates and compels the person to behave in a particular manner [58]. According to Daniel Pink, most people require three major drivers for it—autonomy, mastery, and purpose [81]. Motivation is a key to the effectiveness of the organization [58]. The laboratory management and supervisors are the main drivers to influence the motivation of personnel in laboratories [82]. Thus, laboratory management and supervisors should understand some factors that motivates laboratory personnel. Some people need predominantly some kind of "purpose" stimuli for them to be motivated [58, 59]. Several approaches can be used to motivate workers to perform more effectively [58, 59, 82]. A detailed discussion regarding how laboratory leaders could motivate team members can be found in the book, "Leadership basics for clinical laboratory professionals" written by Yenice and Randell [83]. Organizations need to convince personnel that they are involved in a meaningful job with the possibility to be trained to become experts/specialists in some aspects of the containment laboratory. Assigned goals and timelines will give a sense of personal satisfaction and accomplishment to the personnel if they are able to meet them [58, 82]. Informing personnel about the progress of the BRM system applied in the laboratory is also a possible motivating factor, since some personnel prefer to have an extrinsic driver to increase their productivity. Favorable characteristics of laboratory leaders and environmental factors are also important extrinsic factors for enhancing continuous motivation of personnel [82]. Some examples of extrinsic driving factors are incentives or rewards, creating a "just culture" environment (open and trusted communication environment) with good and effective leaders, creating team cohesion and informing personnel about critical decision-making processes and explaining to them decisions that can impact their tasks. However, one should be careful in using extrinsic motivation drivers for personnel who have already had very high intrinsic motivation [82].

Table 7 includes the BRM operational activities that could be applied as attributes for enhancing continuous motivation of containment laboratory personnel to avoid complacency.

7 Conclusions

The WHO website https://search.bvsalud.org/global-literature-on-novel-coronavirus-2019-ncov/ shows that more than 123,000 Covid-19 related papers have been published by 7 Nov 2020. Based on this data, a vast number of research laboratories around the world are assumed to be involved in SARS-CoV-2 research. Many biorisk management professional, laboratory directors, institutional heads and regulators have some concerns about booming SARS-CoV-2 research in life sciences laboratories since tangible/intangible and unexpected biorisks could be associated with such activities. It is critical for institutions to ensure that an effective BRM system is implemented to mitigate these risks in their containment laboratories. Tables 4, 5, 6, and 7 includes biosafety, biosecurity and bioethics related organizational processes and operational activities which should be managed and led by the leadership of the institution, laboratory management, biorisk management committee and supervisors. Those processes and activities could be recognized as supportive organizational behaviors as described in IAEA's guidance principles for implementing safety and security culture. We propose here to strengthen the existing WHO laboratory BRM paradigms (risk-based approach with ethical responsibility) with supportive organizational behaviors lead by institutional and laboratory leadership as described in the international standard for laboratory BRM, ISO 35001 (*see* conceptual illustration in Fig. 1).

If the laboratory management tabulates their operational activities as presented in Tables 4, 5, 6, and 7, gaps in the operational activities and organizational processes could be identified. The

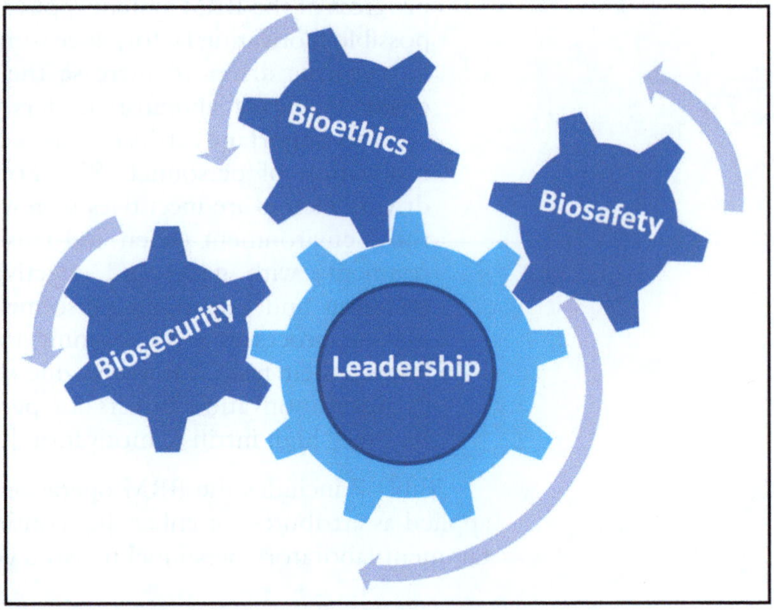

Supportive Organizational behaviour

Fig. 1 Cog wheels showing how organizational behaviors led by institutional and laboratory leadership can strengthen the biorisk management paradigm

operational activities presented in Tables 4, 5, 6, and 7 may not be comprehensive and are not intended as a full list of the attributes for supportive organizational behavior. However, the list may be useful as a starting point for the laboratory management team to prepare a self-evaluation/self-assessment form to gauge the status of their BRM system. The authors believe that assessing or evaluating organizational behavior should be done by the laboratory owners themselves like in the nuclear industry [68]. The IAEA has issued a guidance document for a nuclear facility operator to design their own organizational culture assessment checklist for nuclear power plants and laboratories [68]. The self-assessment action is based on the assumption that operators could recognize the organizational behavior and culture more easily than externals auditors since the operators have in-depth knowledge of the workplace, its people, its processes and its key influencers [83]. Organizational behavior is reflecting the organizational culture. Organizational culture is influenced by the larger societal culture (of the corporate world or the country or both) thus by nature it cannot be standardized easily [84]. If an individual operator designs a self-assessment/self-evaluation form based on the list in this chapter for his/her own laboratory, it could be able to identify and monitor potential organizational culture/behavior issues that might compromise BRM in the laboratory. The laboratory management could find out which operational activities and which organizational processes are attributes for a week BRM system, which stakeholders are involved in managing those activities and processes, how to interact/communicate with stakeholders to enhance an effective BRM, or how to rethink/retool it to achieve an effective system. The self-assessment form for organizational culture and behavior could complement the currently available containment laboratory assessment forms from the WHO Biosafety manual [1], Biosafety branch, Singapore Ministry of Health [43], US Federal select agent program [85], Canadian Public Health Agency of Canada (PHAC) and the Canadian Food Inspection Agency [48], UN Food and Agriculture Organization laboratory mapping tool [86], and European Checklist for Laboratory Biorisk Management [87]. Those assessment forms do not cover the organizational behavior and ethical related questions, but the biosecurity self-scanning tool kit [62] and vulnerability scanning tool kit [61] developed by the Netherlands biosecurity office have cover those questions.

The authors' impression is that assessing the operational activities of the existing BRM system from an organizational behavior perspective can easily be done in the majority of containment laboratories with little or no direct financial cost to the laboratory. Among the operational activities, the most challenging maybe implementing the "just culture" environment in the laboratory. This activity needs solid supports from the top management and organizational health and safety administrators. If the latter group does not believe in a "just culture" concept, and organizational policy of incident reporting and investigation established on a "rule based" concept (i.e., without taking into account human fallibility), the laboratory management alone cannot implement it effectively.

Most importantly, all stakeholders have to understand that biosafety and biosecurity responsibilities lie with them, and they must work together as a team to countermeasure emerging biorisks in containment laboratories. The authors also recommend organizations to equip all researchers involved in security-sensitive BA research activities with appropriate ethical education. This would result in diligently assessing potential benefits (drugs, vaccines, diagnostics, etc.) vs. potential harmful effects (biorisks) in humans, society and environment with maximizing the benefits and minimizing the potential for harmful consequences. Biosafety professional organizations or ad hoc laboratory networks can play a role as trainers/mentors for junior biosafety professional and should be encouraged by national regulatory agencies and international initiatives such as the EU CBRN CoE and Global Health Security Agenda. This comprehensive approach will facilitate a sustainable way to counter biorisks including unexpected laboratory related biorisks and the proliferation of biological weapons that may lead to a negative impact on health, economy, and national security.

Note: The amateur biologists' movement-related biosecurity risks and individual-level behaviors are not discussed in this chapter.

References

1. WHO (2004) Laboratory biosafety manual, 3rd edn. WHO Press, Geneva
2. WHO (2006) Biorisk management. Laboratory biosecurity guidance. WHO Press, Geneva
3. National Institute of Environmental Health Sciences, UK. Bioethics. https://www.niehs.nih.gov/research/resources/bioethics/index.cfm. Accessed 23 Dec 2020
4. WHO (2010) Responsible life sciences research for global health security. Guidance document. WHO Press, Geneva
5. Biosecurity Office, National Institute for Public Health and the Environment, Netherlands. Dual-use research. https://www.bureaubiosecurity.nl/en/policy/dual-use-research. Accessed 23 Dec 2020
6. Biological agents and toxins Act (2005). https://sso.agc.gov.sg/Act/BATA2005. Accessed 25 Nov 2020
7. CDC USDA Federal select agent program. Select agents and toxins list. https://www.selectagents.gov/sat/list.htm?CDC_AA_refVal=https%3A%2F%2Fwww.selectagents.gov%2FSelectAgentsandToxinsList.html. Accessed 25 Nov 2020
8. Australia group. List of human and animal pathogens and toxins for export control. https://www.dfat.gov.au/publications/minisite/theaustraliagroupnet/site/en/human_animal_pathogens.html. Accessed 25 Nov 2020
9. OIE World Organization for Animal Health. Covid-19 portal. Questions and answers on Covid-19. https://www.oie.int/en/scientific-expertise/specific-information-and-recommendations/questions-and-answers-on-2019novel-coronavirus/. Accessed 23 Sept 2020
10. World Health Organization (WHO) (2020) Laboratory biosafety guidance related to coronavirus disease (covid-19) Interim guidance. https://www.who.int/publications/i/item/laboratory-biosafety-guidance-related-to-coronavirus-disease-(covid-19). Accessed 23 Sept 2020
11. Biosafety Branch, Ministry of Health, Singapore. Laboratory Biosafety Resources Related to SARS-CoV2/COVID-19 Virus Samples/Materials (2020) https://www.moh.gov.sg/docs/librariesprovider7/useful-info-and-guidelines-documents/consolidated-laboratory-biosafety-resources-related-to-sars-cov-2-covid-19-virus-materials.pdf. Accessed 23 Sept 2020
12. US Communicable Disease Prevention and Control. Interim laboratory biosafety guidelines for handling and processing specimens associated with coronavirus disease 2019 (Covid-19). https://www.cdc.gov/coronavirus/2019-ncov/lab/lab-biosafety-guidelines.html. Accessed 25 Nov 2020
13. University of South Florida Biosafety Office. SARS-COV-2 (COVID-19) Research laboratory biosafety guidelines. https://www.usf.edu/research-innovation/research-integrity-compliance/documents/biosafety/usf-sars-cov-2-biosafety-guidelines.pdf. Accessed 23 Sept 2020
14. Willemarck N, Brosius B, Van Vaerenbergh B et al (2013) Laboratory-acquired infections in Flinders (2007–2012): an on-line survey. Ref: D/2012/2505/59. https://www.biosafety.be/sites/default/files/2012_lai_reportflanders2007_2012_sbb_2505_59.pdf. Accessed Dec 2020
15. Poh LL, Asok K, Gowri G et al (2004) Laboratory-acquired severe acute respiratory syndrome. N Engl J Med 350:1740–1745
16. Normile D (2004) Mounting lab accidents raise SARS fears. Science 304:659–661
17. Ling AE (2007) Editorial on Laboratory-acquired incidents in Taipei, Taiwan and Singapore following the outbreak of SARS coronavirus. Appl Biosaf 12:17
18. Centres for Disease Control and Prevention (CDC) (2015) Historical CDC laboratory safety activities. https://www.cdc.gov/labs/safety-history.html. Accessed 25 Sept 2020
19. McCarthy M (2014) Biosafety lapses prompt US CDC to shut labs and launch review. BMJ 349:g4615. https://doi.org/10.1136/bmj.g4615. Accessed 25 Sept 2020
20. Briggs PG (1965) Causal factors in microbiology laboratory accidents and infections. US Army Biological Laboratories, Fort Detrick, Frederick, MD
21. WHO (2020) How India scaled up its laboratory testing capacity for Covid-19. https://www.who.int/india/news/feature-stories/detail/how-india-scaled-up-its-laboratory-testing-capacity-for-covid19. Accessed 25 Nov 2020
22. Munir K, Ashraf S, Munir I et al (2020) Zoonotic and reverse zoonotic events of SARS-CoV-2 and their impact on global health. Emerg Microbes Infect 9:2222–2235
23. Oreshkova N, Molenaar RJ, Vreman S et al (2020) SARS CoV-2 infection in farmed

minks, the Netherlands, April and May 2020. Euro Surveill 25:2–8

24. Rosengard AM, Liu Y, Nie YZ et al (2002) Variola virus immune evasion design: expression of a highly efficient inhibitor of human complement. Proc Natl Acad Sci U S A 99: 8808–8813
25. Cello J, Paul AV, Wimmer E (2002) Chemical synthesis of poliovirus cDNA: generation of infectious virus in the absence of natural template. Science 297:1016–1018
26. Tumpey TM, Basler CF, Aguilar PV et al (2005) Characterization of the reconstructed 1918 Spanish influenza pandemic virus. Science 310:77–80
27. Committee on advances in technology and the prevention of their application to next generation biowarfare threats. Development, security, and cooperation policy and global affairs division, Board on Global Health, Institute of Medicine and National Research Council of the National Academies (2006) Advances in technologies with relevance to biology: the future landscape. In: The National Academies of Sciences Engineering Medicine (ed) Globalization, biosecurity, and the future of the life sciences. The National Academies Press, Washinton, DC, pp 139–212
28. Hamilton RA (2020) Understanding the threat of deliberate transmission of Covid-19 and deliberate release of viruses. Presented in Covid-19 and future pandemics: the spectre of bioterrorism webinar organized by United Nation Interregional Crime and Justice Research Institute (UNICRI) and UN Office of Counter-Terrorism Centre (UNCCT) 2 July 2020
29. Meo SA, Abukhalaf AA, Alomar AA et al (2020) Impact of weather conditions on incidence and mortality of Covid-19 pandemic in Africa. Eur Rev Med Pharmacol Sci 24: 9753–9759. https://www.europeanreview.org/wp/wp-content/uploads/9753-9759.pdf. Accessed 23 Dec 2020
30. Ragan I, Hartson L, Pidcoke H et al (2020) Pathogen reduction of SARS-CoV-2 virus in plasma and whole blood using riboflavin and UV light. PLoS One 15:e0233947. https://doi.org/10.1371/journal.pone.0233947. Accessed 25 Nov 2020
31. Wang F, Zhang W (2019) Synthetic biology: recent progress, biosafety and biosecurity concerns, and possible solutions. J Biosaf Biosecur 1:22–30
32. MacIntyre CR (2015) Biopreparedness in the age of genetically engineered pathogens and open access science: an urgent need for a paradigm shift. Mil Med 180:943–949. https://doi.org/10.7205/MILMED-D-14-00482
33. WHO (2016) International Health Regulations (IHA) 2005, 3rd edn. WHO Press, Geneva
34. WHO (2016) Joint external evaluation tool: International Health Regulations (2005), 2nd edn. WHO Press, Geneva
35. UN Security Council Resolution 1540 (2004). https://www.un.org/disarmament/wmd/sc1540/. Accessed 26 Sept 2020
36. UNSCR 1540 Assessment matrix. https://www.un.org/en/sc/1540/documents/Final%20Matrix%20Template%20(E).pdf. Accessed 25 Nov 2020
37. BWC Implementation Support Unit Bureau C.115 UN Office for Disarmament Affairs. Biological weapon convention. https://www.unog.ch/80256EDD006B8954/(httpAssets)/CF1D6A8403271778C1257AAF0055BB6C/$file/BWC-Background_Inf_Nov%202012.pdf. Accessed 23 Dec 2020
38. Meeting of experts, Meeting of the state parties to the convention on the prohibition of the development, production and stockpiling of bacteriological (biological) and toxin weapons and on their destruction. BWC/MSP/2005? MX/WP.35. 24 Jun 2005. Geneva, Switzerland. https://undocs.org/pdf?symbol=en/BWC/MSP/2005/MX/WP.35. Accessed 25 Nov 2020
39. IATA (2020) The IATA dangerous goods regulations manual, 61st edn. IATA, Montreal
40. OIE (2018) Biosafety and biosecurity: standard for managing biological risk in the veterinary laboratory and animal facilities. In: Guidance on biosafety and biosecurity in veterinary laboratories and animal facilities. https://www.oie.int/fileadmin/Home/eng/Health_standards/tahm/1.01.04_BIOSAFETY_BIOSECURITY.pdf. Accessed 24 Dec 2020
41. International Standard Organization (ISO) (2019) Biorisk management for laboratories and other related organisms ISO 35001:2019 (standard). ISO, Geneva
42. WHO (2019) Guidance on regulations for the transport of infectious substances, 2019–2020. WHO Press, Geneva
43. Singapore MOH certification checklist for high containment laboratories. #13 MOH BSL-3 laboratory certification checklist. https://www.moh.gov.sg/biosafety/useful-info/useful-info-and-guidelines. Accessed 25 Nov 2020
44. Federal Select Agent Programme (2012) Inspection checklist for BSL-3 laboratories

(7 CFR 331; 9CFR 121; 42 CFR 73; BMBL 5th ed). Revised July 2012. https://gcbs.sandia.gov/human_capacity_development/hcd_gbrmc_docs_manual_brm/Att%20B-2%20US%20Select%20Agent%20BSL3%20Checklist.pdf. Accessed 6 Oct 2020

45. EU directive 2000/54/EC-Protection of workers from risks related to exposure to biological agents at work (2000)
46. EU Directive 2009/41/EC-Contained use of genetically modified micro-organisms (2009)
47. Council Regulation (EC) No. 428/2009-Setting up a Community regime for the control of exports, transfer, brokering and transit of dual-use items and other directives (2009)
48. Public Health Agency of Canada (PHAC) and the Canadian Food Inspection Agency (CFIA) (2015) Canadian Biosafety Standard (CBS), 2nd edn. https://www.canada.ca/en/public-health/services/canadian-biosafety-standards-guidelines/second-edition.html. Accessed 25 Nov 2020
49. Department of Health and Aging, Office of Gene Technology Regulator, Australia government (2006) Guidelines for certification of a physical containment level 3 laboratory. Version 2.1
50. EU CBRN CoE (2020) How the Centres of Excellence are combating biological and chemical threats globally. https://europa.eu/cbrn-risk-mitigation/news-0/cbrn-coe-biological-and-chemical-projects-brochure-2020-12-18_en. Accessed 23 Dec 2020
51. Clevestig P (2009) Handbook of applied biosecurity for life science laboratories. Stockholm International Peace Research Institute, Stockholm
52. Wagener S, Bollaert C (2013) Biosafety, biosecurity, and bioethics governance in synthetic biology: the "7P" approach. Appl Biosaf 18: 178–186
53. Ritterson R, Casagrande R (2017) Basic scholarship in biosafety is critically needed to reduce risk of laboratory accidents. mSphere 2: e000010–e000017. https://doi.org/10.1128/mSphere.00010-17. Accessed 23 Dec 2020
54. Health & Safety Executive (HSE), UK Government Agency (1999) Reducing error and influencing behaviour. HSG48, 2nd edn. TSO (The Stationery Office), Part of Williams Lea Tag, UK
55. Reason J (2000) Human error: models and management. BMJ 320:768–770
56. Pruchnicki S, Key K, Rai AH (2019) Problem solving/decision making and procedures for unexpected events: a literature review. Technical report, Office of Aerospace Medicine, Washington, DC. http://www.faa.gov/go/oamtechreports. Accessed 25 Nov 2020
57. Lekka C (2011) Highly reliability organizations. A review of the literature. Health and Safety Laboratory for the Health and Safety Executive. Research report. https://www.hse.gov.uk/research/rrpdf/rr899.pdf. Accessed 25 Nov 2020
58. Saravanakumar A (2020) Unit 1 Organizational behaviour. https://www.researchgate.net/publication/340266415_ORGANIZATIONAL_BEHAVIOUR. Accessed 23 Dec 2020
59. Gibson JL, Ivancevich JM, Donnelly JH Jr et al (2011) Organizations: behaviour, structure, processes, 14th edn. McGraw-Hill Irwin, New York. http://dl.motamem.org/organizations_behavior_structure.pdf. Accessed 23 Dec 2020
60. Kuykendall T, Khrpunov I, Lowe J (2018) Final report on the proceedings of the international workshop, Serpong, Indonesia 29--31 January 2018. Research Gate. https://www.researchgate.net/publication/325908800. Accessed 25 Nov 2020
61. Netherlands Biosecurity Office. Vulnerability scanning. https://www.biosecurityvulnerabilityscan.nl/. Accessed 23 Dec 2020
62. Netherlands Biosecurity Office. Biosecurity self-scan toolkit. http://www.biosecurityselfscan.nl/mainMenu.html. Accessed 25 Nov 2020
63. Hystad SW, Bartone PT, Eid J (2013) Positive organizational behavior and safety in the offshore oil industry: exploring the determinants of positive safety climate. J Posit Psychol 9: 42–53
64. (2014) Committee on Establishing and Promoting a Culture of Safety in Academic Laboratory, Research Board on Chemical Sciences and Technology Division on Earth and Life Studies, Board on Human-Systems Integration Division of Behavioral and Social Sciences and Education. In: Safe science: promoting a culture of safety in academic chemical research. National Research Council of the National Academies. The National Academy Press, Washington, DC. Accessed 10 Oct 2020
65. Randell E (2018) Chapter 6. Workplace culture, ethics and laboratory leadership. In: Yenice S, Randell E (eds) Leadership basics for clinical laboratory professionals. Committee on Clinical Laboratory Management, International Federation of Clinical Chemistry and Laboratory Medicine, Washington, DC

66. Franz DR (2019) The role of leaders in laboratory safety: with an example from industry. Biosaf Health 1:4–5
67. Khine-Wamono (2018) Chapter 2. Effective leadership styles. In: Yenice S, Randell E (eds) Leadership basics for clinical laboratory professionals. Committee on Clinical Laboratory Management, International Federation of Clinical Chemistry and Laboratory Medicine, Washington, DC
68. IAEA (2008) IAEA nuclear security series no. 7. Implementing guide. Nuclear security culture. IAEA, Vienna
69. Boysen PG (2013) Just culture: a foundation for balanced accountability and patient safety. Ochsner J 13:400–406
70. Reason J (2000) Safety paradoxes and safety culture. Int J Inj Control Saf Promot 7:3–14
71. Homer PM, Kahle LR (1988) A structural equation test of the value-attitude behaviour hierarchy. J Pers Soc Psychol 54:638–646
72. Committee on the Department of Homeland Security Workforce resilience; Board on Health Sciences Policy; Institute of Medicine (2013) A ready and resilient workforce for the department of homeland security: protecting America's front line. The National Academies Press, Washington, DC
73. WHO, Clinical and Laboratory Standards Institute and CDC (2011) Laboratory quality management system handbook. WHO, Geneva
74. Novossiolova T, Martellini M (2019) Promoting responsible science and CBRN security through codes of conduct and education. Biosaf Health 1:59–64
75. World Conferences on Research Integrity Foundation (WCRIF) (2010). https://wcrif.org/guidance/singapore-statement. Accessed 25 Nov 2020
76. WHO Code of Conduct for responsible research (2017) Geneva. https://www.who.int/about/ethics/code-of-conduct-responsible-research.pdf. Accessed 25 Nov 2020
77. The National Academies of Sciences Engineering medicine (2011) Current condition: Establishing a baseline about education on dual use issues. In: Challenges and opportunities for education about dual use issues in the life sciences The National Academies Press, Washington DC
78. Kelley M (2006) Infectious disease research and dual-use risk. AMA J Ethics 8:230–234
79. International Working Group on Strengthening Culture of Biosafety, Biosecurity, and Responsible Conduct in the Life Sciences (2020). Culture of biosafety, biosecurity, and responsible conduct in the life sciences. Self-assessment framework. Working draft. https://absa.org/wp-content/uploads/2020/02/Culture_of_Biosafety-Biosecurity_Self-Assessment_Framework.pdf. Accessed 6 Oct 2020. Accessed 25 Nov 2020
80. Sedaoui A (2016) On thoughts on safety culture. A questioning attitude: a mark of professionalism and the maturity of a company's safety culture. ICSI issue 25. https://www.icsi-eu.org/documents/157/icsi_thoughts_safety_culture_25_november_2016.pdf
81. Pink D (2009) Drive: the surprising truth about what motivates us. Riverheads Books, New York
82. Yenice S (2018) Chapter 5. The leader as visionary and motivator. In: Yenice S, Randell E (eds) Leadership basics for clinical laboratory professionals. Committee on Clinical Laboratory Management, Washington, DC
83. IAEA (2016) Performing safety culture self-assessments, Safety reports series No. 83. IAEA, Vienna
84. Kastenberg WE (2015) Ethics, risk and safety culture. In: Ahn J, Carson C, Jensen et al (eds) Reflections on the Fukushima Daiichi nuclear accident - toward social-scientific literacy and engineering resilience. Springer International Publishing, Switzerland
85. Federal Select Agent Programme (2012) Inspection checklist for BSL-3 laboratories (7 CFR 331; 9CFR 121; 42 CFR 73; BMBL 5th ed). Revised July 2012. https://gcbs.sandia.gov/human_capacity_development/hcd_gbrmc_docs_manual_brm/Att%20B-2%20US%20Select%20Agent%20BSL3%20Checklist.pdf. Accessed 25 Nov 2020
86. FAO Laboratory mapping tool-Core (LMT-Core) Version 2016. http://www.fao.org/ag/againfo/programmes/en/empres/news_130514.html. Accessed 25 Nov 2020
87. Lloyd G, Lippolito G, Di Caro A et al on behalf of the Laboratory Network Under the EU Funded Joint Action QUANDHIP- Quality Assurance Exercises and Networking on the Detection of Highly Infectious Pathogens (2015) Integrated European checklist for laboratory biorisk management in handling of high consequence risk group 3 and 4 agents (ECL-Biorisk). https://www.emerge.rki.eu/Emerge/EN/Content/Topics/Rules/ECL_Biorisk.pdf?__blob=publicationFile. Accessed 25 Nov 2020

Chapter 24

Biorisk Management for SARS-CoV-2 Research in a Biosafety Level-3 Core Facility

Tessy Joseph, Sabai Phyu, Su Yun Se-Thoe, and Justin Jang Hann Chu

Abstract

The emerging severe acute respiratory syndrome coronavirus 2 (SARS-CoV-2) presents hazards to researchers and other laboratory personnel in research settings where the live virus is stored and handled. The Biosafety Level-3 (BSL-3) Core Facility (CF) at Yong Loo Lin School of Medicine in National University of Singapore (NUS Medicine) has implemented a biorisk management (BRM) system to ensure that biorisk to employees, the public, or the environment are consistently minimized to an acceptable level while working with SARS-CoV-2. This chapter summarizes how a BRM system can be implemented in academic institutions based on international standards in the context of existing local legislations/regulations and institutional policies/guidelines to minimize the risk of laboratory-acquired infections and deliberate misuse of the newly emerged virus, SARS-CoV-2 in BSL-3 laboratories. The BRM programs prioritize performing risk assessments prior to implementation of work processes and reassessing the risk portfolio of the facilities from time to time, determining root causes and prevention of recurrences. Focusing on awareness-raising and educating the laboratory users in biosafety and biosecurity, and identifying opportunities for improvement are the other key factors for a sustainable and successful BRM system in the NUS Medicine BSL-3 CF.

Key words Biorisk management, Risk assessment, SARS-CoV-2, COVID-19, Biosafety, Biosecurity, BSL-3, ISO 35001:2019

1 Introduction

The threat of pandemics is ever-present globally. Massive urbanization and globalization accelerates emerging zoonotic diseases and the rapid spread of infectious diseases beyond the native country, the country where the disease was first discovered or identified. This is well evident in the recent coronavirus disease (COVID-19) pandemic. Although there were multiple layers of mitigation measures to prevent transmission in place by many countries, COVID-19 has widely spread to most countries and territories in few months, after it was first made known to the world. Prior to COVID-19, the world had experienced a series of emerging and

Justin Jang Hann Chu et al. (eds.), *SARS-CoV-2: Methods and Protocols*,
Methods in Molecular Biology, vol. 2452, https://doi.org/10.1007/978-1-0716-2111-0_24,

reemerging infectious disease outbreaks [1]. Therefore, it is important for all countries to be prepared and ready to respond to any future emerging or reemerging infectious diseases.

In Singapore, after experiencing the severe acute respiratory syndrome coronavirus (SARS-CoV) outbreak and a laboratory incident that occurred in 2003 [2], the country has since enhanced its public health and laboratory preparedness toward infectious disease outbreaks. The efforts ranged from the enactment of biosafety and biosecurity related legislations; enhancement of healthcare, diagnostic, and research laboratory capacities and capabilities (including BSL-3 laboratories); and establishment of a robust training program for different stakeholders involved in epidemic and pandemic responses. These efforts helped Singapore to manage several subsequent epidemics, such as the H1N1 influenza; hand, foot, and mouth disease (HFMD); Zika virus (ZIKV) disease outbreaks; and lately the COVID-19 pandemic [1, 3–5].

During epidemics and pandemics, the priority is to identify infected cases, contain disease transmission, provide appropriate treatment to the affected individuals, and reduce mortality rates. Conducting infectious disease research is also critical as it can help to understand the pathogenesis of the disease, develop more efficient diagnostic tools and preventive measures or to innovate treatment strategies that will better support the management of the patients. The World Health Organization (WHO) and national organizations/regulatory agencies, including Singapore's Ministry of Health, recommended the use of Biosafety Level-2 (BSL-2) laboratories for diagnostic work and manipulations of COVID-19 clinical/environmental samples of low-risk activities in a BSL-2 with enhanced (BSL-2 plus) practices [6–9]. However, for high-risk laboratory activities with severe acute respiratory syndrome coronavirus 2 (SARS-CoV-2) such as virus isolation, culturing, manipulation of pure virus, or use of live virus in functional assays and animal experiments, a BSL-3 containment is required in Singapore and in many other countries [9–12].

A laboratory-specific, robust BRM programme is essential to ensure that the research activities and laboratory operations are in compliance to all relevant requirements and the experimental work are carried out safely. This chapter focuses mainly on how a comprehensive BRM system established by the NUS Medicine BSL-3 CF that addresses biosafety and biosecurity programs to manage SARS-CoV-2 research during COVID-19 pandemic. The BSL-3 CF at NUS Medicine had developed its BRM system by adapting the (European Committee for Standardization) Workshop Agreement (CEN CWA) 17593:2011 Laboratory BRM [13] since the CF was established in 2014 and it was then modified based on International Organization for Standardization (ISO) 35001: 2019 standard, the first international standard in 2020 [14].

2 Biorisk Management in BSL-3 Core Facility

A comprehensive and customized BRM system is implemented in the NUS Medicine BSL-3 CF for handling SARS-CoV-2 positive samples. Although WHO and other international bodies published interim guidelines and recommendations for handling SARS-CoV-2 samples, it was necessary to customize the system for individual laboratories based on the risk assessments, available resources and specific needs. The BRM system of BSL-3 CF that is described in this chapter is aimed for continual improvement through a cycle of planning, implementing, reviewing, and improving the processes and undertaking improvement actions to meet the biosafety and biosecurity goals. This is known as the plan–do–check–act (PDCA) framework which is adopted and modified from ISO 35001:2019 standard [14, 15], and the backbone of this standard was the CEN 1753:2011 Laboratory BRM [13]. This approach for BSL-3CF through continual improvement has considered its management structure and oversight model of the organization with the components of leadership, planning, support, operations, and performance evaluation and improvement for the implementation of BRM system [15]. This is to ensure that risks to employees, the public, or the environment are consistently minimized to an acceptable level for all operational activities and experimental work, including the SARS-CoV-2 work. Each component of the BSL-3 CF BRM system is detailed in this chapter.

2.1 Policy and Leadership Commitment

The BSL-3 CF has developed and documented a BRM Policy with the following intended outcome for the implementation and proactively improving the overall biorisk performance:

1. Protecting laboratory personnel, community, and environment from biological hazards.
2. Reducing the risks of accidental or deliberate acts that may cause the release or loss of BAs or sensitive information.
3. Compliance to legal requirements and other (e.g., institutional) requirements.
4. Achievement of safety and health objective.
5. Continual improvement of biorisk performance.

The BSL-3 CF's Policy was developed based on the scope and objectives of the BSL-3 CF BRM system. The leadership of NUS Medicine has established specific policies to ensure that BSL-3 research is conducted in full compliance to the institutional, national [16, 17], and international requirements [14, 18], provided the directions and resources needed for the establishment, implementation, maintenance, and continual improvement of the BRM system. It is also aligned and contextualized to the legal requirements, safety policies of NUS and the NUS Medicine (institutional level) and is appropriate to mitigate all potential risks associated with all Risk Group 3 pathogens including SARS-CoV-2 research. Maintaining a safe and secure containment laboratory also requires knowledge and good decision making of the CF leadership and each staff member [19] which was also ensured by the top management. The BSL-3 CF BRM programme is established to protect the safety and health of all personnel using the facility including visitors, from exposure to biohazards associated with their work and accidental/intentional release of BAs to the environment.

2.2 Roles and Responsibilities, and Authorities

2.2.1 Facility Operator

According to Biological Agents and Toxins Act (BATA), 2005 by the Singapore Ministry of Health (MOH), the roles and responsibilities of the "Operator" (the person who operates the facility or who has the management or control of the facility) include: (1) appoint a Biosafety Committee (BC) comprising appropriate personnel, and a Biosafety Coordinator (BCO); (2) obtain advice of the BC as to the safety measures required for carrying out any activities involving any BA or toxin, and inactivation of BAs; (3) obtain approval to possess BAs or toxins, including SARS-CoV-2 (First Schedule Part II BA); (4) ensure adequate safety and security measures are in place before any work involving BA or toxin can be commenced in the facility; and (5) other duties and obligations which aim to protect the safety of the laboratory personnel, visitors of the CF, the community and the environment [17]. The operator of the NUS Medicine BSL-3 CF is the dean of NUS Medicine and various BRM related roles have been officially delegated to the facility director, the BSL-3 BC, the BSL-3 CF Management, and the CF operation team.

2.2.2 BSL-3 Facility Management

The responsibilities and authorities for relevant roles are assigned and communicated to the facility personnel including those who manage the facility, who perform research (users of the facility) and verify the work (biosafety coordinator) associated with handling and control of biological materials, and so on [17]. In Singapore, some of the roles and responsibilities of personnel working for such facilities (Operator, BC, BCO, etc.) are defined in the BATA.

2.2.3 BSL-3 Biosafety Committee

A Biosafety Committee (BC) has been established to support the BRM system for the BSL-3 CF. The committee was set up according to the requirements stipulated in the BATA [17] and it includes a representative from a cross section of expertise appropriate to the nature of the research activities in the facility. The committee members and the chairperson were appointed by the Facility Operator. The responsibilities of the committee members and the chairperson have been clearly defined in the form of a Terms of

Reference and it is approved by the NUS Medicine's senior management.

The BC act as an independent review group for biorisk issues of the BSL-3 CF and reports to NUS Medicine senior management. The BC meets frequently (monthly or when required, whichever earlier) to discuss biosafety and biosecurity related issues, and all discussions and/or issues raised/addressed in the meeting BSL-3 are formally recorded. All actions allocated are tracked, followed up and closed out effectively. The key responsibilities of the BC include reviewing of policies, programs (safety/security/research), codes of practice and all risk assessments for the BSL-3 CF. The BC is also responsible of informing the Facility Operator of any change of measures, policies, programs, and codes of practice, whenever the BC deems relevant and/or necessary; as well as review and verify if the BA inactivated at the facility has been properly inactivated.

2.2.4 Biosafety Coordinator

Based on the laboratory BRM standard (ISO 35001:2019), the competent individual providing advice and guidance on BRM is recognized as a biological safety officer or biorisk management advisor [14]. This role with similar responsibilities is named as biosafety coordinator (BCO) in Singapore as per the BATA [17]. The BCO is appointed by the facility operator and is a key member of the BC. The BCO is responsible for implementing the measures, policies, programs, and codes of practice which are devised or formulated by the BC. In addition, she/he should also ensure that any changes to the existing measures, policies, programs, or code of practice proposed by the BC are implemented as soon as possible within the timeframe stipulated by the BC.

2.2.5 BSL-3 Facility Operation Team

The BSL-3 CF operation team is responsible for ensuring that all personnel involved in research with BAs including SARS-CoV-2 can demonstrate proficiency in safe laboratory practices prior to work with the BAs. The team also oversees and ensures that:

- The BSL-3 users are provided with appropriate training, advice, and guidance on BRM.
- Overall operations and maintenance of general safety and laboratory equipment.
- The facility air conditioning, mechanical, and ventilation (ACMV) and other engineering and infrastructure related systems are working throughout.
- Administrative (safety and security) control measures and work practices that are suitable for the BAs are implemented.
- Access is restricted and issued with Personal Identification Number (PIN) to only authorized individuals for entering the dedicated room.

- Ensure that BAs are kept in secured place according to BSL-3 CF standard operating procedure (SOP) and up-to-date inventory is maintained.
- Monitor the effectiveness of the BRM program.

Laboratory BRM is a collaborative effort, that to be successful, should have commitment and participation of all persons involved. The appropriate breadth of collaboration depends on the type of hazards in use, risks involved, and the complexity of a given laboratory. In addition to the support from senior management and BC/experts, there should be collaboration with human resources, facility engineering, occupational and environmental health and safety, security, and emergency preparedness personnel. The BSL-3 CF also works together with external collaborators such as first responders (e.g., fire, police, and emergency medical personnel), designated hospital and other health-care facilities (e.g., NUS Occupational Health clinic) as well as contractors involved in facility and equipment maintenance. A detailed description of the typical roles and responsibilities of institutional and laboratory management are captured in the previous chapter.

3 Risk Management of Biohazards

The main component of risk management included identification of hazards specific to all experimental work (i.e., in this case SARS-CoV-2 work), through risk assessments, risk mitigations and communications, determined the legal requirements (e.g., BATA) and other internal (e.g., NUS safety Policies and Directives) and external (e.g., WHO requirements) requirements. Much emphasis was placed on risk assessment and its methodology as mandated by the Singapore Ministry of Manpower's Workplace Safety and Health (WSH) Regulatory Framework and BATA [17, 20, 21]. The risk management has been integrated into daily CF operations and the process of identification of hazards and prioritization of risks and the establishment of risk mitigation protocols are tailored to specific situations that are ongoing in BSL-3 CF.

3.1 Hazard Identification and Risk Assessment for SARS-CoV-2 Work

The risk assessment process included identification of various types of hazards associated with the work in BSL-3 CF. In this chapter, the focus is on biological hazards such as COVID-19 samples and SARS-CoV-2 isolates and excluded details on other types of general laboratory hazards. When the CF had to handle COVID-19 samples in January 2020, it was discussed by the facility team and BC with national regulatory agencies on what type of facilities and risk controls which should be in place to deal with COVID-19 samples for research purposes. Given that SARS-CoV-2 is from the *Coronaviridae* family and has proven similarities in terms of morphology and mode of infection [22–24], risk management methodology used for SARS-CoV and MERS-CoV were initially adopted in most laboratories. Since the infectivity and severity of SARS-CoV-2 was initially uncertain, a high standard of precautions with general BSL-3 containment and special enhanced biosafety practices were followed for all work in BSL-3 CF.

Various guidance documents, including the MOH "Biosafety Guidelines for Laboratories and Personnel Handling Samples or Materials Associated with SARS-CoV-2/COVID-19 virus," and the WHO and CDC published guidance have provided useful biosafety advices on packaging, transporting, shipping, and standard precautions or good laboratory practices in the handling of COVID-19 samples [6, 9, 11, 21, 25]. These were referenced when implementing specific control measures for the handling of COVID-19 samples/SARS-CoV-2 in NUS Medicine BSL-3 CF.

A comprehensive risk assessment was performed by listing all different types of samples received in the BSL-3 CF and all potential scenarios during handling of COVID-19 clinical samples and SARS-CoV-2 virus isolates for various procedures. The type of samples received included respiratory samples, saliva, blood and serum samples, and environmental samples collected on various types of media (e.g., filters). The risk assessment process included the following steps: (1) identifying the hazards associated with the infectious agent or material, (2) identifying the category of hazards including physical situation (e.g., a fire or explosion) and work activities (see examples of activities in the next paragraph), equipment (e.g., use of centrifuge), materials (in this case the principal hazard is SARS CoV-2 [BA], but others includes chemicals and other type of hazards) or people involved (immunocompromised personnel, noncompetent researchers, etc.).

The hazard identification was performed by a team consisting of facility personnel (operation team), scientists, biosafety professionals, and so on using their prior experiences and knowledge, in consultation with local regulatory authorities and international guidelines. The risk assessment team listed various types of routine activities such as vortexing, culturing of virus, centrifugation, use of any equipment, and use of animals and other nonroutine activities such as accidental spillage of biological materials, and hazards associated with each activity were identified and documented. For the evaluation of the risk, a quantitative method using a 3×3 matrix which is used in NUS was adopted. The risk was then calculated as the product of the numbers (1 to 3) assigned to "likelihood of exposure to a hazard" and "potential severity of harm" as shown in Fig. 1 and with the risk levels 4 to 9 requiring additional risk mitigating measures before allowing the activities to be carried out in BSL-3 CF.

The risk assessment for NUS Medicine BSL-3 CF is mainly categorized into four categories: (1) risk assessment for routine

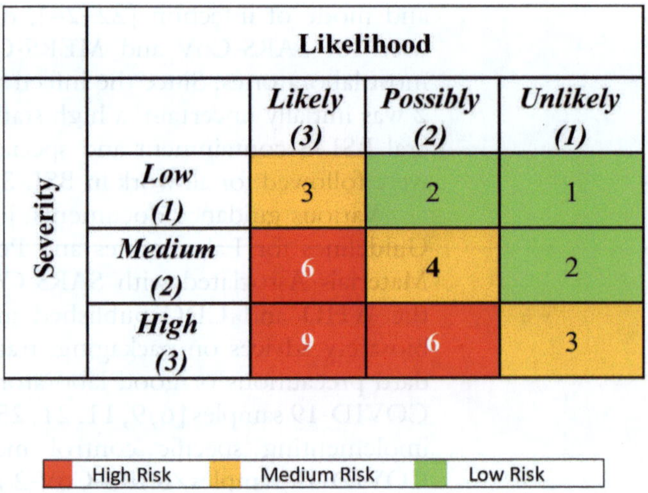

Fig. 1 Risk Assessment template based on a 3 × 3 matrix scheme which use the Likelihood and Severity to assess the risk level

activities led by the researchers (decontamination of BSC after work, labelling and storage of BAs in a freezer, etc.); (2) risk assessment for routine activities performed by the facility operation team (e.g., regular monitoring and maintenance of safety equipment, PPE, etc.), (3) risk assessment for nonroutine activities led by the facility operation team (e.g., for an emergency equipment maintenance work by a contractor), and (4) risk assessment for specific research protocols led by the researchers (e.g., administration of BAs into animals). All these risk assessments are then subjected to BC's review, deliberation, and approval at the regular BC meetings. The biosecurity aspects of the facility's infrastructure and the activities involving SARS-CoV-2 have also been considered in the risk assessment based on the framework established by Infrastructure Protection Act (IPA) and BATA [16, 17]. Dedicated rooms in BSL-3 CF have been assigned for working with SARS-CoV-2 virus. All personnel involved are provided with the risk communication information, training, special precautions, appropriate personal protective equipment (PPE), and OH requirements as per the NUS OH program. Any inadequate protection for the work activities in the laboratory, storage and transportation of BAs, and waste disposal would result in potential consequences of LAIs.

3.2 Legal and Other Requirements that Govern Research Laboratories Handling SARS-CoV-2

In addition to the international standards and guidelines, national legislations, regulations/guidance documents play a major role in influencing the BRM practices for handling this novel virus at laboratory level [17, 21]. The BRM system of BSL-3 CF has adopted all applicable national legal framework supporting BSL-3 laboratories and set the goal of managing potential biorisks and to appropriately address, manage and minimize. The BSL-3 CF also follow all relevant rules and regulations of the university and NUS Medicine. The operation team ensured the strengthening and enforcement of the existing regulations and procedures for all laboratory activities, storage and transportation of SARS-CoV-2 samples and waste management. The legal frameworks in Singapore guides the national authorities, facility operators, BC, laboratory managers and ultimately laboratory workers to take the responsibility in developing and implementing the necessary safeguards. Some of the most relevant legislations, regulations and guidelines are described below.

3.2.1 Workplace Safety and Health Act

The Workplace Safety and Health Act (WSHA) is a legislation relating to the safety, health and welfare of persons at work in any workplaces in Singapore including biomedical research laboratories [20]. It places responsibilities on various stakeholders who have control to ensure safety at the workplace. There are several WSHA subsidiary legislations and several Codes of Practices applicable to all workplaces including research laboratories [20, 26]. The WSHA requires all workplaces to conduct risk assessments to identify the sources of risk and take all reasonably practicable steps to eliminate any foreseeable risk to any person who may be affected by the undertaking in the workplace.

3.2.2 Biological Agents and Toxins Act

The BATA is a risk-based (agent/activity) legislation that sets stringent controls on BAs and toxins that pose high or significant risks to public health and security. It regulates the possession, use, import, transfer and transportation of BAs and toxins, and to provide safe practices in the handling of such BAs and toxins. An approval is required from the MOH for physical possession of all high-risk BAs and toxins. MOH provided guidance to stakeholders on the applicable biosafety regulations and requirements for storage and handling of COVID-19 samples for laboratory diagnosis, clinical trial testing, assay development, kit evaluation, research, and so on. The guidance has been updated several times based on evolving scientific information and situations since it was released in January 2020 [21].

SAR-CoV-2 was listed into First Schedule Part II BA under the BATA as of 30 January 2020 by the MOH, where in BAs listed under this Schedule are generally Risk Group 3 BAs that pose high safety and security risk to laboratory personnel and public, an approval for possession is required from the MOH [12]. The preconditions for possessing and handling high risk activities involving live SARS-CoV-2 virus is a BSL-3 facility that is gazetted as a Protected Place under the Infrastructure Protection Act [16, 17]. The BATA also requires the BSL-3 facility to conduct risk assessment and to implement appropriate measures to safeguard the safety and security of personnel, the public and the environment; and to set training requirements, duties, and obligations for facility operators, BC, BCO, and operation staff of facilities in relation to BAs and toxins [17]. The BATA, however, allowing certain activities such as the use of SARS-CoV-2 clinical and environmental samples for the purposes of diagnosis, autopsy, and surveillance in a non-BSL-3 and nonprotected settings, provided the activities do not involve (a) culturing, isolating, or purifying of live SARS-CoV-2 from COVD-19 samples, or (b) testing of the samples in animals.

3.2.3 Animals and Birds Act

This Act regulates all animal work involving animal pathogens or zoonotic agents (including SARS-COV-2) and animals [27].

3.2.4 Human Biomedical Research Act

The Human Biomedical Research Act (HBRA) regulates research involving the use of human tissues/materials of human origin [28]. To use COVID-19 clinical samples, the principal investigators (PIs) need to obtain approval from Institutional Review Board, a requirement under the HBRA.

3.2.5 Infrastructure Protection Act

Under the BATA, any BSL-3 facility possessing high security risk BAs, such as those listed in First Schedule Part II and Second Schedule, shall be gazetted as a Protected Place and ensures compliance with security requirements under the IPA [16].

3.2.6 Environmental Public Health (Toxic Industrial Waste) Regulations

This Act controls handling, transportation, treatment, and disposal of toxic industrial waste including all type of biological wastes [29].

3.2.7 Singapore Biosafety Guidelines for Research on Genetically Modified Organisms

These Guidelines were developed by the Singapore Genetic Modification Advisory Committee (GMAC) to ensure the safe containment, handling and transport of Genetically Modified Organisms (GMOs) used in research and to provide a common framework for assessment and notification of experiments on GMOs [30]. These Guidelines cover experiments that involve genetic manipulation work and importation of GMOs and/or GMO-derived products for research purposes.

3.2.8 National Advisory Committee for Laboratory Animal Research (NACLAR) Guidelines

All research activities that involve animals and are conducted at BSL-3 Core Facility need to adhere to the Guidelines on the Care and Use of Animals for Scientific Purposes issued by the National Advisory Committee for Laboratory Animals Research (NACLAR) taking into consideration the relevant scientific, ethical, and legal issues [31].

3.3 Risk Mitigation

The risk mitigation strategies for biosafety and biosecurity risks associated with SARS-CoV-2 research in BSL-3 CF are based on risk assessments. In the WHO approach, effective biosafety practices are the foundation of laboratory biosecurity activities: indeed, the implementation of good biosafety practices also addresses certain key dimensions of laboratory biosecurity [32, 33]. This component of BRM is focused on implementing mitigation controls, including training and exercise along with operational issues. The risk assessment process at the planning stage recognized a hierarchy of controls, beginning with the elimination or reduction of hazards (not very relevant and practical in BSL-3 CF for most of the activities involving SAR-CoV-2), then progressed to implement the appropriate engineering and/or administrative controls along with other operational controls to address residual risks, and identified and implemented appropriate PPE to protect the users. For research work involving high risk BAs including SARS-CoV-2, the BSL-3 CF is relying on a combination of engineering controls, additional administrative and work practice controls, and personal protective equipment (PPE) for the safety mitigation measures.

3.3.1 Engineering Controls

Based on the types of samples needed to be handled and aligned to the international [34, 35] and national standards/regulations and guidelines [16, 17], the requirements for the laboratory infrastructure, the primary barriers and other engineering controls have been decided at the design and construction stage of the CF. The facility is designed and fitted with inward airflow and additional engineering controls and precautions such as HEPA filtration on the exhaust air to mitigate the risks posed to personnel, the surrounding community, and the environment.

Biological Safety Cabinets (BSCs) are used as the main safety equipment in BSL-3 CF and all work involving handling of infectious agents/materials including SARS-CoV-2 are performed inside Class II BSCs. Other biosafety engineering controls such as enclosed containers (to contain aerosol, droplet, and leakage, etc.) and centrifuges with sealed rotors and safety cups are also used to protect personnel, the surrounding community, and the environment from possible exposure to hazardous BAs and toxins, depending on the risk assessment.

3.3.2 Administrative Controls

The BSL-3 CF has developed a detailed Laboratory Safety Manual (that is structured in PDCA format) and several facility specific SOPs which include but are not limited to entry and exit procedure, laboratory signage, facility emergency response plans, and designating who is responsible for the day-to-day operation, implementation of procedures and overall oversight of the facility, and so on. Administrative controls also include facility maintenance program. There are written SOPs, on standard microbiological practices such as frequent handwashing, minimizing the generation of aerosols, limiting the use of sharps and safe disposal, routinely decontaminating work areas and equipment, and safely collecting and decontaminating biohazardous wastes. Other examples of administrative controls implemented at CF are SOPs emergency response and contingency plans, biological spill cleanup, and reporting of incidents to relevant authorities. Training is provided to the respective personnel in charge of specific tasks and drills are conducted on a regular basis. The BRM in CF is continuously strengthened by having sufficient preparedness, tabletop exercises, and incorporating biosafety and biosecurity disciplines [36]. Immediate warning, containment, detection, and decontamination measures are necessary parts of preparedness regardless of whether the release of BAs is intentional or accidental.

3.3.3 Biosecurity Risk Controls

Laboratory biosecurity often remains as an undervalued aspect of BRM in many laboratories. Based on the risk assessment for the BSL-3 CF and the pathogens under possession and use in the facility, appropriate control measures were adopted for controlling the identified risks to prevent unauthorized access to BAs and the deliberate release of those agents. The biosecurity control measures of the BSL-3 facility include Information Control, Personnel Control, Physical Security, Transport Control, Pathogen Accountability, and Dual-Use Research Concern. Any genetic modification research work performed in BSL-3 CF need to undergo a review (which includes assessment of safety and dual-use risks) and approval by the BC followed by the endorsement of national Genetic Modification Advisory Committee (GMAC) to ensure appropriate control measures are adopted to mitigate the biosafety and biosecurity risks. The BC and the GMAC committee evaluate every research proposal involving genetic manipulation work or use of GMOs for high-risk biological agents including SARS-COV-2 and provide expert advice for the risk assessment and mitigation measures [30]. Additional administrative controls and policies are always established for a minimum of two personnel to be present in the containment area (i.e., a "buddy system") for biosafety and biosecurity purposes.

3.4 Risk Communication

A culture of effective communication of potential risk and reporting of risk indicators, including incidents and near misses is practiced in the BSL-3 CF and it is done in a nonpunitive manner. The BCO coordinates the facility's safety update program regularly and assists in the development of risk communication documents including incident trends and mitigations, SOPs, laboratory safety manuals, risk mitigation plans, and emergency response plans. The hazard and risk communications are also done by using appropriate signage/labels and trainings. Staff and students working in the BSL-3 CF are encouraged to report issues, including incidents and near misses. Since there is a no "blame culture" in the CF, all users of the CF are very prompt in reporting minor incidents and near misses to the operation team. Any lessons learned from these cases are shared with all CF users without disclosing the names of the personnel and other sensitive information (if any). The PIs/supervisors are also responsible for communicating hazards and risks in the laboratory to their staff/students.

3.5 Risk Assessment Review

The risk assessments are reviewed minimally every 2 years, or earlier such that when there are new activities or laboratory incidents, change/introduction of new regulations, and/or significant alterations to SOPs (disinfection/waste management methodologies, provision/usage of PPE, entry/exit protocols, etc.). A periodic review of the risk assessments is required by the BATA and WSHA. The risk assessments are reviewed by the BC, operation team, the PIs, and their team members to ensure the sustainability and effectiveness of biorisk control measures implemented in CF. During the periodic review, the risk evaluation also considers the experiences, latest information, and knowledge available at that point of time to make necessary adjustments for the control measures.

4 Human Factors

4.1 Behavioral Factors

It is believed that up to 80% of work-related accidents are down to employees behavior or human factor, in the form of acts or omissions [37]. Such behavior can lead to many small factors coming together to produce a negative outcome or accident. There are many reasons why employees cause accidents at work including but not limited to, cutting corners to save time, ergonomic factors, repetitive procedures, health factors, and misunderstanding/miscommunication of risks. Therefore, BSL-3 CF operations team closely monitor individual users and acknowledge their best practices and put in effort to remove or reduce unsafe acts by limiting the working hours and having a "buddy system." Other monitoring approaches include carrying out unannounced/announced inspections, reassessing users' competency for critical SOPs, providing constructive feedback on their performance, and sharing the best practices and lessons learned at various platforms. Such environment with positive organizational culture would help to enhance their awareness in biosafety and biosecurity that gradually lead to instill a safety and security culture at the individual level. More discussions on how positive organizational culture could promote safety and security in laboratories are covered in the previous chapter.

4.2 Personal Heath Factors and BSL-3 Occupational Health Program

Personnel with special health conditions such as immunosuppressive conditions or drug therapy that suppresses the immune system or women that are pregnant or become pregnant are considered in the biorisk assessment since those personnel may suffer from serious consequences if they are exposed to dangerous BAs in workplace. They are required to inform NUS occupational health (OH) physicians about their health status. This risk assessment of such personnel needs to be reviewed by the OH physician, BC and Human Resource (HR) officer before the decision on whether to allow them to work inside the facility is made. Sometimes, the HR officer may also act as a liaison between the employee and their respective PIs or supervisors.

As required by the BATA and NUS OH Program, all personnel working with BSL-3 CF including SARS-CoV-2 need to enroll under NUS BSL-3 OH program. The OH program includes undergoing medical surveillance and vaccination (if it is necessary and available) and providing medical advice with the goal of reducing and ultimately preventing infection/disease. In January 2021, when the first batch of COVID-19 vaccine was available in Singapore, the BSL-3 workers who are working with SARS-CoV-2 were given priority to be vaccinated. The medical surveillance program comprises a baseline medical evaluation prior to BSL-3 work and thereafter, the medical evaluation will be conducted on an annual basis. When the researcher stops working in the BSL-3 CF, they are required to undergo an "exit" medical evaluation to ensure that he/she has not sustained any medical condition associated with the BSL-3 work. Ad hoc medical evaluations are required in the event of accidental exposures in the BSL-3 CF or if any personnel develop a medical condition that may compromise his/her natural individual immunity (such as pregnancy, the use of immunosuppressive drugs such as long-term steroids, or autoimmune/immune diseases) and suffering from a chronic medical condition.

4.3 Competence and Awareness of the Personnel

All personnel who need to work in the BSL-3 CF are required to undergo mandatory BSL-3 training program. The BSL-3 CF operations team ensures that all personnel registering for the BSL-3 training program have basic knowledge and skills based on their prior experiences, education and training. The PIs, supervisor, the BC, and the facility operation team review and agree on the training needs for everyone. The facility has developed a training plan and SOPs to ensure a consistent approach to safety and security training requirements for all BSL-3 users to meet NUS Medicine/NUS policies and national legislations and regulations. It is the policy of the NUS Medicine BSL-3 CF that all researchers applying for BSL-3 training program need to have at least 2 years of BSL-2 experience.

The BSL-3 specific training include competency in the related operational SOPs (common SOPs), research-specific protocols, and emergency response procedures. Additional training is provided for any new procedures, handling of specific BAs (SARS-CoV-2 isolation, SARS-CoV-2 inoculation in animals, inactivation of SARS-CoV-2, etc.), and systems or equipment that require heightened control. The training also includes the main routes of exposure of

SARS-CoV-2 to humans, and common signs and symptoms of the COVID-19 infection [38], laboratory hazards of SARS-CoV-2 and the reporting of any suspected cases of infection. Focused and specialized one-on-one training is provided in some cases. The behavior of each user is also monitored during the training and after the certification for their performance and interactions with the facility, equipment, and other users of the facility.

A survey conducted by Wurtz et al. (2016) revealed that laboratory-acquired infections have been infrequent and even rare in recent years, and human errors (e.g., use of inappropriate procedures and techniques, and inadequate training) represent a very high percentage of the cases compared to failure of containment or safety equipment [39]. Therefore, it is paramount that there must be an initial and follow-on training on laboratory procedures, practices, and techniques described in the BSL-3 laboratory safety manuals and various SOPs [40].

5 Operations/Implementation

5.1 Inventory System and Agent Tracking

A comprehensive program has been established for the accountability of BAs in the BSL-3 CF. The BA inventory is kept up-to-date, complete, and accurate regularly. It includes description of the BAs, its quantities, storage location and use, the person responsible, documentation of every use, and internal and external transfers. In the case of SARS-CoV-2 related projects, there are separate inventory for different types of COVID-19 clinical samples under possession in the freezers and SARS-CoV-2 cultures and are maintained and monitored through periodic audits.

5.2 Transport and Storage of Biological Materials

Samples are packed, placed in secondary containers for transfer within the facility. Trolleys and transfer baskets are also used if more than one sample is transferred at a time. For local transfer and transportation (transfer between buildings by walking and for samples that are sent out via vehicles) of BAs, BATA (Transportation) Regulation is followed [41]. All BAs including COVID-19/SARS-CoV-2 samples are stored in dedicated fridges/freezers which can be locked and with restricted access inside the BSL-3 CF.

5.3 Inactivation of Biological Agents and Removal from the BSL-3 CF

For the inactivation of any BAs including SARS-CoV-2 virus, there is a need to develop specific SOPs and Risk Assessments for review and approval by the BC before the inactivation experiment can be performed. Three independent validation experiments need to be performed by researchers and witnessed by another experienced person (a BSL-3 certified user who can verify the critical steps of inactivation) and submit the results to the BC again for approval before the "inactivated" samples can be removed from the BSL-3 CF to lower containment laboratories. Personnel performing the inactivation must demonstrate proficiency in carrying out the procedure successfully and as written in the inactivation SOP. A system is also in place to track the movement of "inactivated" samples out of the CF.

5.4 Documentation and Document Control

BSL-3 CF maintains all records and documentations associated with its policies, plans, procedures, function, and operations up to 5 years. The facility has developed a BSL-3 Laboratory Safety Manual (which covers the entire BRM elements) and several operations SOPs that include facility and pathogen related procedures. There are several new SOPs developed specifically for working with SARS-CoV-2 virus and various types of COVID-19 samples (clinical and environmental). All documentations of BSL-3 CF are kept securely with controlled access.

5.5 Safe Use and Disinfection/Cleaning Procedures for Equipment and Instruments

5.5.1 Selection of Appropriate Disinfectants

Based on the risk assessment and the BAs handled inside the BSL-3 CF, appropriate disinfectants are used for the respective BAs: For examples, PREempt and 4% Melsept®SF [42–44] are used for most of the BAs including SARS-CoV-2 for decontamination because of their broad spectrum of antimicrobial activity. Ethanol (70%) is also used for disinfecting the workspace (e.g., BSC, microscope, and work benches), reusable PPE, and other small equipment or tools before exiting the BSC/experimental room. All items taken out of a BSC are kept in secondary containers regardless of the disinfectants used for surface decontamination.

5.5.2 Use of Centrifuge

The centrifuges used in BSL-3 facility have aerosol-tight centrifuge buckets or rotor equipped with "O-Ring" sealed aerosol-tight caps/lids and are maintained regularly. These enclosed containers are designed to prevent aerosols from being released during centrifugation. After centrifugation of sample tubes containing infectious BAs/materials the aerosol tight caps/lids of the rotor/buckets are opened only inside a BSC.

5.5.3 Use of Biosafety Cabinet

The facility is provided with Class II, Type A2, Class II, Type B2, and Class III BSCs. For the surface decontamination of BSC, with the cabinet blower running, all containers and equipment surface are decontaminated and removed from the cabinet when work is completed. When biological materials and hazardous agents are removed out from the BSCs, the containers are always decontaminated with disinfectant and placed inside secondary containers. At the end of the experimental work, final decontamination is done for the work surface, the cabinet's sides and rear part, and the interior of the glass window sash.

5.5.4 Use of Autoclave

All biological wastes generated in the BSL-3 CF are decontaminated before they leave the facility unless otherwise stated. Two double-door autoclaves with bioseals are used to decontaminate biological wastes and reusable materials (scrubs, animal cages, etc.) derived from the containment area. Heat resistant autoclave gloves are used when unloading autoclaves to avoid injuries from burns and scalds.

5.5.5 Animal Holding Racks

Only small animals which can be housed in isolation cages are used for experiments in the BSL-3 CF. Bio Containment Unit (BCU) Racks fitted with certified HEPA filters are used for animals infected with SARS-CoV-2. When a cage is removed from the BCU cage rack, both the cages and the port on the rack are automatically sealed from the room. The air from the cages is individually ventilated and HEPA filtered before released into the exhaust system. Preventive maintenance and certification of these racks are conducted annually.

5.5.6 Ultralow-Temperature Freezers

Specimens stored at ultralow temperatures are extremely cold (−70 to −85 °C) and direct contact with the skin may cause severe cold burns. Personnel using the freezers are required to wear thermally resistant gloves on top of the standard BSL-3 PPE (*see* Subheading 5.8), when handling items stored at ultralow temperatures.

5.6 Maintenance, Calibration and Certification of Equipment

The facility has a program in place to ensure all safety equipment and facility engineering systems are inspected, monitored, maintained, calibrated, and certified according to the regulatory/institutional requirements. All BSL-3 facilities in Singapore need to be certified by the MOH-Approved Facility Certifiers and MOH officers on an annual basis. Before the certification, the facility generally will undergo a complete decontamination (by fumigation), followed by maintenance, servicing, and/or calibration work on the facility, the ACMV system, as well as all the safety equipment. Decontamination of the facility (and the equipment within) is necessary to eliminate the risk of workers and certifiers exposure to any biohazards when carrying out the above precertification tasks as well as certification process.

5.7 Biowaste Management

There are different types of wastes generated in the BSL-3 CF. All contaminated biowaste and materials are segregated into solid and liquid wastes, packaged, and decontaminated with appropriate methods approved by the BC. All solid wastes including used disposable laboratory consumables, animal carcasses, sharps containers (with used needles and sharps) and used gloves (outer layer) are double bagged (biohazard bags) within the BSC and the outer surface are disinfected with 4% Melsept. The biohazard bags are then taken out of BSC after the exterior surface is decontaminated for autoclaving. Autoclaved reusable items (e.g., animal cages) are kept in the waste collection room till the biological indicator tubes show negative results before sending out for washing. Other solid wastes that are autoclaved are sent out for disposal through licensed waste collecting agencies [29]. Liquid wastes are decontaminated first by an appropriate chemical disinfectant (compatible with chemicals present in the liquid waste) and allowed appropriate contact time followed by autoclaving prior to discarding. Melsept (4%) is used for decontaminating SARS-CoV-2 liquid wastes and its efficacy on this novel virus has been validated in the BSL-3 CF.

5.8 Clothing and Personal Protective Equipment

Before entering the containment area of the facility, all personnel must change from street cloth to lab scrub suit and put on PPEs which include a first layer of gloves (nitrile gloves), socks, head cover, Powered Air Purifying Respirator (PAPR) and shoes at the clean changing room. All these PPE are stored in the facility (clean) changing area. Before entering any of the BSL-3 suites where pathogens are stored/handled, the second layer of PPE are donned which include shoe covers (knee height over the shoes), a pair of latex gloves and long sleeved disposable surgical gown. When exiting the BSL-3 suites, the PPE decontamination and removal is done according to the standard CF procedure. The disposable gown, shoe covers, and outer layer of gloves are disinfected by spraying PREempt and removed after 5 min in the anteroom before exiting the BSL-3 suites. Shoes are taken off at the clean corridor, just before entering the dirty changing room, and the PAPR, hair cover, and the scrubs are removed in the dirty changing room. The scrubs are then placed in biohazard bags for it to be autoclaved and sent for laundry.

6 Performance Evaluation and Continual Improvement

6.1 Performance Monitoring, Measurement and Evaluation

Based on the PDCA cycle, all processes in BSL-3 CF and components of its BRM are monitored and measured by evaluating all relevant data collected. There are checklists for checking and monitoring various aspects of the BRM (include both biosafety and biosecurity). Some areas and systems are inspected daily (e.g., security system and electrical systems), weekly, monthly, annually, and so on. The performance monitoring is done by external parties as well as internal parties. Internally, the CF operation team is responsible for conducting regular inspections of the facility on various aspect to review and verify that safety and security policies and procedures are implemented, and all relevant parties comply and conform to these policies and procedures for the use of biological agents/toxins and other hazards in the facility. The facility's safety management system is also reviewed yearly by the NUS Safety Office or by the NUS Medicine's Safety Committee.

As required by BATA, BSL-3 CF need to be certified annually by MOH and the certification audit is conducted by an MOH approved third party certifier using a MOH's checklist. The MOH certification audit and review of BRM program is done on-site, covering all safety and security aspects including testing and/or verification of facility infrastructure, containment protection system and safety equipment, review of facility policies, procedures and compliance to legislations, and assessment of staff competency in biosafety and biosecurity procedures and response to emergency incidents. As part of the MOH certification and monitoring of biosecurity of the CF, an in-house Red Teaming exercise also must be done and submit the report to MOH before the annual review by the certifier.

In addition to the outcome of audits and inspections, other parameters such as incident trends and reports, effectiveness of operational control, outcome of training exercises, safety/security/quality service feedback by users, and outcome and after-actions reviews of safety and security emergency drills are also monitored and used for assessing the effectiveness of the biorisk management programme to identify any areas for improvement. The performance evaluation referred in this chapter includes evaluation of results and outcome of audits and inspections (internal and external), incident investigations, analysis of measurement results. Once the data and outcome are documented and analysed, the risk assessment may be revised to implement additional mitigation measures if needed. The stakeholders are retrained or the system/processes are improved appropriately if any gaps are observed in specific procedures/areas. The outcome of performance monitoring and evaluation is brought up to the management during the annual Management Review exercise.

6.2 Management Review of Goals and Objectives of the BRM System

Management review of BRM system is conducted at the end of each year after compiling the results of performance measurement and evaluation for that year. The Management Review committee consists of senior management of the NUS Medicine or representatives of institution and the BSL-3 CF management representative and they review all elements of BRM programme to ensure its continuing suitability, adequacy, and effectiveness, and the outcome of the review is documented and followed by the responsible parties. Management review also includes opportunities for improvement of the BRM, any changes to the system, policies, programs, and objectives to determine any modifications needed or to establish new ones for the next years.

6.3 Continual Improvement Process

Based on the outcome of the Management Review or when scientific knowledge changed, the BSL-3 CF operation team makes necessary changes and improvements for existing programs and policies if necessary. For instance, when the first batch of COVID-19 clinical samples were received in January 2020, a dedicated BSL-3 suite with enhanced practices were identified during the risk assessment stage (PLAN of PDCA) for isolation and culturing of the virus since there was not much information on the biorisk aspects of SARS-CoV-2 available during the early stage of the COVID-19 outbreak. These practices are designed for handling pathogens with unknown risk/higher risk in the BSL-3 CF. Subsequently, SARS-CoV-2 experimental work was shifted to a BSL-3 virus suite when more scientific information such as mode of transmission and pathology about this novel virus was available.

For the SARS-CoV-2 work, many additional mitigations (in addition to the standard practices) were included during the initial risk assessment. Some examples of additional measures included BSL-3 CF use was limited only for SARS-CoV-2 work, personnel working with SARS-CoV-2 were not allowed to have interactions with general community and they were provided with dedicated accommodations, and "batch validation" of samples were required before inactivated samples could be taken out of BSL-3 CF to BSL-2 facilities. After 2 months, when more scientific information on this virus was available, some of these additional controls were removed or relaxed based on the reevaluation of the risk assessment and then, other pathogens related research works could be continued in the CF. Any improvement programs which enhance the BRM performance are communicated to all relevant parties including the BSL-3 CF staff and laboratory personnel, and ensure its timely implementation.

6.4 Some of the Challenges Faced by BSL-3 CF for Starting COVID-19 Research During the Outbreak

1. Managing high volume requests by scientists to transfer clinical samples from hospitals to the CF and start COVID-19 research with limited trained personnel.
2. Limited number of virologists who had working experience in a BSL-3.
3. Developing appropriate SOPs for the novel virus with limited information available, at the beginning of the outbreak.
4. Limited trainers available to provide laboratory safety, security and SARS-CoV-2 specific training within a limited time span.
5. Deciding and implementing additional OH programme for SARS-CoV-2 workers.
6. Ensuring availability of suitable disinfectants that are effective for SARS-CoV-2.
7. Ensuring availability of PPE and other lab consumable items.
8. Selection and validation of appropriate inactivation methods for SARS-CoV-2 and various types of clinical samples, in the initial stage.

These challenges were managed with tremendous support from the NUS Medicine leadership and by carefully analyzing the situation, coordinating and collaborating with various stakeholders. Since there was uncertainty about the risk nature of SARS-CoV-2, more precautions and stringent measures were implemented with the additional resources (manpower, consumables, etc.) provided. The operation team was empowered to lead the discussions and take decisions when necessary to deal with operational crisis and emergency situations related to COVID-19/SARS-CoV-2 work in the BSL-3 CF.

7 Conclusion

Implementing and maintaining a comprehensive biorisk management system requires an organizational culture that holistically ensures the biosafety and biosecurity of infectious BAs and toxins, in addition to complying with relevant legal requirements and policies, and conducting science in a responsible manner [45, 46]. Promoting a positive safety and security culture in laboratories by integrating a risk management process into daily laboratory operations results in the ongoing identification of hazards, prioritization of risks and the establishment of risk mitigation measures tailored to specific situations. To be successful, this process must be collaborative and inclusive of all stakeholders involved [47].

As described in this chapter, a good BRM system is a system that is practical and sustainable, with support from various parties especially senior management. It can be adapted and applied to any organizations dealing with BAs regardless of the type of biological hazards, including emerging or novel pathogens such as SARS-CoV-2. The implementation of BRM system in the NUS Medicine BSL-3 CF has resulted in the following.

- Increased safety and security awareness among BSL-3 CF users.
- A clear understanding of legal requirements among CF staff and users.
- Users are more willing to discuss and share their concerns openly with the facility team which help to identify and address them promptly.
- The facility team is fully empowered and supported by the senior management of the institution.
- Bottom-up approach allowed every stakeholders of the facility to take ownership of their activities.
- Continual improvements of BRM and prevention that could assist in mitigation of biorisks.

In summary, it is possible to continuously improve and further enhance the overall performance of a containment laboratory to minimize the risks and create a positive safety culture through the implementation of an effective and sustainable biorisk management system. As Callihan et al. [47] highlighted, a laboratory that has a robust BRM system could be prepared to address risks associated with emerging infectious diseases because of its personnel have necessary knowledge, skills, and abilities required to deal with unknown risks such as COVID-19. This same approach can be applied to any other containment facilities which are working with highly infectious BAs around the world.

Acknowledgments

The BRM system of NUS Medicine BSL-3 Core Facility is supported by the NUS Medicine's Senior Management, BSL-3 Biosafety Committee, and the BSL-3 Core Facility Operations Team. Key terms and definitions used in this document are adopted from "Section 3—Terms and definitions" of ISO 35001:2019.

References

1. Jain N, Saiju P, Jain R (2020) A comprehensive review on viral zoonosis: emphasising on pathogenesis, diagnosis, treatment, prevention strategies and future perspectives. Int J Pharm Sci Res 11(10):4712–4738
2. Senio K (2003) Singapore SARS case a laboratory accident. Lancet, Infectious Diseases 3. http://infection.thelancet.com
3. Lee SU (2017) Infectious disease in Singapore and Asia: persistent challenges in a new era. Singap Med J 58(4):169–170
4. Wu Y, Yeo A, Phoon MC et al (2010) The largest outbreak of hand; foot and mouth disease in Singapore in 2008: the role of enterovirus 71 and coxsackievirus A strains. Int J Infect Dis 14:e1076–e1081
5. Lin HZ, Tambyah PA, Yong EL, Biswas A, Chan SY (2016) A review of Zika virus infections in pregnancy and implications for antenatal care in Singapore. Singap Med J 58: 171–178
6. WHO Laboratory biosafety guidance related to coronavirus disease (COVID-19) (2020) Interim guidance. https://www.who.int/publications/i/item/laboratory-biosafety-guidance-related-to-coronavirus-disease-(covid-19)
7. University of South Florida Biosafety Office. SARS-CoV-2 (COVID-19) Research Laboratory Biosafety Guidelines. https://www.usf.edu/research-innovation/research-integrity-compliance/documents/biosafety/usf-sars-cov-2-biosafety-guidelines.pdf
8. Hong KH, Lee SW, Kim TS et al (2020) Guidelines for laboratory diagnosis of coronavirus disease 2019 (COVID-19) in Korea. Ann Lab Med 40:351–360. https://doi.org/10.3343/alm.2020.40.5.351. ISSN 2234-3806, eISSN 2234-3814
9. Centres for Disease Control and Prevention, Interim Guidelines for Biosafety and COVID-19. https://www.cdc.gov/coronavirus/2019-ncov/lab/lab-biosafety-guidelines.html. Accessed 22 Jun 2021
10. Kauffer AM, Theis T, Joanna LK, Gray JL, Rawlinson WD (2020) Laboratory biosafety measures involving SARS-CoV-2 and the classification as a Risk Group 3 biological agent. Pathology 52(7):790–795
11. Biological Agents and Toxins Act-Updated Biological Agents and Toxins List Jan 2020. https://www.moh.gov.sg/docs/librariesprovider7/news-updates-documents/list-of-biological-agents-and-toxins.pdf
12. BATA-S 238/2020. Biological agents and toxins (COVID-19 Research Laboratory-exemption) Regulations 2020. https://www-moh-gov-sg-admin.cwp.sg/docs/librariesprovider7/about-bata-documents/covid19-research-lab-exemption.pdf
13. CEN Workshop Agreement 15793 Laboratory Biorisk Management. https://internationalbiosafety.org/wp-content/uploads/2019/08/CWA-15793-English.pdf. Accessed 22 Jan 2021
14. International Organization for Standardization (ISO) 35001:2019. (2019, November 12). https://www.iso.org/standard/71293.html. Accessed 22 Jun 2021

15. Joseph T (2021) Management system approach for addressing biosafety and biosecurity of emerging pathogens in a biosafety level-3 core facility. Appl Biosaf. 26(4): 210-220. https://doi.org/10.1089/apb.2021.0007
16. Infrastructure Protection Act. https://www.police.gov.sg/Advisories/Infrastructure-Protection/Infrastructure-Protection-Act. Accessed 22 Jan 2021
17. Biological Agents & Toxins Act, Biosafety: About BATA. https://www.moh.gov.sg/biosafety/about-bata#legislation. Accessed 22 Jan 2021
18. WHO Laboratory biosafety manual. 4th ed. https://www.who.int/publications/i/item/9789240011311. Accessed 22 Jan 2021
19. Franz DR (2018) Facilities, equipment and procedures: an historic glimpse at high-containment lab safety and security. J Biosaf Biosecur 1(2):98–99
20. Workplace Safety and Health Act: what it covers. https://www.mom.gov.sg/workplace-safety-and-health/workplace-safety-and-health-act/what-it-covers. Accessed 21 Jan 2021
21. MOH Laboratory Biosafety Resources Related to SARS-CoV2/COVID-19 Virus Samples/Materials (2020). https://www.moh.gov.sg/docs/librariesprovider7/useful-info-and-guidelines-documents/consolidated-laboratory-biosafety-resources-related-to-sars-cov-2-covid 19-virus-materials.pdf
22. Chan JF-W, Kok K-H, Zhu Z et al (2020) Genomic characterization of the 2019 novel human-pathogenic coronavirus isolated from a patient with atypical pneumonia after visiting Wuhan. Emerg Microbe Infect 9:221–236
23. Xu J, Zhao S, Teng T, Abdalla AE, Zhu W, Xie L, Wang Y, Guo X (2020) Systematic comparison of two animal-to-human transmitted human coronaviruses: SARS-CoV-2 and SARS-CoV. Viruses 12:244. https://doi.org/10.3390/v12020244
24. Wilder-Smith A, Chiew CJ, Lee VJ (2020) Can we contain the COVID-19 outbreak with the same measures as for SARS? Lancet Infect Dis 20:e102–e107
25. Centres for Disease Control and Prevention, Frequently Asked Questions about Coronavirus (COVID-19) for Laboratories. https://www.cdc.gov/coronavirus/2019-ncov/lab/faqs.html#Specimen-Handling. Accessed 22 Jun 2021
26. Workplace Safety and Health Act: Subsidiary Legislation. https://sso.agc.gov.sg/Act/WSHA2006?ViewType=Sl. Accessed 22 Jan 2021
27. Animals and Birds Act; Search within Legislation. https://sso.agc.gov.sg/Act/ABA1965. Accessed 22 Jan 2021
28. The Human Biomedical Research Act (HBRA) (2015). https://www.research.nhg.com.sg/wps/wcm/connect/d036ee8047fb27ad971cd7cc4956140b/Human+Biomedical+Research+Act+2015.pdf?MOD=AJPERES
29. Toxic Waste Control, National Environment Agency. https://www.nea.gov.sg/our-services/pollution-control/hazardous-waste/toxic-waste-control. Accessed 22 Jan 2021
30. Singapore Biosafety Guidelines for Research on Genetically Modified Organisms (GMOs) 2006. Genetic Modification Committee (GMAC). https://www.gmac.sg/Index_Singapore_Biosafety_Guidelines_for_Research_on_GMOs.html
31. The NACLAR guidelines on the care and use of animals for scientific purposes. https://www.nparks.gov.sg/avs/animals/animals-in-scientific-research/naclar-guidelines/naclar-guidelines. Accessed 22 Jun
32. WHO (2015) Responsible life sciences research for global health security. https://www.who.int/csr/resources/publications/HSE_GAR_BDP_2010_2/en/. Accessed 21 Jun 2021
33. Association of public Health Laboratories (APHL). Biorisk management programme guidance document-biorisk management for clinical and public health laboratories. https://www.aphl.org/programs/preparedness/Biosafety-and-Biosecurity/Documents/APHL_Biorisk_management_program_guidance_document.pdf
34. National Institute of Health, Design Requirement Manual (2016) Division of Technical resources, Office of research facilities (1.5: 3/5/2020). https://www.orf.od.nih.gov/TechnicalResources/Documents/DRM/DRM1.503262020.pdf
35. Standard ANSI Z9.14: Testing and performance verification methodologies for ventilation systems for Biological Safety Level 3 (BSL-3) and animal Biological Safety Level 3 (ABSL-3) facilities. https://www.orf.od.nih.gov/TechnicalResources/Bioenvironmental/Documents/StandardANSIZ914JCHASArticleAutosaved508.pdf
36. An efficient and practical approach to biosecurity, Denmark. Centre for biosecurity and biopreparedness. https://internationalbiosafety.org/wp-content/uploads/2019/08/Efficient__Practical_Approach_to_Biosecurity.pdf

37. Burton S (2012) Behavioural safety - human factors. Paper presented at the international conference on health, safety and environment in oil and gas exploration and production, Perth, Australia, September 2012, Paper Number: SPE-154923-MS. https://doi.org/10.2118/154923-MS
38. Centres for Disease Control and Prevention, Symptoms of Coronavirus. https://www.cdc.gov/coronavirus/2019-ncov/symptoms-testing/symptoms.html. Retrieved 22 Jan 2021
39. Wurtz N, Papa A, Hukic M et al (2016) Survey of laboratory-acquired infections around the world in biosafety level 3 and 4 laboratories. Eur J Clin Microbiol Infect Dis 35:1247–1258
40. Borchardt JK (2006) Lab safety requires training and commitment. Science. https://www.sciencemag.org/careers/2006/08/lab-safety-requires-trainingand-commitment
41. Biological Agents and Toxins Act (Transportation) Regulations. https://www.moh.gov.sg/docs/librariesprovider7/certifiedfacility/bata_transportation_regulation.pdf. Accessed 22 Jun
42. Christel & Elena. (2015) Melsept SF, Surface Disinfectant: Buy Online. https://www.praxisdienst.com/en/Hygiene/Disinfectants/Surface+Disinfectants/Melsept+SF+1000ml.html. Accessed 22 Jun 2021
43. PREempt. https://www.contecinc.com/products/preempt-rtu-disinfectant-solution/. Accessed 22 Jun 2021
44. Government of Canada, Canada.ca, https://www.canada.ca/en/health-canada/services/drugs-health-products/disinfectants/covid-19/list.html#tbl1
45. Perkins D, Danskin K, Rowe AE, Livinski AA (2019) The culture of biosafety, biosecurity, and responsible conduct in the life sciences: a comprehensive literature review. Appl Biosaf 24(1):34–45
46. Guidelines for biosafety laboratory competency CDC and the Association of Public Health Laboratories. U.S. Department of Health and Human Services; Centers for Disease Control and Prevention, Morbidity and Mortality Weekly Report; Supplement/Vol. 60, 15 Apr 2011
47. Callihan DR, Downing M, Meyer E, Ochoa LA, Petuch B, Tranchell P, White D (2021) Considerations for laboratory biosafety and biosecurity during the coronavirus disease 2019 pandemic: applying the ISO 35001: 2019 standard and high-reliability organization principles. Appl Biosaf. https://doi.org/10.1089/apb.20.0068

Chapter 25

Methods of SARS-CoV-2 Inactivation

Enyia R. Anderson, Tessa Prince, Lance Turtle, Grant L. Hughes, and Edward I. Patterson

Abstract

Inactivation methods allow for hazard group 3 (HG3) pathogens to be disposed of and used safely in downstream experiments and assays to be carried out at lower containment levels. Commonly used viral inactivation methods include heat inactivation, fixation methods, ultraviolet (UV) light and detergent inactivation. Here we describe known methods used to inactivate SARS-CoV-2 for safe downstream biological assays.

Key words Inactivation, SARS-CoV-2, TRIzol, Vero, Assay, UV

1 Introduction

SARS-CoV-2 has been classified as a hazard/risk group 3 (HG3/RG3) pathogen. Therefore, inactivation methods are required to allow laboratory work to be carried out at lower containment level settings in a safe manner and samples to be disposed of safely [1–3]. Commonly used viral inactivation methods include heat inactivation, and chemical treatment, such as with fixatives or detergents [4]. Of these, the most appropriate method of inactivation to use depends upon the needs of the downstream biological assays. It has been shown that for SARS-CoV, heat inactivation works through the process of thermal aggregation of membrane proteins, allowing for the preservation of antigens, while detergents disrupt the virus structure but maintain proteins related to the host immune response [5]. Comparatively, fixation of cells infected with SARS-CoV-2 using methanol or paraformaldehyde allows for the preservation of cell morphology while also making the samples noninfectious [6]. Ultraviolet (UV) light can be divided into

Department of Vector Biology and Tropical Disease Biology, Centre for Neglected Tropical Disease, Liverpool School of Tropical Medicine, Liverpool UK

Justin Jang Hann Chu et al. (eds.), *SARS-CoV-2: Methods and Protocols*,
Methods in Molecular Biology, vol. 2452, https://doi.org/10.1007/978-1-0716-2111-0_25,

three spectral ranges: UVA (320–400 nm), UVB (280–320 nm), and UVC (200–280 nm). UVC light has been commonly used to inactivate coronaviruses including SARS-CoV [7]. UVC light acts to cause the formation of pyrimidine dimers in the viral RNA, resulting in inactivation. This method can be used to preserve the structure of proteins and antigens for downstream analysis. Here we describe methods that have been shown to be effective at inactivating SARS-CoV-2, therefore making it safe to carry out downstream biological assays and reduce the risk of laboratory-acquired infection.

Many institutions require that these methods be validated in-house and on specific sample substrates (such as cell-free, cell monolayer, or tissue sample) with the target virus or appropriate surrogate before being put to regular use. The results must show that the inactivation method renders the entire sample noninfectious. Therefore, the entire sample should be tested on cells to observe any potential virus replication. While heat and UV treated samples may be applied to cells directly, reagents used in chemical treatments are typically cytotoxic, requiring removal from the sample prior to cellular assays to confirm inactivation.

Furthermore, we have presented methods for quantifying viable viral titer through 50% tissue culture infectious dose ($TCID_{50}$) and plaque assays. Both assays rely upon visual identification of cytopathic effects (CPE) in stained cell cultures, and therefore a cell line that is permissive to virus infection that results in observable CPE [8]. Hence for evaluating these inactivation assays Vero E6 cells have been used throughout due to the ability of SARS-CoV-2 to replicate in vitro [9]. $TCID_{50}$ assays can be sensitive down to a single infectious virus particle and may be preferred when testing samples with low titer. A plaque assay, on the other hand, allows for more accurate determination of higher viral titers. Both of these assays can be used to measure the reduction of viable virus particles, thus indicating the effectiveness of the methods of inactivation.

2 Materials

2.1 Inactivation Assays

1. Vero E6 cells, confluently seeded onto 96-well plate.
2. Vero E6 cells, confluently seeded onto 24-well plate.
3. Centrifugal filtration columns.
4. PBS.
5. Benchtop centrifuge.
6. Vortexer.
7. Sample or SARS-CoV-2 stock.
8. Inactivation reagent/equipment:

(a) Water bath or heat block.
(b) 0.5% detergent solution: Triton X-100/NP-40/SDS diluted in PBS.
(c) TRIzol/TRIzol LS.
(d) 100% ice-cold methanol.
(e) 4% paraformaldehyde.
(f) CL1000 UV Crosslinker—see Manufacturer's Operating Instructions and Service Manual before using the UVP Crosslinker for the first time.

2.2 50% Tissue Culture Infectious Dose Assay ($TCID_{50}$)

1. Vero cell maintenance medium: DMEM supplemented with 2% heat inactivated FBS and 0.05 mg/mL gentamicin.
2. 96-well cell culture plate.
3. Vero E6 cells seeded onto 96-well culture plate.
4. Crystal violet solution: 0.25% crystal violet, 30% methanol, top up to final volume with water.
5. 10% formalin.
6. 37 °C incubator with 5% CO_2 inflow.
7. Sample or SARS-CoV-2 stock.

2.3 Plaque Assay

1. Vero cell maintenance medium.
2. 96-well dilution plate.
3. 24-well treated cell culture plates seeded with Vero E6 cells.
4. 2% low melting point agarose solution.
5. 10% formalin.
6. Crystal violet solution.
7. 37 °C incubator with 5% CO_2 inflow.
8. Sample or SARS-CoV-2 stock.

3 Methods

All protocols should be carried out at the appropriate biosafety level for your region.

3.1 Heat Inactivation

1. Add at least 300 μL of sample or SARS-CoV-2 stock solution to a microcentrifuge tube.
2. Incubate in heat block or water bath at 80 °C for 1 h (*see* **Note 1**). For a negative control, leave the sample at room temperature (RT) for 1 h.
3. Remove sample from heat and mix prior to serial dilutions for quantification assay.

3.2 Detergent Inactivation

1. To 250 μL of sample or SARS-CoV-2 stock, add 50 μL of 0.5% detergent solution. Mix gently by flicking the tube to avoid the formation of bubbles. For negative control, add 50 μL of PBS to the sample instead of detergent solution.
2. Incubate sample at RT for 30 min.
3. After 30 min has elapsed, transfer the sample to a 15 mL or 50 mL tube. Use either PBS or DMEM to top up the total volume to 15 mL or 50 mL, depending on the critical micelle concentration (CMC) of the detergent used (*see* **Note 2**). Gently mix.
4. Transfer diluted sample to a centrifugal filter tube. Centrifuge according to manufacturer's protocol. Discard waste and transfer additional diluted sample as necessary (*see* **Note 3**).
5. Centrifuge filter tubes until ≤300 μL of the sample remains. Transfer the sample to a new tube and top up the volume to 300 μL if needed with Vero cell maintenance media.
6. Mix sample prior to serial dilutions for quantification assay.

3.3 TRIzol Inactivation

1. Add 25 μL of sample or SARS-CoV-2 stock, to 75 μL of TRIzol reagent (*see* **Note 4**). Mix by vortexing tube. For negative control, add 75 μL of PBS to the sample instead of TRIzol.
2. Incubate sample at RT for 5 min.
3. After 5 min has elapsed transfer the sample to a 15 mL or 50 mL tube. Use either PBS or DMEM to top up the total volume to 15 mL or 50 mL, depending on the compatibility of phenols with your centrifugal filter tube (*see* **Note 5**). Gently mix.
4. Transfer diluted sample to a centrifugal filter tube. Centrifuge according to manufacturer's protocol. Discard waste and transfer additional diluted sample as necessary.
5. Centrifuge filter tubes until ≤300 μL of sample remains. Transfer the sample to a new tube and top up volume to 300 μL if needed with Vero cell maintenance media.
6. Mix sample prior to serial dilutions for quantification assay.

3.4 Inactivation with Fixatives (Methanol or Paraformaldehyde)

1. This method is used to inactivate virus in cells. In a microcentrifuge tube, pellet an infected cellular sample by centrifuging for 5 min at 1000 × *g*. Remove supernatant without disturbing the pellet.
2. Add 1 mL of 100% ice-cold methanol or 4% paraformaldehyde and pipette up and down to resuspend the pellet. Incubate at RT for 30 min. For a negative control, resuspend cells in 1 mL of PBS.

3. Pellet the sample by centrifuging for 5 min at 1000 × *g*. Remove the fixative solution without disturbing the pellet.
4. Wash the pellet twice by resuspending in 1 mL of PBS, centrifuging for 5 min at 1000 × *g* after each wash.
5. Resuspend the pellet in 300 μL of Vero cell maintenance media.

3.5 Inactivation Using UV Irradiation Using the CL1000 UVP Crosslinker (See Note 6)

1. Transfer 1 mL/10 mm^2 growth area aliquots of sample into individual wells of a 6, 12, or 24-well tissue culture plate or into individual tissue culture dishes (i.e., 1 mL/6-well, 0.5 mL/12-well, 0.25 mL/24-well, 0.8 mL/35 mm^2 dish, 2 mL/60 mm^2 dish, 5 mL/100 mm^2 dish).
2. Place the 24-well plate, with the lid removed, inside the crosslinker such that the bottom of the well is exactly 6 cm (measure using a ruler) from the UV bulbs on the ceiling of the chamber. This can be achieved by placing the plate in receptacle containing ice or using ice-blocks to elevate the plate.
3. For each new virus solution, the method will need to be optimized to determine the best energy exposure for inactivation. Perform the inactivation in separate plates at a range of energy exposures, for example., 0.01–1 J/cm^2. By using a systematic range of energy exposures, you will be able to determine the appropriate energy exposure to inactivate your sample type.
4. Switch on the power supply to the UVP Crosslinker.
5. To set the UV energy exposure, press **ENERGY** on the touchpad and enter the desired exposure value (**NB**, enter values as microjoules × 100, for example, for 0.72 J/cm^2 enter 7200). The LED display will flash the value you have entered. Check the entry is correct and then press **ENTER** on the touchpad.
6. Press **START** on the touchpad. After a slight delay, the LED will begin to countdown.
7. The unit will automatically stop at the end of the exposure cycle and will beep five times. Exposure is now complete. Check that the countdown recorder has reached zero to ensure the cycle has reached completion.
8. Open the UVP Crosslinker, remove the tissue culture receptacle and replace the lid. A small aliquot of the inactivated sample should be taken for plaque assay or $TCID_{50}$ to ensure that no infectious virus remains. If multiple aliquots are inactivated, pool and use pooled inactivated isolate to perform plaque assay/$TCID_{50}$ to confirm inactivation.

3.6 50% Tissue Culture Infectious Dose ($TCID_{50}$) Assay

1. On the day of the inactivation, experiment have a 96-well cell culture plate seeded with Vero E6 cells at >80% confluence (*see* Note 7).

2. Prepare a separate 96-well plate for serial dilutions. Leaving the first-row blank, prepare at least 3 rows (each sample run in triplicate) of 7 dilutions for each replicate, by adding 90 μL of Vero cell maintenance medium to each well (*see* **Note 8**).
3. Add 100 μL of the sample to the first row in triplicate. Transfer 10 μL from the first row to the second and mix thoroughly by pipetting up and down 20 times. Dispose of pipette tips. Repeat until all dilutions are complete from 10^{-1} to 10^{-7}.
4. Transfer the entire sample from the dilution wells into corresponding Vero E6 wells (*see* **Note 9**). Change pipette tips after each sample. Incubate plate at 37 °C with 5% CO_2.
5. Monitor cells for CPE for 4 days (*see* **Notes 10** and **11**).
6. On day 4, fix cells with 10% formalin and incubate for 30 min at RT.
7. Remove supernatant and stain cells with crystal violet solution. Wash stain from plates with water to view results.
8. Note the wells with CPE to calculate virus titer (*see* **Note 12**).

3.7 Plaque Assay

1. Seed a 24-well plate with 500 μL of 10^5 Vero E6 cells/mL and incubate at 37 °C with 5% CO_2 to become confluent overnight (*see* **Note 13**).
2. In a 96-well plate, add 225 μL of Vero cell maintenance media to each well in the dilution series (*see* **Note 14**).
3. Dilute sample 1 in 10 by adding 25 μL of sample to the 225 μL of media. Continue the 1:10 serial dilution until 10^{-6}. When transferring sample, mix by pipetting up and down 20 times. Discard pipette tip between each dilution.
4. Remove all Vero cell growth media from the 24-well plate.
5. Add 100 μL of sample dilution from 10^{-6} dilution to 10^{-1} to the wells. Rock the plate left to right, then front and back to spread the virus evenly over the cells (*see* **Note 15**).
6. Place in incubator at 37 °C with 5% CO_2 for 60 min, rock the plate every 15 min.
7. Make up the overlay by adding 2% low melting point agarose to Vero cell maintenance media in a 1:4 ratio. Make enough overlay to add 1 mL per well (24 mL per plate) (*see* **Note 16**).
8. Add the overlay to the wells. Ensure that the overlay is not too warm when added to the well, nor is the overlay placed on the cells too forcefully as this will damage the monolayer of cells.
9. Incubate the plate for 3 days at 37 °C with 5% CO_2.
10. After the incubation period, place 2 mL of 10% formalin onto the agar overlay and leave at RT to fix the cells for 30 min. After

30 min, pour the formalin from the plate and remove the overlay plugs with a flat edged spatula.

11. Cover wells with crystal violet solution, incubate stain for a few minutes and then wash off crystal violet with water. Count plaques to determine virus titer.

4 Notes

1. Any range of temperature and time can be evaluated.
2. The CMC of detergents may vary greatly (https://assets.thermofisher.com/TFS-Assets/LSG/Application-Notes/TR0019-Remove-detergent.pdf).
3. Gently mix the sample in the filter tube after each addition of sample and prior to removing the sample to dislodge any aggregates blocking the filter.
4. There are multiple TRIzol reagents. Determine which is most suitable for your sample. For stock virus, either TRIzol or TRIzol LS will work.
5. TRIzol may have a phenol content of 30–60%.
6. Here we demonstrate the inactivation of the virus using the CL-1000 UVP Crosslinker (UVP, Upland, CA, USA), however, many different models may be used. The majority of UV inactivation studies have been performed using mercury bulb UV crosslinkers (using 254 nm wavelengths), though newer models are becoming available with wavelengths at 222 nm. The UVP crosslinker consists of 5 × 254 nm shortwave tubes, which is the recommended wavelength for neutralizing viruses, in particular SARS-CoV [2]. As UV lamps age, their irradiance output decreases. The CL-1000 UVP Crosslinker has an inbuilt UV sensor that continually measures the UV energy and an inbuilt microprocessor that automatically adjusts the UV output to compensate for variations in UV intensity that occurs as the UV tubes age so that the machine can be set to deliver a fixed amount of energy, rather than relying on exposure time. This provides a built-in safety measure against the ageing of the UV bulbs.
7. To prepare a 96-well plate to be confluent the next day, seed wells with 100 μL of Vero cell growth media containing 2×10^5 Vero E6 cells/mL the day proceeding the experiment.
8. Different numbers of replicates can be used to calculate $TCID_{50}$. We typically use 3 to allow for one condition to be performed in triplicate and be spaced 1 well apart on a 96-well plate.

9. The Vero cell growth media on the cells can either be removed prior to adding the sample or exchanged for Vero cell maintenance media when virus titer is expected to be low or longer time points are required to see CPE.
10. Cytotoxicity will usually be apparent within 1 day. CPE seen on the same day will likely be due to cytotoxicity, meaning that the inactivation reagent was not completely removed from the sample. CPE seen on day 1 post-inoculation may be due to cytotoxicity or very high virus titer.
11. Multiple passages can be used to ensure that no viable virus is present. In these cases, transfer half of the supernatant from the current plate to wells in a new plate with confluent Vero E6 cells. Typical schedules are 2 passages of 4 days or 3 passages of 3 days. If any new samples become positive for CPE in the new plate, these can be reported as having viable SARS-CoV-2, with no concentration calculated by $TCID_{50}$.
12. Spearman–Karber and Reed–Muench methods are both common for calculating $TCID_{50}$ titers.
13. Different plates may be used (6-well and use 2 mL of cells, 12-well and use 1 mL of cells, 24-well and use 0.5 mL of cells, 48-well and use 0.25 mL of cells, 96-well and use 0.1 mL of cells) to space plaques farther apart.
14. Microcentrifuge tubes may be used instead of a 96-well plate. This may be more convenient when a small number of samples are used. Dilutions may be vortexed to mix rather than pipetting.
15. It is important to make sure the sample is in the middle of the well. Rocking the plate in a circular motion will push the sample to the edge, making it difficult to count plaques.
16. Warm media prior to adding an overlay to avoid the overlay solidifying before being added to cells.

References

1. Patterson EI, Warmbrod KL, Bouyer DH, Forrester NL (2019) Evaluation of the inactivation of Venezuelan equine encephalitis virus by several common methods. J Virol Methods 254: 31–34
2. Darnell MER, Subbarao K, Feinstone SM, Taylor DR (2004) Inactivation of the coronavirus that induces severe acute respiratory syndrome, SARS-CoV. J Virol Methods 121(1):85–91
3. Patterson EI, Prince T, Anderson ER, Casas-Sanchez A, Smith SL, Cansado-Utrilla C et al (2020) Methods of inactivation of SARS-CoV-2 for downstream biological assays. J Infect Dis 222(9):1462–1467
4. Song H, Li J, Shi S, Yan L, Zhuang H, Li K (2010) Thermal stability and inactivation of hepatitis C virus grown in cell culture. Virol J 7:40
5. Kampf G, Voss A, Scheithauer S (2020) Inactivation of coronaviruses by heat. J Hosp Infect 105(2):348–349

6. Rabenau HF, Cinatl J, Morgenstern B, Bauer G, Preiser W, Doerr HW (2004) Stability and inactivation of SARS coronavirus. Med Microbiol Immunol 194:1–6
7. Heβling M, Hones K, Vatter P, Ligenfelder C (2020) Ultraviolet irradiation doses for coronavirus inactivation – review and analysis of coronavirus photoinactivation studies. GMS Hyg Infect Control 15:Doc08
8. Smither S, Lear-Rooney C, Biggins J, Pettitt J, Lever M, Olinger G Jr (2013) Comparison of the plaque assay and 50% tissue culture infectious dose assay as methods for measuring filovirus infectivity. J Virol Methods 193(2):565–571
9. Ogando NS, Dalebout TJ, Zevenhoven-Dobbe JC, Limpens R et al (2020) SARS-coronavirus-2 replication in Vero E6 cells: replication kinetics, rapid adaptation and cytopathology. J Gen Virol 101(9):925–940

INDEX

Justin Jang Hann Chu et al. (eds.), *SARS-CoV-2: Methods and Protocols*,
Methods in Molecular Biology, vol. 2452, https://doi.org/10.1007/978-1-0716-2111-0,

MIX
Papier aus verantwortungsvollen Quellen
Paper from responsible sources
FSC® C105338

If you have any concerns about our products,
you can contact us on
ProductSafety@springernature.com

In case Publisher is established outside the EU,
the EU authorized representative is:
Springer Nature Customer Service Center GmbH
Europaplatz 3, 69115 Heidelberg, Germany

Printed by Libri Plureos GmbH
in Hamburg, Germany